Green Technology and Smart Materials for Engineering Applications

Sustainability and technological advancements are rapidly gaining traction on a global scale and are becoming increasingly prominent across a wide range of industries, as evidenced by current market trends. Companies are making significant investments in the research and development of smart materials that can adapt to their surroundings in real time, thereby improving their performance and productivity.

The heightened interest in environmentally friendly and intelligent materials can be attributed to the recent surge in research, conferences, and patent applications in this field, highlighting the need for a comprehensive resource that can provide a thorough explanation of these developments. Green Technology and Smart Materials for Engineering Applications provides a thorough examination of smart materials, including their unique properties and applications in sustainable construction. It explores the versatile uses of green materials in different industries, emphasizing sustainable manufacturing practices and resource-efficient materials.

The integration of eco-design and innovation is looked at for the creation of sustainable materials, highlighting the importance of green and smart materials in optimizing energy consumption. Additionally, the book offers insights into reuse and recycling techniques that promote circular economy principles and sustainable business models related to green and smart materials. Real-world examples and success stories are used to illustrate the environmental and economic impacts of implementing smart and green materials in various industries.

This book is intended to serve as a comprehensive reference guide for academics, researchers, scholars, and professionals working in the fields of manufacturing processing, material science, and environmental engineering. The primary objective of this book is to highlight sustainability by emphasizing the diverse array of green and smart materials utilized in various engineering applications. Through this publication, readers will gain valuable insights into the importance of incorporating environmentally friendly materials into their work, ultimately contributing to a more sustainable future.

Innovations in Smart Manufacturing for Long-Term Development and Growth
Series Editors: Atul Babbar, Gursel Alici, Yu Dong, Ankit Sharma

Fabrication Techniques and Machining Methods of Advanced Composite Materials
Vikas Dhawan, Atul Babbar, Inderdeep Singh and Jonathan M. Weaver

Advances in Pre- and Post-Additive Manufacturing Processes
Innovations and Applications

Naveen Mani Tripathi and Ankit Sharma

Handbook of 3D Printing in Pharmaceutics
Innovations and Applications

Prakash Katakam, Ranvijay Kumar, Nishant Ranjan and Atul Babbar

Intelligent Machinery Fault Diagnostics and Prognostics
The Future of Smart Manufacturing

Deepam Goyal, Ankit Sharma and Mohamad Abou Houran

For more information about this series, please visit: www.routledge.com/Innovations-in-Smart-Manufacturing-for-Long-Term-Development-and-Growth/book-series/CRCISMLDG

Green Technology and Smart Materials for Engineering Applications

Edited by Jashanpreet Singh, Hitesh Vasudev,
Ankit Sharma, and Murali Mohan Cheepu

CRC Press
Taylor & Francis Group
Boca Raton London New York

CRC Press is an imprint of the
Taylor & Francis Group, an **informa** business

Designed cover image: Shutterstock

First edition published 2025
by CRC Press
2385 NW Executive Center Drive, Suite 320, Boca Raton FL 33431

and by CRC Press
4 Park Square, Milton Park, Abingdon, Oxon, OX14 4RN

CRC Press is an imprint of Taylor & Francis Group, LLC

ISBN: 978-1-032-84920-1 (hbk)
ISBN: 978-1-032-85002-3 (pbk)
ISBN: 978-1-003-51607-1 (ebk)

DOI: 10.1201/9781003516071

Typeset in Times New Roman
by Deanta Global Publishing Services, Chennai, India

This book is dedicated to our families, friends, teachers, and research supervisors.

Contents

Preface

The book *Green Technology and Smart Materials for Engineering Applications* elaborates emerging areas like sustainability, energy efficiency, and cutting-edge functionality where smart and green materials are useful. The goal of this textbook is to demonstrate the field of material science, with a particular emphasis on sustainability in the environment for the long term with minimum pollution and energy consumption. This book highlights a wide range of green and smart materials used in engineering applications. The challenges and risks associated with the suitability and their usage are also explored in this book. This book represents a torch for future researchers working in this field.

Green Technology and Smart Materials for Engineering Applications may help in spreading information which is crucial for a sustainable future with its focus on the positive effects of materials on the environment and the economy. The book sheds light on the environmental and economic implications of smart and green material implementations in many industries by offering real-life examples and successful implementations. This book thoroughly analyses some smart and environmentally friendly materials used in engineering including shape memory alloys, self-healing materials, piezoelectric materials, conductive polymers, photocatalytic materials, bio-based and recycled materials, energy-efficient materials, nanomaterials, biomimetic materials. Lastly, this book will result in several benefits, including reduced environmental impact, better environmental quality, safety, cost savings, increased performance, process efficiency, and energy efficiency. This might be useful in light of the worldwide push to conserve natural resources and lessen human impact on the environment.

This book is written for academicians, researchers, scholars, and industries in the field of material science and engineering. Readers would have a better grasp of the real advantages and sustainable practices in engineering and industry. This book also provides a few case studies and practical applications that can be helpful for the future.

About the Editors

DR. JASHANPREET SINGH

Among the *World's Top 2% Scientists* (2022, 2023, and 2024)

Dr. Jashanpreet Singh is currently working as an Assistant Professor at the University Center for Research and Development & Department of Mechanical Engineering at Chandigarh University, Mohali, India. He obtained his PhD degree in the field of Slurry Erosion and Thermal Spray Coatings from Thapar Institute of Engineering and Technology, Patiala in 2019. He has teaching experience of more than three years and industry experience of more than two years. His area of research is tribology-solid particle erosion, power plant engineering, ash conveying, thermal spray coatings, CFD simulation, composites, and materials characterization. He has contributed comprehensively to thermal spray coatings and tribology in reputed journals including *Wear, Tribology International, Ceramics International, Applied Surface Science Advances, Particulate Science and Technology, Engineering Failure Analysis, ASME: Journal of Tribology, ASCE Journal of Pipeline Systems – Engineering and Practice, Coatings, Proceedings of IMechE Part J: Journal of Engineering Tribology, Part E: Journal of Process Mechanical Engineering, Surface Review and Letters, Materials Science & Engineering B, Industrial Lubrication and Tribology*, under the flagship of various publishers like Elsevier, Taylor & Francis, Sage, Springer Nature, World Scientific, Emerald Publishing, etc. He has authored 85+ research publications of which 46 publications are indexed in SCIE/ESCI (Web of Science) and 75 are indexed in the SCOPUS database in various international journals, books and conferences of repute. He has authored 1 book with CRC Press and book chapters in six different CRC Press books. He is serving as a reviewer for various journals such as *Ceramics International, Surface Coatings and Technology, Tribology in Industry, Surface Topography: Metrology and Properties, Vacuum, IMechE Part E: Journal of Process Mechanical Engineering, Material Research Express, Surface Review and Letters, International Journal of Interactive Design and Manufacturing, Industrial Lubrication and Tribology, Materials Today: Proceedings*, etc. He is currently an editor of Elsevier Book. He has served as guest editor for the *International Journal of Interactive Design and Manufacturing* (Springer Nature), *Recent Patents on Mechanical Engineering* (Springer Nature), *Recent Advances in Computer Science and Communications* (Bentham Science), *Global Clinical Engineering Journal* and *Recent Patents of Engineering* (Bentham Science). He is also a section editor (Special Issues) for *Recent Advances in Computer Science and Communications* (Bentham Science) for the section "Advanced Applications of Artificial Intelligence in Manufacturing Technologies". He is listed in the *World's Top 2% List of Scientists* by Stanford University (USA) from 2022 to

2024. He received "Trusted Reviewer Status" from the Institute of Physics (IOP) in 2020. He has been awarded an "Outstanding Reviewer Award 2021" by the Institute of Physics (IOP) for the journal: *Surface Topography: Metrology and Properties.* He is a life member of AICTSD, IAENG, and NITTSD.

PROF. (DR.) HITESH VASUDEV

Among the *World's Top 2% Scientists* (2023 and 2024)

Dr. Hitesh Vasudev is a Professor in the School of Mechanical Engineering, Lovely Professional University, Phagwara, Punjab, India. Dr. Vasudev is also deputed as HOD of Division of Research and Development at Lovely Professional University, Phagwara. He received his PhD degree from Guru Nanak Dev Engineering College, Ludhiana-India in 2018 under the guidance of Prof. (Dr.) Harmeet Singh, GNDEC, Ludhiana and Dr. Lalit Thakur, NIT, Kurukshetra. His research areas include Surface Engineering/Thermal Spraying (HVOF, Flame Spray, Cold and Plasma Spray, and Microwave Claddings). He is currently working on the development of nanostructured, multi-modal, and high entropy alloy coatings for high-temperature oxidation and corrosion resistance – thermal barrier coatings (TBCs) and microwave processing of materials. Recently, he has been listed in the *World Top 2% Scientists* published by Stanford University and Elsevier – 2023. He has contributed extensively to thermal spray coatings in reputed journals which include *Surface Coatings and Technology, Materials Today Communications, Engineering Failure Analysis, Journal of Cleaner Production, Surface Topography: Metrology and Properties*, and *Journal of Failure Prevention and control, International Journal of Surface Engineering*, and *Interdisciplinary Materials Science* under the flagship of various publication groups such as Elsevier, Taylor & Francis, Springer nature, IGI Global and In-Tech Open. Moreover, he is a dedicated reviewer of reputed journals such as *Surface Coatings and Technology, Journal of Thermal Spray and Technology, Ceramics International, Journal of Material Engineering Performance, Engineering Failure Analysis, Surface Topography: Metrology and Properties, Material Research Express, Engineering Research Express*, and *IGI Global Journals*, etc. He has been awarded a "Top Cited Paper Awards India 2022" by @IOP Publishing, United Kingdom, in the Review Category. He has won the "Research Excellence Award" 2019, 2020, 2021 and 2023 at the Lovely Professional University. He has 40+ SCI indexed papers, 100 SCOPUS-Indexed papers, 15 conference papers, 5 books (1 self-authored and four edited books), and 25 book chapters. He has granted a patent titled "High-Temperature Oxidation and Erosion Resistant Alloy-718/Al_2O_3 Composite Coatings". He has organized national conferences and international conferences. He has received SEED MONEY for a project titled "Fabrication of coatings for Biomedical Applications", Project Number: LPU/DRD/SEED/SAC/140222/24804 from Lovely Professional University, India. He has also received a grant from the Scientific Engineering and Research Board (SERB), New Delhi, Government of India, to conduct a conference in 2023.

DR. ANKIT SHARMA

Dr. Ankit Sharma holds the position of Head of the Chitkara University Publications, Division of the Research and Innovation Network (CURIN) Department, Chitkara University, Punjab, India. He received his doctorate in Mechanical Engineering from the Thapar Institute of Engineering and Technology, India. He has more than 10 years of experience in academics, research, consulting, training, and industry. Dr. Sharma is also organizing an international conference – ICEMSMCI 2023 and serving as the general chair and convenor. He has authored numerous national and international publications in SCIE, SCOPUS, and Web of Science-indexed journals. He has filed/published 30+ international patents. He is also the Book Series Editor entitled "Innovations in Smart Manufacturing for Long-Term Development and Growth" of CRC Press, a Taylor & Francis publisher. He has been the editor of CRC Press, Taylor & Francis, and Elsevier's books related to advanced manufacturing processes and several other assignments are under the pipeline. He is the managing and special issue editor of several international journals which are SCIE, ESCI, and SCOPUS-indexed. The list of journals is the *International Journal on Interactive Design and Manufacturing*, Springer Publications; *Journal of Adhesion Science & Technology*, Taylor & Francis; *Machines*, *MDPI*, etc. He has delivered several invited seminar and keynote talks on international (USA, China, India) platforms and awarded with best research paper awards. He is also a review board member of several international journals such as *Measurement* (Elsevier), *Sage Part E*, *International Journal of Advanced Manufacturing Technology* (Springer), *Silicon* (Springer), *Sustainability*, etc.

DR. MURALI MOHAN CHEEPU

Dr. Murali Mohan Cheepu is currently a Research Manager at STARWELDS Inc. Busan, Republic of Korea. He has several years of industry experience, most notably at JSW Steel Ltd, India, as a Junior Manager and at STARWELDS Inc., as a Research Manager, specializing in the development of emerging materials, wire arc additive manufacturing processes for large-scale components, welding process design, automation, digitalization, and artificial intelligence. He received his PhD in Mechatronics Engineering at Kyungsung University in 2019. His master's degree was in Welding Engineering from the National Institute of Technology Tiruchirappalli. As well, he has a bachelor's degree in Mechanical Engineering from Andhra University, India. He also carried out his post-doctoral research at Pukyong National University, Korea, in 2019. Dr. Murali has extensive research experience in additive manufacturing, welding and joining technologies, development of welding filler materials, mechanical behavior materials, and machine learning and AI of the WAAM process. He has published numerous papers in scientific and conference proceedings, edited books, and patents, and continues to serve as the editorial board member of several journals. In addition, he received the "Best Researcher

Award" and "Young Fellow Award" from the Korean Welding and Joining Society in 2020 and 2023, and the "Outstanding Young Manufacturing Engineer Award" from SME in 2023. He was also awarded the "Outstanding Contributor Award" from the Korean Society for Precision Engineers in 2023. Dr. Murali is an IEEE senior member and member of KWJS, HKSME, AWS, KSME, and SME.

List of Contributors

Shuaibu Abdullahi, Department of Chemistry, Bayero University, Kano, Nigeria; and
Department of Chemistry, Modibbo Adama University, Yola, Nigeria

Mohamed Aboughaly, Faculty of Energy Systems and Nuclear Science, University of Ontario Institute of Technology, Oshawa, ON, Canada

A.J Aliyu, Airforce Institute of Technology, Kaduna, Nigeria

B.M. Aliyu, Department of Scientific and Industrial Research, National Research Institute for Chemical Technology, Zaria, Nigeria

Abdullah Al Noor, School of Bioengineering and Biosciences, Lovely Professional University, Punjab, India

Shashi Bahl, Department of Mechanical Engineering, I.K. Gujral Punjab Technical University Jalandhar Kapurthala, India

K.C. Bright, Department of Physics, Mar Ivanio's College (Autonomous), Thiruvananthapuram, Kerala, India

Subhendu Chakroborty, School of Applied Sciences, Chandigarh University, Unnao, India

Kritika Charak, University Institute of Biotechnology, Chandigarh University, Mohali, Punjab, India

Arish Dasan, FunGlass-Centre for Functional and Surface Functionalized Glass, Alexander Dubček University of Trenčín, Trenčín, Slovakia

Lipsa Das, Department of Electronics Engineering, Amity University, Greater Noida, India

Opeoluwa O. Fasanya, Industrial and Environmental Pollution Department, National Research Institute for Chemical Technology, Zaria, Nigeria

Ashish Goyal, Mechanical Engineering Department, Manipal University, Jaipur, Rajasthan India

Nitin Goyal, Mittal School of Business, Lovely Professional University, Phagwara, India

Niyati Gupta, University Institute of Biotechnology, Chandigarh University, Mohali, Punjab, India

Sunilkumar Harsur, Dr. D. Y. Patil Institute of Technology, Pimpri, Pune, India

Pallavi Jadhav, Dr. D. Y. Patil Institute of Technology, Pimpri, Pune, India

Vishal Jagota, Chitkara University Institute of Engineering and Technology, Chitkara University, Rajpura, Punjab, India

M. Abila Jeba Queen, Department of Physics, Holy Cross College (Autonomous), Nagercoil, India

Y. Julit, Department of Physics, Holy Cross College (Autonomous), Nagercoil, India

Gitanjali Kale, Dr. D. Y. Patil Institute of Technology, Pimpri, Pune, India

Gurmeet Kaur, Department of Computer Science and Engineering, Punjabi University, Patiala, Punjab, India

Gursimran Kaur, Mittal School of Business, Lovely Professional University, Phagwara, India

Hardeep Kaur, University Institute of Biotechnology, Chandigarh University, Mohali, Punjab, India

Sukhman Kaur, University Institute of Biotechnology, Chandigarh University, Mohali, Punjab, India

Karuna Khajuria, Department of Health and Medical Education, Jammu and Kashmir, India

Rajnish Kler, Department of Commerce, Motilal Nehru College, University of Delhi, Delhi, India

Anish Kumar, School of Bioengineering and Biosciences, Lovely Professional University, Punjab, India

Ashish Kumar, Chitkara University Institute of Engineering and Technology, Chitkara University, Punjab, India

Ashish Kumar Agrahari, Department of Physics and Astronomy, Clemson University, South Carolina, USA

Ashiwani Kumar, Mechanical Engineering Department, Feroze Gandhi Institute of Engineering and Technology, Raebareli, India

Hitesh Kumar, Department of Mechanical Engineering, Rayat Bahra Institute of Engineering and Nano Technology, Hoshiarpur, Punjab, India

Mukesh Kumar, Mechanical Engineering Department, MNIT Jaipur, Rajasthan, India

Rajeev Kumar, Mechanical Engineering Department, Lovely Professional University, Phagwara, Punjab, India

Sanjay Kumar, Mechanical Engineering Department, Dr BR Ambedkar National Institute of Technology, Jalandhar, Punjab, India

Rakesh Kumar Phanden, Federation University, Ballarat, Australia

Praveen Kumar Verma, School of Mechanical Engineering, Lovely Professional University, Phagwara, Punjab, India

Magaji Ladan, Department of Chemistry, Bayero University, Kano, Nigeria

Sawanta S. Mali, School of Chemical Engineering, Chonnam National University, Gwangju, South Korea

Kamalkishor Maniyar, Dr. D. Y. Patil Institute of Technology, Pimpri, Pune, India

Roshan Marode, Universiti Teknologi Petronas, Malaysia

Amrinder Mehta, Research and Development Cell, Lovely Professional University, Phagwara, India; and School of Mechanical Engineering, Lovely Professional University, Phagwara, India

Chandra Mohan, Department of Chemistry, School of Basic & Applied Sciences, K R Mangalam University, Gurugram, India

Ganesh D. Mote, Pharmaceutical Chemistry Department, Annasaheb Dange College of B. Pharmacy, Ashta, Sangli, Maharashtra, India

Nigar K. Mujawar, Department of Pharmaceutics, Womens College of Pharmacy, Peth Vadgaon, Kolhapur, Maharashtra, India

H. Musa, Department of Chemistry, Bayero University, Kano, Nigeria

J.M Nasir, Department of Chemistry, Bayero University, Kano, Nigeria

Prashant Patil, Dr. D. Y. Patil Institute of Technology, Pimpri, Pune, India

Gaurav Prashar, Department of Mechanical Engineering, Rayat Bahra Institute of Engineering and Nano Technology, Hoshiarpur, Punjab, India

Jaiinder Preet Singh, School of Mechanical Engineering, Lovely Professional University, Phagwara, Punjab, India

Saloni Raheja, Mittal School of Business, Lovely Professional University, Phagwara, India

Samrat Ray, International Institute of Management Studies, Pune, India

Shivaji B. Sadale, Department of Technology, Shivaji University, Kolhapur, Maharashtra, India

Vinay Saini, Department of Mechanical and Manufacturing Engineering, University of Calgary, Canada

Ankur Salunke, Dr. D. Y. Patil Institute of Technology, Pimpri, Pune, India

Avinash Salunke, Dr. D. Y. Patil Institute of Technology, Pimpri, Pune, India

Ankitha Sharma, Mittal School of Business, Lovely Professional University, Phagwara, India

Parveen Sharma, School of Mechanical Engineering, Lovely Professional University, Phagwara, India

Rohit Sharma, School of Mechanical Engineering, Lovely Professional University, Phagwara, India

Vikas Sharma, University Centre for Research and Development, Chandigarh University, Mohali, Punjab, India

Nitin Sherje, Dr. D. Y. Patil Institute of Technology, Pimpri, Pune, India

Jashanpreet Singh, University Centre for Research and Development, Chandigarh University, Mohali, Punjab, India

Amanjot Singh Syan, Canadore Stanford International College, Ontario, Canada

Neelofar Sohi, Department of Computer Science and Engineering, Punjabi University, Patiala, Punjab, India

Ruchira Srivastava, Department of Mechanical Engineering, Amity University, Greater Noida, India

Shyam Sunder Agrawal, Mittal School of Business, Lovely Professional University, Phagwara, India

Karunesh Tiwari, Department of Physics, Mai Nefhi College of Science, Himbriti, Mai Nefhi, Maekel Region, Eritrea

Ayushi Thakur, Department of Mechanical Engineering, Amity University, Greater Noida, India

Babaso V. Udugade, Department of Pharmaceutical Chemistry, Ashokrao Mane College of Pharmacy, Peth Vadgaon, Kolhapur, Maharashtra, India

Hitesh Vasudev, Research and Development Cell, Lovely Professional University, Phagwara, India; and School of Mechanical Engineering, Lovely Professional University, Phagwara, India

Sunil Verma, Department of Management Information System, Webster University, Tashkent, Uzbekistan

Rajan Vohra, School of Electronics and Electrical Engineering, Lovely Professional University, Phagwara, Punjab, India

Lalitkumar Wadhwa, Dr. D. Y. Patil Institute of Technology, Pimpri, Pune, India

Abdul Wahab Hashmi, School of Mechanical Engineering, Shandong University of Technology, Zibo, China

Pushparaj Warke, Dr. D. Y. Patil Institute of Technology, Pimpri, Pune, India

J.M Yelwa, Department of Scientific and Industrial Research, National Research Institute for Chemical Technology, Zaria, Nigeria; and Department of Chemistry, Bayero University, Kano, Nigeria

Noushi Zaidi, Department of Basic Science, College of Science, King Faisal University, Al-Ahsa, Saudi Arabia

1 Introduction to Green Technology and Smart Materials

Jashanpreet Singh

Description	Abbreviation
Additive manufacturing	AM
Calcium–silicate–hydrate	C–S–H
Electric vehicles	EVs
Electroactive Polymer	EAP
European Union's Circular Economy Action Plan	EUCEAP
Federal investment tax credit	FITC
Green technology and smart materials	GTSM
Intergovernmental panel on climate change	IPCC
Light-emitting diode	LED
Magneto-Rheological	MR
Nitinol	NiTi
Photochromic material	PCM
Self-healing material	SHM
Shape memory alloy	SMA
Shape memory polymers	SMP
Surface acoustic wave	SAW
Tetracycline ribosomal protection protein	TRPP
Thermo-chromic material	TCM

1.1 INTRODUCTION TO GREEN TECHNOLOGY

Nowadays, there is a need for effective solutions to environmental problems such as global warming, depletion of resources and pollution. Today, it has been acknowledged that green technology and smart materials are vital in the development of a more sustainable society [1,2]. Green technology is a technology that is supposed to reduce the negative effect of human beings on this planet and lower the hazard of climate change [3]. On the other hand, smart materials are defined mainly as materials

DOI: 10.1201/9781003516071-1

that exhibit the ability to change shape or other characteristics, as a response to applied signals like heat, pressure or an electric field [4]. Synergistically, both fields have all the potential to reimagine industries in construction, manufacturing, energy and transportation. The main concepts of green technology and smart materials are depicted in the Venn diagram in Figure 1.1 below.

Clean technology or environmental technology are other terms used to describe green technology [5]. This technology is applied to address the needs of the population with the least negative effects on the natural environment. The core idea underlying green technology is to develop constructions and products that release the lowest emissions of pollutants, preserve natural resources and consume minimal energy. This may include solar power and wind power as types of energy, energy-saving building constructions, water purification methods or technologies for recycling waste, among others [6–8]. The impact of green technology is considerable if it attempts to avoid the negative impact of human activities on the surrounding environment. For many decades, industrialization and mass production have been two of the main factors that are contributing to environmental issues, for example, deforestation and pollution of water and air resources, as well as dehydration of natural resources [6,]. From an international viewpoint, green technology is a pioneer in preventing global warming. According to the Intergovernmental Panel on Climate Change (IPCC), it is still possible to avoid such consequences if global warming does not exceed 1.5°C above pre-industrial levels [12–14]. This can only be done by reducing the emission of greenhouse gases, which can only be achieved with the adoption of green technologies in different areas of the economy.

This chapter describes the fundamental concepts of green technology and smart materials and then outlines some of their benefits and drawbacks. Further, a few

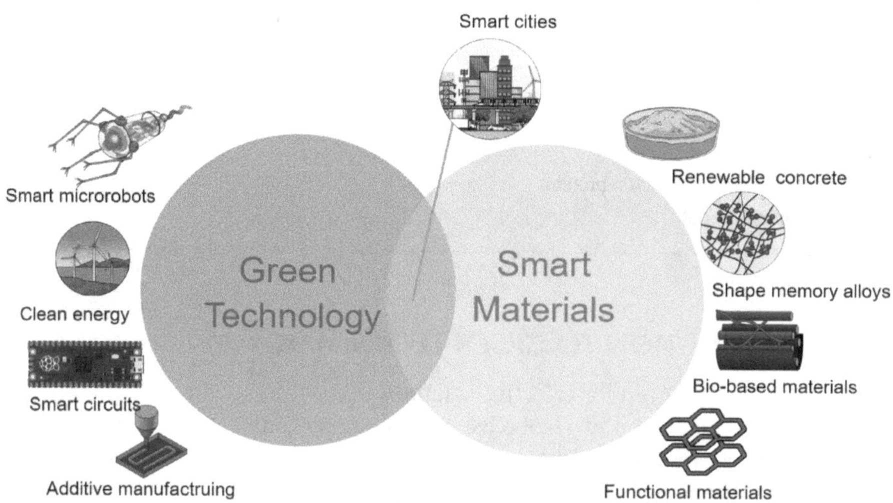

FIGURE 1.1 Schematic diagram on the synergy of green technology and smart materials (Credit statement: created by author).

smart materials which use less energy and have the potential to contribute to building smart cities are also described in this chapter. Other topics that are covered in this chapter include technological solutions to environmental challenges, how smart materials work and applying technology to various industries to improve a product's quality by reducing wastage. Through the use of many types of smart materials, smarter and more energy-efficient cities can be developed. Conventional smart materials include shape memory alloys, smart self-healing materials and photo-/thermochromic materials [15,16]. The environmental effects of buildings can be reduced through the development of more smart materials by exploiting their physicochemical properties and their responsiveness to external stimuli, such as temperature increase, additional sunshine or the appearance of cracks. Further, one traditional material that is mentioned as being 'smarter' and 'greener' through the inclusion of sensors is concrete [17].

1.2 IMPORTANCE OF GREEN TECHNOLOGY

1.2.1 Renewable Energy

Renewable energy is essential to the global energy supply because it decreases the demand for fossil fuels and minimizes carbon emissions. It will also become more economical as consumers demand better technologies, which brings down the cost of production of energy and improves the efficiency of energy retrieval and storage. One of the major fields of green technology is the utilization of renewable energy sources. However, it cannot be said that such use of green technology is all-pervasive and probably the most well known. While examples of fossil energy sources are coal, oil and gas, renewable energy like solar, wind or hydro energy are in vast abundance and emit a minimal volume of greenhouse gases compared to fossil fuel energy sources [18,19]. Figure 1.2 represents a schematic diagram for the use of renewable energy in smart cities. Clean energy technologies include solar power systems that use solar energy, windmills that use wind energy, and hydroelectric power stations that use water energy [20]. Renewable energy is the best way to reduce the use of fossil fuels and thereby reduce the emission of carbon into our atmosphere [21]. At the same time, it is becoming more economical because in recent years new technologies that allow for cutting the costs of manufacturing and improving the efficiency of storage and generation of energy have been developed [22,23].

1.2.2 Energy Efficiency

Energy efficiency is another important characteristic of green technology. Energy conservation systems and products refer to those used during construction in order to ensure the minimal possible use of energy for executing a specific task [24]. Some examples include light-emitting diode (LED) lights, efficient electrical appliances and improved building materials that can be used when developing buildings and that are efficient in minimizing heat gain from the surroundings [25]. Energy saving is a part of reducing the organizational energy need and this results in a low carbon

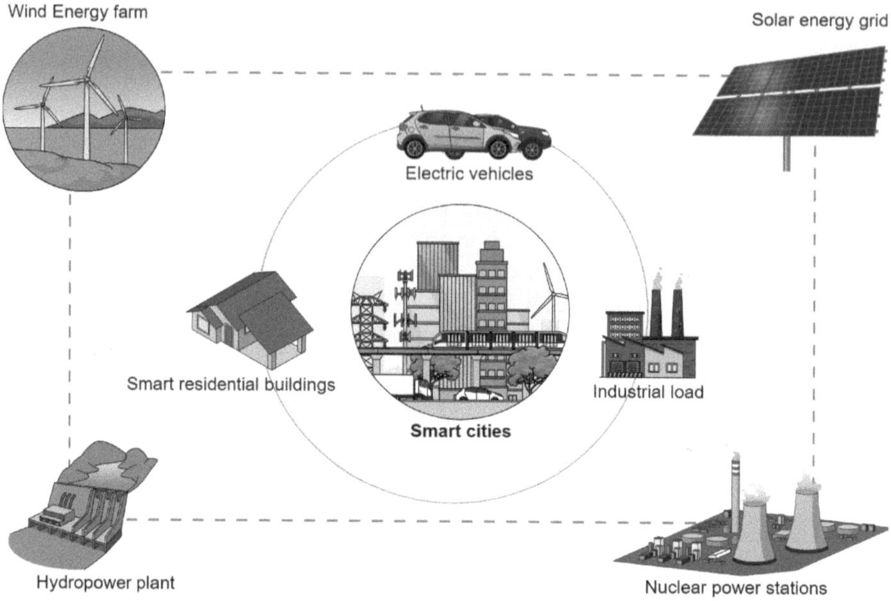

FIGURE 1.2 Schematic diagram for the use of renewable energy in smart cities (Credit statement: created by author).

footprint on the power networks [26]. Energy efficiency technologies are therefore as equally crucial as renewable energy in the effort to increase the sustainability of energies.

1.2.3 Sustainable Transportation

Green technology is also applied to transport, which is strongly indicted for emitting greenhouse gases into the atmosphere. Technological advancements in electric vehicles (EVs), hybrid cars and hydrogen fuel cells are reducing the effects of transport on the environment [27], as shown in Figure 1.3. Also, other comprehensive land uses that include the improvement of transport networks, like better means of transport, including public transport, cycling and walking infrastructure, are also embraced.

1.3 SMART MATERIALS

Smart, functional, adaptive or responsive materials are those whose characteristics vary based on heat, light, pressure and other such signals, or changes due to the influence of electric fields [28–31]. These materials can undergo other phase changes reversibly to adapt themselves to the altering environmental conditions at a particular time. This is because smart materials are a new generation materials and have immense potential for bringing out revolutionary changes and modernization

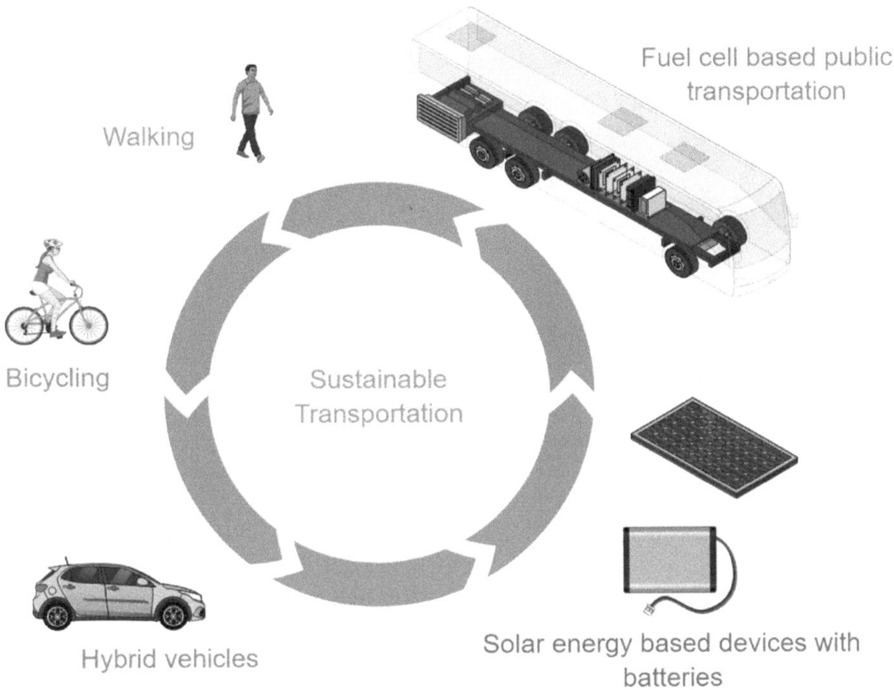

FIGURE 1.3 Schematic diagram for sustainable transportation in smart cities (Credit statement: created by author).

in many fields, namely medical, electrical and manufacturing sectors. Smart materials are grouped into types based on their characteristics or the services they provide. Some of these distinctly developed intelligent materials are as follows:

1.3.1 Magneto-Rheological (MR) Fluids

Magneto-rheological fluids can be classified as smart materials that show unique characteristics with respect to the alteration of rheological properties when a magnetic field is applied [32–34]. The mechanism of MR fluid is shown in Figure 1.4. These include a non- colloidal based suspension in a low viscosity carrier fluid, such as mineral oil or silicone oil, of larger ferromagnetic particles, especially carbonyl iron [35]. MR fluids change their viscosity when exposed to magnetic fields and thus find applications in automobile engineering, such as designing a shock absorber that adapts to the available magnetic fields. This property enables MR fluids to transform from a free-flowing liquid to a semi-solid state in a very short time, thus making it suitable for several functions. The process includes the formation of linear structural phases from ferromagnetic particles causing an increase in the yield stress and change in the rheological properties of the fluid. Such behavior is often described using the Bingham plastic model in which the fluid possesses a yield value of stress.

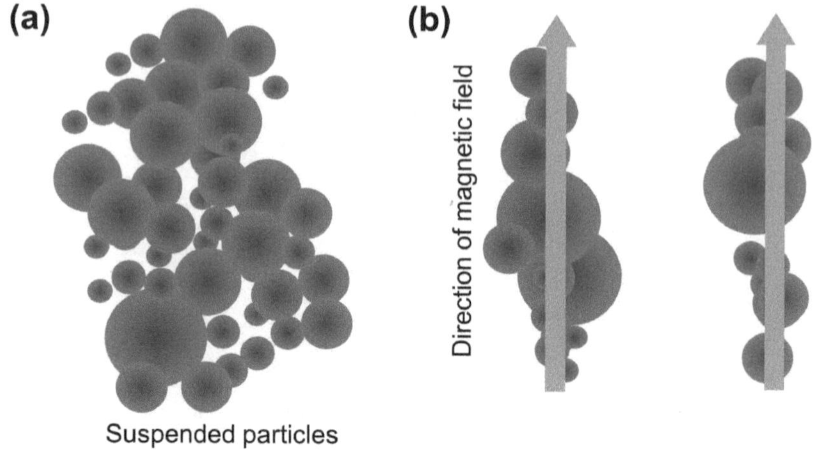

FIGURE 1.4 Mechanism of MR fluids with (a) no magnetic fluid and (b) with an applied magnetic field (Credit statement: created by author).

Above this value, the fluid flows and shows solid-like behavior at rest, but flows like a liquid under stress [36–38].

MR fluids are applied to a wide variety of fields, such as automotive, aerospace, civil engineering, manufacturing and industrial, robotics, etc. [39]. In automotive applications, MR fluids are used in high-performance suspension and dampers wherein the damping coefficients depend on the road conditions as well as on the behavior of the vehicle. This capability enhances vehicle stability and comfort together with safety by increasing traction required while manuvering. In the case of aerospace applications, MR fluids play a role in shock absorbing functions, and vibration control technologies that enable the tracking of the control of the dynamic loads and mitigating effects of vibrations on instruments and people. Furthermore, it is also used in prosthetic or rehabilitation devices where the ability to control the motion of the device has to be adjusted by the resistance of the fluids and their ability to change resistance following natural movements. However, a number of issues affect the highly attractive characteristics and realizable applications of MR fluids. The major limitation is associated with sedimentation: the ferromagnetic particles eventually settle at the bottom of the carrier fluid thus weakening the performance of the device. Presently, investigators and researchers are working on several additives and formulations with a view to increasing MR fluid stability while still yielding the optimal rheology. Still, a major challenge is maintaining a fine particle size and uniformity of the particles. Although finer particles are helpful in improving the stability of the fluid and reducing the achievable yield stress under the magnetic stimulation. For the best performance in these applications, a number of factors also have to be well balanced. A new method in material science is to use dimorphic MR fluids in which spherical particles contain nanowires or other shapes to reduce sedimentation rates as well as increase the magnitude of the yield stress. These advancements are pushing the producers of MR fluid to create better formulations, which will suit specific industrial applications. Further, in present development more efforts have been directed toward improvements in thermal stability and the working range of the MR fluids for different severe applications.

MR fluids show how smart materials can bring change in several fields because of their flexibility. They are also able to undergo phase transition between solid and liquid phases, provided they receive stimuli from their environs; thus improving performance in addition to giving a new perspective to engineering as compared to design. However, more studies are currently being carried out to address these problems and to increase the functionality of these materials. Even further, MR fluids have a great potential to play a large role in future technologies in numerous areas.

1.3.2 Electroactive Polymers (EAPs)

Electroactive polymers (EAPs) are part of the largest group of smart materials that undergo mechanical strain when exposed to electric currents, and they can be applied as actuators and artificial muscles [40–42]. These polymers can be deformed rather easily, at times to as much as 380% of their original length; quite apart from this they offer considerable force, which, of course, is in sharp contrast to some of the ceramic actuators which offer negligible deformation and no force at all. Figure 1.5 shows a schematic diagram of 'on-demand' molecular release activated by electrical stimulation of EAPs [43].

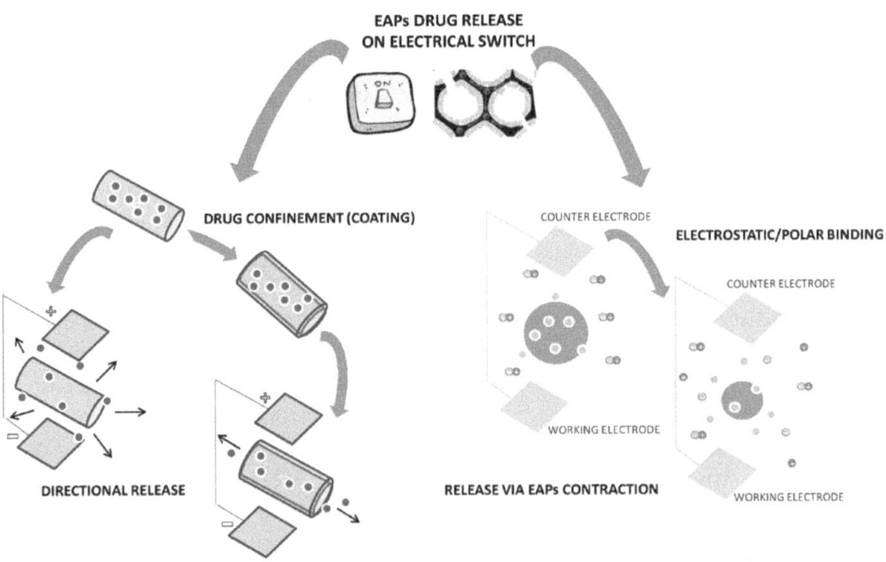

FIGURE 1.5 Schematic diagram of 'on-demand' molecular release activated by electrical stimulation of EAPs [43] (Credit statement: Creative Commons International Attribution 4.0).

EAPs are generally classified into two main categories: the first comprises *aqueous* and *polar* covalent linkages, while the other consists of *ionic* and *dielectric* bonds. *Ionic EAPs* attain their actuation from the movement of ions within the polymer substrate [44–46]. They are useful at low voltage (1–2 V) for bending movement such as in soft robotics and artificial muscles [47]. However, they must be provided with water to function properly and this makes them inconvenient for specific environments. On the other hand, *Dielectric EAPs* exploit forces that are generated by an external electric field, which turns the material into a different form [48,49]. This type in general requires high voltage, but it does not require any other power in the form of voltage to maintain its form and constant voltage; hence its use in robots and sensors [50]. The low density, high tracking rate and mechanical elasticity of the EAPs make the offering applicable in innovative applications in biomedical applications, such as prosthetics and tissue regeneration and in smart structures and robotics. However, some difficulties are still associated with the questions of the EAPs' scalability, stability and the control mechanisms required for the wide applications. The EAPs are expected to make a substantial input on the emergence of the next generation smart material with possible biological behavior and compatibility with stimuli.

1.3.3 Self-Healing Materials (SHMs)

SHMs are a new and strange type of smart material with the capacity to restore itself in the event it gets harmed [51]; this makes it stronger, particularly for their use in coating for electronic devices and construction materials [52,53]. Such materials possess the ability to reform itself similar to the way tissue in living things is repaired. The material can self-heal and is designed to respond to damage without more input from the programmer. A schematic diagram of the self-healing phenomenon and its applications is shown in Figure 1.6 [54]. Self-healing can be categorized into two forms, namely intrinsic self-healing and extrinsic self-healing. *Intrinsic SHMs* contain within themselves the ability to repair fractured bonds as soon as they are damaged [55,56], while *extrinsic SHMs* are embedded with microcapsules of healing agents that are released as soon as the structure cracks or gets damaged [57,58]. In several cases on polymer-based SHMs, the microcapsules carrying a healing agent are designed to break up upon the failure of the material. This results in the healing agent infiltrating the crack zone and subsequently solidifying. Perhaps, this can help to achieve better mechanical properties and protection in damaged condition. This has given rise to innumerable possibilities in the usage of SHMs in aerospace structures, automobile industries in construction, electronics, etc. It also assists in reducing the costs of maintenance and is very effective in preventing major damage to crucial structures. Apart from that, the use of SHMs in consumer products like phones and automotive paint can help to extend the durability of the product, reducing the amount of waste produced [59]. Nevertheless, it has some notable difficulties concerning its applicability at larger scales or its profitability. Recent investigations are directed at increasing the effectiveness of the healing process, reducing the time required for healing, and the stability of performance with reference to the

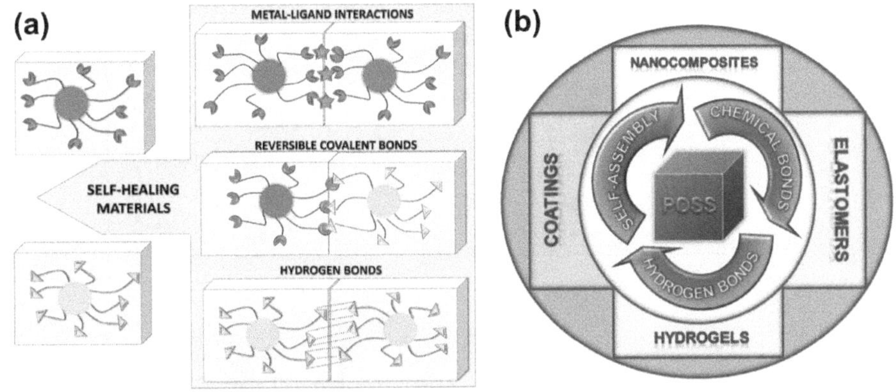

FIGURE 1.6 Schematic diagram of (a) the self-healing phenomenon and (b) applications of polyhedral oligomeric silsesquioxanes (POSS) based SHM [54] (Credit statement: Creative Commons International Attribution 4.0).

characteristics of the environment. Still more improvement is being made, thus it can be said that SHMs are going to revolutionize several industries with the need for efficient, durable and resource-efficient functional materials. Looking into the future of SHMs, very high possibilities exist for their use for the improvement of structures to withstand detrimental processes causing their degradation.

1.3.4 Shape Memory Alloys (SMAs)

Shape memory alloys can change their shape back to an initial shape set during the manufacturing of the product under certain factors, such as temperature conditions [60,61]. They are used in medical applications and actuators. Many potential applications of SMAs especially those based on NiTinol alloys include biomedical, aerospace, construction civil engineering industries, etc. [62–65]. SMAs work on the super-elasticity phenomenon given in Figure 1.7. NiTinol possesses several crystal structures at different temperatures. The crystal lattice at lower temperatures may have two mirror images of the constituent cell in the martensite phase, which are energetically equivalent [66]. It may also undergo large deformation at high temperatures, other than the martensite–austenite transition point, and this is known as the austenite phase. It should be noted that NiTinol has the properties of super-elasticity if it is in the austenite phase. Large transformation of an SMA is accomplished by mechanical loading, not via elastic straining, by the changing of the elementary cells' place. As long as the temperature of the material does not rise beyond the martensite–austenite transition point, the deformation will be retained. This type of material is almost immediately restored to its memory state when this happens. This is so since during the austenite phase there is a limited number of possible arrangements of the crystal lattice [67]. This becomes apparent when the temperature goes above the martensite–austenite transition temperature and the entire material adopts

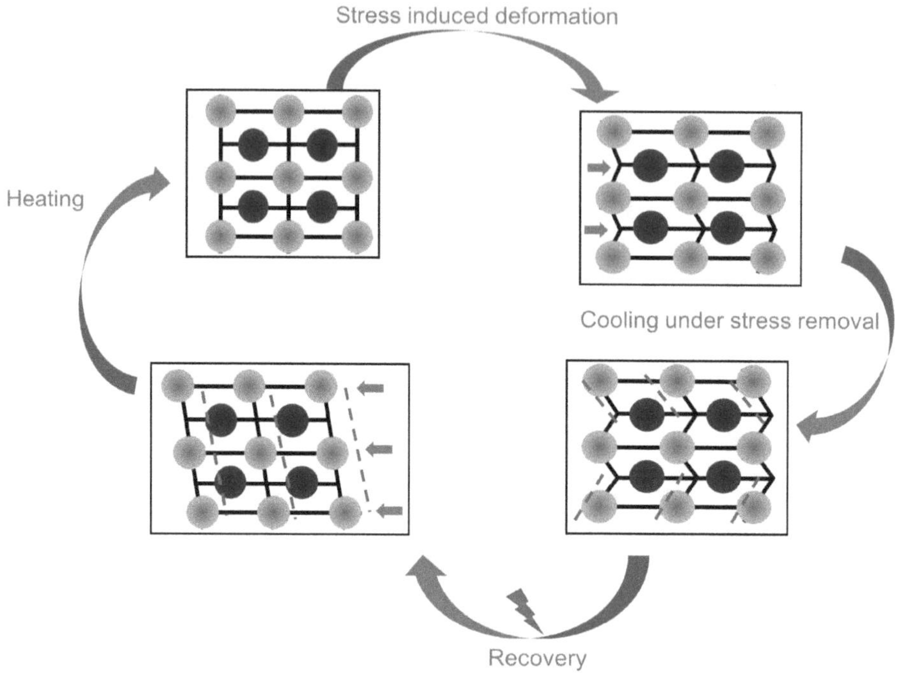

FIGURE 1.7 Schematic diagram of the super-elasticity phenomenon in SMAs (Credit statement: created by authors).

this form with each elementary cell transforming at once. Metallurgical manufacturing of the wire involves a solidification and cooling stage, which can put a seal on the memory shape. NiTinol has an additional interesting property that may be applied to green technology or city design [68]. In normal life, external forces might distort their arteries. On the other hand, super-elastic stents do not harm the human arteries and can undergo transformation to extremely low or high diameters [69].

1.3.5 Piezoelectric Materials

Piezoelectric material is one of the most wonderful types of smart material, which has the capability to convert mechanical energy to electrical energy and vice versa [70,71]. This is because of what is called the piezoelectric effect in which a section of the material develops electricity, producing an electric charge when pressure is applied to it in the manner shown in Figure 1.8. In the other case, when the electric field is applied these materials deform or in some cases, a minor deformation occurs. This direct and reverse energy conversion makes the piezoelectric material favorable for several applications, including its use in sensors, actuators and energy harvesting. It is applied for small and general purposes, and in other devices, such

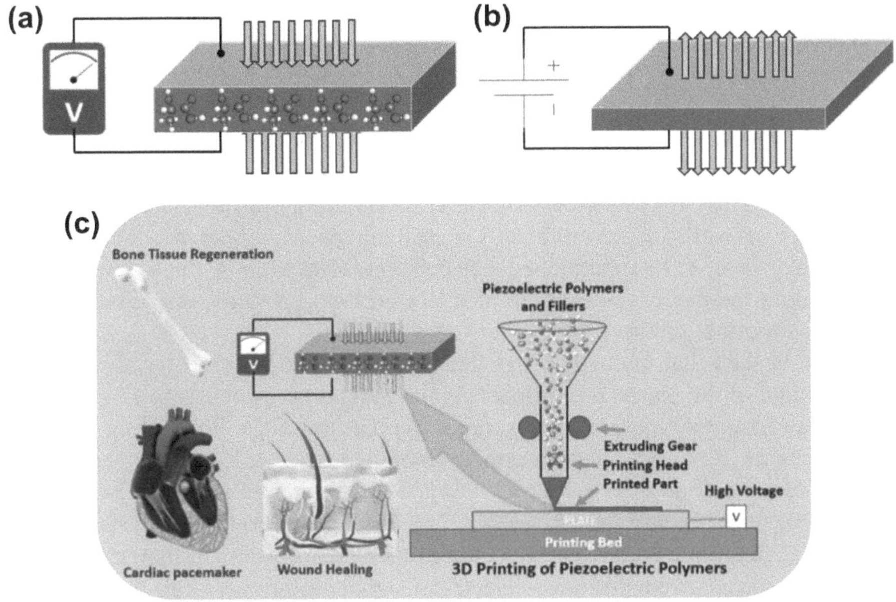

FIGURE 1.8 Schematic diagram of (a) the direct and (b) the converse piezoelectric effect, and (c) applications of 3D printed piezoelectric polymers [74] (Credit statement: Creative Commons International Attribution 4.0).

as microphones or speakers, as well as in devices of larger scale, such as pressure sensors, and also in medical applications and in robots.

Piezoelectric material can be divided into single crystals, such as barium titanate, ceramics including lead zirconate titanate and polymers, such as polyvinylidene fluoride [72,73]. Both possess distinct pros, for example, piezoceramics has high piezoelectric coefficients and chemical stability and it is more or less performance-driven. Piezoelectric polymers are quite unlike these, as they are flexible in form and also light in mass, and it can be used in wearable technologies and soft robotics. There have been enhancements in the properties of nanostructured piezoelectric materials that has created opportunity for the invention of several devices capable of charging through vibrations or movements of a person.

Therefore, with regard to the further development of technologies, the prospects for the application of piezoelectric materials are quite promising. Further advancements in mechanical and material engineering technologies are expected to bring even more efficient and versatile piezoelectric systems to meet the needs of the future generation and its applications across several fields of endeavor [74]. From improving consumer electronics up to a better healthcare system, and even powering the green environment, piezoelectric materials are a prime example of smart materials shaping the lives of today's citizens. The other characteristic which supports the classification of such materials as smart materials is the ability of these materials to function intelligently and adapt to the different stimuli received from the environment. More importantly these materials fit perfectly with the need for a more responsive and interactive technological landscape.

In the area of energy harvesting, piezoelectric material has the very important function of converting kinetic energy, which includes kinetic energy generated from walking or from the running on a treadmill, into electricity [75,76]. This capability is valid only in self-charging devices and wireless sensors that rarely need to be recharged. The currently emerging trend in the design of other materials under investigation is work on new fabrication technologies, polymer nano-composites with conductive filler to enhance piezoelectric phenomena [77]. Such development is pertinent in the process of making ideas for energy generation tangible at a time when it is important to move away from fossil fuel [78].

Also, it is seen that biomedical applications of piezoelectric material prove the huge potential of the material to impact significantly on the concerns of biomedical science. For example, these energy sources can be used in power-storing implantable devices or in physiological parameter-measuring devices, such as sensors for detecting heartbeats or muscle contractions [79,80]. Typical examples of 3D-printed piezoelectric polymers are shown in Figure 1.8. Bulk-specific piezoelectric polymers exhibit biocompatibility; they can be used in medicine as implant materials or tissue engineering scaffolds to induce electrical signals to promote cell proliferation. Despite this, there are challenges that make it difficult to apply piezoelectric materials widely.

1.3.6 Thermo-chromic and Photochromic Materials

Thermo-chromic and photochromic materials can be considered as two crucial examples of smart materials that change their properties depending on the environmental factors, temperature and light, respectively.

- Thermo-chromic materials (TCM) are those that alter their color with temperature and this may be as a result of phase change, change in the ligand field or change in the number of solvent molecules surrounding the compound []. These materials can be categorized into four main types: inorganic, organic, polymeric and hybrid sol-gel systems. Each one of them has its peculiarities and application in practice. For instance, inorganic LCMs like vanadium dioxide (VO_2) show phase transformation from semiconductor to metal, accompanied by a color change which can be used for smart windows for building energy efficiency []. On the other hand, organic TCMs exist and are often used to incorporate leuco dyes that change their color reversibly by responding to fluctuations in temperature [85] (Figure 1.9). They feature in mood rings and thermo-chromic inks for fabrics and packaging material. This is so because TCMs have variegated applications and their applications encompass energy-saving coatings for construction and sensors maintaining changes in temperature under diverse circumstances.
- Photochromic materials (PCMs) undergo reversible transitions from one isomeric form to another when the material is exposed to a defined wavelength

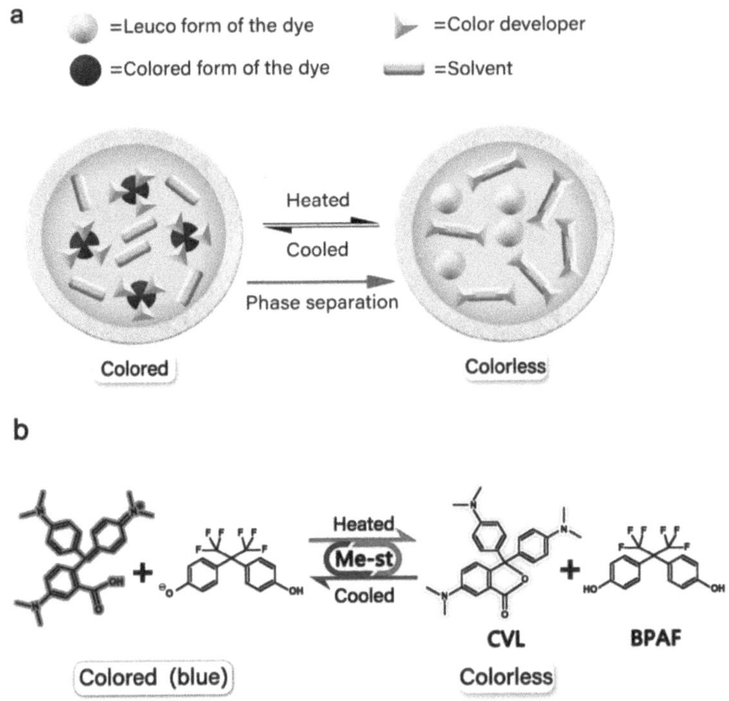

FIGURE 1.9 Schematic illustration of the thermos-chromic behavior of RT microcapsules (a) leuco dye (CVL), color developer (BPAF) and solvent configuration; (b) chemical reaction of the CVL dye and color-forming reaction of CVL with BPAF [81] (Credit statement: Creative Commons International Attribution 4.0).

of light, normally ultraviolet (UV) or visible light []. This makes them easily switchable in terms of color or in terms of opacity and can be used in products, such as darkening glasses (telescopic sunglasses), or smart windows that change shades according to the intensity of light. Concerning the working principles of photochromism, the change that is often associated with molecular constitution involves photoisomerization or redox [88–90]. Organic compounds, such as spiropoxazines and diarylethenes, revert from color to color on exposure to light and from color to colorless on exposure to light. In the visible region of the spectrum between the grounded and the excited state the incoming photons of energy equal to the light-induced excitation of the electrons cause the material to partly admit these photons and display a shade of color typical to the material makeup. Also, while luminescence is connected with radiative de-excitation along with the outgoing photons, in such a case the de-excitation can be non-radiative, by means of phonon oscillations. As Figure 1.10(a) indicates, this relates to two groups of compounds and all of the listed properties are similar for both. One group is used to name other organic dyes, such as anthocyanin

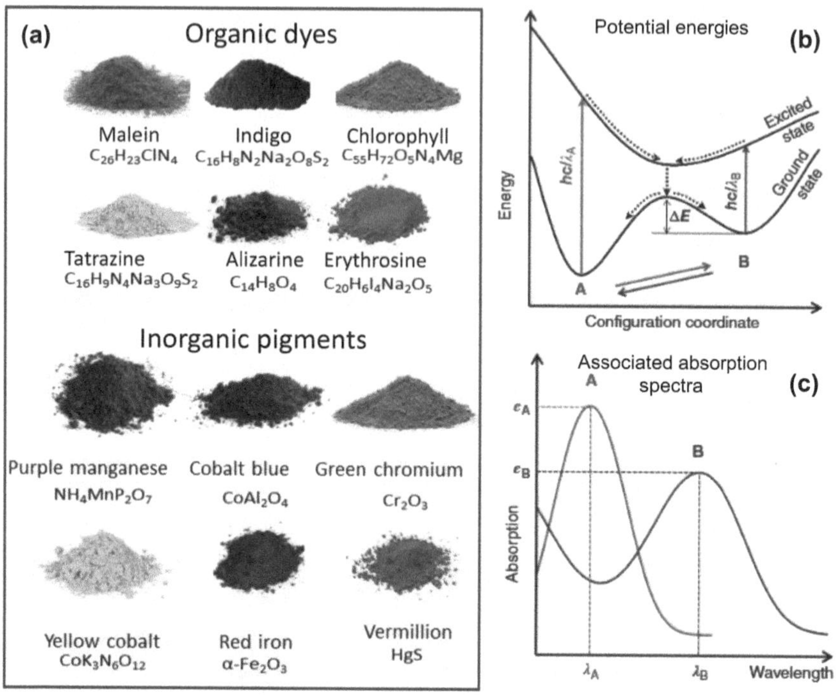

FIGURE 1.10 (a) Various organic and inorganic pigments, and (b, c) graphs representing electromagnetic excitation between two molecular states (A & B) result in a bidirectional reaction [92] (Credit statement: Creative Commons International Attribution 4.0).

dye or the malachite green dye, due to solubility in dispersive media. The second group is formed by inorganic pigments: smalt, cinnabar, vermilion or ochre, which are not soluble in the dispersive phase [91]. Despite this, a few of the organic compounds described above can be used as pigments but not as dyes. It is also important to know there are organic dyes, and inorganic pigments can demonstrate the ability to change their color depending on the length of the exposure to light. Therefore, these PCMs are used in the manufacturing of eyeglasses, data storage media and some parts of molecular switches' high-performance electronics. Although great strides have been made with these polymers, questions remain about the stability of such systems and the fatigue properties of the materials when exposed to prolonged use in the presence of light. Schemes of any distortions that allow a material to undergo a reversible change for the removal from a thermally stable state A to a metastable state B produced by electromagnetic radiation are called photochromic [92], as illustrated in Figure 1.10(b,c). The B state especially takes more time to reach the most generally typical stable state of A, either by its own transformation or by the application of some activation process like heating, exposure to a certain medium, exposure to different wavelengths of light, etc. In general, photochromes are classified depending on which type of stimulus helps to initiate the B → A revert status. There are various ways in which the system could return to its original state provided that the transition from A → B is always photoactivated; for instance, a low ΔE and high higher potential energy barriers.

Both TCMs and PCMs can be viewed as integral parts of smart materials and technologies to store and deliver useful energy to users. Not only can they facilitate products that possess such flexibility to respond to outside signals for developing artistic products but also play roles in the energy conservation of various applications in terms of sustainability. For instance, a thermo-chromic coating may be applied on building surfaces in order to reflect sun rays thereby reducing the amount of heated conditioned air required. Similarly, it can be stated that PCMs can be used to save energy because they can control the extent of light coming into a particular room thereby reducing the extent to which artificial light is called into use. The current literature review established that the current research on both TCMs and PCMs still holds prospects for the revelation of new potentialities for their utilization in numerous areas.

1.3.7 Renewable Concrete

A new approach to reduce the CO_2 emission rate and increase building durability should be undertaken to implement the concept of autonomous reparability and healability directly into cementitious materials. Because of its manufacture and relatively desirable mechanical properties, concrete has been the most prevalent material in the building industry up to this point [93]. Its elements include sand, coarse filler, water and cement. A major component of the inorganic composite that shares credit for the composite rigidity is the calcium–silicate–hydrate (C–S–H) domain, which is formed after the concrete been dried and cured [94,95]. The tensile properties of concrete, such as elongation and tensile strength, to a large extent are limited, even though they possess very high compressive strength. Therefore, during the service period of bridges, roads and buildings that are constructed from concrete, they are likely to crack. One way of ensuring structures are strong despite the presence of water is self-healing concrete that involves sealing off the damage zone without outside interference [96–98]. The same self-healing capacity of cementitious materials for cracks up to 50 μm could be observed here without any obvious change [99]. As illustrated in Figure 1.11, several processes can be involved with this so-called autogenous healing. This includes the incorporation of dissolved calcium hydroxide, swelling of the C–S–H phases, the physical closures of cracks that usually arise due to the presence of small particles in water and delayed cement hydration of residual reagents [100]. In any case, water must get to the damaged area to react with either the matrix of the concrete or to transport the necessary chemicals.

Combing with autogenous or bacteria-based self-healing concretes, carbon, cellulose, thermoplastic or glass fibers enhance the concrete with a higher capability to withstand larger deformation and control fracture propagation [50,]. Another

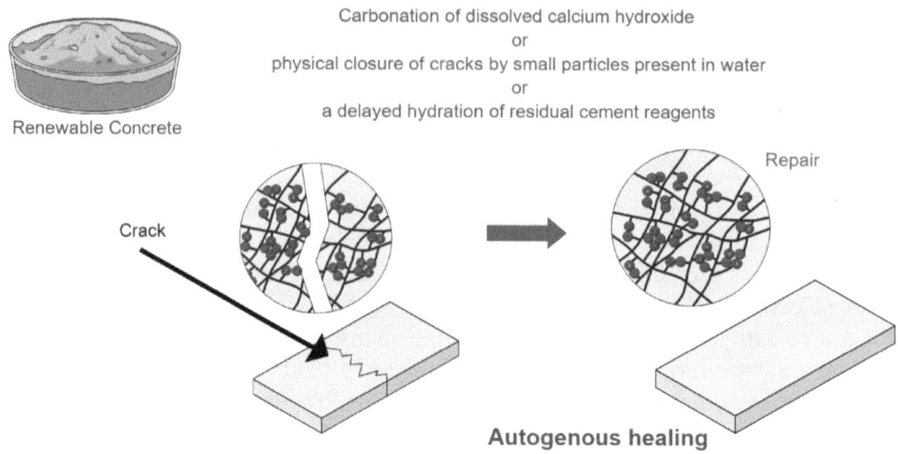

FIGURE 1.11 Schematic diagram of the autogenous healing process in renewable concrete (Credit statement: created by authors).

conventional way to improve the fracture toughness of fiber-reinforced composites is fiber bridging, which takes place when fibers stop cracking [104–108]. For an autogenous self-healing, crack width is a problem, but with the help of fibers it becomes possible, though it can only heal minor fractures [109]. This, together with enhanced tensile properties, significantly reduces the risk of failure in the mentioned states [110–118].

Another method of getting concrete to self-heal is by incorporating functional polymers into the cementitious mix [102]. Some of the reversible mechanisms that functional polymers use for healing include hydrostatic swelling of superabsorbent polymers in water [119], thermo-activated viscosity change, e.g., covalent adaptable polymer networks [120], and thermo-activated expansion/contraction of SMPs [121]. It has also been suggested that the incorporation of certain polymers into the concrete enhances its ductility, fracture toughness and permeability as well as the self-healing properties of concrete [119,121]. Various materials as well as advancements in construction applications have tapped into chemical and physical self-healing mechanisms, including microcapsules, vascular systems, dynamic polymer networks and others.

1.3.8 Innovative Concrete

Concrete structures range from buildings, tunnels and bridges and these all require a huge amount of energy to produce on a global scale. Another approach might be to decrease the overall energy use of the construction industry by improving cement-making technologies and fine-tuning all stages of construction and large structure creation [122,123]. This objective is highly investigated and with a great interest in research and development worldwide. Third, we have to make all reinforced concrete structures to last as long as possible. This can only be achieved if these buildings are maintained and preserved in the best way possible. Appropriate application of self-healing concrete solutions is required in these fields, and as important is monitoring the signs of degradation of the structure's integrity for performing the corresponding maintenance, repair or consolidation work. The monitors installed on the structure and over the bridge deck comprise a battery of sensors placed directly on the concrete surface, which is presently used to monitor the early signs of degradation of bridges and tunnels [124,125]. The sensors used on the structures include those for deformation, temperature, vibration and humidity. For the purposes of reading as well as transmitting data without a wire, these sensors are connected to an electrical box. I have seen panels of photovoltaic material installed on these boxes to power the systems within them. These solutions are widely promoted by many providers and are known well. However, these technologies cannot be used to measure other parameters of the structures, such as internal concrete temperature or deformation. It is not optimal to obtain these two values from observations of the surface. Applications such as structural health monitoring and other smart concrete applications might require Surface Acoustic Wave (SAW) sensors being incorporated into concrete structures [126,127]. Other extensions of SAW solutions for the measurement of other parameters, including deformation and corrosion, can be implemented in the future. Besides, such sensors may help in the formulation of high-performance concrete, which may help in minimizing the adverse impact of construction on the environment.

1.4 APPLICATIONS OF SMART MATERIALS IN DIFFERENT FIELDS

Smart materials possess characteristic features that make them suitable for use in various fields, cutting across disciplines. Some of the applications are presented below.

1.4.1 Medical Science

Smart structures in the form of advanced materials continue to advance the field of medical devices because they feature increased responsiveness and adaptability to care delivery and treatments for patients. These materials can react to the presence of or changes in external stimuli, including temperature, pH, and mechanical stress; all of which are currently being employed in many ways in the biomedical field [128–130]. Smart medical applications include smart implants, drug delivery systems and tracking sensors, where the implants monitor the patient's health in real time. For example, SMAs are used in stents placed within blood vessels to increase the opening of the stent. The most significant application is likely to be in drug delivery where smart materials can be made to release a therapeutic agent into the body at a controlled rate and to a specific organ. The strategy of the drug action decreases side effects and increases therapeutic effects on chronic diseases or localized illnesses [131]. Also, tetracycline ribosomal protection proteins (TRPPs) are associated with tissue engineering as stimuli–responsive materials that offer scaffolds that reflect the microenvironments required for tissue development. These materials can change mechanical properties in order to help with the formation of tissues and cells for healing injured tissues/organs [132]. Moreover, the self-sensing properties of smart materials are also transforming health monitoring because biosensors, which are being integrated into smart materials, can identify biomarkers or physiological alterations in real time. Perhaps, this capability can assist patients by providing an early diagnoses and intercessions; therefore bringing about overall health improvements for the patients [43,127]. Another application includes artificial muscles associated with electroactive polymers, which aims at imitating human muscle movements. These materials are used when designing compact prosthetic parts and also in muscle movement technology for the disabled people [133]. Also, materials that are able to perform self-healing, mimicking biological functions, are studied solely for use in implants and wound remedies, as damaged body parts are able to cure and so have a more prolonged life cycle and capability [134–136]. Altogether, the integration of smart materials into miscellaneous medical devices enhances these devices and opens the possibilities for creating the idea of a unique approach to treatment called personalized medicine.

1.4.2 Aerospace and Defense

A few of the uses of smart materials in the aerospace industry are vibration control, noise-reduced floors, and structural health monitoring. In defense, they are used for better armor and to create adaptable and responsive systems that can alter their properties in external stimuli. Such materials as SMA, EAP and piezoelectric each has its own prospects [43,63,74]. For instance, SMA can change size with the surrounding temperature, and this makes it possible to produce structures that can change state, for instance, between the wings of an aircraft to minimize drag and hence expenditure on fuel. EAPs can also be harnessed for actuating or sensing real-time control in strategic areas including smart structures like the flying capabilities of an aircraft's landing gear or the propeller of a helicopter's vibration isolating mechanism. Furthermore, piezoelectric material is applied to structural health and energy harvesting activities, where the piezoelectric material turns the mechanical energy of the vibration into electrical energy to be used to power the onboard equipment. In defense-related applications, smart materials are used primarily in personal protection, where advanced polymers provide enhanced ballistic protection and lower weight body armor and helmets [57]. Besides, these materials can be applied in hardware systems of camouflage that can effectively function in various climates. The conveyance of nanomaterials propels these applications and provides better strength and performance at lower weights. In the future, research in the aerospace and defense industries can focus on smart materials as innovative more effective systems for use on the complications of modern events and in aerospace engineering.

1.4.3 Construction

Smart products in construction, for instance, self-healing concrete, reduces costs relating to repair and maintenance and also enhances the aspects of durability, stability and workability of structures [137]. These electrically active materials, which can alter their characteristics based on factors such as temperature, stress and electric fields, are gradually being introduced into a number of applications in the field of engineering [138]. Among the areas of extensive usage are smart buildings, where it is used to address provisions for vibration and noise to increase inhabitant comfort and safety. For instance, piezoelectric material is capable of converting mechanical energy from vibrations into electrical energy that is used in monitoring the structure or power of the sensors installed in the structure [71]. In addition, SMAs are employed in seismic protection systems, where they store or release energy during an earthquake and reduce structural failure [67]. Smart materials are also useful in road construction; for example, in pavement, TCMs allow for the clearing of snow using a low voltage current to reduce risks in winter. Moreover, these materials provide structural health monitoring by incorporating sensors that identify cracks and other defects that may likely compromise structures; hence, this enables their timely maintenance and repair [83]. This practice of prevention not only extends the expected life spans of structures but also enables structures to withstand acts of nature. Luckily, as the research advances, incorporating more smart materials into construction will lead to more research that will in the end lead to better methods of construction, overcoming modern construction challenges by embracing sustainability and energy utilization.

1.4.4 Energy

Amorphous silicon or first generation solar cells have been the most well known for many decades but they are characterized by its efficiency and cost [139]. A recently discovered and promising photovoltaic material, because of high efficiency and low cost of production, is a new class of material with a perovskite structure. Perovskite materials are generally characterized by a given crystal structure that helps in the absorption of light and also the conversion of such light energy into electricity, which most suitable for application in solar cells [140]. Perovskite solar cells can attain stoichiometric efficiency or even surpass the silicon PV counterparts. Moreover, carbon nanotubes could be manufactured employing more energy-effective methods than other materials, thus, they are more eco-friendly. However, in the last few years, significant progress has been made toward optimizing the stability and performance of perovskite solar cells, which were once considered a significant factor in the commercial application of perovskite solar cells [141]. As such, these solar cells could be employed in solar power systems that would form part of larger systems.

It is especially pertinent to discuss energy storage as one of the key challenges affecting the renewable energy industry. Solar and wind-generated energy sources are relatively volatile because they are entirely reliant on either the existence of sunlight or wind [142]. Thus the call for storage of such power for whenever the sun is not out or the wind is not blowing. Smart material science advancements will gradually become a key input in the production of future energy storage solutions [143]. Among the most significant developments in energy storage technology is the solid oxide batteries, in which the electrolyte is not in the liquid form. They are safer, have a larger storage capacity and have higher energy density than traditional lithium-ion batteries. The strategic components that have been introduced to improve the efficiency and durability of solid-state batteries for the renewable energy supply are advanced ceramics and polymers.

1.5 SYNERGY BETWEEN GREEN TECHNOLOGY AND SMART MATERIALS (GTSM)

Green technology and smart materials possess the potential for enhanced sustainability and success in the future. Smart materials can enhance the innovative concept of green technology while green technology enlightens experimental results of smart materials in environment-friendly applications. For instance, concerning the utilization of smart materials, these can be used in passive energy structures, which can change the conditions of the environment. Two applications of smart and green technology include smart windows that are capable of adjusting their level of transparency in relation to sunlight, energy harvesting materials that are able to convert vibrations into electrical energy and self-healing coatings for structures. Solar panels in energy conversion and storage or wind energy turbines through the integration of more efficient renewable energy technologies as well as smart materials are used in storage systems. Some of the examples of advanced materials used in energy conversion include piezoelectric material; piezoelectric material can be used to transform mechanical vibrations that can be obtained from wind or wave.

Green technology integrated with smart materials is a shift in the technological developmental patterns of sustainable architecture and construction by invoking new solution proposals based on the environment. Hence, the link between these two domains must become more significant as international interest in climate change is increasing. Green technology can also mean various processes and products that are involved in the reduction of the human impact on the environment, for instance, using environmentally friendly power, climate-friendly structures and green manufacturing. At the same time, it is possible to note that smart materials can actively respond to their surroundings and have a number of characteristics that can significantly enhance the significance of their effect in construction and manufacturing.

Smart materials are comprised of a range of applied sensitive materials, such as thermo-biometals, that alter in size in relationship to heat and light-emitting concrete that captures light throughout the day and 'glows' during the night. In addition to improving the efficiency of construction, these materials support energy saving and the innovation of traditional energy sources. For instance, cement that emits light used in urban construction in the production of roads, can help save on the costs of street lighting and therefore help in cutting down on energy and carbon emissions. Furthermore, smart materials can assist buildings to self-heal and improve the lifespan of buildings as well as reduce waste.

The incorporation of smart materials in green technology enhances the performance of energy-related structures. Ideas such as green roofs, where a plant system absorbs the building's heat load, make sound sense as they not only increase the environmental advantages of the property, but also freshens internal air. Also, with regard to putting sensors into structures, this enables the monitoring of such factors as energy use and environmental conditions and adjustment of such factors to guarantee the highest efficiency at any given time. It also helps with the creation of sustainable urban habitats and smart cities that entail infrastructural growth that is sustainable both in terms of its impact on the physical environment and in terms of the engineering technology applied.

Further, smart materials integrated into the circular economy bring innovation to green technology approaches. When it comes to designing for production, there is more likelihood that the materials that are used in the production of these commodities will not only be protected but also preserved in that more of the commodities that are made will have the capacity to be recycled or even be biodegradable. This approach is well illustrated by smart materials like biopolymers or those derived from renewable raw materials as they are green replacements for the frequently harmful normal materials. This is a good example of how advances in material science can take forward the sustainability agenda when these materials are developed.

Nevertheless, smart materials are not only sustainable but also improve economic sustainability since cost is likely to rise in the future. The structures that are made from such materials do not require much energy for either heating or cooling, hence the energy bills are low. They also have a long life expectancy, hence reducing maintenance expenses, and they give buildings a longer life span, therefore making them economically reasonable structures in the long run.

Green technology when used in tandem with smart materials is further enhanced by innovative technologies such as AI and the IoT. Such technologies can improve design solutions, and enable better joining of smart systems in a building. For example, AI can employ measurements collected by smart sensors installed in the materials used in construction to reduce energy consumption. The said integration can not only leverage organizational performance optimization but also contribute to the inception of a sustainable urban strategy.

The mutual relationship between green technology and smart materials is a trending and standout theme in the current world of design and construction. By adopting the newest environmental conditions such as those operating based on the opening and closing of new materials, as well as the newest technological developments toward the protection of the environment, we get an opportunity to develop and come up with the better and ideal structures. Such a combination is important in addressing the new challenges that are associated with climatic change, as well as in promoting a more sustainable urban form in the future.

While the discovery of these advancements goes on in both fields, it will only get better and provide the kind of innovative solutions that will assist in the construction of a more environmentally sensitive world.

The combination of green technology and smart materials is quite effective in the development of green and resilient cities. The applicative potentials of smart materials in enhancing the efficiency and sustainability of systems in cities enable them to offer a better quality of life as well as an efficient use of resources. This is due to the growing innovation and increasing research and development that is carried out in these fields, and therefore we shall see advanced solutions taking place in future of urban living.

1.6 CHALLENGES AND FUTURE DIRECTIONS

However, many issues have to be faced so that further potentials of green technology and SMs can be explored.

1.6.1 Scalability and Cost

At present, scaling up for mass production and commercialization is not a core focus of the technology's development. Thus, despite successful experiments carried out with the help of green technologies and smart material usage, the challenge of expanding them and applying them to large commercial use is more considerable and can be very expensive. For instance, although sources such as wind and solar power have been cheaper in the recent past, the costs of developing infrastructure can be very expensive for some types of geography. Likewise, smart materials may be costly to develop and also may entail substantial modifications to the existing system architectures of products.

1.6.2 Resource Availability

Smart materials, as well as environmental technologies in general, are frequently dependent on scarce resources like metals and minerals. That demand creates problems connected with supply chain and ore extraction, which can be critical for the environment. It will therefore be important to either invent new materials or find ways to reprocess existing ones in order to take on this challenge.

1.6.3 Regulatory and Policy Barriers

Through the analysis of green technology and smart materials, it is said that government regulations and policies are central. Now and then, existing rules and regulations may slow down the process of integrating new technologies since some regulatory frameworks become obsolete or may be extremely prescriptive. There is a need for the formulation of policies that foster innovation, but at the same time, environmental and safety legal mandates must be met.

1.7 ROLE OF POLICY AND REGULATION IN PROMOTING GREEN TECHNOLOGY

Technological progression in green technology, building integrated photovoltaics and smart materials are crucial to the development of green building systems, but they are complemented by government policies and laws. This section analyzes how the existing policy environment can facilitate the construction and application of these technologies.

1.7.1 Regulations for Energy Efficiency and Sustainable Materials

Two other policies that can be named include policies that set standards for the mandatory use of energy-efficient products and encourage the use of sustainable materials. Policies at the industry structural level such as those that recommend the use of materials that have efficient-energy resource utilization can encourage the utilization of green technology in the construction industries. Similarly, the measures limiting the emissions of car exhausts, for example, may promote the sales of electrical and hybrid vehicles. They can also control aspects like the life cycle of the material or product or its recyclability among other aspects concerning the manufacture of the material. For example, the European Union's Circular Economy Action Plan (EUCEAP) is 'A plan toward reduction of wastes and increased usage of recycled materials in various sectors.'

1.7.2 Challenges in Policy Implementation

As seen there are benefits to promoting GTSM but the policy together with its application has its challenges, including resistance from industries that have heavily invested in traditional technologies, high costs of implementation of green technologies and the challenges of overhauling regulations, among others. In order to solve these challenges, which are currently being grappled by many governments, they have to convince the stakeholders, provide the standard regulatory instructions and guidelines, as well as make sure they are using the correct responsive policies in light of current technological advancements.

1.8 CONCLUSION AND FUTURE OUTLOOK

It's worth mentioning that green technology and smart materials occupy the same level of invention, separation with innovation, and care about the environment. The combination of green technology and smart materials offers realistic solutions to some of the greatest challenges that presage the future of the earth and the specters of climate change, scare resources and pollution. During their evolution, these branches can become of much greater importance in shaping a less wasteful and more environment-friendly world.

Sustainability may also continue to diversify as well as grow together with GTSM in the future. Going forward it could be possible to solve such problems as scalability, resource availability and regulation to fully harness this freedom to create a better world to live in. Employment of the GTSM is now a giant leap up on the ladder toward combating environmental problems globally. These areas include, for instance, renewable energy, energy efficiency, sustainable transport and sustainable buildings. Some of these technologies are rapidly transforming the world and making it sustainable. In the future, the use of nano-scale and biological techniques in the material sciences will determine the development of a new generation of products, especially in such parameters as power consumption, useful life and cost.

However, the application of green technologies and smart materials only improves if and when political actions, as well as regulation and the market, improve as well. It therefore means that each government, industry, consumer as well as every world citizen has to make positive effort to support sustainable development. Following these progresses, one can construct a future that is rich in information technology applications yet safe for the environment.

REFERENCES

1. Nilimaa J. Smart materials and technologies for sustainable concrete construction. *Dev Built Environ.* 2023;15:100177.
2. Hui CX, Dan G, Alamri S, et al. Greening smart cities: An investigation of the integration of urban natural resources and smart city technologies for promoting environmental sustainability. *Sustain Cities Soc.* 2023;99:104985.
3. Pratap B, Kumar S, Nand S, et al. Wastewater generation and treatment by various eco-friendly technologies: Possible health hazards and further reuse for environmental safety. *Chemosphere.* 2023;313:137547.
4. Scalet G. Programmable materials: Current trends, challenges, and perspectives. *Appl Mater Today.* 2024;40:102372.
5. Aydin M, Degirmenci T. The impact of clean energy consumption, green innovation, and technological diffusion on environmental sustainability: new evidence from load capacity curve hypothesis for 10 European Union countries. *Sustain Dev.* 2024;32:2358–2370.
6. Rani GM, Pathania D, Umapathi R, et al. Agro-waste to sustainable energy: A green strategy of converting agricultural waste to nano-enabled energy applications. *Sci Total Environ.* 2023;875:162667.
7. Tayebi ST, Sambucci M, Valente M. Waste management of wind turbine blades: A comprehensive review on available recycling technologies with a focus on overcoming potential environmental hazards caused by microplastic production. *Sustainability.* 2024;16:4517.
8. Khawaja MK, Alkayyali K, Almanasreh M, et al. Waste-to-energy barriers and solutions for developing countries with limited water and energy resources. *Sci Total Environ.* 2024;926:172096.
9. Arachi D, Prakash DRM. Environmental degradation-conservation strategies. In Mukul Machhindra Barwant, Vishnu Kiran Manam, and Aruna Kumari Nakkella (editors), .*Environment Conservation Challenges Threats in Conservation of Biodiversity.* 2022;235. Science Publications.
10. Kolawole AS, Iyiola AO. Environmental pollution: threats, impact on biodiversity, and protection strategies. In Izah SC, Ogwu, MC, editors, *Sustainable Utilization and Conservation of Africa's Biological Resources and Environment*, 377–409. Singapore: Springer; 2023.
11. Subramaniam S, Raju N, Ganesan A, et al. Artificial intelligence technologies for forecasting air pollution and human health: A narrative review. *Sustainability.* 2022;14:9951.
12. Hansen JE, Sato M, Simons L, et al. Global warming in the pipeline. *Oxford Open Clim Chang.* 2023;3:kgad008.
13. Reddy SJ. A note on the reality of "global warming" in the context of climate change. *Res Contempl Multidiscip Approach.* 2024;18.

14. Cointe B, Guillemot H. A history of the 1.5° C target. *Wiley Interdiscip Rev Clim Chang.* 2023;14:e824.
15. Gudimetla A, Kumar P, Prasad SS, et al. Towards smart materials: Enhancing the efficiency of the materials. *Model Charact Process Smart Mater.* 2023;1–30.
16. Kumari H, Chahal S, Kumar P, et al. Photocatalytic degradation of RB dye via cerium substituted SnO2 photocatalysts. *Mater Sci Eng B* [Internet]. 2023;296:116654.
17. Tian Q, Zhou J, Hou J, et al. Building the future: Smart concrete as a key element in next-generation construction. *Constr Build Mater.* 2024;429:136364.
18. Ngobeh JM, Sannoh M, Thullah J. A comparative analysis of the sustainable growth of global hydro, solar, and wind power systems (Renewable Energy Systems). *Open J Energy Effic.* 2023;12:49–61.
19. Algarni S, Tirth V, Alqahtani T, et al. Contribution of renewable energy sources to the environmental impacts and economic benefits for sustainable development. *Sustain Energy Technol Assessments.* 2023;56:103098.
20. Ogunjo S, Olusola A, Olusegun C. Potential of using floating solar photovoltaic and wind farms for sustainable energy generation in an existing hydropower station in Nigeria. *Clean Technol Environ Policy.* 2023;25:1921–1934.
21. Kumar P, Singh J, Singh S. Neural network supported flow characteristics analysis of heavy sour crude oil emulsified by ecofriendly bio-surfactant utilized as a replacement of sweet crude oil. *Chem Eng J Adv* [Internet]. 2022;11:100342. https://doi.org/10.1016/j.ceja.2022.100342.
22. Singh R, Toseef M, Kumar J, et al. Benefits and Challenges in additive manufacturing and its applications. In Kaushal S, Singh I, Singh S, et al., editors, *Sustainable Advanced Manufacturing and Materials Processing.* 1st ed., 137–157. Boca Raton: CRC Press; 2022. https://www.taylorfrancis.com/books/9781003269298/chapters/10.1201/9781003269298-8.
23. Singh J, Singh R, Singh S, et al. Reducing scrap due to missed operations and machining defects in 90PS pistons. *Int J Interact Des Manuf.* 2022;17:2527–2539. https://doi.org/10.1007/s12008-022-01071-0.
24. Vandenbogaerde L, Verbeke S, Audenaert A. Optimizing building energy consumption in office buildings: A review of building automation and control systems and factors influencing energy savings. *J Build Eng.* 2023;76:107233.
25. Olajiga OK, Ani EC, Sikhakane ZQ, et al. A comprehensive review of energy-efficient lighting technologies and trends. *Eng Sci Technol J.* 2024;5:1097–1111.
26. Luo S, Yuan Y. The path to low carbon: The impact of network infrastructure construction on energy conservation and emission reduction. *Sustainability.* 2023;15:3683.
27. Markard J, Wells P, Yap X-S, et al. Unsustainabilities: A study on SUVs and space tourism and a research agenda for transition studies. *Energy Res Soc Sci.* 2023;106:103302.
28. Balcerak-Woźniak A, Dzwonkowska-Zarzycka M, Kabatc-Borcz J. A comprehensive review of stimuli-responsive smart polymer materials—recent advances and future perspectives. *Materials (Basel).* 2024;17:4255.
29. Chen Q, Kalpoe T, Jovanova J. Design of mechanically intelligent structures: Review of modeling stimuli-responsive materials for adaptive structures. *Heliyon.* 2024;10(14):e34026.
30. Wang X, Qin Q, Lu Y, et al. Smart triboelectric nanogenerators based on stimulus-response materials: From intelligent applications to self-powered systems. *Nanomaterials.* 2023;13:1316.
31. Singh J, Kumar S, Mohapatra SK, et al. Shape simulation of solid particles by digital interpretations of scanning electron micrographs using IPA technique. *Mater Today Proc.* 2018;5:17786–17791. https://doi.org/10.1016/j.matpr.2018.06.103.

32. Bhat A, Rao VS, Jayalakshmi NS. Review of the evolution of magnetorheological fluid-based rehabilitative devices: From the perspective of modelling, sensors and control strategies. *IEEE Access*. 2023;11:88759–88777.
33. Bendezu MAV. *Investigation of the Rheological Behavior of Polyborosiloxane with Magnetically Responsive Particles*. Peru: Pontificia Universidad Catolica del Peru; 2024.
34. Wiener T, Offenzeller C, Martetschläger D, et al. Phenomenological studies on magnetic and mechanical remanence effects in magnetorheological fluids. *Smart Mater Struct*. 2024;33:45004.
35. Fenyk A, Horak W, Zieliński M. Investigation the Effect of MR fluid composition on properties at low strain ranges. *Materials (Basel)*. 2023;16:5730.
36. Malkin AY, Derkach SR, Kulichikhin VG. Rheology of gels and yielding liquids. *Gels*. 2023;9:715.
37. Moyers-Gonzalez M, Hewett JN, Cusack DR, et al. Non-isothermal thin-film flow of a viscoplastic material over topography: Critical Bingham number for a partial slump. *Theor Comput Fluid Dyn*. 2023;37:151–172.
38. Kumar L. An algebraic thixotropic elasto-visco-plastic constitutive equation describing pre-yielding solid and post-yielding liquid behaviours. *J Nonnewton Fluid Mech*. 2023;321:105105.
39. Masa'id A, Lenggana BW, Ubaidillah U, et al. A review on vibration control strategies using magnetorheological materials actuators: Application perspective. *Actuators*. 2023;12:113.
40. Ahmad D, Parancheerivilakkathil MS, Kumar A, et al. Recent developments of polymer-based skins for morphing wing applications. *Polym Test*. 2024;135:108463.
41. Cazacu M, Dascalu M, Stiubianu G-T, et al. From passive to emerging smart silicones. *Rev Chem Eng*. 2023;39:941–1003.
42. Jyothish KJ, Mishra S. A Survey on Robotic Prosthetics: Neuroprosthetics, Soft Actuators, and Control Strategies. *ACM Comput Surv*. 2024;56:1–44.
43. Guarino V, Zuppolini S, Borriello A, et al. Electro-Active Polymers (EAPs): A promising route to design bio-organic/bioinspired platforms with on demand functionalities. *Polymers (Basel)*. 2016;8:185.
44. Bernat J, Gajewski P, Kołota J, et al. Review of soft actuators controlled with electrical stimuli: IPMC, DEAP, and MRE. *Appl Sci*. 2023;13(3):1651.
45. Choi H, Kim Y, Kim S, et al. Ions-silica percolated ionic dielectric elastomer actuator for soft robots. *Adv Sci*. 2023;10:2303838.
46. Biswal DK. Application of electroactive polymer actuator: A brief review. *Biomimicry Mater Appl*. 2023;127–146.
47. Qiu Y, Ashok A, Nguyen CC, et al. Integrated sensors for soft medical robotics. Small. 2024;20(22):2308805.
48. Sîrbu I-D, Preninger D, Danninger D, et al. Electrostatic actuators with constant force at low power loss using matched dielectrics. *Nat Electron*. 2023;6:888–899.
49. Gour S, Kumar D. Unequal-biaxial taut states of electro-active polymeric composites. *J Appl Mech*. 2024;91(5):051004.
50. Singh A, Kaushik V, Chahal S, et al. Efficient degradation of methylene blue dye and antibacterial performance of shape controlled RuO2 nanocomposites. *ChemistrySelect*. 2021;6:10038–10050. https://www.scopus.com/inward/record.uri?eid=2-s2.0-85116495488&doi=10.1002%2Fslct.202102546&partnerID=40&md5=f86b39f7294c94f8be0f7194778c3bb6.

51. Hashmi AW, Singh J, Ahmad S. Future Perspective of self-healing nano-coatings for corrosion protection. In Hardeep Kaur, Rakesh Kumar Phanden, Rajeev Kumar Singh, and Basant Singh Sikarwar (editors), *Discov Disruption, Futur Implic Nanomater.* 2024;597–624.
52. Bouiti K, Labjar N, Benmessaoud M, et al. Nano-hybrid smart coatings: Advancements in self-healing and responsive functionalities. In Ashish Kumar and Abhinay Thaku (editors), *Nano-hybrid Smart Coatings: Advancements in Industrial Efficiency and Corrosion Resistance*;1469:279–302. ACS Publications; 2024.
53. Song M-M, Wang Y-M, Liang X-Y, et al. Functional materials with self-healing properties: A review. *Soft Matter.* 2019;15:6615–6625.
54. Nowacka M, Kowalewska A. Self-Healing Silsesquioxane-Based Materials. *Polymers (Basel).* 2022;14(9):1869.
55. Li B, Cao P-F, Saito T, et al. Intrinsically self-healing polymers: from mechanistic insight to current challenges. *Chem Rev.* 2022;123:701–735.
56. Chen C, Shen T, Yang J, et al. Room-Temperature intrinsic self-healing materials: A review. *Chem Eng J.* 2024;498:155158.
57. Agrawal N, Arora B. Self-healing polymers and composites: extrinsic routes. *Mini Rev Org Chem.* 2022;19:496–512.
58. Utrera-Barrios S, Verdejo R, López-Manchado MA, et al. Evolution of self-healing elastomers, from extrinsic to combined intrinsic mechanisms: A review. *Mater Horizons.* 2020;7:2882–2902.
59. Wang J, Tang J, Chen D, et al. Intrinsic and extrinsic self-healing fiber-reinforced polymer composites: A review. *Polym Compos.* 2023;44:6304–6323.
60. Ahmad S, Hashmi AW, Singh J, et al. Innovations in additive manufacturing of shape memory alloys: Alloys, microstructures, treatments, applications. *J Mater Res Technol.* 2024;32:4136–4197.
61. Singh J, Hashmi AW, Ahmad S, et al. Critical review on biodegradable and biocompatibility magnesium alloys: Progress and prospects in bio-implant applications. *Inorg Chem Commun.* 2024;169:113111.
62. Nair VS, Nachimuthu R. The role of NiTi shape memory alloys in quality of life improvement through medical advancements: A comprehensive review. *Proc Inst Mech Eng Part H J Eng Med.* 2022;236:923–950.
63. Sadashiva M, Sheikh MY, Khan N, et al. A review on application of shape memory alloys. *Transform [TMT].* 2021;21:24.
64. Behera A, Sahoo AK, Mohapatra SS. Nickel–titanium smart hybrid materials for automotive industry. In Sabu Thomas, Ajit Behera and Tuan Anh Nguyen (editors), *Nickel-Titanium Smart Hybrid Materials*, 271–295. Elsevier; 2022.
65. Twala B, Kaur H, Thakur A, et al. A review of shape memory alloy applications and future trends in the aerospace industry. SSRN 4813587.
66. Rogovoy AA, Stolbova OS. Microstructural model of magnetic and deformation behavior of single crystals and polycrystals of ferromagnetic shape-memory alloy. In Altenbach H, Eremeyev V, editors, *Advances in Linear and Nonlinear Continuum and Structural Mechanics*, 397–442. Cham: Springer; 2023.
67. Biffi CA, Fiocchi J, Tuissi A. Relevant aspects of laser cutting of NiTi shape memory alloys. *J Mater Res Technol.* 2022;19:472–506.
68. Ratheesh A, Elias L, Aboobakar Shibli SM. Tuning of electrode surface for enhanced bacterial adhesion and reactions: A review on recent approaches. *ACS Appl Bio Mater.* 2021;4:5809–5838.
69. Yin M, Chen T, Liu P, et al. Application and progress of NiTi alloys in vascular interventional medical devices. *J Mater Res Technol.* 2024;32:4246–4264.

70. Nguyen QH, Ta QTH, Tran N. Review on the transformation of biomechanical energy to green energy using triboelectric and piezoelectric based smart materials. *J Clean Prod.* 2022;371:133702.
71. Mahapatra S Das, Mohapatra PC, Aria AI, et al. Piezoelectric materials for energy harvesting and sensing applications: roadmap for future smart materials. *Adv Sci.* 2021;8:2100864.
72. Nivedhitha DM, Jeyanthi S. Polyvinylidene fluoride, an advanced futuristic smart polymer material: A comprehensive review. *Polym Adv Technol.* 2023;34:474–505.
73. Habib M, Lantgios I, Hornbostel K. A review of ceramic, polymer and composite piezoelectric materials. *J Phys D Appl Phys.* 2022;55:423002.
74. Ali F, Koc M. 3D printed polymer piezoelectric materials: Transforming healthcare through biomedical applications. *Polymers (Basel).* 2023;15(23):4470.
75. Covaci C, Gontean A. Piezoelectric energy harvesting solutions: A review. *Sensors.* 2020;20:3512.
76. Wang L, Fei Z, Qi Y, et al. Overview of human kinetic energy harvesting and application. *ACS Appl Energy Mater.* 2022;5:7091–7114.
77. Kumar A, Grover N, Manna A, et al. Investigating the influence of WEDM process parameters in machining of hybrid aluminum composites. *Adv Compos Lett.* 2020;29:2633366X20963137.
78. Kumar P, Singh J, Singh JP. Design and performance improvement of downdraft co-current gasifier by adjustment of gasification parameters and properties of biomass materials. *AIP Conf Proc.* AIP Publishing; 2024.
79. Roy S, Azad ANMW, Baidya S, et al. Powering solutions for biomedical sensors and implants inside the human body: A comprehensive review on energy harvesting units, energy storage, and wireless power transfer techniques. *IEEE Trans Power Electron.* 2022;37:12237–12263.
80. Al-Suhaimi EA, Aljafary MA, Alfareed TM, et al. Nanogenerator-based sensors for energy harvesting from cardiac contraction. *Front Energy Res.* 2022;10:900534.
81. Crosby PHN, Netravali AN. Green thermochromic materials: A brief review. *Adv Sustain Syst.* 2022;6:2200208.
82. Hakami A, Srinivasan SS, Biswas PK, et al. Review on thermochromic materials: Development, characterization, and applications. *J Coatings Technol Res.* 2022;19:377–402.
83. Hakami A. *Microencapsulation of Thermochromic Materials for Thermal Storage and Energy Efficiency of Buildings.* University of South Florida; 2021.
84. Hossain S, Sadoh A, Ravindra NM. Principles, properties and preparation of thermochromic materials. *Material Science & Engineering International Journal* 2023;7(3):146–156.
85. Liu H, Deng Y, Ye Y, et al. Reversible thermochromic microcapsules and their applications in anticounterfeiting. *Materials (Basel).* 2023;16(14)–5150.
86. Kayani ABA, Kuriakose S, Monshipouri M, et al. UV photochromism in transition metal oxides and hybrid materials. *Small.* 2021;17:2100621.
87. Leistner A, Pianowski ZL. Smart photochromic materials triggered with visible light. *European J Org Chem.* 2022;2022:e202101271.
88. Irie M. *Diarylethene Molecular Photoswitches: Concepts and Functionalities.* 2021. Wiley.
89. Xiao Y, Ko C-C. Photochromic reactions in coordination compounds. In Bahnemann, DW, Patrocinio AOT, editors, *Springer Handbook of Inorganic Photochemistry*, 547–579. Cham: Springer; 2022.
90. Jago D, Gaschk EE, Koutsantonis GA. History and fundamentals of molecular photochromism. *Aust J Chem.* 2023;76(10):635–654.

91. Pfaff G. *Inorganic Pigments*. Berlin, Boston: Walter de Gruyter GmbH & Co KG; 2023.
92. Badour Y, Jubera V, Andron I, et al. Photochromism in inorganic crystallised compounds. *Opt Mater X*. 2021;12:100110. https://www.sciencedirect.com/science/article/pii/S2590147821000395.
93. Kulkarni SD, Manjunatha, Chandrasekhar U, et al. Design and optimization of polyvinyl-nitride rubber for tensile strength analysis. *Int J Interact Des Manuf.* 2024;18:5893–5908.
94. Moshiri A. Molecular crosslinking of calcium-silicate-hydrate with organosilanes: Toward engineering of cementitious matrices with high thermal resistance. 2021. PhD dissertation, University of Houston, Houston, United States.
95. Duque-Redondo E, Bonnaud PA, Manzano H. A comprehensive review of CSH empirical and computational models, their applications, and practical aspects. *Cem Concr Res*. 2022;156:106784.
96. Wang Y. *Damage Detection in Reinforced Concrete and Self-Healing Concrete Structures Using Non-Destructive Testing Techniques*. Cardiff University; 2020.
97. Cappellesso V, di Summa D, Pourhaji P, et al. A review of the efficiency of self-healing concrete technologies for durable and sustainable concrete under realistic conditions. *Int Mater Rev*. 2023;68:556–603.
98. Way AC. *SLS Reliability Considering Autogenous Self-Sealing in Tension Governed Reinforced Concrete Water Retaining Structures*. Stellenbosch: Stellenbosch University; 2021.
99. Fernandez CA, Correa M, Nguyen M-T, et al. Progress and challenges in self-healing cementitious materials. *J Mater Sci*. 2021;56:201–230.
100. Zhang Y, Wang R, Ding Z. Influence of crystalline admixtures and their synergetic combinations with other constituents on autonomous healing in cracked concrete—A Review. *Materials (Basel)*. 2022;15:440.
101. Dallaev R. Advances in materials with self-healing properties: A brief review. *Materials (Basel)*. 2024;17:2464.
102. Rauf M, Khaliq W, Khushnood RA, et al. Comparative performance of different bacteria immobilized in natural fibers for self-healing in concrete. *Constr Build Mater.* 2020;258:119578.
103. Prashar G, Vasudev H. Surface topology analysis of plasma sprayed Inconel625-Al_2O_3 composite coating. *Mater Today*. 2022;50:607–611. https://linkinghub.elsevier.com/retrieve/pii/S221478532102109X.
104. Monazami M, Gupta R. Investigation of mechanical behavior and fracture energy of fiber-reinforced concrete beams and panels. *Cem Concr Compos*. 2022;133:104656.
105. Ch CP, Ali AYA. Impact of hybrid fibers on the crack healing of concrete. *Экономика строительства*. 2023;3:104–109.
106. Sharma V, Kumar V. Comparison of machining characteristics of metal matrix composites using CO_2 laser curve cutting process. *Proc Inst Mech Eng Part E J Process Mech Eng*. 2018;232:349–368.
107. Sharma V, Kumar V. Application of Box-Behnken design and response surface methodology for multi-optimization of laser cutting of AA5052/ZrO2 metal – matrix composites. *Proc Inst Mech Eng Part L J Mater Des Appl*. 2018;232:652–668.
108. Sharma V, Kumar V, Bist A. Investigations on morphology and material removal rate of various MMCs using CO 2 laser technique. *J Brazilian Soc Mech Sci Eng*. 2020;42:1–14.

109. Cuenca E, D'Ambrosio L, Lizunov D, et al. Mechanical properties and self-healing capacity of ultra high performance fiber reinforced concrete with alumina nano-fibres: Tailoring ultra high durability concrete for aggressive exposure scenarios. *Cem Concr Compos.* 2021;118:103956.
110. Nicolay P, Schlögl S, Thaler SM, et al. Smart materials for green (er) cities, a short review. *Appl Sci.* 2023;13:9289.
111. Sharanabasava H, Raviprakash M, Prasad CD, et al. Microstructure, mechanical and wear properties of SiC and Mo reinforced NiCr microwave cladding. *Adv Mater Process Technol.* 2023;10(4):362–3633.
112. Poojari M, Hanumanthappa H, Prasad CD, et al. Computational modelling for the manufacturing of solar-powered multifunctional agricultural robot. *Int J Interact Des Manuf.* 2023;18:1–12.
113. Yedida VVS, Vasudev H. A review on the development of thermal barrier coatings by using thermal spray techniques. *Mater Today Proc.* 2022;50:1458–1464.
114. Kattimani MA, Venkatesh PR, Masum H, et al. Design and numerical analysis of tensile deformation and fracture properties of induction hardened inconel 718 superalloy for gas turbine applications. *Int J Interact Des Manuf.* 2023;18:1–11.
115. Prashar G, Vasudev H. A review on the influence of process parameters and heat treatment on the corrosion performance of Ni-based thermal spray coatings. *Surf Rev Lett.* 2022;29:2230001.
116. Majji BGR, Vasudev H, Bansal A. A review on the oxidation and wear behavior of the thermally sprayed high-entropy alloys. *Mater Today Proc.* 2022;50:1447–1451.
117. Dutta V, Thakur L, Singh B, et al. A study of erosion–corrosion behavior of friction stir-processed chromium-reinforced NiAl bronze composite. *Materials (Basel).* 2022;15:5401. https://www.mdpi.com/1996-1944/15/15/5401.
118. Sunitha K, Vasudev H. A short note on the various thermal spray coating processes and effect of post-treatment on Ni-based coatings. *Mater Today Proc.* 2022;50:1452–1457. https://linkinghub.elsevier.com/retrieve/pii/S2214785321058132.
119. Jung A, Endres MB, Weichold O. Influence of environmental factors on the swelling capacities of superabsorbent polymers used in concrete. *Polymers (Basel).* 2020;12:2185.
120. Schmidleitner C, Kriehuber MU, Schlögl S, et al. Frontal polymerization of covalent adaptable networks. *Polym Chem.* 2025;16:963–971.
121. Chen Y, Chen C, Rehman HU, et al. Shape-memory polymeric artificial muscles: Mechanisms, applications and challenges. *Molecules.* 2020;25:4246.
122. Afzal M, Liu Y, Cheng JCP, et al. Reinforced concrete structural design optimization: A critical review. *J Clean Prod.* 2020;260:120623.
123. Murtagh N, Scott L, Fan J. Sustainable and resilient construction: Current status and future challenges. *J Clean Prod.* 2020;268:122264.
124. Kot P, Hashim KS, Muradov M, et al. How can sensors be used for sustainability improvement? In Ren J, editor, *Methods in Sustainability Science*, 321–344. Elsevier; 2021.
125. Sharma VB, Tewari S, Biswas S, et al. Recent advancements in AI-enabled smart electronics packaging for structural health monitoring. *Metals (Basel).* 2021;11:1537.
126. Shi X. Online monitoring of reinforced concrete corrosion using sensors. In Pacheco-Torgal F, Melchers RE, Shi X et al., editors, *Eco-Efficient Repair and Rehabilitation of Concrete Infrastructures*, 85–111. Elsevier; 2024.
127. Hafizh M, Soliman MM, Qiblawey Y, et al. Surface Acoustic Wave (SAW) sensors for hip implant: A numerical and computational feasibility investigation using finite element methods. *Biosensors.* 2023;13:79.
128. Bratek-Skicki A. Toward a new class of stimuli-responsive polymer-based materials– Recent advances and challenges. *Appl Surf Sci Adv.* 2021;4:100068.

129. Singh J, Singh JP, Kumar S, et al. Short review on hydroxyapatite powder coating for SS 316L. *J Electrochem Sci Eng*. 2023;13:25–39. https://www.scopus.com/inward/record.uri?eid=2-s2.0-85148748776&doi=10.5599%2Fjese.1611&partnerID=40&md5=9dbb686d916cc26266b626a0873c5c0a.
130. Singh J. Slurry erosion performance analysis and characterization of high-velocity oxy-fuel sprayed Ni and Co hardsurfacing alloy coatings. *J King Saud Univ – Eng Sci*. 2021;35:415–429. https://doi.org/10.1016/j.jksues.2021.06.009.
131. Placha D, Jampilek J. Chronic inflammatory diseases, anti-inflammatory agents and their delivery nanosystems. *Pharmaceutics*. 2021;13:64.
132. Mubeen B, Ansar AN, Rasool R, et al. Nanotechnology as a novel approach in combating microbes providing an alternative to antibiotics. *Antibiotics*. 2021;10:1473.
133. Prajapati AK, Komath M, Subhash NN. Prosthetics and orthotics for persons with movement disabilities in India in the postpandemic milieu. *Prosthet Orthot Int*. 2022;48:616–622.
134. Antil P, Kumar Antil S, Prakash C, et al. Multi-objective optimization of drilling parameters for orthopaedic implants. *Meas Control*. 2020;53:1902–1910.
135. Venkateswaran MR, Khosravi A, Zarepour A, et al. Self-healing materials in biomedicine and the circular economy. *Environ Sci Nano*. 2024;11:2771–2802.
136. Singh H, Singh S, Prakash C. Current trends in biomaterials and bio-manufacturing. In: Prakash C, Singh S, Singh R, et al., editors, *Biomanufacturing*, 1–34. Cham: Springer International Publishing; 2019. https://doi.org/10.1007/978-3-030-13951-3_1.
137. Makul N. Advanced smart concrete-A review of current progress, benefits and challenges. *J Clean Prod*. 2020;274:122899.
138. Sharma V, Singh JP, Hussan K, et al. Machine learning approach for predicting the machining characteristics of additive manufactured ABS material. *Proc Inst Mech Eng Part C J Mech Eng Sci*. 2024;239:09544062241283330.
139. Kim JY, Lee J-W, Jung HS, et al. High-efficiency perovskite solar cells. *Chem Rev*. 2020;120:7867–7918.
140. Guo Z, Jena AK, Kim GM, et al. The high open-circuit voltage of perovskite solar cells: A review. *Energy Environ Sci*. 2022;15:3171–3222.
141. Yoo JJ, Seo G, Chua MR, et al. Efficient perovskite solar cells via improved carrier management. *Nature*. 2021;590:587–593.
142. Adedeji PA, Akinlabi SA, Madushele N, et al. Neuro-fuzzy resource forecast in site suitability assessment for wind and solar energy: A mini review. *J Clean Prod*. 2020;269:122104.
143. Ma W, Zhang Y, Pan S, et al. Smart fibers for energy conversion and storage. *Chem Soc Rev*. 2021;50:7009–7061.

2 Green Technology and Smart Materials for Biomedical Material Applications

Praveen Kumar Verma and Hitesh Vasudev

2.1 INTRODUCTION

Modern medical practice has come to rest heavily upon the use of biomedical implants, which serve as treatment for several diseases (e.g., joint replacements and cardiovascular devices) [1, 2]. These implants aim to improve the patient's quality of life by decreasing pain and increasing functionality. Nevertheless, the stuff used in their manufacture and coatings applied to their surfaces play a pivotal role in the success and lifetime of these implants [3–6]. Conventional materials, such as metals, ceramics, and synthetic polymers, possess superior mechanical strength and durability; therefore, these types of materials have been extensively employed for biomedical applications [7–9]. However, they tend to have significant barriers concerning corrosion resistance, biocompatibility, and an increased potential for adverse biological reactions. Some conventional coatings, for example, might cause inflammatory responses, cause harmful compounds to leak out, or not have the bioactivity needed to encourage cellular interactions [10]. This has created a pressing demand for creative solutions that address environmental issues as well as enhance the functionality of these coatings [11]. Researchers are focusing more on green technology and smart materials as viable paths for the creation of cutting-edge biomedical coatings in response to these difficulties [12]. Figure 2.1 shows the integrated green technology flows.

Sustainable development and a reduction in ecological footprint are the goals of green technology, which emphasizes the use of eco-friendly materials and processes. This corresponds to the creation of coatings made from non-toxic substances, renewable resources, and environmentally friendly synthesis techniques in the context of biological applications. In terms of implant coatings, for example, bio-based polymers like chitosan, gelatin, and alginate have become quite attractive because of their natural source, biocompatibility, and intrinsic bioactive qualities [14].

DOI: 10.1201/9781003516071-2

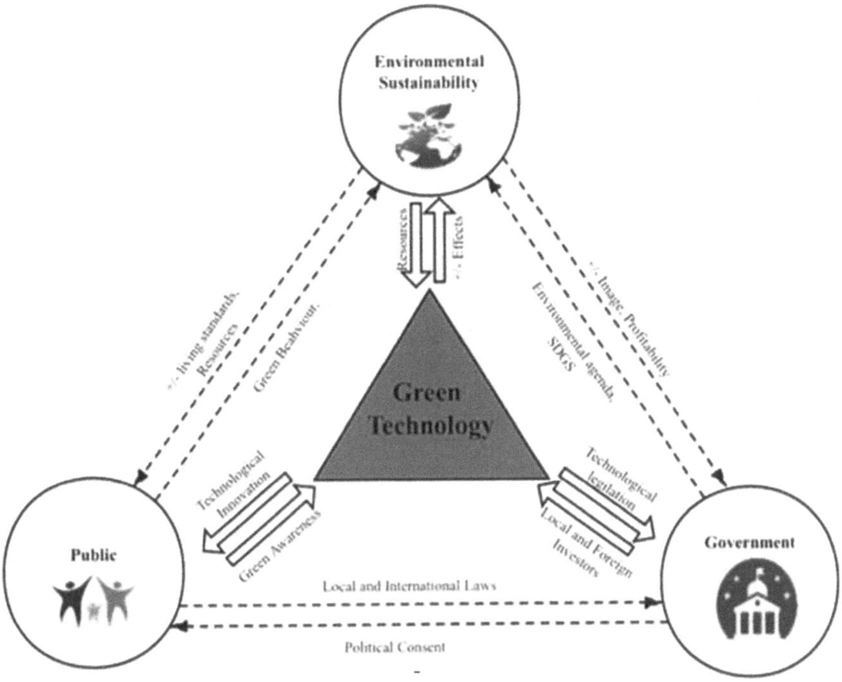

FIGURE 2.1 A framework of integrated green technology flows [13] (CC BY 4.0 Attribution).

These materials show beneficial interactions with biological systems and reduce the environmental impact of standard manufacturing techniques, improving the overall efficacy of biomedical implants. Some of the recent advancements in green synthesis techniques include; sol-gel processing, electrospinning, and hydrothermal synthesis, among which have facilitated the synthesis of elaborate coatings with specific characteristics. These, in turn, open opportunities for better integration of the implant with the surrounding tissues through using bioactive substances that stimulate cell attachment on the implant surface, cell division, and cell specialization. It is also important that new technology for using environmentally friendly solvents and procedures correlate well with a shift toward sustainability in the biomedical field, which makes it a favorable replacement for the older manner of doing things. Parallel to the progress of ecologically friendly technology is another realm of this rapidly expanding field – the incorporation of smart materials into biomaterials' coatings. Smart materials differ from ordinary materials since they quickly respond to numerous external triggering factors such as temperature, pH, as well as mechanical stress. In essence, these coatings can use their responsiveness to better adapt to the physiological environment, and, thus, enhance their working and operation. For example, stimulus-responsive hydrogels for drug delivery may be created to release a therapeutic compound in response to specific biological signals to deliver medication to the implant site with great accuracy. This is helpful for extended-release implants where systemic side effects are minimized, and the effectiveness of the afforded medicines is enhanced. Also, smart materials may show that coatings should be equipped with the inherent capability of repairing themselves when they are damaged, thereby increasing the expected lifetime of the implants [15].

This feature is particularly invaluable for load-carrying implant fixtures that could debond over time. These examples of these materials can further be explained by the incorporation of smart materials in biomedical coatings and the introduction of shape memory alloys and polymers. As a result of burying common issues such as corrosion, wear, and mechanical failure, in addition to boosting the mechanical efficiency of implants, these enhancements also co-operate to form a more successful common benefit. Nevertheless, the following barriers remain as challenges to green technology and smart material for biomedical coatings even though significant advancement has been made. One of them is the low compatibility of green synthesis technologies for large-scale production. Many greener synthetic processes show great potential in the lab, however, cost, productivity, and reproducibility are critical issues that need to be taken into consideration when it comes to the real world pushing a process forward [16].

Moreover, adapting intelligent materials into coatings often demands complex fabrication techniques, which can complicate the fabrication process. Also, before new coatings can be approved for clinical purposes the approval of materials and coatings in biomedical systems can be both a lengthy and demanding process that involves safety testing as well as effectiveness and biocompatibility. These difficulties express the significance of the cooperation of material scientists, biological engineers, and regulatory specialists erasing barriers of academic disciplines to ensure that new coatings can be translated from a laboratory prototype to clinical application. The combination of smart materials and green technology appears to be one of the promising prospects for further research and application with reference to biomedical implants' further development. This is really the key to innovation in this field – while there is always a demand for better, safer, and more effective medical solutions to a variety of ailments, recent years have also seen people wake up to the importance of environmental conservation. To enhance the effectiveness and durability of biomedical implants, green coatings will have to be sourced with the help of biomedical engineers, companies, and doctors. Meanwhile, the use of green technology for biomedical implant coatings based on smart materials shows great potential to improve patients' status and reduce negative effects on the environment. By applying sustainability and responsiveness paradigms, researchers can guide the biomedical industry in the direction of a new generation of implants that will benefit both the human body and the global environment [17, 18]. Figure 2.2 shows the recent trends of hydrogel research in the biomedical field.

This study intends to evaluate recent accomplishments, problems, and prospects in the sector, shining light on the transformative impact of sustainable practices and new materials in determining the future of biomedical applications. By combining these state-of-the-art technologies, the field of biomedical implants stands to be completely transformed and materials that are safe, effective, and environmentally friendly will be encouraged to be developed [20].

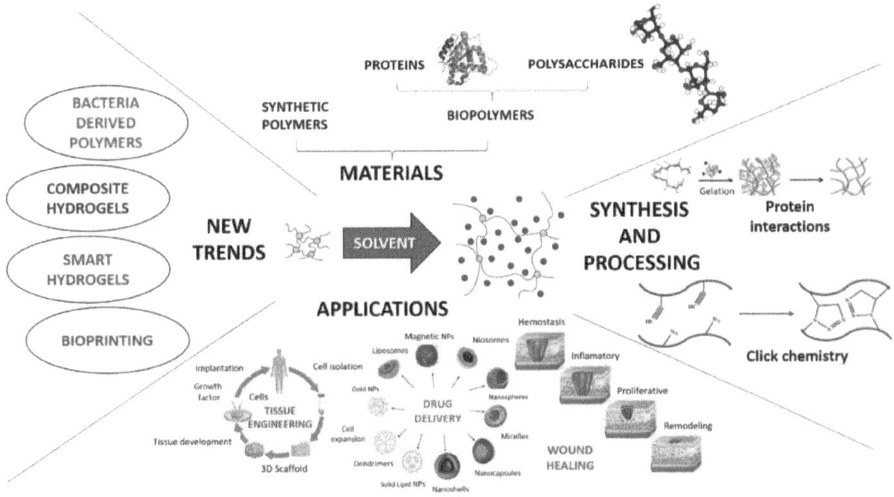

FIGURE 2.2 General perspective of the present and recent trends of hydrogel research in the biomedical field [19] (CC BY 4.0 Attribution).

2.2 GREEN TECHNOLOGY IN BIOMEDICAL COATINGS

In biomedical coatings, green technology focuses on environmentally friendly and sustainable ways to enhance implant functionality and patient safety. These coatings are developed using biodegradable ingredients, renewable resources, and energy-efficient procedures, decreasing environmental impact while boosting biocompatibility. The goal of innovations in this sector is to provide coatings that minimize the usage of hazardous compounds, minimize corrosion, and encourage tissue regeneration. Green coatings allow implants to function for a long time by combining intelligent materials including polymers, magnesium alloys, and bioactive ceramics [21–25]. Furthermore, these technologies support the development of more environmentally friendly biomedical solutions that improve patient outcomes and environmental health in line with global sustainability goals [26].

2.2.1 Need for Advanced Coatings in Biomedical Implants

The biomedical implant coating has functions such as increasing biocompatibility, decreasing friction, sterilizing bacteria, and incorporating a controlled drug release mechanism. Additional characteristics of these coatings may also influence the implants' durability and their functionality in general. However, traditional coating materials have many drawbacks, including low adhesion, poor mechanical properties, and toxicity effects. For instance, most of the ordinarily worn layers could corrode and launch unsafe material into the body, boosting the likelihood of issues. There is, therefore, a dire need for the existence of new coating material originating from environmental and health issues, and the improvement of the excellent

TABLE 2.1
Application of Advanced Coatings in Biomedical Implants

Coating Material	Properties	Application
Hydroxyapatite (HA)	Biocompatible, osteoconductive	Bone implants
Graphene Oxide	Antimicrobial, enhances mechanical strength	Orthopedic and dental implants
Zirconia (ZrO_2)	High hardness, wear-resistant	Dental and bone implants
Magnesium Alloys (Mg)	Biodegradable, corrosion control	Biodegradable implants
Niobium-based Thin Films (Nb)	High biocompatibility, corrosion resistance	Dental implants
Titanium Dioxide (TiO_2) Doped	Antimicrobial, bioactive	Dental implants

performance in biomedical implants [27]. Table 2.1 shows the application and properties of advanced coatings in biomedical implants.

2.2.2 Green Synthesis Methods

Environment-friendly approaches to chemical synthesis have therefore emerged as core strategies for the development of green materials for biomedical applications. This technique focuses on maximizing the use of renewable materials and minimizing the damage that the traditional chemical synthesis processes are known to cause to the environment. Green synthesis directs reagents that do not pose severe toxicity, materials that are biodegradable products, and energy conservation through sustainable processes. But other green synthesis methods are available which can also bring out the sol-gel processing, one of which is the preparation of thin film coatings and nanoparticles from precursor solutions. Through this approach, a solution is transformed into a sol-gel phase; this forms a gel network that can accommodate medications or any bioactive agents. Some advantages of using sol-gel technology include low processing temperature that makes it possible to achieve thickness consistency and the ability to control the chemical and physical properties of the final product [28].

Another important green synthesis method is hydrothermal synthesis, which employs high temperature and pressure to increase the rate of reaction of the precursor material in an aqueous solution. This is particularly advantageous for the synthesis of metal oxides and ceramics which are prominently used for biomedical coatings. Particle form and size may be well defined by employing hydrothermal synthesis which in turn results in enhanced mechanical strength and bioactivity among others [29]. Figure 2.3 shows the schematic diagram of porous nanomaterials' biomedical applications.

Another cutting-edge method for creating nanomaterials using natural resources like bacteria, fungi, and plants is biological synthesis. These microbes could produce biopolymers with advantageous qualities for coating applications or aid in the

FIGURE 2.3 Schematic diagram of porous nanomaterials biomedical applications [31]. (CC BY 4.0 Attribution).

reduction of metal ions into nanoparticles. For example, phytochemical-rich plant extracts can function as stabilizers and reducing agents to produce biocompatible nanoparticles appropriate for use in biomedicine.

In general, green synthesis techniques are essential for creating biocompatible, sustainable coatings for biomedical implants. These techniques meet the rising need for environmentally friendly medical solutions by reducing their negative effects on the environment and improving material performance [30]. Figure 2.4 illustrates the application of bioinspired items in product design.

2.3 SMART MATERIALS IN BIOMEDICAL COATINGS

More specifically, in the field of biomedical coatings, smart materials are becoming revolutionary structures because of their very nature that enables them to respond rapidly and efficiently to external signals. In this case, the use of this responsiveness results in improved performance of the biomedical implants by enhancing their usefulness, robustness, and compatibility with biological systems. The use of intelligent materials and biomedical coatings represents a significant advancement for biomedical coatings that eliminates several constraints arising from conventional coating material. There are potential enhancements of functionality for smart materials in biomedical applications such as enhanced biocompatibility, self-healing properties, as well as programmed drug delivery. By using these materials in coatings, scientists

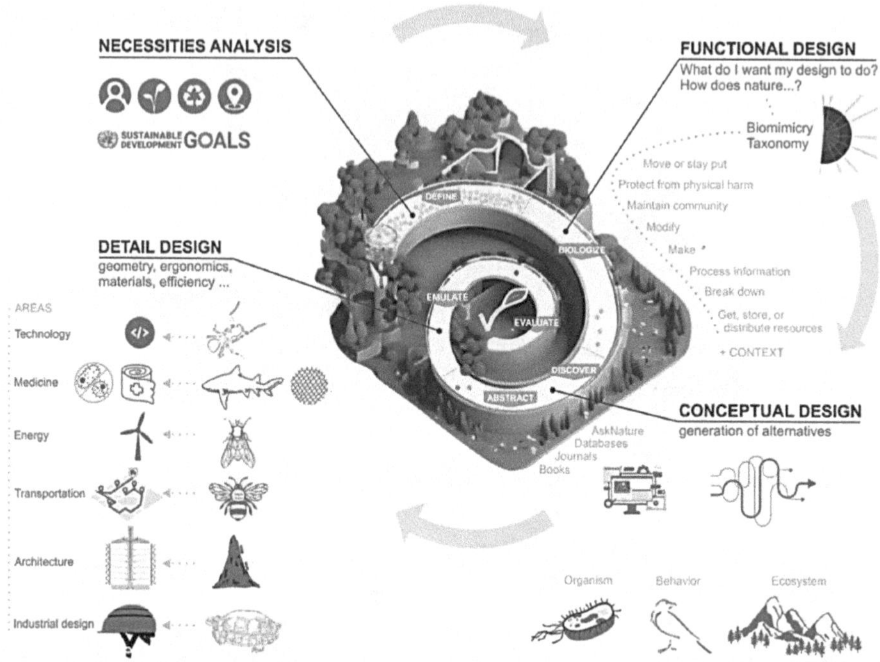

FIGURE 2.4 Potential application of bioinspired items in product design [32]. (CC BY 4.0 Attribution).

can design implants that meet the mechanical requirements and at the same time can adapt to the environment of the body thereby enhancing tissue integration and healing [33].

2.3.1 Applications of Smart Materials in Biomedical Coatings

There are numerous important uses of a wide variety of smart materials in biomedical coatings. Here, one of the best-known examples is the use of stimuli-responsive hydrogels. That is, they can expand or contract in response to changes in pH, temperature, or ionic strength. Controlled drug delivery to deliver a drug from one place to another at some rate, e.g., hydrogels can be designed that hold a therapeutic compound until it is released in response to specific environmental changes. An implant's hydrogel surface can then sense pH or temperature changes that indicate an inflammatory process, and release anti-inflammatory drugs. This targeted drug delivery improved the treatment therapeutic outcomes and minimized systemic side effects [34].

2.3.2 Self-Healing Coatings

These new creative applications of smart materials in biomedical coatings include also the development of self-healing coatings. The coatings were designed to jerk into action when the implant is damaged and to keep going for a longer period of operation, which should all add up to an extended life for the implants themselves. Chemists and engineers have found different methods to produce materials with build-in self-healing capabilities, such as introducing dynamic covalent connections that can adapt themselves under mechanical stress or embedding microcapsules that store healing agents and break open and release the chemical as soon as a crack arises. Self-healing in coatings gives implants longer life, and less risk of corrosion and failure from wear [35].

2.3.3 Shape Memory Alloys and Polymers

In biomedical applications, shape memory alloys (SMAs) and shape memory polymers (SMPs) have also attracted interest as a class of smart materials. When certain stimuli, like heat, are applied to certain materials, they can revert to their original shape. SMAs and SMPs can be used in the context of biomedical coatings to produce coatings that modify their characteristics in response to external stimuli. For instance, a shape memory polymer coating can improve tissue integration or enable controlled medication release by altering its porosity in response to temperature changes. Shape memory materials are very promising for usage in dynamic situations such as the human body because of their versatility [36]. Figure 2.5 presents the biomedical applications of shape memory polymers.

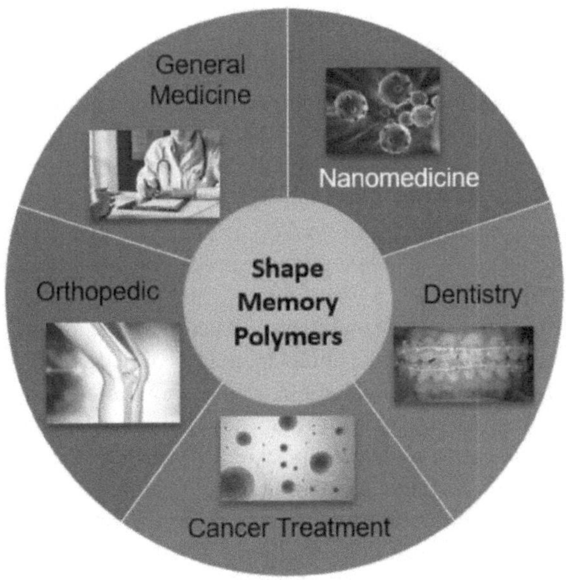

FIGURE 2.5 Biomedical applications of shape memory polymers (SMPs) [37]. (CC BY 4.0 Attribution).

2.3.4 Enhanced Biocompatibility and Functionality

The biocompatibility of biomedical coatings is also significantly enhanced with the addition of smart materials. Traditional coatings often have difficulty integrating well with biological systems, meaning that implants may be rejected or cause inflammation, regulatory or structural consequences; however, smart materials can be engineered to promote favorable cell–material interactions. For instance, the surface properties of smart coatings can be modified to enhance tissue integration through increased cell adhesion and protein adsorption. Moreover, smart materials are customizable as they can be programmed and thus their surface properties could match special biomedical requirements [38].

2.4 RECENT ADVANCES AND TRENDS

Green technology along with smart materials has seen huge development in the biomedical coating field. The emerging data reveal that polymers obtained from biological resources like gelatin, and chitosan have extensively been used to synthesize coatings that possess ideal antibacterial and biocompatibility properties. Other studies have also focused on drug-delivery stimuli-responsive hydrogels, capable of releasing the medicine in conditions of pH or temperature changes. These developments demonstrate that composite material of sophisticated material science and ecological technology can be incorporated into biomedical implants more securely and dependably [39].

2.4.1 Innovations in Green and Smart Coatings

Smart and green advancements in coatings are reshaping cutting-edge materials for various applications, especially biomedical implants, electronics, and protective layers. These advancements solve the two related issues namely functionality and performance in ambient environment through the synthesis of smart and eco-friendly materials. With green coatings, traditional oppressive materials have their effects on the environment reduced through elements that are non-hazardous and ingredients derived from renewable sources, besides the efficient use of energy in the manufacturing of the coating solutions. Researchers are again and again attracted to natural resources of bio-based polymers like chitosan, gelatin and alginate. But these materials, especially those with biomedical applications, show interesting behaviors when interacting with biological systems. Further, they decrease dependence on products produced from petroleum to a significant extent. Owing to the recent advances in green chemistry approaches, such as hydrothermal synthesis and sol-gel processing, it is now possible to formulate superior coatings containing bioactive ingredients that have minimal or no effect on the environment. These methods help to control the thickness of deposited films and the reproducibility of fabrication of nanoparticles to achieve target characteristics, and also improve the efficiency and biocompatibility of coatings utilized in biomedical applications [40].

However, smart coatings represent a significant breakthrough in material technology since they have functionality that allows them to respond dynamically to changes in their surroundings. These coatings are susceptible to changes in temperature, pH, moisture content, mechanical stress, and other factors that may cause them to release embedded therapeutic compounds. Stimuli-responsive hydrogels, for example, can be engineered to expand or contract in response to pH or temperature changes to enable controlled drug delivery at the implant site. This capacity lowers systemic adverse effects, improves patient comfort and safety, and boosts the therapeutic efficacy of medicines. Another significant development is the development of coatings with the capacity to repair themselves when damaged. This feature is especially useful in load-bearing applications where the integrity of the underlying material and the coating may be compromised by mechanical wear and corrosion. Self-healing systems can be designed using a range of strategies, including the incorporation of microcapsules carrying healing chemicals that are released upon cracking or the utilization of dynamic covalent connections that realign in response to mechanical stress. These advancements significantly lengthen the coatings' lifespan and improve the overall durability of the materials [41]. Figure 2.6 shows the alternative model for transforming the second-order factors into the first-order factors of green innovation and business sustainability.

Green and smart coatings are revolutionizing the electronics industry by improving the sustainability and performance of electrical gadgets. Eco-friendly coatings that have superior mechanical qualities and electrical conductivity can be made using conductive polymers and nanoparticles. Flexible electronics can be made more environmentally friendly and perform better by applying these coatings Additionally, the smart coatings also encompass self-cleaning surfaces, which are very useful in several industrial industries. They can protect surfaces from being dirty and reduce

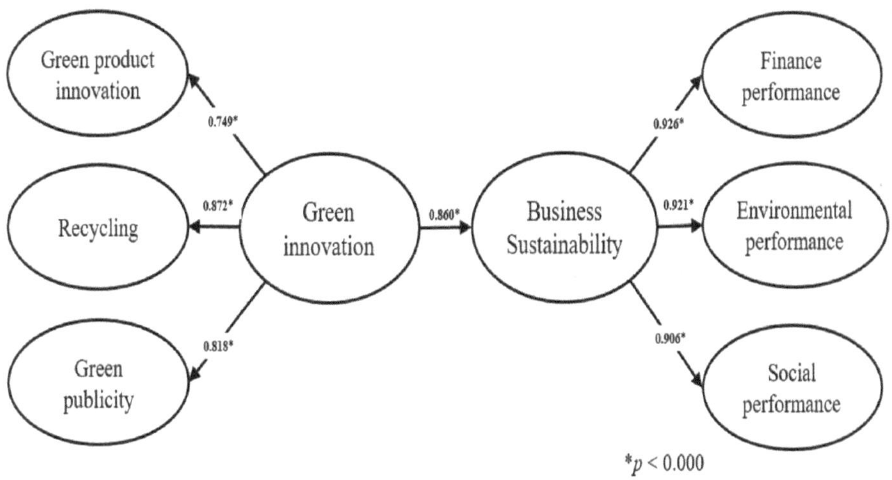

FIGURE 2.6 An alternative model for transforming the second-order factors into the first-order factors of green innovation and business sustainability [42]. (CC BY 4.0 Attribution).

the amount of money to be used in cleaning through the addition of photocatalytic elements or superhydrophobicity. Nanotechnology improves the performance of intelligent and eco-friendly coatings because it allows the synthesis of new material elements that cannot be obtained with the help of other coatings. Sustainable and intelligent coatings have been progressing; nevertheless, there are challenges, namely the integration and the adherence to the norms and standards. Many of the new materials and methods are still under the research and development category. When such discoveries are made, questions of production viability and economic feasibility in getting the product to the market come into play. Also, the governing of such innovative material can take time and be cumbersome hence requiring more effort when used in biological processes that will require more safety and efficacy results before general use. Smart and green coatings have been revolutionary to the enhancement of materials research through sustainability and functionality. To address the need for advanced coating solutions for sophisticated applications without negative impacts on the environment, current techniques are being integrated with environmentally sustainable processing technologies to develop the future's coatings. For these new coatings to meet their full potential and move beyond existing barriers, it may require interdisciplinary cooperation between engineers, material scientists, and regulatory professionals as this industry continues to expand. Sophisticated and eco-responsible products are likely to become crucial to competitive, high-performance, value-added solutions impacting numerous sectors, boding well for coatings [43].

2.5 GREEN TECHNOLOGY: A SUSTAINABLE APPROACH

Green technology refers to the application of environmentally friendly substances and processes that lessen their adverse effects on the environment and encourage sustainability. When discussing green technology concerning biomedical applications, non-toxic materials, renewable resources, and ecologically friendly synthesis methods are used to create coatings. The development of new coatings with improved characteristics has been made easier using green synthesis techniques including hydrothermal and sol-gel processing. These techniques allow for the addition of bioactive substances that support cellular adhesion and proliferation while also lessening the environmental impact of conventional production processes. Green technology also makes it possible to create coatings that are more suited to the biological environment, which lowers the possibility of host tissue rejection and inflammation [44]. Figure 2.7 shows the green environment perspective of the proposed sustainable inventory system.

2.6 CHALLENGES IN CURRENT RESEARCH AND DEVELOPMENT

Although using green technology and smart materials for biomedical coatings has made significant strides, there are still several obstacles to overcome. Scalability is a major obstacle to large-scale production of green synthesis technologies. Laboratory-scale procedures might work well, but transferring these techniques to industrial

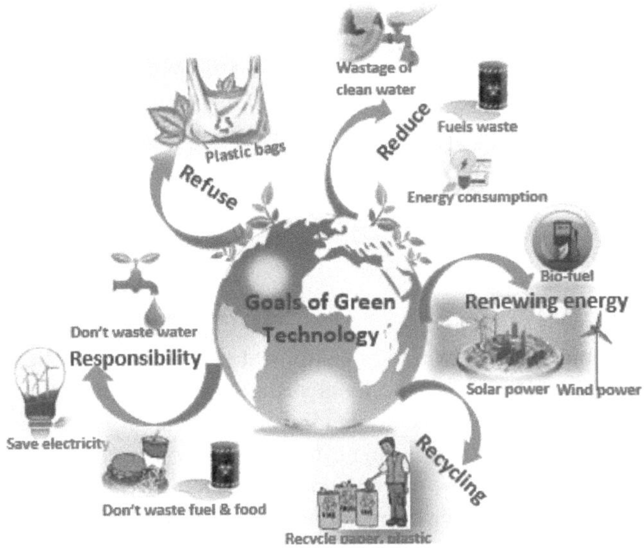

FIGURE 2.7 Green environment perspective of the proposed sustainable inventory system [45]. (CC BY 4.0 Attribution).

settings calls for rigorous assessment of cost, effectiveness, and consistency. In addition, in many cases for the incorporation of smart material into coatings, it is necessary to have complex fabrication processes, which may make the processes to be more intricate. Moreover, there are some challenges in the biomedical industry regarding the approval of new coatings and materials through the regulatory body. As previously stated, there are numerous safety, efficacy, and biocompatibility assessments that the researchers have to go through before new coatings for clinical application are approved. This problem raises awareness of an interdisciplinary issue regarding how material scientists, biomedical engineers, and regulatory specialists must collaborate to make novel coatings that can be effectively translated from the laboratory to clinical practice [46–48].

2.7 FUTURE PERSPECTIVES

A hybrid system developed with green technology and smart materials is another reasonable direction to be explored more and expanded in the future based on the improvements of biomedical implants. The need for environmental consideration and the resultant changes as well as the growing need for better and safer medication has spurred innovation in this field. For enhancing the potential of sustainable coatings that can enhance the work and durability of biomedical implants, collaboration between researchers and industrialists, as well as healthcare professionals will be critical [49, 50].

The studies of biomedical implant coatings demonstrates that current innovations in the biomedical field and biomedical implant coatings are extensible for environmental and patient benefits. It is to provide a comprehensive evaluation of the subject matter in terms of current trends, challenges, and prospects. This chapter successfully assessed the transformative impact of sustainable techniques and innovative materials on the advancement of biomedical applications. It is about time society embraced biomedical implants, which will be created using smart materials and environmentally sustainable technology [51, 52].

2.8 CONCLUSION

Smart materials and green technology to innovate biomedical applications as such constitute a radical shift toward functional sustainability in healthcare. Green technology is the practice of environment-friendly measures, such as the use of renewable resources, biodegradable materials, and efficient energy procedures, in the development of biomedical devices to be more efficient and safer with minimal effects on the environment. These developments are supposed to replace materials and processes that may be injurious to the lives of humans and those of other living things in the environment. In the same manner, bioactive ceramics, shape memory alloys, and stimulus-sensitive polymers are smart materials that are developed basically to alter their properties in response to real-time physiological stimuli such as pH, temperature, or stress to release drugs and limit microbial growth or promote tissue healing. For example, magnesium alloys can degrade in the body and cause a few complications that require subsequent surgeries to remove implanted devices, while bioactive ceramics promote increased bone contact. Incorporating these technologies, scientists are developing new biomedical coatings, enhancing implant function, and making the environment better. In keeping with worldwide sustainability agendas, green, intelligent materials facilitate implant longevity, reducing infection incidence and accelerating patients' healing times. It ensures that the future medical equipment needed to respond to clinical as well as ecological realities of the advanced world is environmentally optimal and functionally enhanced.

REFERENCES

1. Singh J, Singh JP, Kumar S, Gill HS. Short review on hydroxyapatite powder coating for SS 316L. *J Electrochem Sci Eng*. 2023;13(1):25–39.
2. Singh J, Singh S. A review on Machine learning aspect in physics and mechanics of glasses. *Mater Sci Eng B*. 2022; 284:115858.
3. Kumar S, Singh J, Mohapatra SK. Role of particle size in assessment of physico-chemical properties and trace elements of Indian fly ash. *Waste Manag Res J a Sustain Circ Econ*. 2018, 11;36(11):1016–22.
4. Dutta V, Thakur L, Singh B, Vasudev H. A study of erosion–corrosion behaviour of friction stir-processed chromium-reinforced NiAl bronze composite. *Materials*. 2022;15(15):5401.

5. Kumar P, Singh J, Singh S. Neural network supported flow characteristics analysis of heavy sour crude oil emulsified by ecofriendly bio-surfactant utilized as a replacement of sweet crude oil. *Chem Eng J Adv.* 2022;11:100342.
6. Prashar G, Vasudev H. A review on the influence of process parameters and heat treatment on the corrosion performance of Ni-based thermal spray coatings. *Surf Rev Lett.* 2022;29(01):2230001.
7. Sharanabasava H, Raviprakash M, Prasad CD, Ramesh MR, Phanibhushana MV, Vasudev H, et al. Microstructure, mechanical and wear properties of SiC and Mo reinforced NiCr microwave cladding. *Adv Mater Process Technol.* 2023. https://doi.org/10.1007/s40033-024-00670-3
8. Ahmad S, Hashmi AW, Singh J, Arora K, Tian Y, Iqbal F, et al. Innovations in additive manufacturing of shape memory alloys: Alloys, microstructures, treatments, applications. *J Mater Res Technol.* 2024. https://doi.org/10.1016/j.jmrt.2024.08.213
9. Singh J, Hashmi AW, Ahmad S, Tian Y. Critical review on biodegradable and biocompatibility magnesium alloys: Progress and prospects in bio-implant applications. *Inorg Chem Commun.* 2024;21:113111.
10. Singh J. Slurry erosion performance analysis and characterization of high-velocity oxy-fuel sprayed Ni and Co hardsurfacing alloy coatings. *J King Saud Univ - Eng Sci.* 2021. https://doi.org/10.1016/j.jksues.2021.06.009
11. Sajjad R, Chauhdary ST, Anwar MT, Zahid A, Khosa AA, Imran M, Sajjad MH. A review of 4D printing–technologies, shape shifting, smart polymer based materials, and biomedical applications. *Adv Ind Eng Polym Res.* 2024;7(1):20–36.
12. Su M, Song Y. Printable smart materials and devices: Strategies and applications. *Chem Rev.* 2021;122(5):5144–64.
13. 3Ikram M, Sroufe R, Awan U, Abid N. Enabling progress in developing economies: A novel hybrid decision-making model for green technology planning. *Sustainability.* 2021;14(1):258.
14. Greco F, Mattoli V. Introduction to active smart materials for biomedical applications. In *Piezoelectric Nanomaterials for Biomedical Applications* 2012 Mar 31 (pp. 1–27). Berlin, Heidelberg: Springer Berlin Heidelberg.
15. Costa CM, Cardoso VF, Martins P, Correia DM, Gonçalves R, Costa P, et al. Smart and multifunctional materials based on electroactive poly (vinylidene fluoride): recent advances and opportunities in sensors, actuators, energy, environmental, and biomedical applications. *Chem Rev.* 2023;123(19):11392–487.
16. Sahayaraj AF, Dhamotharan S, Sandeep D, Ramachandran P, Jenish I, Divakaran D, Suyambulingam I, Sanjay MR, Siengchin S. Sustainable smart polymer composite materials: A comprehensive review. In *E3S Web of Conferences* (Vol. 428, p. 02014). EDP Sciences, 2023.
17. Thakur A, Srivastava R, Bahadur PS, Rana A, Exploring the various biomedical applications of smart materials. In *Discovery, Disruption, and Future Implications of Nanomaterials* (pp. 249–274). IGI Global, 2024.
18. Cecen B, Hassan S, Li X, Zhang YS, Smart biomaterials in biomedical applications: Current advances and possible future directions. *Macromol Biosci.* 2024;24(3):2200550.
19. Sánchez-Cid P, Jiménez-Rosado M, Romero A, Pérez-Puyana V. Novel trends in hydrogel development for biomedical applications: A review. *Polymers.* 2022;14(15):3023.
20. Fathi-karkan S, Shamsabadipour A, Moradi A, Rahdar A, Pourmadadi M, Romanholo Ferreira LF. Four-dimensional printing techniques: A comprehensive review of biomedical and tissue engineering developments. *BioNanoScience.* 2024:14(4):4189–218.
21. Majji BGR, Vasudev H, Bansal A. A review on the oxidation and wear behavior of the thermally sprayed high-entropy alloys. *Mater Today Proc.* 2022;50:1447–51.

22. Kattimani MA, Venkatesh PR, Masum H, Math MM, Bahadurdesai VN, Mustafkhadri S, et al. Design and numerical analysis of tensile deformation and fracture properties of induction hardened inconel 718 superalloy for gas turbine applications. *Int J Interact Des Manuf.* 2023;18(8), 5939–49.
23. Singh R, Toseef M, Kumar J, Singh J. Benefits and challenges in additive manufacturing and its applications. In: Kaushal S, Singh I, Singh S, Gupta A, eds. *Sustainable Advanced Manufacturing and Materials Processing* (1st ed., pp. 137–57). CRC Press, 2022.
24. Singh J, Singh R, Singh S, Vasudev H, Kumar S. Reducing scrap due to missed operations and machining defects in 90PS pistons. *Int J Interact Des Manuf.* 2022. https://doi.org/10.1007/s12008-022-01071-0
25. Poojari M, Hanumanthappa H, Prasad CD, Jathanna HM, Ksheerasagar AR, Shetty P, et al. Computational modelling for the manufacturing of solar-powered multifunctional agricultural robot. *Int J Interact Des Manuf.* 2023;1–12. https://doi.org/10.1007/s12008-023-01291-y
26. Singh S, Melnik R. Coupled multiphysics modelling of sensors for chemical, biomedical, and environmental applications with focus on smart materials and low-dimensional nanostructures. *Chemosensors.* 2022;10(5):157.
27. Pepelanova I. Tunable hydrogels: introduction to the world of smart materials for biomedical applications. In *Tunable Hydrogels* 2021 (pp. 1–35). Springer.
28. Vasanthanathan A, Menaga S, Rosemi K. A comprehensive review of smart systems through smart materials. *Curr Mater Sci.* 2019;12(1):77–82.
29. Ramakrishnan T, Kumar SS, Chelladurai SJ, Gnanasekaran S, Sivananthan S, Geetha NK, Arthanari R, Assefa GB. Recent developments in stimuli responsive smart materials and applications: an overview. *J Nanomater.* 2022;2022(1):4031059.
30. Kumar A, Kumar P, Srivastava AK, Goyat V, eds. *Modeling, Characterization, and Processing of Smart Materials.* IGI Global, 2023 Aug 7.
31. Parvin N, Kumar V, Mandal TK, Joo SW. Advancements in nanoporous materials for biomedical imaging and diagnostics. *J Funct Biomat.* 2024;15(8):226.
32. 17Aguilar-Planet T, Peralta E. Innovation inspired by nature: Applications of biomimicry in engineering design. *Biomimetics.* 2024;9(9):523.
33. Sharma SS, Bashir S, Kasi R, Subramaniam RT. The significance of graphene based composite hydrogels as smart materials: A review on the fabrication, properties, and its applications. *FlatChem.* 2022;33:100352.
34. Bocchetta G, Fiori G, Sciuto SA, Scorza A. Performance of smart materials-based instrumentation for force measurements in biomedical applications: A methodological review. *InActuators* 2023;12(7):261.
35. Tripathy DB, Gupta A. Nanocomposites as sustainable smart materials: A review. *J Reinfor Plast Compos.* 2024:07316844241233162.
36. Dutschk V, Karapantsios T, Liggieri L, McMillan N, Miller R, Starov VM. Smart and green interfaces: From single bubbles/drops to industrial environmental and biomedical applications. *Adv Colloid and Interface Sci.* 2014;209:109–26.
37. Rokaya D, Skallevold HE, Srimaneepong V, Marya A, Shah PK, Khurshid Z, Zafar MS, Sapkota J. Shape memory polymeric materials for biomedical applications: An update. *J Compos Sci.* 2023;7(1):24.
38. Abdelhamid HN. Smart materials for bioimplant. In *Surface Engineering of Biomaterials* 2024 Mar 20 (pp. 85–104). CRC Press.
39. Kumar A, Mishra RK, Verma K, Aldosari SM, Maity CK, Verma S, Patel R, Thakur VK. A comprehensive review of various biopolymer composites and their applications: From biocompatibility to self-healing. *Mater Today Sustain.* 2023;23:100431.

40. Cheiliakh O, Cheylyakh Y, Mak-Mak N, Malysheva I, Burova D, Mikula J. Modern trends in application of smart materials in biomedical engineering. In Arras, Peter and Luengo, David (editors), *Teaching and Subjects on Bio-medical Engineering.* 2021:272–284.
41. Qiu X, Hu S. "Smart" materials based on cellulose: A review of the preparations, properties, and applications. *Materials.* 2013;6(3):738–81.
42. Li L, Msaad H, Sun H, Tan MX, Lu Y, Lau AK. Green innovation and business sustainability: New evidence from energy intensive industry in China. *Int J Environ Res Public Health.* 2020;17(21):7826.
43. Alemu Reta B, K MB, Tesfaye T. Smart and intelligent biomaterials for novel applications–a review. *Int J Polym Mater.* 2024:1–9. https://doi.org/10.1080/00914037.2024.2316191
44. Verma PK, Singh S, Kapoor M, Singh S. A review on the surface topography and corrosion behavior of Mg-alloy coatings for biomedical implants. *Resul Sur Inter.* 2024:100227. https://doi.org/10.1016/j.rsurfi.2024.100227
45. Bhavani GD, Meidute-Kavaliauskiene I, Mahapatra GS, Činčikaitė R. A sustainable green inventory system with novel eco-friendly demand incorporating partial backlogging under fuzziness. *Sustainability.* 2022;14(15):9155.
46. Tripathy DB, Gupta A. Nanocomposites as sustainable smart materials: A review. *J Reinf Plast Compos.* 2024:07316844241233162.
47. Rezaei N, Akbarzadeh I, Kazemi S, Montazeri L, Zarkesh I, Hossein-Khannazer N, Hassan M, Vosough M. Smart materials in regenerative medicine. *Mod Med Lab J.* 2021;4(1):39–51.
48. Sunitha K, Vasudev H. A short note on the various thermal spray coating processes and effect of post-treatment on Ni-based coatings. *Mater Today Proc.* 2022;50:1452–57.
49. Yedida VVS, Vasudev H. A review on the development of thermal barrier coatings by using thermal spray techniques. *Mater Today Proc.* 2022;50:1458–64.
50. Singh SP, Verma PK, Singh S, Kapoor M, Balu R. Microstructural characterization and electrochemical behaviour of HVOF sprayed WC-Co-Cr coatings on 316L boiler steel stainless steel. *Resul Sur Inter.* 2024;16:100248.
51. Amukarimi S, Ramakrishna S, Mozafari M. Smart biomaterials—a proposed definition and overview of the field. *Curr Opin Biomed Eng.* 2021;19:100311.
52. Khan A, Alamry KA. Recent advances of emerging green chitosan-based biomaterials with potential biomedical applications: A review. *Carbohyd Res.* 2021;506:108368.

3 Recent Progress in Shape Memory Alloys and Additive Manufacturing

Jashanpreet Singh and Abdul Wahab Hashmi

Description	Abbreviation
Additive manufacturing	AM
Build direction	BD
Digital image correlation	DIC
Direct energy deposition	DED
Direct metal laser sintering	DMLS
Electron beam freeform fabrication	EBF3
Laser additive manufacturing	LAM
Laser metal deposition	LMD
Laser powder bed fusion	LPBF
Nitinol	NiTi
Powder bed fusion	PBF
Scanning electron microscopy	SEM
Selective laser sintering	SLS
Selective laser melting	SLM
Electron beam melting	EBM
Shape memory alloy	SMA
Wire arc additive manufacturing	WAAM

3.1 INTRODUCTION

Nowadays, 3D modelling is used to make parts or products straight from digital product data. This technology is called additive manufacturing (AM) or 3D printing. When it first came out in the late 1980s, AM technologies were used to make models and prototypes [1]. Householder [2] has published the first US patent on AM technology. After more than 40 years, AM is now a useful tool in the materials science field. In contrast to the traditional manufacturing process of removing materials, AM technologies use a layer-by-layer approach to build solid parts by superimposing layers one on top of the other [3]. By layer melting, sintering, or bonding powder or wire

DOI: 10.1201/9781003516071-3

materials [4], this method makes both a 3D model of the part and a finished part. In each stage of development, a slice CAD model is shown in a cross-section. This enables AM to make many different kinds of materials with complicated structures, like metals, ceramics, plastics, polymers, and organic molecules that have different properties [4]. It can also be used in health, medical tools, products, aircraft, and aerospace [5–8]. After rapid research in recent decades, AM is now a state-of-the-art technology with its own methods for selecting materials, printing precisely, rapidly, and affordably, and producing high-quality goods [9]. Presently, several AM technologies are under development, including electro-hydrodynamic jet, microfluidic control, laser metal deposition (LMD), 3D printing, selective laser sintering (SLS), and laser additive manufacturing (LAM) [4, 5, 10]. One AM process that uses an intense laser beam to produce heat is LAM. More complex and precise components can be manufactured, along with high-performance parts that can be quickly shaped and densified, thanks to the relatively concentrated heat source and high energy density [11, 12]. Many advantages of LAM are the capacity to manufacture direct high-precision components, quicker research and development and production procedures, and the ability to make intricate forms that aren't conceivable with normal techniques [13, 14].

However, the different AM techniques have unique advantages, features, and applications. Table 3.1 represents the brief comparative details of different types of AM technologies with reference to the kind of mechanisms involved, primary benefits of the distinct technologies, common drawbacks, and common applications. These technologies include but are not limited to material extrusion, which is well recognised for its capability of generating complex designs in a very short time, even for more sophisticated systems, such as powder bed fusion (PBF) and direct energy deposition (DED), which are well noted for their accuracy and ability to work with difficult materials. These materials have different properties that qualify them for application in various industrial fields including aerospace, automobiles, machinery, medical, etc., which underscores the versatility of AM technologies. The cited papers shed more light on the working dynamics and prospects of each of the AM techniques, thus underlining the irrevocable role of AM in the current manufacturing environments. A conceptual network of SMA phases, crystal structures, and additive manufacturing is presented in Figure 3.1, which classifies them according to their relevant specifications. The information in this chart functions as a pragmatic manual for the advancement of SMA applications.

Utilising additive manufacturing, research has created multifunctional porous nitinol to promote osseointegration and bone tissue regeneration. To emulate the properties of trabecular bone, AM integrates a biofunctionalised surface and a microarchitecture that has been systematically designed to optimise pore size, surface area, and curvature. By controlling the release kinetics of $rhBMP_2$, which is accomplished via polydopamine-immobilised $rhBMP_2$, the biofunctionalised surface facilitates enhanced cell adhesion, proliferation, and osteogenic properties [24]. A thermal treatment protocol for SMAs composed of AMed nitinol (NiTi) is presented by Tareq et al. [25]. Following a one-hour solution cycle at 900°C, the protocol stipulates a 30-minute ageing cycle at 450°C. By generating ultrafine metastable Ni-rich

TABLE 3.1
Various AM techniques with characteristics and applications

S. No.	AM Technique	Description Of Process Or Features	Benefits or Advantages	Challenges Or Drawbacks	Applications	Ref.
1	Material Extrusion	Deposition through a nozzle; layer-by-layer solidification	Facilitates intricate geometries, minimal waste	Surface roughness, limited material compatibility	Prototyping, small parts manufacturing	[15]
2	Powder Bed Fusion (PBF)	Uses a laser or electron beam to melt powder	High geometric precision, suitable for complex structures	Porosity, residual stress, high energy consumption	Aerospace, medical implants	[16, 17]
3	Directed Energy Deposition (DED)	Uses a multi-axis nozzle and either a laser or an electron beam for deposition. capacities	Rapid deposition speeds and massive build	Flaws on the surface, poor resolution	Maintenance and repair of substantial structural parts	[18, 19]
4	Binder Jetting	Powder materials may be more easily joined with the help of binding agents	Flexible material choices with little waste	Need for post-processing and reduced mechanical strength	Artistic concepts and working models	[20]
5	Sheet Lamination	Materials joined in layers using glue or welding	Quick processing time,	Low dimensional precision, and limited material processing options	Items for decoration and packaging	[21]
6	Hybrid Technologies	Combining AM with conventional manufacturing for synergy	Enhances microstructure and material characteristics	Increased operating expenses and complicated setup	Top-quality components needed for exacting requirements	[22, 23]

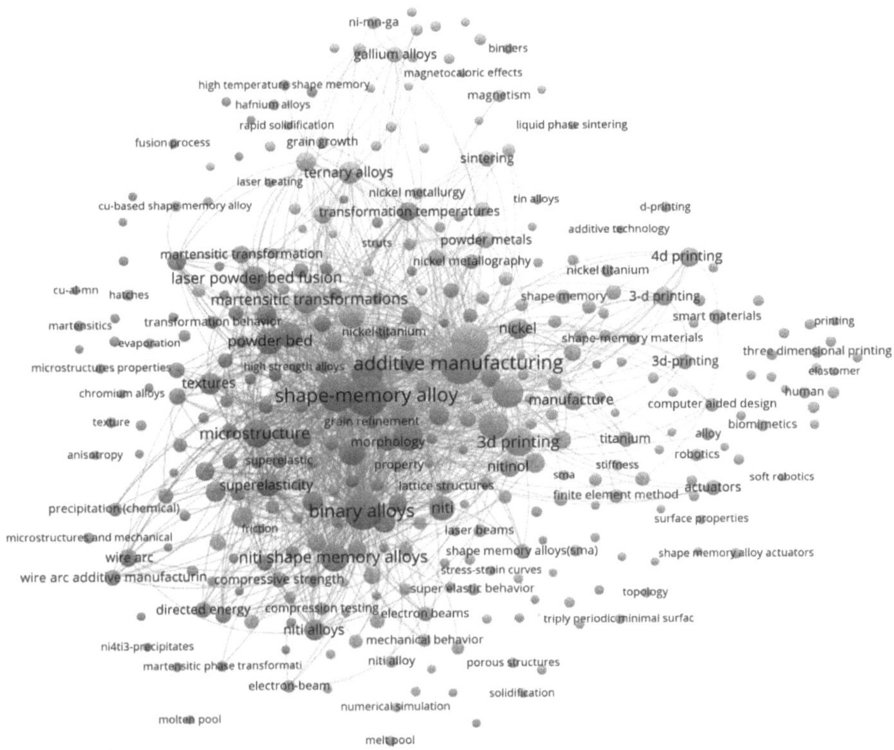

FIGURE 3.1 Conceptual network of SMA phases, crystal structures, and additive manufacturing [5] (Credit statement: Under Creative Commons Attribution 4.0).

precipitates and achieving a homogenised microstructure, this treatment promotes the formation of advantageous phase transformation characteristics and enhances super-elasticity, as evidenced by recovery ratios exceeding 90% and recoverable strains surpassing 6.5%. Andani et al. [26] employ porosity to achieve an equivalent rigidity of NiTi components to that of human bone. With 58% porosity, they simulate and fabricate porous NiTi structures to reduce stiffness from 69 GPa to 20.5 GPa, demonstrating that it is possible to accomplish the desired levels of stiffness by adjusting porosity. Concerning obstacles to bone–implant stability, Rahmanian et al. place particular emphasis on mismatching rigidity and bone–implant integration. They investigate the use of nitinol additive manufacturing to fabricate implants with a customised surface and rigidity for bone attachment [27]. Jamshidi et al. [28] studied the application of laser powder bed fusion (LPBF) in the production of nitinol stents, optimising the elastic moduli and strut diameters to enable customised designs. The research underscores the impact of process parameters on the properties of stents and emphasises the potential of additive manufacturing in the production of personalised stents that yield enhanced outcomes. By means of SLM, Lu et al. [29] investigate the microstructure, shape memory characteristics, and biocompatibility of porous NiTi scaffolds. A comparison is made between additively manufactured SS316L and NiTi in terms of their biocompatibility and mechanical properties as they pertain to biomedical implants. The authors emphasise the superior strength and biocompatibility of NiTi, which is particularly advantageous for surgical applications [30–32]. Constant et al. [33] studied nitinol's mechanical stability and biocompatibility in relation to intra-articular prosthetic devices. Incubation in synovial fluid caused changes to the surface's morphology but had no discernible impact on the surface's composition or mechanical characteristics. So, nitinol makes a good case for being used in these kinds of contexts [30].

This chapter analyses the literature on AM-made nitinol (NiTi) alloys, identifies open questions, and charts recent growth patterns. In order to get a clearer picture, we also examine the microstructural properties, dynamic properties, and strengthening mechanism of AMed aluminium alloys, as well as the process utilised to manufacture them. In conclusion, this study discusses and summarises the present issues, trends in development, and potential future applications of aluminium metal produced using AM technology.

3.2 ALLOYING ELEMENTS FOR SHAPE MEMORY ALLOYS

This section summarises various alloy elements which can be added to shape memory alloys (SMAs) and offers an overview of their unique features, effects on the behaviour of SMAs, and presents the most popular applications. Combination of various elements can improve certain properties of the composites, including service temperature change, mechanical properties, corrosion resilience, as well as operational stability under different circumstances [34–44]. Different types of alloying elements are introduced to SMAs in order to improve the performance of these materials and their utilisation in numerous specific assignments. Copper (Cu) reduces transformation temperatures and improves super-elasticity for NiTiCu alloys used in applications where low activation temperature is desirable [45]. Iron (Fe) enhances the corrosion protection and high-temperature stability of the NiTiFe alloys for aerospace actuators and other high-temperature applications [46–48]. Manganese (Mn) suppresses hysteresis effects and enhances shape memory performance; the latter is important for using SMPs in medicine where temperature control is critical [49]. NiTiAl alloys exhibit enhanced mechanical strength and resistance to attrition due to additions of aluminium (Al), thus the alloys can be applied where high performance is desirable, such as in robotics or medical applications [50, 51]. Vanadium (V) improves elevated temperature strength and mechanical characteristics; therefore, it is used in fire-protectant products and vehicles [52, 53]. Hysteresis and work output of shape memory super-elastic NiTiZr alloys useful in surgery and space exploration are increased by the element zirconium (Zr) [54–56]. Tantalum (Ta) has very high corrosion resistance and enhanced strength, and a higher fatigue life among the high-temperature SMAs used in orthodontic wires and various medical applications [57, 58]. A 3D microstructure produced from gallium (Ga) allows for low-temperature driven actuation and is best suited for self-healing materials and micro actors [59, 60]. Hafnium (Hf) enhances high-temperature characteristics, phase stability, and surface quality, making it suitable for use where high-temperature stability is desired [61, 62]. Finally, silicone (Si) improves the mechanical properties and mechanical attrition of NiTiSi alloys in demanding fields such as robotics and aerospace applications [63–65].

3.3 ADDITIVE MANUFACTURING IN SHAPE MEMORY ALLOYS

This section outlines a summary of the important work reported in the current manuscript for SMAs employing various AM techniques. There are several techniques related to AM of SMAs like selective laser melting (SLM), direct metal laser sintering (DMLS), electron beam melting (EBM), etc. It identified that selective laser melting of NiTi alloys could provide high-resolution fabrication with controllable microstructures for application in biomedical implants and aerospace parts [66]. CuAlNi-based alloy processed by DMLS was studied to investigate the influence of heat treatment on the properties required for actuator and robotics purposes [67–69]. FeMnSi alloys were studied using electron beam melting on two corrosion characteristics and mechanical strength for application in structure for marine [70–72]. Binder Jetting of NiTiCu alloys focused on enhancing properties of printed parts for application in consumer electronics and automotive sensors by optimising processing parameters was explored [73–75].

SLS produced NiTiNb alloys with high ductility and low modulus for superior mechanical compatibility of biomedical uses for implants and surgical instruments [46, 76]. EBM also showcases the improved high-temperature characteristics of NiTiHf alloys and makes it suitable for aerospace engine parts and automotive actuator applications [46, 76]. DED of NiMnGa alloys improved the magnetic characteristics and transformation rates of the same alloys for use in magnetic sensors and smart actuators [59, 77, 78]. The super-elasticity of SLS of CuZnAl alloys increased shape memory recovery and fatigue life, making it suitable for wearable devices and telecommunication [79, 80]. Lastly, in the Binder Jetting of CoNiAl alloys [81, 82], work was done on non-conventional systems' thermal characteristics that can be useful for heat-exchange and high-temperature applications. These findings are consistent with the specific use of different AM technologies to improve the functional properties of SMAs for specific applications.

The emergence of the advanced direct metal deposition (DMD), also sometimes referred to as additive manufacturing, of shape memory alloys has unveiled a wide-spectrum opportunity for biomedical application due to its properties and biocompatibility. Biocompatibility is clearly seen in NiTi alloys synthesised with selective laser melting, and these are useful in applications for orthodontic wires and stents [83, 84]. Cryomilled NiTiCu alloys possess a medium degree of biocompatibility and can be used in implants requiring a relatively low activation temperature, for which EBM is appropriate [73, 85]. Likewise, NiTi alloys when produced by DMLS are characterised by high biocompatibility and therefore are suitable for use in surgical instruments and bone plates [86, 87]. On the other hand, CuAlNi alloys using Binder Jetting have relatively low biocompatibility, thus not suitable for implant applications and commonly used in brace construction [88–90].

FeMnSi alloy treated by the DED process has moderate biocompatibility, which is used in surgical instruments and custom clamps [70, 71], while NiTiHf alloys, having high biocompatibility fabricated by SLM, are widely used in high-temperature medical devices and prosthetic components [91]. The low biocompatibility of NiMnGa alloys also makes them useful in magnetic dental implant applications and magnetic actuators for robotic surgery because of their magnetic response [78]. A CuZnAl alloy made by EBM is moderately biocompatible and is used in dental retention bar and clip attachments and vascular surgeries in the form of clips [80]. Finally, CoNiAl alloys produced by the Binder Jetting process, which have relatively lower biocompatibility, are applied to the external fixations and the supportive rehabilitation equipment [82]. These enhancements in AM processing have enabled the variety usage of SMAs to various biomedical gadgets which enhance their prospects in medical therapies [92].

3.4 NITINOL (NITI) BIOIMPLANT FABRICATION AND APPLICATIONS

Nickel-titanium (NiTi) is an alloy with super-elastic and shape memory properties that is frequently used in bio-implant manufacturing. Fabrics made from nitinol are flexible when used in medical applications [93]. The conventional processes used to direct the shape and structure of nitinol bio-implants are directed energy deposition and powder bed fusion [94]. Micro-roughening and bioactive coatings have been designed to enhance biological compatibility and retention by surrounding living tissues, therefore extending nitinol's uses in the human body for a longer time [95–99]. Some other uses of nitinol in the biomedical area include stents, orthopaedic implants like bone screws and plates, and dental applications, which use the shape memory and super-elasticity of nitinol [100]. Current developments are directed toward improving nitinol's mechanical characteristics and resistance to fatigue failure by adjusting the chemical composition of the alloy and improving the technological process [101, 102]. Research on anti-bacterial silver has proved its biocompatibility and efficiency, while more research is being done to overcome the problems related to the release of nickel ions over the long run [102, 103]. Future directions will focus on advanced bio-implants based on nitinol for improved and tailored structures, such as nitinol functionally graded materials together with the utilisation of new 3D fabrication techniques for tailor-made bio-implant structures [94, 104].

3.5 CHARACTERISTICS OF NITI SMAS

This section provides the main selection characteristics of SMAs used in additive manufacturing, indicating their main properties and advantages, possible difficulties, and typical uses. SMAs when used in additive manufacturing are guided by a number of critical parameters that improve their applicability in many high-performance applications. Super-elasticity is one such characteristic where SMAs are capable of withstanding and self-healing large deformations; this makes them ideal for use in areas such as medical stents and orthodontic wires [105–107]. Nonetheless, achieving the desired microstructure during processing becomes a real issue. The other important requirement is biocompatibility. SMAs coexist in the human body and can serve as effective long-term medical and dental implants [108, 109]. For these reasons, the finest material purity should be achieved and products should undergo testing that is extremely rigorous to prevent negative outcomes. They also possess high damping capacity, thus effectively damping mechanical energy in applications such as in the automotive and aerospace industries due to its ability to reduce vibrations [110, 111]. This, however, is only possible by optimising on the material composition in a way that enhances its use in countless ways. Their ability to resist fatigue, thus allowing them to endure cyclic deformation without failure, makes SMAs suitable for aerospace parts and surgical instruments, but, of course, the high cost of SMA materials and the required high standards of fabrication can be problematic [112, 113].

Furthermore, cooling rates in SMAs are slow, a feature that ensures that the material has a low coefficient of thermal expansion; that is, it does not expand or contract when exposed to heat or cold and is thus ideal for use in manufacturing mechanisms such as micro-electro-mechanical systems (MEMS) and composites [114, 115]. However, this property needs to be selective, for each type of use calls for a distinct property. While this causes problems in the control of composition and thermal treatment to achieve the desired actuation capability based on the environmental temperature changes, the temperature-induced actuation capability of SMAs is exceptional for actuators and temperature-sensitive applications [116]. Last, SMAs exhibit good specific strength, which is a valued characteristic for aerospace and automotive applications, as the former requires lightweight and strong materials for construction [66, 117]. However, stringent processing methods must be used to achieve greater utilisation of this strength.

3.5.1 Various Mechanical Properties of SMAs

When it comes to the AM of SMAs their mechanical characteristics and performance depend on many factors that improve the usage of these materials in diversified fields. Chemical composition which consists of copper (Cu), iron (Fe), manganese (Mn), aluminium (Al), vanadium (V), zirconium (Zr), tantalum (Ta), gallium (Ga), hafnium (Hf) and silicon (Si) is added to NiTi to enhance the transformation temperatures, mechanical properties, and specified application properties in load-bearing and structural applications [118, 119]. Shape memory alloys custom-made through AM exhibit enhanced mechanical properties, including fatigue, strength, and wear resistance, categories crucial in cyclic operations in aerospace, medical, and automotive applications, among others [112–123]. Stress-strain characteristics of SMAs also depend on the composition of the SMA and the processing. For example, two NiTi alloys NiTi-1 (Ni = 55.6%) have a plateau of 467 MPa and fail at 583 Mpa, and NiTi-2 (Ni = 55.2%) changes at 110 MPa and reaches failure at 505 MPa, demonstrating the importance of optimisation for use in structures [15]. Often in techniques like laser powder bed fusion where scan strategies play a part in determining the surface finish and crack initiation, surface morphology, as analysed through scanning electron microscopy (SEM), is critical to producing complex functional components without defects [124, 125]. The deformation behaviour of FeMnSi SMAs as a function of their orientation in the frames of AM processes is a factor that determines mechanical performance efficiently and affects most of all tailored components used in aerospace, automobiles, and robotic engineering, where accurate mechanical attributes are critical [72].

3.5.2 Microstructure and Phases

SEM of as polished cross-sections NiTi-1 (Ni = 55.6%) and NiTi-2 (Ni = 55.2%) is shown in Figure 3.2. Specimens were placed with their bottom edges towards the print bed throughout the printing process. The rate of porosity of the NiTi-2 samples is 12%, whereas that of the NiTi-1 specimens is 5%. It seems like the pores on the second specimen are longer and more crooked. This indicates that the densification process during sintering was not fully successful [11]. Backscatter electron contrast allows for the separation of two phases from the NiTi matrix. By using EDX, one of the phases may be recognised as TiNi oxide, which is somewhat darker (Figure 3.2c, g). Another, with a much higher oxygen content, is a Ti carbide that looks much darker. Figure 3.2(c) shows that the specimens' exteriors are coated with a fragile carbide and oxide layer. The NiTi2 specimens seem to have a thinner shell in the low-magnification photograph. Zykova et al. [126] illustrated the application of "ColdArc" Wire Arc Additive Manufacturing (WAAM) in the production of Ti-6Al-4V components. At room temperature, the resulting components display an entirely austenitic microstructure and demonstrate super-elastic behaviour. This study demonstrates how WAAM can be utilised to fabricate intricately shaped components from advanced functional materials like Ti-6Al-4V. As depicted in Figure 3.3, tensile testing samples were acquired through wire cutting from the deposited layers. These samples were subsequently assessed via digital image correlation (DIC) and SEM analysis. The results of the SEM analysis unveiled super-elastic behaviour and irrecoverable strain accumulation in WAAM Ti-6Al-4V components. The microstructural evolution of the as-deposited NiTi components is explained by Zeng et al. [127]. Specifically, the deposited layers undergo a transition from columnar grains with an inclination in the initial deposited layer to equiaxed grains at the interface of subsequent deposited layers [127].

3.5.3 Super-Elasticity

Super-elasticity, also referenced to as pseudoelasticity, is one of the unique features of SMA, which means that SMAs can achieve really large amounts of elastic deformation and still remain undamaged. It is characterised by a shift in phase from austenite to martensite in the material whenever the same is subjected to stress greater than the transformation temperature. When this stress is relieved, the alloy reverts back to its fully austenitic state, and so returns to its original shape. Super-elasticity also indicates that SMAs can withstand strains up to 10% with their material

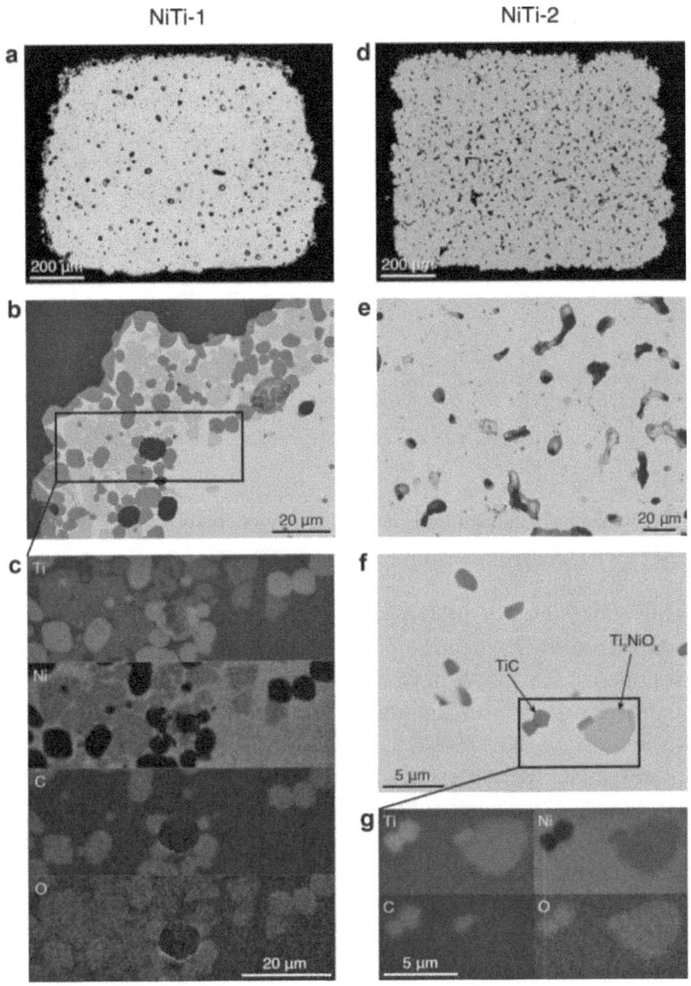

FIGURE 3.2 SEM and EDX Analysis of NiTi-1 (Ni = 55.6%) and NiTi-2 (Ni=55.2%) alloys: cross-sectional phase identification [15]. (Credit statement: Under Creative Commons Attribution 4.0).

properties intact, making them suitable for products such as biomedical, eyeglasses frame hinges, and flexible electronics, since resilience is central to their application.

3.5.4 Deformation Behaviour

Ferretto et al. [72] studied the deformation mechanics and mechanical properties of FeMnSi-based SMAs fabricated through LPBF with regard to the effect of sample orientation on texture during AM. Using an argon environment to reduce oxygen content, the examined samples were manufactured on a Sisma MySint 100 LPBF

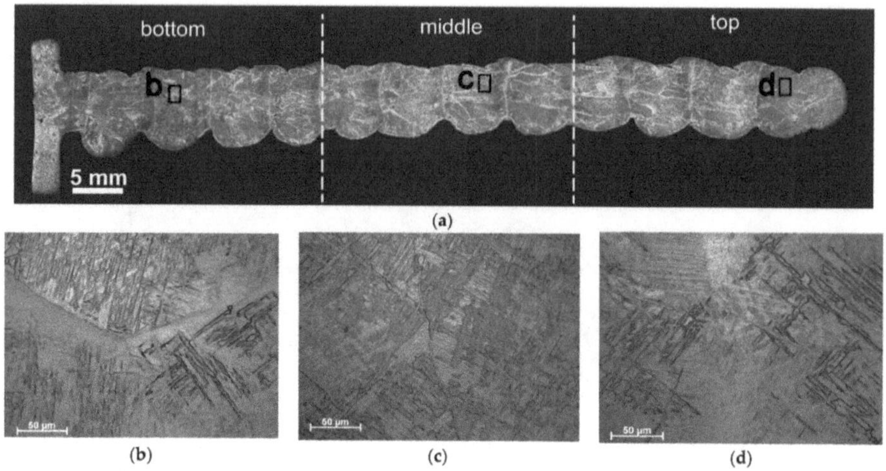

FIGURE 3.3 Macroscopic and microstructural overview of WAAM NiTi specimens [126] (Credit statement: Under Creative Commons Attribution 4.0).

machine having a fibre laser of 1070 nm. Based on previous research, precise parameters were established for the scanning strategy, layer thickness, and seam spacing, while the laser operated at 175 W power and 300 mm/s. The specimens of horizontal cylindrical bars were cast with their axis lying perpendicular to the build direction (BD), while those of the vertical cylindrical bars had their respective main axis lying parallel to the BD with reference to the build plate. All rods were annealed at 800°C for two hours subsequent to fabrication in order to eliminate any remaining bcc-(ferrite). Heat-treated cylindrical rods were subsequently machined into round specimens measuring 6 mm in diameter and 25 mm in gauge length. By utilising this method, it was possible to generate tensile specimens in which the loading direction (LD) was oriented differently in relation to the BD. For horizontal specimens, this meant perpendicularity, while vertical specimens exhibited parallel LD. Ferretto et al. [72] carried out the engineering stress-strain analysis of both specimens. The vertical sample yields at a lower stress than the horizontal sample. Both kinds of samples return to linearity at stresses much below their respective yield strength for this material; the vertical sample departs from linearity thereafter at 140 MPa. Figure 3.4 shows that the increase in the horizontal sample work hardening behaviour appears at higher stress levels of about 200 MPa based on the work hardening rate curves.

3.6 APPLICATIONS OF NITI SMAS

NiTi SMAs have characteristics that allow them to be applied in a vast number of fields, and in this chapter they are discussed in relation to their ability alter their phases with the help of thermal or mechanical triggers and get back to their original shape, which is among the strengths that allow dozens of inventive applications.

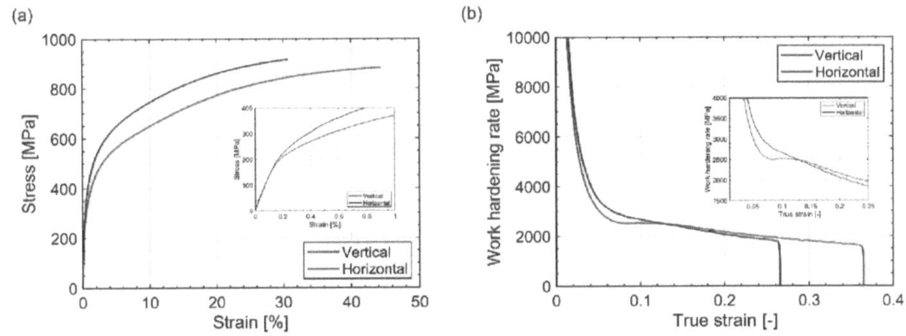

FIGURE 3.4 Stress-strain and work hardening rate curves for vertical and horizontal samples [72] (Credit statement: Under Creative Commons Attribution 4.0).

3.6.1 Orthodontic Applications

Specific applications of nickel-titanium compounds use characteristics of pseudo-elastic as well as the shape memory alloys, which have established extensive usage in hardwing dentistry for the future time-dependent and adaptation capability [108, 128]. Various orthopaedic applications of NiTi alloys are shown in Figure 3.5. NiTicable wires are applied in multibracket fixed orthodontic treatment, by initiating the constant forces it assists in extensive movement of the teeth which is paramount to orthodontic treatments for long-term results. Palatal arches with NiTi composition apply softer and gentler forces, just like brackets with lightweight memory wires. This not only decreases the patients' discomfort but also ensures the corrective forces' role without affecting their effectiveness. The alloy property enables these endodontic files to be highly flexible and have the capacity to "bounce" back to their original shape in a timely manner. Various NiTi endodontic files are shown in Figure 3.6. As a result precise cleaning and debridement is highly facilitated thus significantly reducing the risk of canal perforation. On the upside, NiTi alloys have such advantages as high fatigue resistance threshold, braceability beyond the ASTM-provided static recovery limit, super-elasticity, a lower probability of overfilling, as well as a risk of hypersensitivity from nickel hypersensitivity patients. Thus, the materials are used in certain cases when typical properties provide a definite superiority over other materials.

The use of NiTi alloys in medical applications benefits from shape memory, and pseudoelastic properties, which allows for bone treatment improvement and new joint solutions. Spinal and intramedullary implants that have NiTi as the base material can conform to body temperature, and spine structures become stable and thus improves the bone fracture healing process through adapting to the anatomy forms. With regard to scoliosis, it must be noted that NiTi is an excellent material when applied in corrective devices, which is crucial for correcting the spine position. These corrective systems are less invasive compared to common methods. These features are responsible for the application of force, guaranteeing that the latter is unceasing and appropriate which makes the healing process faster and better. Even

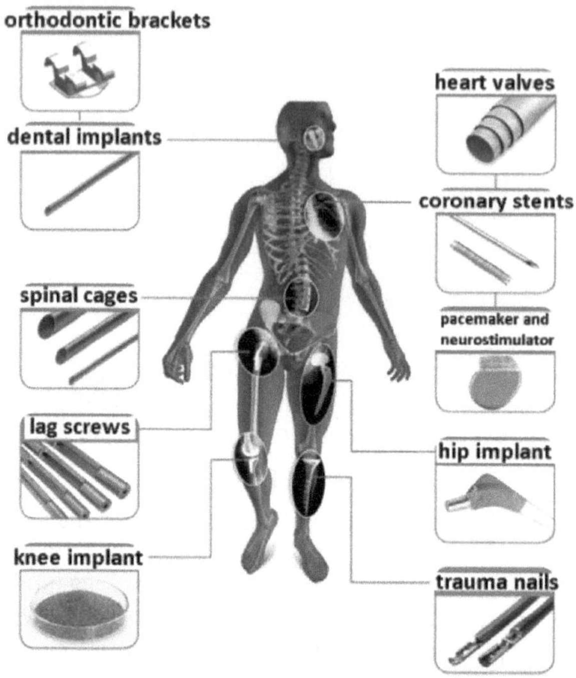

FIGURE 3.5 Orthopaedic applications of NiTi alloys [129] (Credit statement: Reproduction permissions from Elsevier).

so, NiTi does present certain constraints for its application in orthopaedics, including the high cost of the material and the necessity to precisely control the temperature of the procedure. Second, in spite of NiTi having all the benefits of flexibility and biocompatibility, its wear resistance and strength may be low as compared to those of most of the traditional orthopaedic materials like stainless steel and Ti, which may hamper its use in highly load-bearing applications.

3.6.2 Vascular Applications

The use of NiTi alloys in vessels is because of their most outstanding capacity to restore their shape and their pseudoelastic properties under load, which is crucial for minimally invasive operations [131]. By virtue of its characteristics NiTi is used in manufacturing stents and stent-grafts, which are widely used in the procedures of vasal operations that intend to normalise the blood circulation in the arteries that are narrowed or blocked. NiTi stents have a self-expanding nature which means that stents expand with human body temperature and don't require balloon expansion, hence causing minimal trauma and aiding in the healing of the site of the injury. Venous filters, for example, the Simon filter partiality make use of NiTi's shape memory effect and adapt to the micro-anatomy of the veins to perform filtration

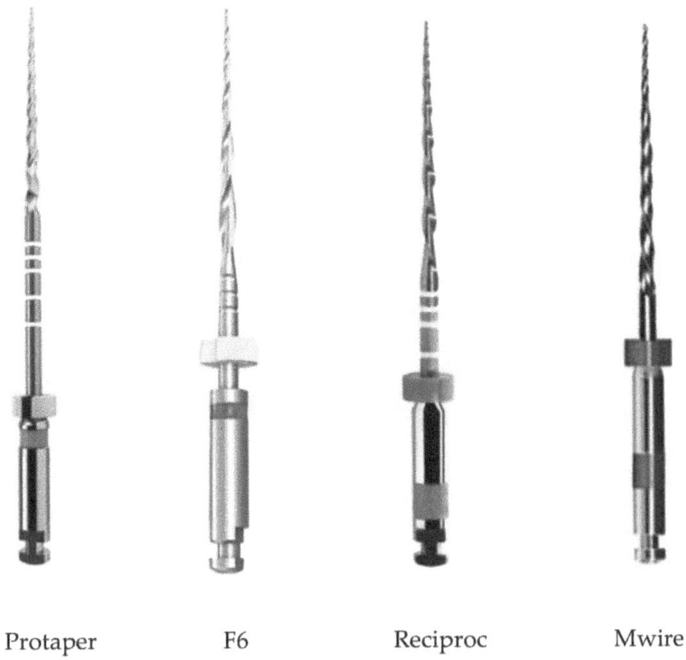

FIGURE 3.6 NiTi endodontic files [130]. (Credit statement: Under Creative Commons Attribution 4.0).

without disrupting flow. These applications demonstrate the delivery of NiTi as a crucially efficient and secure tool for vascular interventions. Nevertheless, mechanical stresses and dynamic vascular environment are wearisome, as they lead to the possibility of stent fractures or migration, as shown in Figure 3.7. Moreover, NiTi being flexible and unobtrusive permits small incisions, that might decrease post-operative complications. However, the devices need fine precision both at production and deployment and this will influence their performance and handling.

3.6.3 Neurosurgical Applications

The neurosurgical uses of nickel-titanium alloys exploit special characteristics such as the pseudoelastic and shape memory effects in these alloys so that they can be applied in treating intricate and complex anatomies in sensitive areas. Coils used for the aneurysm treatment from a platinum and NiTi blend are used to create the formation of blood clots in aneurysm, so they won't be burst due to the softness of NiTi, which does not harm the vessel. Self-expanding neurosurgical stents made from NiTi that are permanently implanted to regulate blood circulation in cerebral arteries where there are close spaces also hold up the walls of the artery firmly but without much pressure. They are a major aid in procedures not only in preventing strokes but also for stabilising arterial walls and enabling the endovascular coiling

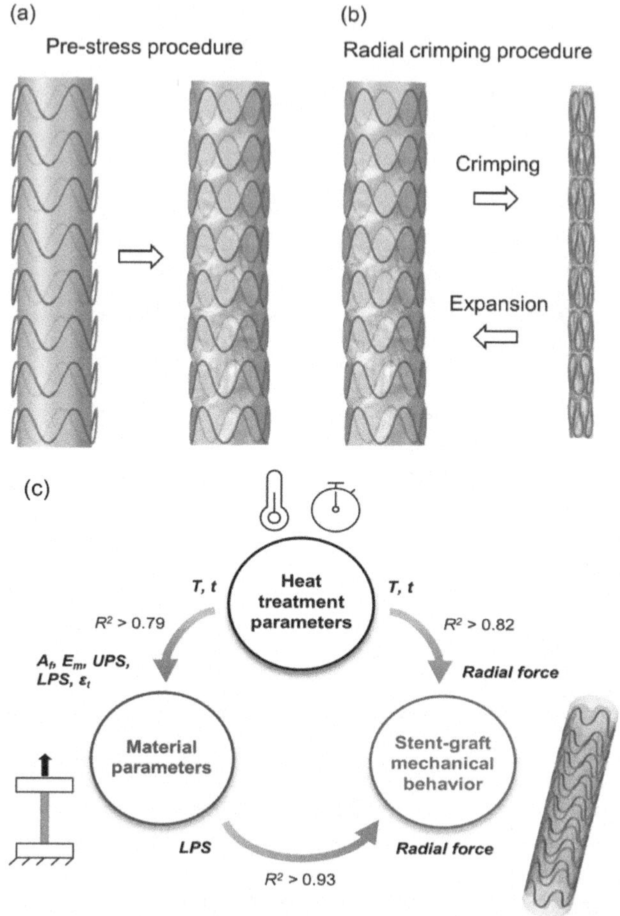

FIGURE 3.7 Design and analysis of NiTi vascular stents [132]. (Credit statement: Under Creative Commons Attribution 4.0).

of aneurysms. Likewise, many of the advantages of NiTi in neurosurgery, however, come with obstacles. The complexity of the build-up of cerebral vessels requires that NiTi devices be very accurate in design and placement to avoid off-target effects or complications. Furthermore, the in-depth comprehension of long-term interactions between NiTi and neural tissue is an active field of research with respect to inflammation and biocompatibility, implying the requirement of using a thorough clinical examination and optimum devices.

3.6.4 Surgical Instruments

The use of the nickel-titanium (NiTi) alloys on surgical instruments benefits from the material's pseudoelasticity and shape memory to offer increased yield flexibility and reliability in multiple minimally invasive procedures. Since these surgical tools are NiTi, their shape may be maintained at a certain structural state or altered to take on the necessary form under the parameters mentioned before, allowing for very precise and less intrusive procedures [133]. Devices like this, called guidewires and baskets, are used to remove kidney and bladder stones. They are constructed of NiTi, which contracts to a prior shape, making manipulation simpler and the removal of stones easier. Catheters and snares made from NiTi have a great ability to manoeuvre and reduce the tendency to kink in which complex treatment of vasculature is achieved.

Nevertheless, it is quite evident that alongside the many benefits that NiTi holds there are some restrictions related to its use in surgical instruments. NiTi may be more expensive than traditional materials in the long run; consequently, its use can be limited to niche applications where it has only its unique functionalities in its favour. In addition, the use of specialised machines to manage and process NiTi increases the complexity of surgical investigation. The fourth and final point is that while NiTi has excellent biocompatibility in general, a small number of patients could develop nickel hypersensitivity, which must be taken into consideration while selecting materials that are going to be in contact with the patient's tissues.

3.7 PERFORMANCE IMPROVEMENT OF AMED STRUCTURES USING SMAS

This section summarises key research on the effects of different additive manufacturing techniques and thermomechanical treatments on the microstructure and phase transformations of SMAs, particularly focusing on NiTi alloys. Each study provides insights into how processing conditions influence the functional properties of SMAs, crucial for industrial and biomedical applications. Table 3.2 presents a list of major works dealing with the influence of thermal treatments and post-processing treatments on microstructure, phase transformation and properties of SMAs. Techniques used such as ageing, solution treatments, and the consequences of these treatments on the properties of NiTi as well as other SMA systems are elaborated hence developing probable instruments for use in engineering.

Some prior research on the microstructure of shape memory alloys fabricated through additive manufacturing have produced striking observations on the potential of optimising the materials consumption and utilization. For instance, Mohri et al. [134] sought to present data on how the microstructure of SMA could be improved based on AM technology. The studies supported that thermomechanical training and heat treatment improved simultaneously both the pseudoelasticity and the shape memory effects, thus emphasising the appropriate heat and mechanical processing needed for improving those functional properties of SMAs [135, 136]. Research studies on 3D-printed NiTi, alloys provided super-elastic recovery and stress-free recovery with a variation according to composition and processing of alloy reveal the possibility of attaining SMA properties through enhanced integration of 3D printing technology [15]. Further, explorations of the effects of size and geometry on transformation temperatures and mechanical response show that geometry has a critical

TABLE 3.2
Summary of Research on Thermal Treatments and Post-Processing of SMAs

Ref.	Method	Main Findings	Implications
[141]	Examination of fatigue under cyclic loading	Discussed the impact of microstructure and heat treatment on SMA performance under cyclic loads	Provides guidelines for enhancing SMA fatigue resistance through controlled thermal histories
[142]	Ageing treatments on EBAM-fabricated NiTi alloys	Demonstrated improved super-elasticity through ageing, showing significant recovery rates	Suggests ageing as a critical post-processing step to optimise the functional properties of EBAM NiTi
[143]	Post-heat treatments on EBF3 NiTi alloys	By using a range of post-heat treatments, the martensitic transition may be controlled, and functional attributes intensified for targeted uses	Emphasises the adaptability of thermal post-processing to modify SMA characteristics
[144]	Study on Ni concentration in FeMnAl SMAs	Ni has no effect on particle development, and regulated heat cycling improves super-elasticity	SMAs might be optimised for industrial applications via careful heat control are being considered
[145]	L-PBF of NiTi SMAs	Texturing and heat treatment increase the transition temperatures and strength	Post-processing is critical for AM SMAs to get their required mechanical and phase properties

role, thus calling for efficient part design in terms of size in AM to allow for the desired performance in SMAs [137].

Research on WAAM of NiTi-alloys focused on microstructural development and phase transformation of deposited layers; the work highlighted WAAM's capability to fabricate NiTi SMAs with desired microstructures [138]. In the same way, electron beam freeform fabrication (EBF3) studied the phase constitution, microstructure, and martensitic transformation of NiTi SMAs and the impact of solidification in the mechanical and functional performance, observing the potential of the method to reproduce geometries of NiTi SMAs [139]. Lastly, studies on electron beam additive manufacturing of NiTi alloys established the links between microstructure, martensitic transformation, and mechanical properties; reaffirming that SMA with desired crystallography and mechanical properties can be produced through EBAM [140]. These studies therefore emphasise how the microstructural features could be controlled through several AM techniques for improvement of SMA performance and functionality.

3.8 CONCLUSION

This chapter elaborated on the advances in the AM techniques for shape memory alloys. It can be concluded that AM can revolutionise the capabilities and the performance of a NiTi SMAs in diversified fields. Modern methods of powder bed fusion by lasers and electron beams enable the manufacturing of complex structures and proforma, which are impossible in conventional manufacturing processes. This capability enhances not only the mechanical properties and performance of SMA components but also establishes tailored designs for the required application, especially in biomedical applications. However, there are still some problems, such as phase stability problems, the problem of material defects, and the problem of process optimisation. Clearly, continuous research is requisite for overcoming these challenges and for realising the full potential of SMAs in the area of additive manufacturing. Due to the strengthening of the field, the developments brought by the innovations would also introduce new uses in aerospace, robotics, and other applications that greatly integrate SMAs in engineering and technology in the future. The future development of SMAs in additive manufacturing suggests great prospects for their further evolution and readiness to become the basis for new sophisticated solutions in numerous spheres and fields of activity.

REFERENCES

1. Campbell I, Bourell D, Gibson I. Additive manufacturing: rapid prototyping comes of age. *Rapid Prototyp J.* 2012;18:255–258.
2. Householder DA. The progress of knowledge: a reexamination of the fundamental principles of American copyright law. *Loy LA Ent LJ.* 1993;14:1.
3. Rasiya G, Shukla A, Saran K. Additive manufacturing-a review. *Mater Today Proc.* 2021;47:6896–6901.
4. Singh R, Toseef M, Kumar J, et al. Benefits and Challenges in Additive Manufacturing and Its Applications. In: Kaushal S, Singh I, Singh S, et al., editors. *Sustain Adv Manuf Mater Process* [Internet]. 1st ed. Boca Raton: CRC Press; 2022. p. 137–157. Available from: https://www.taylorfrancis.com/books/9781003269298/chapters/10.1201/9781003269298-8.
5. Ahmad S, Hashmi AW, Singh J, et al. Innovations in additive manufacturing of shape memory alloys: alloys, microstructures, treatments, applications. *J Mater Res Technol.* 2024. https://doi.org/10.1016/j.jmrt.2024.08.213
6. Kumar S, Singh J, Mohapatra SK. Role of particle size in assessment of physico-chemical properties and trace elements of Indian fly ash. *Waste Manag Res J Sustain Circ Econ* [Internet]. 2018;36:1016–1022. Available from: http://journals.sagepub.com/doi/10.1177/0734242X18804033.
7. Sharma V, Singh JP, Raman R, et al. Machine learning approach for prediction analysis of aluminium alloy on the surface roughness using CO_2 laser machining. *Proc Inst Mech Eng Part E J Process Mech Eng.* 2024;09544089241231093.
8. Subramaniam S, Raju N, Ganesan A, et al. Artificial intelligence technologies for forecasting air pollution and human health: a narrative review. *Sustainability.* 2022;14(16):9951.

9. Singh JIP, Sharma V, Singh S, et al. Impact of molding temperature, fiber loading and chemical modifications on the physicomechanical, and microstructural morphology properties of Woven Kenaf Fiber/PLA composites for non-structural applications. *J Nat Fibers*. 2024;21:2326586.
10. Sharma V, Singh JP, Hussan K, et al. Machine learning approach for predicting the machining characteristics of additive manufactured ABS material. *Proc Inst Mech Eng Part C J Mech Eng Sci*. 2024;09544062241283330.
11. Angjellari M, Tamburri E, Montaina L, et al. Beyond the concepts of nanocomposite and 3D printing: PVA and nanodiamonds for layer-by-layer additive manufacturing. *Mater Des*. 2017;119:12–21.
12. Singh J, Singh R, Singh S, et al. Reducing scrap due to missed operations and machining defects in 90PS pistons. *Int J Interact Des Manuf* [Internet]. 2022. Available from: https://doi.org/10.1007/s12008-022-01071-0.
13. Singh J. Sustainability Issues in Laser-Based Additive Manufacturing. Handb Laser-Based Sustain Surf Modif Manuf Tech. Boca Raton: CRC Press; 2023. p. 167–189.
14. Singh J. Applications of laser technology for processing and post-treatments of powders and coatings. *Handb Laser-Based Sustain Surf Modif Manuf Tech*. 2023;135–166. https://doi.org/10.1201/9781003347408-7
15. Wagner MA, Ocana-Pujol JL, Hadian A, et al. Filament extrusion-based additive manufacturing of NiTi shape memory alloys. *Mater Des*. 2023;225:111418.
16. Yakout M, Elbestawi MA, Veldhuis SC. A review of metal additive manufacturing technologies. *Solid State Phenom*. 2018;278:1–14.
17. Frazier WE. Metal additive manufacturing: a review. *J Mater Eng Perform* [Internet]. 2014;23:1917–1928. Available from: http://link.springer.com/10.1007/s11665-014-0958-z.
18. Megahed M, Mindt H-W, N'Dri N, et al. Metal additive-manufacturing process and residual stress modeling. *Integr Mater Manuf Innov*. 2016;5:61–93.
19. Alagha AN, Hussain S, Zaki W. Additive manufacturing of shape memory alloys: a review with emphasis on powder bed systems. *Mater Des*. 2021;204:109654.
20. Blakey-Milner B, Gradl P, Snedden G, et al. Metal additive manufacturing in aerospace: a review. *Mater Des*. 2021;209:110008.
21. Gibson I, Rosen D, Stucker B, et al. Sheet lamination. *Addit Manuf Technol*. 2021;253–283. https://doi.org/10.1007/978-3-030-56127-7_9.
22. Chaurasia JK, Jinoop AN, Paul CP, et al. Effect of deposition strategy and post processing on microstructure and mechanical properties of serviced Inconel 625 parts repaired using laser directed energy deposition. *Opt Laser Technol*. 2024;168:109831.
23. Pragana JPM, Sampaio RF V, Bragança IMF, et al. Hybrid metal additive manufacturing: a state–of–the-art review. *Adv Ind Manuf Eng*. 2021;2:100032.
24. Gorgin Karaji Z, Speirs M, Dadbakhsh S, et al. Additively manufactured and surface biofunctionalized porous nitinol. *ACS Appl Mater Interfaces*. 2017;9:1293–1304.
25. Tareq S, Rahman T, Poudel B, et al. Heat treatment protocol for additively manufactured nitinol shape memory alloys in biomedical applications. *Mater Sci Eng A*. 2024;897:146274.
26. Taheri Andani M, Haberland C, Walker JM, et al. Achieving biocompatible stiffness in NiTi through additive manufacturing. *J Intell Mater Syst Struct*. 2016;27:2661–2671.
27. Rahmanian R, Andani MT, Walker J, et al. Modeling and Validation of Additively Manufactured Porous Nitinol Implants. 2014. Available from: https://search.crossref.org/search/works?q=Rahmanian+R%2C+Andani+MT%2C+Walker+J%2C+et+al.+Modeling+and+Validation+of+Additively+Manufactured+Porous+Nitinol+Implants.+2014.&from_ui=yes#.

28. Jamshidi P, Panwisawas C, Langi E, et al. Development, characterisation, and modelling of processability of nitinol stents using laser powder bed fusion. *J Alloys Compd.* 2022;909:164681.
29. Lu HZ, Ma HW, Luo X, et al. Microstructure, shape memory properties, and in vitro biocompatibility of porous NiTi scaffolds fabricated via selective laser melting. *J Mater Res Technol.* 2021;15:6797–6812.
30. T G, Badhai J, Karthik S, et al. Biocompatibility and Mechanical Behaviour Studies on Wire-Arc Additive Manufactured Stainless Steel 316L and Shape Memory Alloy (NiTi) Materials for Biomedical Implants. 2023. https://doi.org/10.1115/smasis2014-7653.
31. Singh J. Slurry erosion performance analysis and characterization of high-velocity oxy-fuel sprayed Ni and Co hardsurfacing alloy coatings. *J King Saud Univ - Eng Sci* [Internet]. 2021; Available from: https://doi.org/10.1016/j.jksues.2021.06.009.
32. Singh A, Kaushik V, Chahal S, et al. Efficient degradation of methylene blue dye and antibacterial performance of shape controlled RuO2 nanocomposites. *ChemistrySelect* [Internet]. 2021;6:10038–10050. Available from: https://www.scopus.com/inward/record.uri?eid=2-s2.0-85116495488&doi=10.1002%2Fslct.202102546&partnerID=40&md5=f86b39f7294c94f8be0f7194778c3bb6.
33. Singh J, Singh JP, Kumar S, et al. Improvement of erosion performance of WOKA HVOF coating reinforced by ceramics and carbide powdered materials: a case study. *AIP Conf Proc.* AIP Publishing; 2024.
34. Kumar M, Singh J, Kumar S, et al. *Future Perspective on Multilayer Thermal Barrier Coatings for the Protection of Gas Turbine and Its Components BT - Recent Developments in Mechanics and Design.* In: Singh DK, Hegde S, Mishra A, editors. Singapore: Springer Nature Singapore; 2024. p. 209–216.
35. Dutta V, Thakur L, Singh B, et al. A study of erosion–corrosion behaviour of friction stir-processed chromium-reinforced NiAl Bronze composite. *Materials* (Basel) [Internet]. 2022;15:5401. Available from: https://www.mdpi.com/1996-1944/15/15/5401.
36. Kumar A, Grover N, Manna A, et al. Investigating the influence of WEDM process parameters in machining of hybrid aluminum composites. *Adv Compos Lett.* 2020;29:2633366X20963137.
37. Kattimani MA, Venkatesh PR, Masum H, et al. Design and numerical analysis of tensile deformation and fracture properties of induction hardened inconel 718 superalloy for gas turbine applications. *Int J Interact Des Manuf.* 2023;1–11. https://doi.org/10.1007/s12008-023-01452-z.
38. Prashar G, Vasudev H. Surface topology analysis of plasma sprayed Inconel625-Al_2O_3 composite coating. *Mater Today Proc* [Internet]. 2022;50:607–611. Available from: https://linkinghub.elsevier.com/retrieve/pii/S221478532102109X.
39. Yedida VVS, Vasudev H. A review on the development of thermal barrier coatings by using thermal spray techniques. *Mater Today Proc.* 2022;50:1458–1464.
40. Prashar G, Vasudev H. A review on the influence of process parameters and heat treatment on the corrosion performance of Ni-based thermal spray coatings. *Surf Rev Lett.* 2022;29:2230001.
41. Rani N, Chahal S, Mahadevan SK, et al. Development of hierarchical magnesium oxide anchored cerium oxide nanocomposites with improved magnetic properties and photocatalytic performance. *Nanotechnology* [Internet]. 2020;31. Available from: https://www.scopus.com/inward/record.uri?eid=2-s2.0-85087528231&doi=10.1088%2F1361-6528%2Fab96e8&partnerID=40&md5=f0d1cd83633ad8184d627e716ab962f7.

42. Ankita A, Chahal S, Singh S, et al. Europium-doped cerium oxide nanoparticles: investigating oxygen vacancies and their role in enhanced photocatalytic and magnetic properties. *Environ Sci Pollut Res Int* [Internet]. 2024;31:1276–1287. Available from: https://www.scopus.com/inward/record.uri?eid=2-s2.0-85182736017&doi=10.1007%2Fs11356-023-30686-3&partnerID=40&md5=b13f37d2fc45ce1dc37c13738421c65d.
43. Sharma V, Kumar V, Bist A. Investigations on morphology and material removal rate of various MMCs using CO 2 laser technique. *J Brazilian Soc Mech Sci Eng*. 2020;42:1–14.
44. Yang KZ, Pramanik A, Basak AK, et al. Application of coolants during tool-based machining–a review. *Ain Shams Eng J*. 2023;14:101830.
45. Liang W, Zhou M, Ke F. Shape memory effect in Cu nanowires. *Nano Lett*. 2005;5:2039–2043.
46. Xi R, Jiang H, Li G, et al. In-situ alloying of NiTiNb ternary shape memory alloys via laser powder bed fusion using pre-alloyed NiTi and elemental Nb powders: microstructure, phase transformation behavior and functional properties. *Addit Manuf*. 2024;79:103933.
47. Zhang Y, Jiang S, Tang M, et al. Mechanisms for influence of post-deformation annealing on microstructure of NiTiFe shape memory alloy processed by local canning compression. *J Mater Process Technol*. 2021;291:116998.
48. Liu X, Li H, Zhang Y, et al. Cryogenic temperature deformation behavior and shape memory mechanism of NiTiFe alloy. *Intermetallics*. 2023;162:107997.
49. Mediha KÖK. Investigation of thermal properties, chemical analysis and biocompatibility of high temperature oxidized NiTiMn shape memory alloy. *J Phys Chem Funct Mater*. 2018;1:25–35.
50. Cheng Q, Guo N, Fu Y, et al. Study on preparation and stiffness of NiTi shape memory alloy/2A12 aluminum alloy laminated structure. *Int J Adv Manuf Technol*. 2022;122:4179–4189.
51. Singh V, Singh J, Vasudev H, et al. Review on tribo-erosion analysis in Ni, Al_2O_3, and TiO2 based thermal spray coatings. *AIP Conf Proc*. AIP Publishing; 2024.
52. Arun A, Rajkumar K, Santosh S. Fiber laser cutting study on ternary NiTiV shape memory alloy. *Mater Manuf Process*. 2023;38:1745–1754.
53. Aksu Canbay C, Keskin A. Effects of vanadium and cadmium on transformation temperatures of Cu–Al–Mn shape memory alloy. *J Therm Anal Calorim*. 2014;118:1407–1412.
54. Li Y, Ru Z, Cui Y, et al. Phase stability and hardness of some ternary Ti–Zr based shape memory alloys. *Int J Smart Nano Mater*. 2011;2:272–282.
55. Narasimha GB, Murigendrappa SM. Effect of zirconium on the properties of polycrystalline Cu-Al-Be shape memory alloy. *Mater Sci Eng A*. 2019;755:211–219.
56. Sharma V, Kumar V. Application of Box-Behnken design and response surface methodology for multi-optimization of laser cutting of AA5052/ZrO2 metal–matrix composites. *Proc Inst Mech Eng Part L J Mater Des Appl*. 2018;232:652–668.
57. Niendorf T, Krooß P, Batyrsina E, et al. Functional and structural fatigue of titanium tantalum high temperature shape memory alloys (HT SMAs). *Mater Sci Eng A*. 2015;620:359–366.
58. Ercan E, Dagdelen F, Qader IN. Effect of tantalum contents on transformation temperatures, thermal behaviors and microstructure of CuAlTa HTSMAs. *J Therm Anal Calorim*. 2020;139:29–36.
59. Laitinen V, Salminen A, Ullakko K. First investigation on processing parameters for laser powder bed fusion of Ni-Mn-Ga magnetic shape memory alloy. *J Laser Appl*. 2019;31. https://doi.org/10.2351/1.5096108.

60. Mingear J, Hartl D. Liquid metal-induced corrosion of nickel-titanium alloys by gallium alloys for liquid metal-enabled shape memory applications. *Corros Sci.* 2020;167:108524.
61. Zhao T, Li Y, Liu Y, et al. Nano-hardness, wear resistance and pseudoelasticity of hafnium implanted NiTi shape memory alloy. *J Mech Behav Biomed Mater.* 2012;13:174–184.
62. Buytoz S, Dagdelen F, Qader IN, et al. Microstructure analysis and thermal characteristics of NiTiHf shape memory alloy with different composition. *Met Mater Int.* 2021;27:767–778.
63. Sawaguchi T, Kikuchi T, Ogawa K, et al. Development of prestressed concrete using Fe–Mn–Si-based shape memory alloys containing NbC. *Mater Trans.* 2006;47:580–583.
64. Khodaverdi H, Mohri M, Ghorabaei AS, et al. Effect of low-temperature precipitates on microstructure and pseudoelasticity of an Fe–Mn–Si-based shape memory alloy. *Mater Charact.* 2023;195:112486.
65. Singh J, Singh S. Support vector machine learning on slurry erosion characteristics analysis of Ni- and Co-alloy coatings. *Surf Rev Lett* [Internet]. 2023; Available from: https://www.worldscientific.com/doi/10.1142/S0218625X23400061.
66. Elahinia MH, Hashemi M, Tabesh M, et al. Manufacturing and processing of NiTi implants: a review. *Prog Mater Sci.* 2012;57:911–946.
67. Moskvichev EN, Smolin AY, Shamarin NN. Fabrication of promising Cu-Al-Ni alloys by electron-beam additive manufacturing. *AIP Conf Proc.* AIP Publishing; 2022.
68. Pérez-Cerrato M, Fraile I, Gómez-Cortés JF, et al. Designing for shape memory in additive manufacturing of Cu–Al–Ni shape memory alloy processed by laser powder bed fusion. *Materials* (Basel). 2022;15:6284.
69. Agrawal A, Dube RK. Methods of fabricating Cu-Al-Ni shape memory alloys. *J Alloys Compd.* 2018;750:235–247.
70. Gärtner E, Meyenborg I, Toenjes A. Adapting Fe–Mn–Si–Cr shape memory alloy for laser powder bed fusion by adjusting the Mn content. *Prog Addit Manuf.* 2023;46:102071.
71. Del-Río L, Nó ML, Gómez R, et al. Additive manufacturing of Fe-Mn-Si-based shape memory alloys: state of the art, challenges and opportunities. *Materials* (Basel). 2023;16:7517.
72. Ferretto I, Kim D, Lee WJ, et al. Shape memory and mechanical properties of a Fe-Mn-Si-based shape memory alloy: effect of crystallographic texture generated during additive manufacturing. *Mater Des.* 2023;229:111928.
73. Qadır R, Mohammed S, Kök M, et al. A review on NiTiCu shape memory alloys: manufacturing and characterizations. *J Phys Chem Funct Mater.* 2021;4:49–56.
74. Chen L, Oliveira JP, Yan X, et al. Microstructure and phase transformation behavior of NiTiCu shape memory alloys produced using twin-wire arc additive manufacturing. *Addit Manuf Front.* 2024;3:200132.
75. Mousapour M, Partanen J, Salmi M. NiTiCu alloy from elemental and alloyed powders using vat photopolymerization additive manufacturing. *Addit Manuf.* 2023;78:103853.
76. Liu S, Han S, Zhang L, et al. Strengthening mechanism and micropillar analysis of high-strength NiTi–Nb eutectic-type alloy prepared by laser powder bed fusion. *Compos Part B Eng.* 2020;200:108358.
77. Ituarte IF, Nilsén F, Nadimpalli VK, et al. Towards the additive manufacturing of Ni-Mn-Ga complex devices with magnetic field induced strain. *Addit Manuf.* 2022;49:102485.
78. Caputo MP, Berkowitz AE, Armstrong A, et al. 4D printing of net shape parts made from Ni-Mn-Ga magnetic shape-memory alloys. *Addit Manuf.* 2018;21:579–588.

79. Zhuo L, Song B, Li R, et al. Effect of element evaporation on the microstructure and properties of CuZnAl shape memory alloys prepared by selective laser melting. *Opt Laser Technol.* 2020;127:106164.
80. Dong B, Xia Y, Cai X, et al. Addition of Sc in wire-based directed energy deposition of Al–Mg–Zn–Cu alloy: microalloying to refine grains and improve mechanical properties. *Addit Manuf.* 2023;67:103494.
81. Sanin V V, Kaplansky YY, Aheiev MI, et al. Structure and properties of heat-resistant alloys NiAl–Cr–Co–X (X= La, Mo, Zr, Ta, Re) and fabrication of powders for additive manufacturing. *Materials* (Basel). 2021;14:3144.
82. Im HJ, dos Santos JCP, Campbell CE, et al. Microstructure and properties of Co-Ni-Al-W γ/γ' superalloy fabricated via laser fusion of elemental powders. *Addit Manuf.* 2023;76:103790.
83. Shayesteh Moghaddam N, Saedi S, Amerinatanzi A, et al. Achieving superelasticity in additively manufactured NiTi in compression without post-process heat treatment. *Sci Rep.* 2019;9:41.
84. Khorasani AM, Goldberg M, Doeven EH, et al. Titanium in biomedical applications—properties and fabrication: a review. *J Biomater Tissue Eng.* 2015;5:593–619.
85. Sharma N, Singh G, Hegab H, et al. Tribo-corrosion characterization of NiTiCu alloy for bio-implant applications. *Mater Res Express.* 2019;6:96526.
86. Prasad K, Bazaka O, Chua M, et al. Metallic biomaterials: current challenges and opportunities. *Materials* (Basel). 2017;10:884.
87. Abdudeen A, Abu Qudeiri JE, Kareem A, et al. Latest developments and insights of orthopedic implants in biomaterials using additive manufacturing technologies. *J Manuf Mater Process.* 2022;6:162.
88. Zhang H. *Surface Modification by Filtered Cathodic Vacuum Arc and Nanomechanical Properties of Thin-film Media, Cu-Al-Ni Shape-memory Alloy, and Surface-textured Silicon.* Berkeley: University of California; 2009.
89. Manik R, Behera A. Manufacturing of 3D print biocompatible shape memory alloys. Smart 3D nanoprinting. Boca Raton: CRC Press; 2022. p. 281–298.
90. Stevens E, Kimes K, Salazar D, et al. Mastering a 1.2 K hysteresis for martensitic para-ferromagnetic partial transformation in Ni-Mn (Cu)-Ga magnetocaloric material via binder jet 3D printing. *Addit Manuf.* 2021;37:101560.
91. Elahinia M, Moghaddam NS, Amerinatanzi A, et al. Additive manufacturing of NiTiHf high temperature shape memory alloy. *Scr Mater.* 2018;145:90–94.
92. Singh H, Singh S, Prakash C. *Current Trends in Biomaterials and Bio-manufacturing BT - Biomanufacturing.* In: Prakash C, Singh S, Singh R, et al., editors. Cham: Springer International Publishing; 2019. p. 1–34. Available from: https://doi.org/10.1007/978-3-030-13951-3_1.
93. Tyber J, McCormick J, Gall K, et al. Structural engineering with NiTi. I: Basic materials characterization. *J Eng Mech.* 2007;133:1009–1018.
94. Elahinia M, Moghaddam NS, Andani MT, et al. Fabrication of NiTi through additive manufacturing: a review. *Prog Mater Sci.* 2016;83:630–663.
95. Safavi MS, Bordbar-Khiabani A, Khalil-Allafi J, et al. Additive manufacturing: an opportunity for the fabrication of near-net-shape NiTi implants. *J Manuf Mater Process.* 2022;6:65.
96. Singh J, Singh S, Vasudev H, et al. Neural computing and Taguchi's methodbased study on erosion of advanced Mo2C–WC10Co4Cr coating for the centrifugal pump. *Adv Mater Process Technol.* 2024;10:2569–2581.
97. Singh J, Vasudev H, Chohan JS. Review on computational fluid dynamics based analysis of surface erosion in hydraulic machinery. *Int J Interact Des Manuf* [Internet]. 2023; Available from: https://link.springer.com/10.1007/s12008-023-01336-2.

98. Singh J, Kumar S, Kumar R, et al. Machine learning regression tools for erosion prediction of WC-10Co4Cr thermal spray coating. *Results Surf Interfaces* [Internet]. 2023;13:100156. Available from: https://www.sciencedirect.com/science/article/pii/S2666845923000636.
99. Thakur L, Vasudev H, Singh J, et al. *Artificial Intelligence and Machine Learning in the Thermal Spray Industry: Practices, Implementation, and Challenges.* Boca Raton: CRC Press; 2023.
100. Sathishkumar M, Kumar CP, Ganesh SSS, et al. Possibilities, performance and challenges of nitinol alloy fabricated by directed energy deposition and powder bed fusion for biomedical implants. *J Manuf Process.* 2023;102:885–909.
101. Habijan T, Haberland C, Meier H, et al. The biocompatibility of dense and porous nickel–titanium produced by selective laser melting. *Mater Sci Eng C.* 2013;33:419–426.
102. Pourshams M, Cullaz T, Rocco PA, et al. Enhancing NiTi implants through binder jetting: microstructure, impurity control, and post-processing optimization. *Int Conf Shape Mem Superelastic Technol. ASM International.* 2024;84840:16–18. https://doi.org/10.31399/asm.cp.smst2024p0016.
103. Farber E, Zhu J-N, Popovich A, et al. A review of NiTi shape memory alloy as a smart material produced by additive manufacturing. *Mater Today Proc.* 2020;30:761–767.
104. Antil P, Kumar Antil S, Prakash C, et al. Multi-objective optimization of drilling parameters for orthopaedic implants. *Meas Control.* 2020;53:1902–1910.
105. Maroof M, Sujithra R, Tewari RP. Superelastic and shape memory equi-atomic nickel-titanium (Ni-Ti) alloy in dentistry: a systematic review. *Mater Today Commun* [Internet]. 2022;33:104352. Available from: https://www.sciencedirect.com/science/article/pii/S235249282201193X.
106. Zhao Z, Jiang D, Xiao Y, et al. Intrinsic response of nanocrystalline superelastic NiTi shape memory alloy. *Extrem Mech Lett.* 2023;60:101988.
107. Song D, Yu C, Zhang C, et al. Superelasticity degradation of NiTi shape memory alloy in wide ranges of temperature and loading level: experimental observation and micromechanical constitutive model. *Int J Plast.* 2023;161:103487.
108. Huang C-L, Chang C-J, Chen K-C, et al. Effect of thermocycling-induced stress on properties of orthodontic NiTi wires. *J Dent Sci.* 2022;17:1788–1795.
109. Van Humbeeck J. Shape memory alloys: a material and a technology. *Adv Eng Mater.* 2001;3:837–850.
110. Saedi S, Acar E, Raji H, et al. Energy damping in shape memory alloys: a review. *J Alloys Compd.* 2023;956:170286.
111. Sun F, Feng X, Wang H, et al. Brief overview of functionally graded NiTi-based shape memory alloys. *Adv Eng Mater.* 2023;25:2201434.
112. Tang D, Hu Y, Yang L. New insights into the mechanical properties, functional fatigue, and structural fatigue of Ni-Ti alloy porous structures. *Metals* (Basel). 2023;13:931.
113. Chen P, Cai X, Liu Y, et al. Combined effects of grain size and training on fatigue resistance of nanocrystalline NiTi shape memory alloy wires. *Int J Fatigue.* 2023;168:107461.
114. Gangil N, Siddiquee AN, Maheshwari S. Towards applications, processing and advancements in shape memory alloy and its composites. *J Manuf Process.* 2020;59:205–222.
115. Trehern W, Hite N, Ortiz-Ayala R, et al. NiTiCu shape memory alloys with ultra-low phase transformation range as solid-state phase change materials. *Acta Mater.* 2023;260:119310.
116. Chikhareva M, Vaidyanathan R. A Thermal, mechanical, and materials framework for a shape memory alloy heat engine for thermal management. *Nanomaterials.* 2023;13:2159.

117. Jalali M, Mohammadi K, Movahhedy MR, et al. SLM additive manufacturing of NiTi porous implants: a review of constitutive models, finite element simulations, manufacturing, heat treatment, mechanical, and biomedical studies. *Met Mater Int.* 2023;29:2458–2491.
118. Tadaki T, Otsuka K, Shimizu K. Shape memory alloys. *Annu Rev Mater Sci.* 1988;18:25–45.
119. Saud SN, Hamzah E, Abubakar T, et al. A review on influence of alloying elements on the microstructure and mechanical properties of Cu-Al-Ni shape memory alloys. *J Teknol.* 2013;64:51–56.
120. Mohammadgholipour A, Billah AHMM. Mechanical properties and constitutive models of shape memory alloy for structural engineering: a review. *J Intell Mater Syst Struct.* 2023;1045389X231185458.
121. Majji BGR, Vasudev H, Bansal A. A review on the oxidation and wear behavior of the thermally sprayed high-entropy alloys. *Mater Today Proc.* 2022;50:1447–1451.
122. Sunitha K, Vasudev H. A short note on the various thermal spray coating processes and effect of post-treatment on Ni-based coatings. *Mater Today Proc* [Internet]. 2022;50:1452–1457. Available from: https://linkinghub.elsevier.com/retrieve/pii/S2214785321058132.
123. Poojari M, Hanumanthappa H, Prasad CD, et al. Computational modelling for the manufacturing of solar-powered multifunctional agricultural robot. *Int J Interact Des Manuf.* 2023;1–12. https://doi.org/10.1007/s12008-023-01291-y.
124. Wang X, Zhang Y, Zhang Z, et al. Development of Fe-Ni-Co-Al-based superelastic alloys. *Crit Rev Solid State Mater Sci.* 2023;49:1–26.
125. Zhang C, Ozcan H, Xue L, et al. On the effect of scan strategies on the transformation behavior and mechanical properties of additively manufactured NiTi shape memory alloys. *J Manuf Process.* 2022;84:260–271.
126. Zykova A, Savchenko N, Nikolaeva A, et al. Effect of "ColdArc" WAAM regime and arc torch weaving on microstructure and properties of as-built and subtransus quenched Ti-6Al-4V. *Materials* (Basel). 2024. https://doi.org/10.3390/ma17102325.
127. Zeng Z, Cong BQ, Oliveira JP, et al. Wire and arc additive manufacturing of a Ni-rich NiTi shape memory alloy: microstructure and mechanical properties. *Addit Manuf.* 2020;32:101051.
128. Singh J, Kumar S, Kumar R. Support vector machine-based prediction of wear in Ni-20Cr2O3 thermal spray coating under bottom ash slurry conditions. *J Tribol.* 2023;39:84–105.
129. Kityk A, Protsenko V, Danilov F, et al. Enhancement of the surface characteristics of Ti-based biomedical alloy by electropolishing in environmentally friendly deep eutectic solvent (Ethaline). *Colloids Surfaces A Physicochem Eng Asp.* 2021;613:126125.
130. Sánchez P, Vidi B, Rico C, et al. Relevant aspects in the mechanical and aging degradation of NiTi alloy with R-phase in endodontic files. *Materials* (Basel). 2024. https://doi.org/10.3390/ma17133351.
131. Singh J, Hashmi AW, Ahmad S, et al. Critical review on biodegradable and biocompatibility magnesium alloys: progress and prospects in bio-implant applications. *Inorg Chem Commun.* 2024;113111. https://doi.org/10.1016/j.inoche.2024.113111.
132. Carbonaro D, Villa E, Gallo D, et al. Designing the mechanical behavior of NiTi self-expandable vascular stents by tuning the heat treatment parameters. *J Mech Behav Biomed Mater* [Internet]. 2024;158:106653. Available from: https://www.sciencedirect.com/science/article/pii/S1751616124002856.
133. Singh J, Gill HS. Evolution of nickel-based superalloy claddings for against high-temperature oxidation and wear. *Therm Claddings Eng Appl.* CRC Press; 2024. p. 160–181.

134. Mohri M, Ferretto I, Khodaverdi H, et al. Influence of thermomechanical treatment on the shape memory effect and pseudoelasticity behavior of conventional and additive manufactured Fe–Mn–Si–Cr–Ni-(V,C) shape memory alloys. *J Mater Res Technol.* 2023;24:5922–5933.
135. Sharanabasava H, Raviprakash M, Prasad CD, et al. Microstructure, mechanical and wear properties of SiC and Mo reinforced NiCr microwave cladding. *Adv Mater Process Technol.* 2023. https://doi.org/10.1007/s40033-022-00445-8.
136. Kulkarni SD, Manjunatha KV, Chandrasekhar U, et al. Design and optimization of polyvinyl-nitride rubber for tensile strength analysis. *Int J Interact Des Manuf.* 2024;18:5893–5908.
137. Farjam N, Nematollahi M, Andani MT, et al. Effects of size and geometry on the thermomechanical properties of additively manufactured NiTi shape memory alloy. *Int J Adv Manuf Technol.* 2020;107:3145–3154.
138. Yu L, Chen K, Zhang Y, et al. Microstructures and mechanical properties of NiTi shape memory alloys fabricated by wire arc additive manufacturing. *J Alloys Compd.* 2022;892:162193.
139. Li B, Wang L, Wang B, et al. Solidification characterization and its correlation with the mechanical properties and functional response of NiTi shape memory alloy manufactured by electron beam freeform fabrication. *Addit Manuf.* 2021;48:102468.
140. Dutkiewicz J, Rogal Ł, Kalita D, et al. Microstructure, mechanical properties, and martensitic transformation in NiTi shape memory alloy fabricated using electron beam additive manufacturing technique. *J Mater Eng Perform.* 2021;1–13. https://doi.org/10.1007/s11665-021-06241-x.
141. Cebrian Carcavilla A, Zaki W. Fatigue of shape memory alloys with emphasis on additively manufactured NiTi components. *Appl Mech Rev.* 2022;74:40801.
142. Pu Z, Du D, Zhang D, et al. Improvement of tensile superelasticity by aging treatment of NiTi shape memory alloys fabricated by electron beam wire-feed additive manufacturing. *J Mater Sci Technol.* 2023;145:185–196.
143. Li B, Wang B, Wang L, et al. Effect of post-heat treatments on the microstructure, martensitic transformation and functional performance of EBF3-fabricated NiTi shape memory alloy. *Mater Sci Eng A.* 2023;871:144897.
144. Peng H, Yong L, Zuo Y, et al. Effect of cyclic heat treatment on abnormal grain growth in Fe-Mn-Al-based shape memory alloys with different Ni contents. *J Mater Sci Technol.* 2023;153:8–21.
145. Saghaian SE, Nematollahi M, Toker GP, et al. Enhancing shape memory response of additively manufactured NiTi shape memory alloys by texturing and post-processing heat treatment. *Shape Mem Superelasticity.* 2023;9:192–206.

4 Emergence of Piezoelectric Materials in Smart Materials

Mohamed Aboughaly

Description	Abbreviation
Active fibre composites	AFC
$BaTiO_3$	BT
Bismuth sodium titanate	BNT
Carbon monoxide	CO
Cellulose nanofibers	CNF
Hexafluoropropylene	HFP
Lead zirconate titanate	PZT
Macro-fiber composite	MFC
Piezo response force microscopic devices	PFM
Polycaprolactone	PCL
Polylactic acid	PLA
Polyvinylidene fluoride	PVF
Polyvinylidene difluoride	PVDF
Polyvinylidene fluoride trifluoro	P(VDF-TrFE)
Triboelectric nanogenerators	TENG
Zinc oxide	ZnO

4.1 INTRODUCTION

Piezoelectric materials are used in various industries including the ceramic industry due to their excellent heat and corrosion-resistant properties. A common piezoelectric material is lead zirconate titanate, which is used in various engineering applications [1, 2]. Other common piezoelectric materials include titanate, barium, and lead titanate compounds. Lead zirconate titanate is used to manufacture ultrasound ceramics and other sensor devices and actuators as well as ceramic capacitors and ferroelectric chips. Piezoelectric materials have shown excellent applications in the manufacturing of digital sensors and use in damage detection applications due to

 DOI: 10.1201/9781003516071-4

their applicability under quality factor (Q) and electromechanical coupling factor (k), and their abilities to withstand stress and strain and generate electricity under high mechanical pressure [3–6]. Also, polymer composites including polyvinylidene compounds combined with fluoride and trifluoroethylamine copolymers are widely used in flexible and wearable devices [2–8].

These copolymers have shown excellent response sensitivity in biomedical and intelligent wearable devices. They also enhance flexibility, can withstand the high mechanical pressure of an electric field and generate electricity with mechanical pressure. Fabricated nanocomposite films synthesized from optimized polymer modulated porous polyvinylidene fluoride integrated with hexafluoropropylene (PVDF-HFP)-based polymer membranes filled with titania (TiO_2) were able to withstand higher levels of shear stress up to 4 KPa with less deformation compared to various nanocomposite films [4, 6, 9, 10]. Also, high polymer piezoelectric composite compounds based on PVDF materials are used as actuators to generate control forces against various sensor structures. Piezoelectric materials convert electrical energy to mechanical energy and are integrated into various wearable devices [11–13]. Piezoelectric materials are able to convert electrical energy to mechanical energy and vice versa [11–13]. The group of piezoelectric materials include mono- and polycrystals, polymers and polymer composites. Piezoelectric materials are integrated into various analytical devices such as sorption detectors, water detectors, tracing of metals, virus detection such as COVID-19, and air pollution detection devices. Also, the advancement of piezoelectric materials has enabled their use for detecting carbon monoxide (CO) with high precision at ppm level in ambient atmosphere. New piezoelectric quartz crystals are used in various carbon monoxide sorption detectors to monitor carbon monoxide at ppm level [3, 4, 10]. Palladium (II) to acetamide at a ratio of 1 to 10 moles was found to have the highest accuracy for carbon monoxide detection [14–16]. Also, piezoelectric based micro-electromechanical sensors provide active binding sites for the detection of viruses [11, 17, 18]. Green materials have been used in the development of multifunctional components in energy, infrastructure, and construction sectors.

Green materials have been employed in smart material development for various applications, including energy equipment, electronics and building materials due to their sustainability and durability. These materials are regarded as safer and more eco-friendly, making them suitable for intelligent applications and helping to overcome the limitations of traditional materials. Piezoelectric materials exhibit outstanding properties for engineering applications, such as actuators and sensors in industrial equipment. They are known for their efficient mechanical-to-electrical energy conversion capabilities, and those with high piezoelectric charge and voltage coefficients are extensively used in renewable energy applications [19]. Piezoelectric materials are considered a fundamental component of energy harvesting and are used to enhance mechanical movements and vibrations as well as applied stress applications using displaced ions that result in net electric charge caused by the unit cell permanent dipole moment [20].

Piezoelectric materials are either natural or synthesized materials, such as quartz compounds, including SiO_2, tourmaline mineral groups, and organic compounds, including bone, wood, rubber, and hair. While natural materials include synthetic ceramics, polymers, and composite materials. The most common ceramic piezoelectric materials are lead, barium, and strontium-based titanate compounds. These materials are used in relators, relaxer-ferroelectric ceramics and crystals [20–22]. Below are the different categories of piezoelectric materials as shown in Figure 4.1.

Perovskite lead zirconate titanate, referred to as PZT, is a commonly used polycrystalline piezoelectric material doped with niobium or lanthanum to create hard or soft piezoelectric ceramic materials [24–27]. PZT materials have been created in thin sheets on substrates enhancing coupling in response to the growing use of piezoelectric ceramic compounds in micro-device systems [28–31]. The current disadvantage of piezoelectric ceramics is their high brittleness and density. Innovative PZT thin sheets enhance flexibility and utilization of texturing and applying epitaxial thin films to enhance flexibility in different applications that use piezoelectric ceramics in micro-devices [28–34]. The biotoxicity is controlled by encapsulation by polylactic acid, protein-based polymers, polysaccharides, such as chitosan and amino acids [35–38]. Piezoelectric ceramics have superior properties as well as single crystals due to their microstructure compared to other materials [29, 39].

Also, the sensing capabilities of piezoelectric materials achieve mobility and are used to develop micro-devices. Ambulatory locomotion offers the benefit of moving

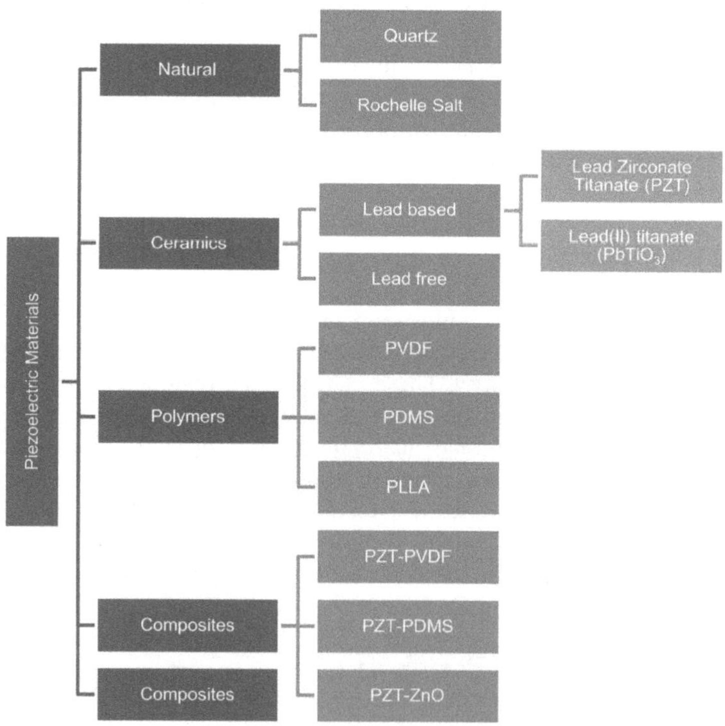

FIGURE 4.1 Different categories of piezoelectric materials.

across rough surfaces more effectively than traditional wheeled mechanisms. To design biologically inspired flapping-wing microrobots for exploration, similar to insects, high-density actuation power is necessary. This power can be achieved through the use of piezoelectric materials integrated with green materials. Examples of green materials used in piezoelectric materials are biodegradable polymers, such as polylactic acid (PLA) or polycaprolactone (PCL), which can be used to create microrobots that degrade safely in the environment after their useful life, reducing waste. Also, bio-based composites and recycled materials are used for the construction of microrobots. Below are the categories of piezoelectric sensing applications used in electronic devices as shown in Figure 4.2.

4.2 GREEN MATERIALS USED IN PIEZOELECTRIC COMPONENTS

Green materials used in piezoelectric compounds have shown high tailor ability and improved piezoelectric response due to their high strength that allows superior sensing and better energy harvesting devices. Natural materials such as quartz, topaz and Rochelle salt are widely used in piezoelectric devices due to their electricity generation capabilities and ability to generate electric charge under mechanical

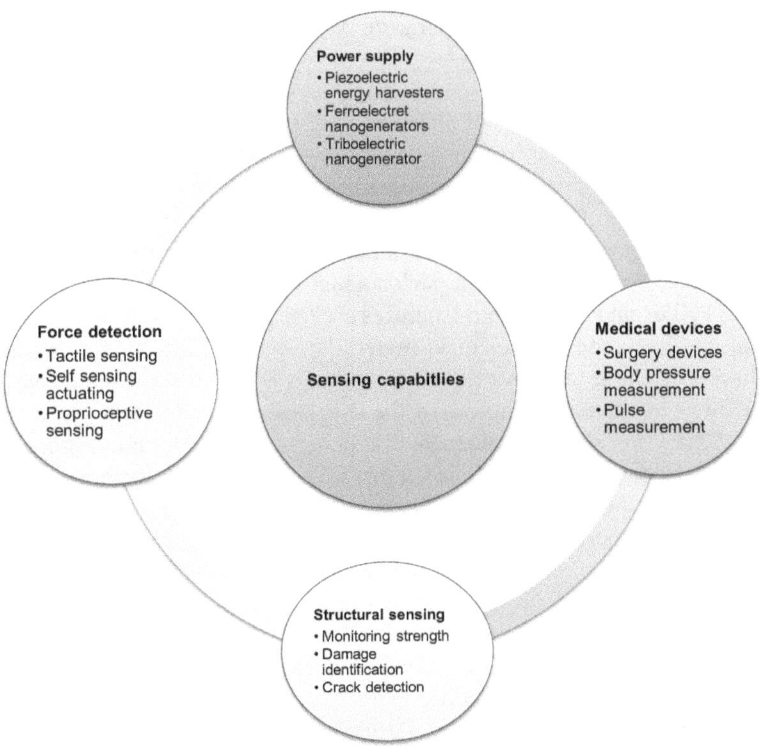

FIGURE 4.2 Classification of sensing capabilities for micro-devices.

stress and improving various engineering applications such as sensors, actuators and energy harvesting devices. Natural materials such as quartz, topaz, and Rochelle salt have been implemented in piezoelectric devices and their applications. The current disadvantages of such materials are the low electromechanical coupling and low piezoelectric properties. Lead zirconate titanate (PZT) is a synthetic polycrystalline ceramic material that has niobium dopes forming soft PZT devices and structures. The current disadvantages of PZT are its high brittleness, density, and high lead content. However, recently developed piezoelectric polymer polyvinylidene fluoride (PVF) compounds have high flexibility and low materials density, which limits its current industrial applications. However, PVDF has lower electromechanical coupling compared to PZT. Green alternatives include non-lead piezoelectric compounds and nanocomposites and foams [32, 33, 40].

Also, electrochemically poled cellulose nanofibers (CNF) thin films have been used as piezoelectric material thin films that are able to function under humid conditions resulting in enhanced crystal alignment. PVDF copolymer layers have also been used to develop low-cost and printable eco-friendly piezoelectric-powered nodes [41]. The current limitations that limit the high performance of piezoelectric materials, such as PVDF, PZT and other single-crystal structures, are a low Curie temperature range within 100°C to 300°C [41]. The piezoelectric material's Curie temperature identifies the thermal instability phase and depolarization, which identifies the highest operating temperature of the material. The Curie temperature could be increased by the addition of different compounds such as langasite (LGS), lithium niobate, and aluminum nitride compounds that raise the Curie temperature up to 1000°C. Also, the addition of different compositions of lithium niobate, langasite, and aluminum nitride compounds (AIN) which increases the Curie temperature allowing its usage in high-temperature applications such as the design of aerospace materials, processing of materials and power plants [42]. Also, single-crystal piezoelectric materials have high electromechanical coupling properties almost ten times higher than that of PZT. Also, high-performance conventional piezoelectric materials, such as PZDF, and crystalline structures are ten times larger than PZT. However, some high-performance piezoelectric materials, such as PZT, single crystals, and PVDF, have low Curie temperatures of less than 350°C that limit their applications [6, 9]. Below is a schematic diagram of the use of electrochemically poled cellulose nanofiber (CNF) thin sheets in piezoelectric generators, as shown in Figure 4.3 [43].

The piezoelectric material's Curie temperature defines the temperature of the material's thermal phase instability and depolarization which is used to identify the upper material operating temperature [44–46]. Conventional PZT materials and their derivatives produce a major challenge due to their lead toxicity which limits their biocompatibility. The presence of lead in PZT derivatives poses potential risks in biological environments when used in implanted sensors, as well as environmental concerns due to regulatory restrictions on hazardous materials, which led to the advancement of lead-free piezoelectric materials, including titanate perovskite, bismuth sodium titanate (BNT), alkaline niobite perovskite-based materials, including potassium, bismuth, sodium niobate, and bismuth non-perovskite-based materials [47, 48]. Currently, the brittle nature of lead-free piezoelectric materials limits their

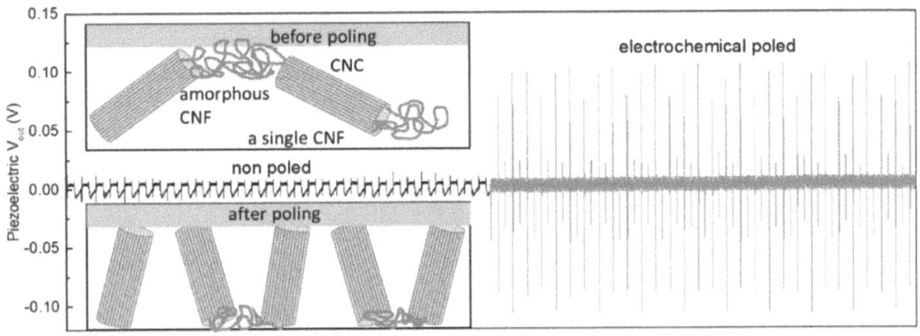

FIGURE 4.3 Green and biomass-based piezoelectric generators based on electrochemical poled nanocellulose fibers [43]. (Credit: Creative Commons 4.0 Attribution).

applications and usage. Durable and sustainable piezoelectric materials are currently able to retain the piezoelectric performance of ceramic-based materials [44–46].

Composite piezoelectric devices that consist of mono lithic piezoceramics using a flexible polymer matrix includes active fiber composites, macro-fiber composites, or piezoelectric particles. Green materials were used in the development of nanowires, nanotubes, and nanofibers that are integrated into flexible and stretchable thin films for the creation of composite piezoelectric devices. Engineered protein filaments developed by bacteria are modified to conduct electricity and synthesized to produce green nanowires and nanotubes, which are used as piezoelectric materials. For example, monolithic piezoceramics are integrated within various polymer matrices and active fiber composites (AFC), zinc oxide compounds (ZnO), and macro-fiber composites (MFCs). Zinc oxide is considered one of the most used piezoelectric nanomaterials used to synthesize nanowires as well as PZT (Lead zirconate titanate) and BT ($BaTiO_3$) nanostructures [26, 49, 50].

Conventional polymeric piezoelectric materials, such as PVDF, improve the flexibility and strength of piezoelectric materials as well as active and macro-fiber composites. Also, nanofiber and nanowire piezoelectric materials are integrated into stretchable and flexible thin films to be used in various micro and nanoscale devices. Also, protein nanowires with conductive ions could be used in piezoelectric materials and to harness energy from moisture in the air [51, 52]. Modification of green materials produced by bacteria could be used to create electronic components that improve the sustainability of piezoelectric materials. Naturally occurring bacteria are also used to synthesize conductive filaments known as nanowires to allow the transfer of electrons across their membranes. Also, piezoelectric materials are used with electromechanical coupling components including transducers, actuators, and portable and kinetic sensors [51–53]. Below is a schematic diagram of the fabrication process of piezoelectric polymer (PVDF-TrFE) thin film generators including silver electrodes fabricated using double-sided graphene transfer [54–56], as shown below in Figure 4.4.

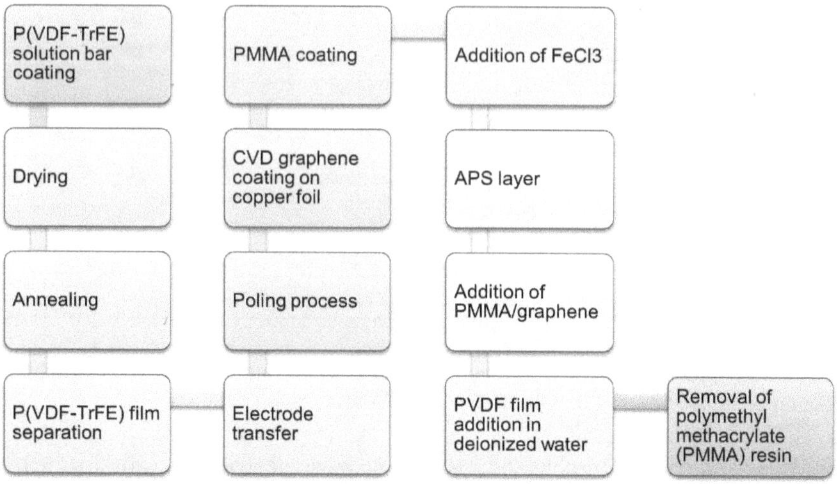

FIGURE 4.4 Schematic of the fabrication process for P(VDF-TrFE) based thin films on silver piezoelectric electrodes.

Conventional polymeric piezoelectric materials, such as PVDF, offer high robustness and flexibility but their applications are limited by poor piezoelectric coupling. Voided piezoelectric nanomaterials, including nanowires, nanofibers, and thin films, are used in both nano- and microscale devices. Zinc oxide is a commonly used compound in these piezoelectric nanomaterials [4, 9]. Void-charged polymer electrets, which exhibit piezoelectric responses, contain permanent charge voids that allow the generation of charges when subjected to mechanical fields. These piezo electrets are typically made from polypropylene, which includes air-filled voids introduced during the fabrication process of PVDF [39, 57, 58]. Such materials are characterized by low density and higher piezoelectric coefficients compared to PVDF. Variants of lead zirconate titanate (PZT) are produced by doping with niobium to create soft PZT or with lanthanum to produce hard PZT. Additionally, the piezoelectric polymer polyvinylidene fluoride has a highly flexible microstructure with low density and lead-free chemical composition [59–61].

Piezoelectricity occurs related to the net polarity of certain materials that enhances the piezoelectric electromechanical coupling properties, which is due to the dipole orientation that exhibits high electromechanical couplings that are ten times higher than in single-crystal piezoelectric materials [59–61]. Single-crystal piezoelectric materials are more expensive and have high manufacturing costs due to the long processing time required for their growth. Additionally, they are less durable than polycrystalline counterparts, as they are more susceptible to mechanical fractures and degradation under operational stress, limiting their applicability in certain high-performance and long-term use cases [62–64]. The current limitations of piezoelectric materials, such as PZT and PVDF compounds, are that they have low Curie temperatures, lower than 350°C, which limits their industrial applications. The Curie temperature of piezoelectric material identifies its depolarization and thermal instability, which defines the material's highest operating temperature [65, 66]. The addition of compositions, such as compounds of lithium niobate, langasite and aluminum nitride compounds, increases the Curie temperature, which allows its usage in high-temperature storage and industrial applications [16, 49, 50].

Compositions like aluminum nitride, langasite, and lithium niobate are utilized due to their ability to raise the Curie temperature, making them suitable for high-temperature applications, including aerospace systems, power plants, and material processing. The main disadvantage for high-temperature structures is that those materials have lower piezoelectric coupling than PZT and other industrial materials. A major advantage of green materials integration of improvement of biocompatibility of PZT materials leads to the improvement of PZT materials in biological environments. Barium nitrate, bismuth sodium nitrate, alkaline niobite perovskite-based, and bismuth non-perovskite-based materials are used in PZT but require improvement in performance. The performance of piezoelectric materials is massively improved using lead-free compositions [16, 65]. Also, propylene is used as a base polymer in piezo-electrets and contains air-filled voids integrated using fabrication processes. Ferro-electret foams have also shown lower density and very high piezoelectric coefficients higher than 5^{-10} and high tailorability and improvement in piezoelectric responses, as well as high strength that allows for superior sensing and electro-ceramics; they are also easily recycled which minimizes environmental impact [67–69]. Due to the high environmental concerns and toxicity of lead, ultra-high piezoelectric devices using piezoelectric materials free of lead are essential in the substation of integrated lead zirconate-based titanate (PZT) ceramics in electronic devices [16]. Piezoelectric actuators change electric into mechanical signals. Piezoelectric sensors measure various forces, including acceleration and pressure, and they improve signals in various medical devices. They are used to develop low-power sensors and phones as well as manufacturing piezoelectric resonators that govern electrical circuit frequencies [49, 67]. Also, piezoelectric nanomaterials have shown excellent properties in electric generators that can self-power using biowaste and therefore eliminate the current limitations in charging batteries.

Green materials such as electrochemically poled cellulose nanofibers (CNF) were used in piezoelectric materials and were able to substitute polyvinylidene difluoride (PVDF) copolymers to manufacture low-cost printable eco-friendly piezoelectric-powered electronic nodes [49, 67]. Generators based on poled low-cost and printable CNF show high piezoelectric voltages similar to PVDF copolymers [70–72].

Organic materials with high molecular structures such as sugar cane reveal piezoelectricity under applied pressure that achieves a direct piezoelectric effect. The converse piezoelectric effect, which occurs under mechanical deformation, is utilized in various applications such as sensors and energy transducers, while the direct piezoelectric effect is employed in actuators [73–76].

Biomolecules like dipeptides and amino acids have non-centrosymmetric piezoelectricity. The larger molecules' hierarchical hydrogen bond structure is essential to piezoelectric action. The hierarchical structures of proteins, polysaccharides, long-chain peptides, and synthetic polymers—such as helical, fibrous, and spatial folding—determine the piezoelectric properties of these macromolecules [5, 6, 10].

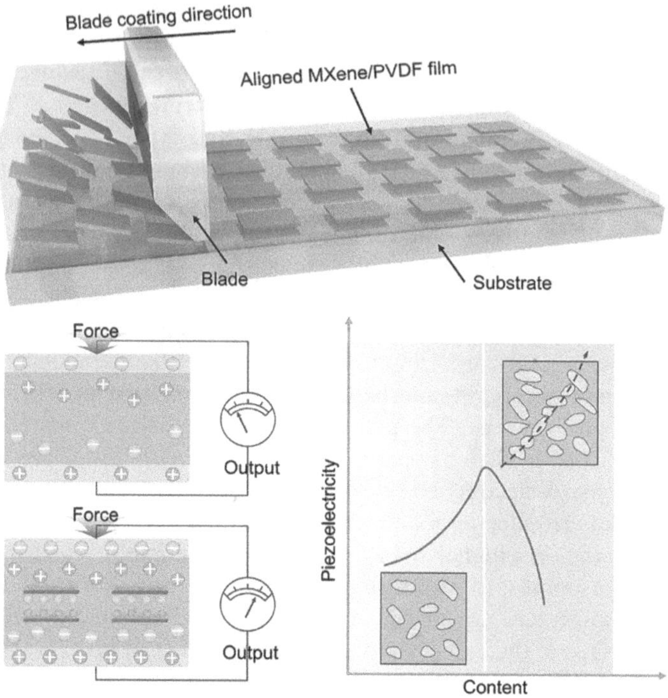

FIGURE 4.5 Mxene-coated PVDF film on dielectric micro capacitor [81]. (Credit: Creative Commons 4.0 Attribution).

The most important parameters for piezoelectric components are the calculated power density (Pd), structural integrity, mechanical stress (σ), electromechanical coupling factor (Q), and electromechanical coupling factor (k). The operating temperature has an impact on material selection [43, 80]. Piezoelectric materials can be classified into four categories: single crystals, ceramics, polymers, and composites. Their properties are characterized by several key factors and parameters, including the electromechanical coupling factor (k), the piezoelectric voltage constant (g), the mechanical quality factor (Q), and the dielectric constant. A schematic diagram of mxene-coated PVDF film on a dielectric micro capacitor as shown in Figure 4.5.

4.3 NANOTECHNOLOGY AND GREEN COMPONENTS INTEGRATED IN PIEZOELECTRIC MATERIALS

With the use of green energy nanotechnology, the mechanical effects of fatigue, torsion, bending, compression, tension, and shear forces have changed the piezoelectric effect. The regulation of energy barriers at contact surfaces, which governs the movement of charge carriers, is a result of the interaction between piezoelectric and semiconducting materials in piezotronics. Utilizing piezoelectric materials, biomechanical energy of compression, tension, shear, and torsion is captured. Electric generators based on piezoelectric nano surfaces have the potential to self-develop into biowaste, doing away with the present battery charging constraints [49, 67, 68].

Many natural materials have a piezoelectric effect including berlinite, sugar cane, quartz, topaz, salt compounds, tourmaline, and apatite crystals. Also, compounds of barium and lead zirconate titanate are considered natural chemicals that have a piezoelectric effect. Also, several green materials were able to clarify how polarization affects piezoelectric catalytic activity and improve it using materials like silk fibroin and collagen-like peptides [73, 74]. Also, raw piezoelectric materials present in cellulose as well as polylactic acid (PLLA) are used to synthesize piezoelectric polymer compounds that are biodegradable which are synthesized by polymerization of lactic acid monomers [82, 83].

Green biomass materials have been used in the synthesis of capacitors where the piezoelectric material is charged by the capacitive piezoelectric current. Such biomass materials are chitosan and chitin, cellulose and its derivatives as well as alginate and other biomass materials. Another excellent biomass-based supercapacitance material is carbon aerogels. Also, the development of porous carbon materials from biomass is used to develop biochar and porous carbon materials with high surface area above 1190 $m^2\ g^{-1}$ w [76, 84]. The supercapacitor has demonstrated high cycle stability, retaining 98% of its performance after over 15,000 cycles and maintaining a current density of 10 A g^{-1} in sodium acetate. Additionally, it achieved an energy density of 15 Wh kg^{-1} at a current density of 0.2 A g^{-1} [76, 77, 84]. The usage of porous carbon materials has shown high storage capacity up to 76% in comparison to non-aqueous-based binders. The usage of carbon aerogels with mesoporous structures is used in synthesis due to their excellent porous structure. Carbon aerogels have excellent electric storage capabilities and are used for storage in hydrogen systems due to their high porosity and corrosion resistance characteristics [84, 85]. Lead-free perovskite-structured ferroelectrics such as $BaTiO_3$, $Bi_{0.5}Na_{0.5}TiO_3$, and $NaTaO_3$ are considered excellent lead-free piezoelectric materials [86, 87]. Energy harvesters based on piezoelectric nano-generation systems are promising due to their high electricity generation capabilities. Green and renewable materials used in nanogenerators include natural materials with high cellulose and sugar contents are used in harvesting electric energy through triboelectric nanogenerators (TENGs) using biocompatible and eco-friendly natural materials [86, 88]. Natural organic piezoelectric materials include biomolecular compounds such as amino acids and peptides, as well as long-chain proteins, polysaccharides, single crystals, and polymers. [88–91]. The high-performance piezoelectric energy generators are assisted by better dielectric charging efficiencies of inorganic structures. Lead-free compositions from polycrystalline and single crystals such as langasite and ferroelectric ceramics with tungsten-bronze structures are reported to be excellent piezoelectric materials [92 –94]. Below is a flowchart of different organic materials used with carious piezoelectric energy harvesters as shown below in Figure 4.6.

Energy harvesting materials are designed to operate in harsh environments and endure temperatures up to 60°C. Piezoelectric harvesters, crafted from sustainable materials, transform mechanical stress into electrical energy. Meanwhile, electrostatic, triboelectric, and electromagnetic harvesters generate electricity through

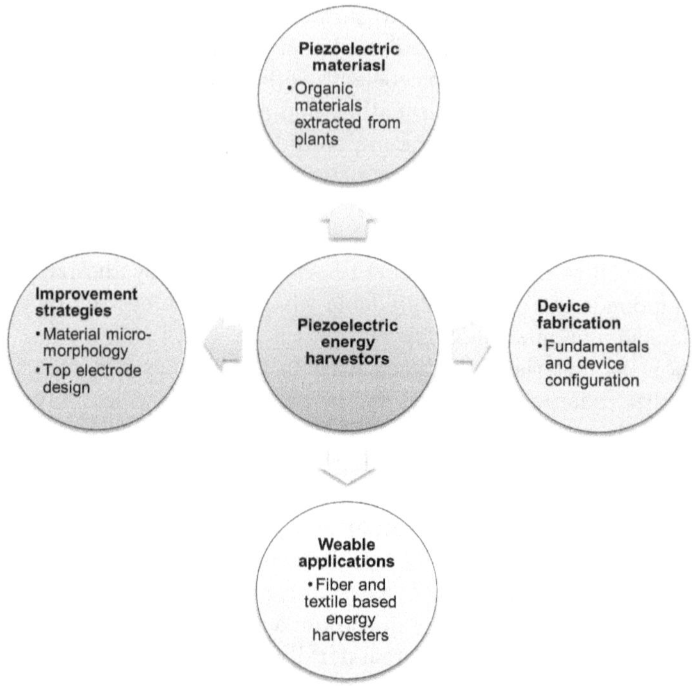

FIGURE 4.6 Integration of green materials in piezoelectric energy harvesters.

variations in capacitance, frictional contact, and magnetic or electrostatic induction [89, 90]. The piezoelectric characteristics of electrochemically poled cellulose nanofiber (CNF) thin films, combined with their crystal alignment, result from the alignment and fragmentation of CNF segments [41]. Cellulose-based materials are considered promising because of their shear piezoelectric properties, which result from dipole moments of hydroxyl groups in various biomass materials.

Nanocellulose, including nanocrystals (CNCs) and nanofibers, possesses unique properties, making piezoelectric materials highly effective in electromechanical coupling systems, such as actuators, transducers, and kinetic sensors. Among piezoelectric polymers, the most effective are fluoro-based polymers based on polyvinylidene fluoride and polyvinylidene fluoride trifluoro ethylene (PVDF-TrFE) [7, 8, 95, 96]. Also, plant-based materials such as cellulose have excellent piezoelectric properties. Cellulose is a key structural component that is extracted as tiny crystals called nanocellulose used to generate voltage under pressure [7, 95, 96]. The cellulose nanofiber piezoelectric performance in thin films is significantly enhanced by electric polling under room humidity conditions. Electric polling also improves the alignment of electric dipoles in ferroelectric materials, further boosting the piezoelectric response of CNF films [36–99].

Biodegradable piezoelectric materials integrate mechanical–electrical coupling with environmentally friendly properties, reducing the need for excess materials and minimizing infection risks. Organic biodegradable piezoelectric compounds, which include amino acids, proteins, synthetic polymers, and polysaccharides, merge these coupling characteristics with biodegradability, thus preventing unnecessary material retention [7, 35, 96]. Below are also different classes of piezoelectric materials as shown in Figure 4.7.

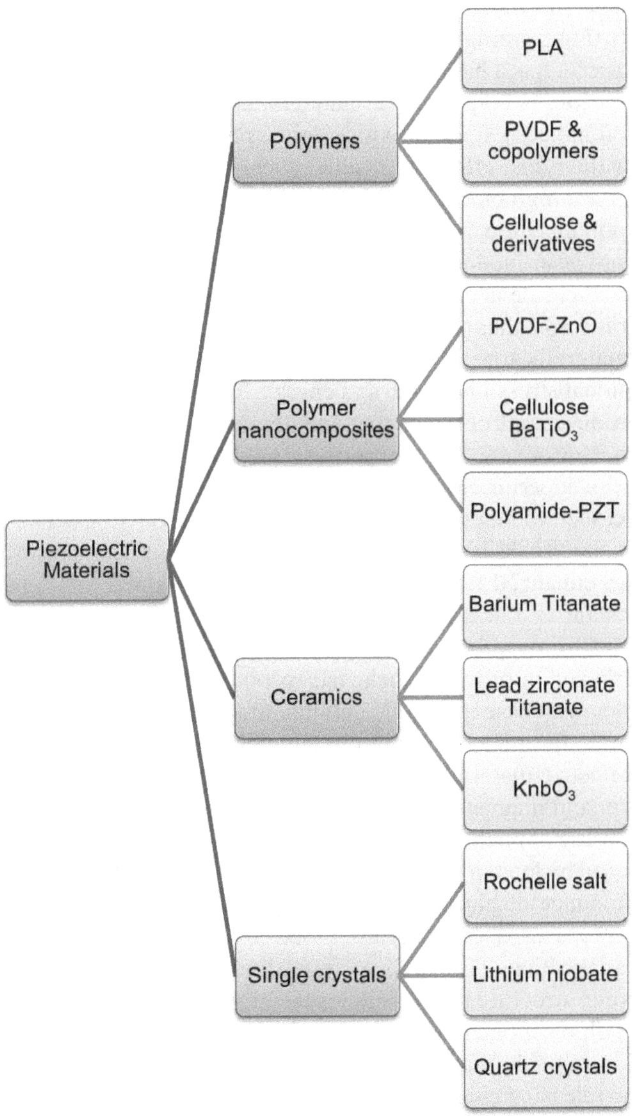

FIGURE 4.7 Different classes of piezoelectric materials.

4.4 IMPROVEMENTS IN ENGINEERING APPLICATIONS USING GREEN PIEZOELECTRIC MATERIALS

Novel piezoelectric materials derived from eco-friendly sources are employed to create bioelectronics that seamlessly integrate with biological systems. These materials are used for applications such as sensing physiological signals and stimulating healing tissues. The generation of piezoelectricity from both synthetic and natural bioresorbable polymer composites is applied in various engineering fields.

Biodegradable piezoelectric polymers include natural materials like collagen, which is a soft tissue similar to cellulose that contributes to piezoelectricity. Also, polylactic acid, which is a biodegradable piezoelectric polymer synthesized by lactic acid polymerization, is considered a biodegradable piezoelectric polymer. Various piezoelectric biomolecules, including proteins, peptides, and amino acids, are recommended for their cost-effectiveness and superior properties in biomedical applications due to their high biocompatibility and biodegradable and durable properties. Piezoelectric polymers are categorized by their electric dipole orientation and structural topology. These biodegradable piezoelectric materials enhance the biocompatibility of medical devices, facilitating the creation of implantable devices that naturally degrade and thus minimize the need for surgical removal. Additionally, piezoelectric materials are employed in digital sensors for damage detection by converting mechanical stress into electrical energy. This capability is also harnessed in actuators to produce control forces that counteract internal structural stresses.

Piezoelectric transduction is a promising method for the storage of mechanical energy due to its superior electromechanical and piezoelectric coupling compared to electrostatic, electromagnetic, and triboelectric transductions. The advancement of micro- and nanoscale biomaterials enables the manufacturing of piezoelectric generators with enhanced features including factors of electromechanical coupling known as constant (k_s) at a lower cost [16, 21, 39, 105]. The green piezoelectric materials synthesize recyclable and highly functional composite materials that are highly tailorable and improve piezoelectric response and allow superior sensing and electro-ceramics with ease to recycle and manufacture composites with comparable performance and minimize their environmental impact [27, 62, 106]. Green and sustainable piezoelectric materials can synthesize functional materials such as dielectric, conductive, and magnetic ceramics for improving the environmental impact in various engineering applications [15, 25, 59]. Also, green materials overcome many drawbacks caused by lead in piezoelectric materials such as more durability, sustainability and avoidance of high toxicity and brittleness of lead compounds. A major disadvantage of green-based materials is the low Curie temperature which limits its industrial applications [14–109]. Also, piezoelectric green materials possess an asymmetric green structure that facilitates the alignment of opposite ions and generates an electric dipole moment which creates domain structures. Also, the integration of green and sustainable materials in piezoelectric components enhances material polarization by releasing charges in a phenomenon known as direct piezoelectricity.

Also, recently, piezoelectric materials have provided an alternative to e-waste disposal concerns and improved the recyclability and biocompatibility of electronic devices [36–99]. Also, advancement in piezoelectric materials is utilized to develop electronic implantable parts in various medical devices that are used in medical tasks and absorbed in body fluids or discarded to harmless materials, thus avoiding the usage of complex devices and the need for retrieval surgery [14, 83, 107, 108]. The crystalline structure within the piezoelectric materials is categorized as either quartz (SiO_2), Rochelle salts or tourmaline. Quartz has very high performance as well as stiffness, high-temperature resistivity and durability. However, quartz isn't suitable for damage detection and structural control applications [107, 108, 110].

Eco-friendly, biodegradable, and biocompatible functional materials are used in developing electromechanical coupling materials in numerous applications with high features of degradability and enhanced mechanical properties to be used in wearable and implantable bioelectronics [111–114]. Biodegradable piezoelectric polymers have been created to enable the development of innovative physiological sensors, transducers, biomedical devices, and medical equipment [38–119]. Also, nanocellulose properties are modified by chemical modification that increases the applicability of piezoelectric materials in biomedical applications [3, 4]. Similarly, biodegradable piezoelectric polymers have been used in medical and physiological sensors biomedical devices and transducers, regenerative and biomedical devices as well as the fabrication of electronic parts of implantable devices for advanced biomedical applications [14, 15, 36, 37, 83]. The following piezoelectric polymers have shown high performance including fluoropolymers based on polyvinylidene difluoride (PVDF) as well as copolymer compounds including polyvinylidene fluoride trifluoro (PVDF-TrFE) [3, 4, 9, 10].

The latest generation of eco-friendly piezoelectric materials offers benefits including lower cost, water solubility, and overall environmental sustainability. Devices made from piezoelectric biomaterials have been shown to produce high-output voltages. The highest piezoelectric coefficient observed in the nanoscale of biomaterials is 38 pm/V, as measured by piezo response force microscopic devices (PFM) [41]. Poling of polymer sheets with cellular framework has shown strong piezoelectric activity which reports biodegradable polylactic acids that reached higher than 600 pC/N based on positive piezoelectric effect [3, 4, 9, 10]. The new green piezoelectric materials such as biodegradable polylactic acid (PLA) materials are beneficial for physiological sensing and have shown high performance [5, 6, 10, 117]. For example, biodegradable polylactic acid (PLA) exhibits strong piezoelectric activity reaching 600 pC/N (column/newton).

4.5 CONCLUSION

The advancement in piezoelectric materials has significantly shifted the focus toward more sustainable and green-based materials. These materials can address the limitations associated with traditional substances by integration of piezoelectric materials such as polyvinylidene fluoride (PVDF or lead zirconate titanate (PZT). Green materials and nanocomposites and piezoelectric foams offer promising enhancement in flexibility, durability, and recyclability. Integration of nanotechnology such as green nanowires and protein-based filaments can expand the capabilities of piezoelectric devices and allow their application in high-temperature and flexible environments. The piezoelectric transduction materials are recommended due to their superior electromechanical properties including coupling and piezoelectric coefficients in micro and nanomaterials which have caused significant advancement in the design of piezoelectric generators enhancing their performance through improved electromechanical coupling factors. These materials over advantage of overcoming the brittleness and toxicity of lead-based piezoelectric materials as well as limitations such as lower Curie temperature and non-symmetric structure that affect their application. The integration of green technologies and smart materials with a focus on biodegradability and desired mechanical properties has a great potential for future wearable and implanted bioelectronic devices These materials offer a sustainable path forward improving both the performance and environmental impact of various engineering applications. Green and smart materials with biodegradability and desired mechanical properties including piezoelectric polymers and nanocellulose are used for electromechanical coupling in various biomedical applications.

REFERENCES

1. S. S. Kang, K. Choi, J. Do Nam, and H. J. Choi, "Magnetorheological elastomers: Fabrication, characteristics, and applications," 2020. https://doi.org/10.3390/ma13204597.
2. S. Bahl, H. Nagar, I. Singh, and S. Sehgal, "Smart materials types, properties and applications: A review," in *Materials Today: Proceedings*, 2020. https://doi.org/10.1016/j.matpr.2020.04.505.
3. R. A. Surmenev *et al.*, "Hybrid lead-free polymer-based nanocomposites with improved piezoelectric response for biomedical energy-harvesting applications: A review," 2019. https://doi.org/10.1016/j.nanoen.2019.04.090.
4. F. Dai and Q. Geng, "Organic biodegradable piezoelectric materials and their potential applications as bioelectronics," *Soft Science*, 2023. https://doi.org/10.20517/ss.2022.30.
5. L. Z. Cheong, W. Zhao, S. Song, and C. Shen, "Lab on a tip: Applications of functional atomic force microscopy for the study of electrical properties in biology," 2019. https://doi.org/10.1016/j.actbio.2019.08.023.
6. S. Ahmad, A. W., Hashmi, J. Singh, K. Arora, Y. Tian, F. Iqbal, M. Al-Dossari, M. I. Khan, "Innovations in additive manufacturing of shape memory alloys: Alloys, microstructures, treatments, applications," *J Mater Res Technol*, vol. 32, 2024, 4136–4197.
7. B. Duan *et al.*, "Bioinspired PVDF piezoelectric generator for harvesting multi-frequency sound energy," *Adv Electron Mater*, vol. 9, no. 8, 2023. https://doi.org/10.1002/aelm.202300348.
8. J. Huang *et al.*, "Intelligent device for harvesting the vibration energy of the automobile exhaust with a piezoelectric generator," *Micromachines (Basel)*, vol. 14, no. 2, 2023. https://doi.org/10.3390/mi14020491.
9. L. Persano, S. K. Ghosh, and D. Pisignano, "Enhancement and function of the piezoelectric effect in polymer nanofibers," *Acc Mater Res*, vol. 3, no. 9, 2022. https://doi.org/10.1021/accountsmr.2c00073.
10. M. P. Nikiforov *et al.*, "Double-layer mediated electromechanical response of amyloid fibrils in liquid environment," in *ACS Nano*, 2010. https://doi.org/10.1021/nn901127k.
11. S. Qian, X. Wang, and W. Yan, "Piezoelectric fibers for flexible and wearable electronics," 2023. https://doi.org/10.1007/s12200-023-00058-3.

12. W. Q. Liao *et al.*, "A molecular perovskite solid solution with piezoelectricity stronger than lead zirconate titanate," *Science (1979)*, vol. 363, no. 6432, 2019. https://doi.org/10.1126/science.aav3057.
13. H. Zhou *et al.*, "Stretchable piezoelectric energy harvesters and self-powered sensors for wearable and implantable devices," 2020. https://doi.org/10.1016/j.bios.2020.112569.
14. T. Liu, Y. Wang, M. Hong, J. Venezuela, W. Shi, and M. Dargusch, "Advances in biodegradable piezoelectrics for medical implants," 2023. https://doi.org/10.1016/j.nantod.2023.101945.
15. M. Ali, S. M. Hoseyni, R. Das, M. Awais, I. Basdogan, and L. Beker, "A flexible and biodegradable piezoelectric-based wearable sensor for non-invasive monitoring of dynamic human motions and physiological signals," *Adv Mater Technol*, vol. 8, no. 15, 2023. https://doi.org/10.1002/admt.202300347.
16. D. Wang *et al.*, "Ultrahigh piezoelectricity in lead-free piezoceramics by synergistic design," *Nano Energy*, vol. 76, 2020. https://doi.org/10.1016/j.nanoen.2020.104944.
17. S. O. Oyedepo, J. O. Dirisu, N. E. Udoye, and O. S. I. Fayomi, "Progresses on green and smart materials for multifaceted applications," in *Handbook of Smart Materials, Technologies, and Devices*, 2021. https://doi.org/10.1007/978-3-030-58675-1_41-1.
18. Y. Ren, X. Sun, and J. Liu, "Advances in liquid metal-enabled flexible and wearable sensors," 2020. https://doi.org/10.3390/mi11020200.
19. A. C. Aydin, and O. Çelebi̇, "Piezoelectric materials in civil engineering applications: A review," 2023. https://doi.org/10.1021/acsomega.3c00672.
20. A. Aabid *et al.*, "Review a review of piezoelectric material-based structural control and health monitoring techniques for engineering structures: Challenges and opportunities," 2021. https://doi.org/10.3390/act10050101.
21. A. Jbaily, and R. W. Yeung, "Piezoelectric devices for ocean energy: A brief survey," 2015. https://doi.org/10.1007/s40722-014-0008-9.
22. Q. Xu *et al.*, "Construction of bio-piezoelectric platforms: from structures and synthesis to applications," 2021. https://doi.org/10.1002/adma.202008452.
23. C. Ribeiro, V. Sencadas, D. M. Correia, and S. Lanceros-Méndez, "Piezoelectric polymers as biomaterials for tissue engineering applications," 2015. https://doi.org/10.1016/j.colsurfb.2015.08.043.
24. W. Qian, W. Yang, Y. Zhang, C. R. Bowen, and Y. Yang, "Piezoelectric materials for controlling electro-chemical processes," 2020. https://doi.org/10.1007/s40820-020-00489-z.
25. H. Shaukat, A. Ali, S. Bibi, W. A. Altabey, M. Noori, and S. A. Kouritem, "A review of the recent advances in piezoelectric materials, energy harvester structures, and their applications in analytical chemistry," 2023. https://doi.org/10.3390/app13031300.
26. C. Yang, J. Ji, Y. Lv, Z. Li, and D. Luo, "Application of piezoelectric material and devices in bone regeneration," 2022. https://doi.org/10.3390/nano12244386.
27. H. Shaukat *et al.*, "Piezoelectric materials: Advanced applications in electro-chemical processes," 2023. https://doi.org/10.1016/j.egyr.2023.03.077.
28. J. Wu, "Perovskite lead-free piezoelectric ceramics," *J Appl Phys*, vol. 127, no. 19, 2020. https://doi.org/10.1063/5.0006261.
29. T. Granzow, S. Glinsek, and E. Defay, "Piezoelectric Ceramics," in *Encyclopedia of Smart Materials*, 2021. https://doi.org/10.1016/B978-0-12-815732-9.00043-7.
30. J. Park, D. G. Lee, S. Hur, J. M. Baik, H. S. Kim, and H. C. Song, "A review on recent advances in piezoelectric ceramic 3D printing," 2023. https://doi.org/10.3390/act12040177.
31. D. Sun, Y. Lu, and T. Karaki, "Review of the applications of 3D printing technology in the field of piezoelectric ceramics," 2023. https://doi.org/10.1016/j.recm.2023.02.001.

32. Q. Guo *et al.*, "Piezoelectric ceramics with high piezoelectricity and broad temperature usage range," *Journal of Materiomics*, vol. 7, no. 4, 2021. https://doi.org/10.1016/j.jmat.2020.11.012.
33. H. Han and J. Ko, "Power-generation optimization based on piezoelectric ceramic deformation for energy harvesting application with renewable energy," *Energies (Basel)*, vol. 14, no. 8, 2021. https://doi.org/10.3390/en14082171.
34. Z. Jin, B. Lu, Y. Xu, and X. Zhang, "Lead-free piezoelectric ceramics – An electroactive material that provides electrical stimulation cues for bone regeneration," 2023. https://doi.org/10.2298/PAC2303214J.
35. S. S. Anandakrishnan *et al.*, "Toward ecofriendly piezoelectric ceramics—reduction of energy and environmental footprint from conceptualization to deployment," 2023. https://doi.org/10.1002/gch2.202300061.
36. K. Kim *et al.*, "Biodegradable, electro-active chitin nanofiber films for flexible piezoelectric transducers," *Nano Energy*, vol. 48, 2018. https://doi.org/10.1016/j.nanoen.2018.03.056.
37. E. J. Curry *et al.*, "Biodegradable piezoelectric force sensor," *Proc Natl Acad Sci U S A*, vol. 115, no. 5, 2018. https://doi.org/10.1073/pnas.1710874115.
38. S. W. Ibrahim, T. I. Hamad, and J. Haider, "Biological properties of polycaprolactone and barium titanate composite in biomedical applications," 2023. https://doi.org/10.1177/00368504231215942.
39. J. Li *et al.*, "3D interpenetrating piezoceramic-polymer composites with high damping and piezoelectricity for impact energy-absorbing and perception," *Compos B Eng*, vol. 232, 2022. https://doi.org/10.1016/j.compositesb.2022.109617.
40. E. B. Jo *et al.*, "The 0-3 Lead Zirconate-Titanate (PZT)/Polyvinyl-Butyral (PVB) composite for tactile sensing," *Sensors*, vol. 23, no. 3, 2023. https://doi.org/10.3390/s23031649.
41. A. Sultana *et al.*, "Toward high-performance green piezoelectric generators based on electrochemically poled nanocellulose," *Chemistry of Materials*, vol. 35, no. 4, 2023. https://doi.org/10.1021/acs.chemmater.2c03020.
42. J. G. Sun, T. N. Yang, C. Y. Wang, and L. J. Chen, "A flexible transparent one-structure tribo-piezo-pyroelectric hybrid energy generator based on bio-inspired silver nanowires network for biomechanical energy harvesting and physiological monitoring," *Nano Energy*, vol. 48, 2018. https://doi.org/10.1016/j.nanoen.2018.03.071.
43. A. Sultana, Md. M. Alam, E. Pavlopoulou, E. Solano, M. Berggren, X. Crispin, and D. Zhao, "Toward high-performance green piezoelectric generators based on electrochemically poled nanocellulose," *Chemistry of Materials*, vol. 35, no. 4, pp. 1568–1578, 2023. https://doi.org/10.1021/acs.chemmater.2c03020
44. T. C. Kaspar *et al.*, "Tuning piezoelectric properties through epitaxy of La2Ti2O7 and related thin films," *Sci Rep*, vol. 8, no. 1, 2018. https://doi.org/10.1038/s41598-018-21009-5.
45. P. Andersson Ersman *et al.*, "Integration of screen printed piezoelectric sensors for force impact sensing in smart multifunctional glass applications," *Adv Eng Mater*, vol. 24, no. 11, 2022. https://doi.org/10.1002/adem.202200399.
46. J. Gao, D. Xue, W. Liu, C. Zhou, and X. Ren, "Recent progress on BaTiO3-based piezoelectric ceramics for actuator applications," 2017. https://doi.org/10.3390/act6030024.
47. X. Guo *et al.*, "Piezoelectric properties of 0-3 composite films based on novel molecular piezoelectric material (ATHP)2PbBr4," *Materials*, vol. 15, no. 18, 2022. https://doi.org/10.3390/ma15186378.
48. B. Y. Lee *et al.*, "Virus-based piezoelectric energy generation," *Nat Nanotechnol*, vol. 7, no. 6, 2012. https://doi.org/10.1038/nnano.2012.69.
49. P. K. Singh *et al.*, "Trends in piezoelectric nanomaterials towards green energy scavenging nanodevices," 2023. https://doi.org/10.1016/j.mtsust.2023.100583.

50. J. González *et al.*, "Advanced cellulose–nanocarbon composite films for high-performance triboelectric and piezoelectric nanogenerators," *Nanomaterials*, vol. 13, no. 7, 2023. https://doi.org/10.3390/nano13071206.
51. C. Liu *et al.*, "Nanowire-bacteria hybrids for unassisted solar carbon dioxide fixation to value-added chemicals," *Nano Lett*, vol. 15, no. 5, 2015. https://doi.org/10.1021/acs.nanolett.5b01254.
52. K. K. Sakimoto, C. Liu, J. Lim, and P. Yang, "Salt-induced self-assembly of bacteria on nanowire arrays," *Nano Lett*, vol. 14, no. 9, 2014. https://doi.org/10.1021/nl502946j.
53. S. Bairagi, Shahid-ul-Islam, M. Shahadat, D. M. Mulvihill, and W. Ali, "Mechanical energy harvesting and self-powered electronic applications of textile-based piezoelectric nanogenerators: A systematic review," 2023. https://doi.org/10.1016/j.nanoen.2023.108414.
54. A. Sultana *et al.*, "Toward high-performance green piezoelectric generators based on electrochemically poled nanocellulose," *Chemistry of Materials*, vol. 35, no. 4, 2023. https://doi.org/10.1021/acs.chemmater.2c03020.
55. D. G. Cahill *et al.*, "Nanoscale thermal transport," 2003. https://doi.org/10.1063/1.1524305.
56. R. J. Warzoha *et al.*, "Applications and impacts of nanoscale thermal transport in electronics packaging," *Journal of Electronic Packaging, Transactions of the ASME*, vol. 143, no. 2, 2021. https://doi.org/10.1115/1.4049293.
57. J. A. Krishnaswamy, F. C. Buroni, R. Melnik, L. Rodriguez-Tembleque, and A. Saez, "Design of polymeric auxetic matrices for improved mechanical coupling in lead-free piezocomposites," *Smart Mater Struct*, vol. 29, no. 5, 2020. https://doi.org/10.1088/1361-665X/ab7e35.
58. S. Gupta *et al.*, "Multifunctional and flexible polymeric nanocomposite films with improved ferroelectric and piezoelectric properties for energy generation devices," *ACS Appl Energy Mater*, vol. 2, no. 9, 2019. https://doi.org/10.1021/acsaem.9b01000.
59. C. C. Hsiao, A. S. Siao, and J. C. Ciou, "Improvement of pyroelectric cells for thermal energy harvesting," *Sensors*, vol. 12, no. 1, 2012. https://doi.org/10.3390/s120100534.
60. J. Najd, E. Zappino, E. Carrera, W. Harizi, and Z. Aboura, "A variable kinematic model for the prediction of capacitance variations in embedded PZT sensors," *J Intell Mater Syst Struct*, vol. 34, no. 7, 2023. https://doi.org/10.1177/1045389X221121974.
61. G. Jian, H. Qingxian, L. Sheng, D. Zhou, and F. Qiuyun, "Effect of solid content variations on PZT slip for tape casting," *Processing and Application of Ceramics*, vol. 6, no. 4, 2012. https://doi.org/10.2298/pac1204215j.
62. A. Megdich, M. Habibi, and L. Laperrière, "A review on 3D printed piezoelectric energy harvesters: Materials, 3D printing techniques, and applications," 2023. https://doi.org/10.1016/j.mtcomm.2023.105541.
63. C. C. Jin, D. M. Liu, and L. X. Zhang, "An emerging family of piezocatalysts: 2D piezoelectric materials," 2023. https://doi.org/10.1002/smll.202303586.
64. J. Li *et al.*, "Piezoelectric materials and techniques for environmental pollution remediation," 2023. https://doi.org/10.1016/j.scitotenv.2023.161767.
65. J., Singh, A.W., Hashmi, S., Ahmad, Y. Tian, "Critical review on biodegradable and biocompatibility magnesium alloys: Progress and prospects in bio-implant applications," *Inorg Chem Commun*, vol. 169, p. 113111, 2024.
66. F. Tian, Y. Liu, R. Ma, F. Li, Z. Xu, and Y. Yang, "Properties of PMN-PT single crystal piezoelectric material and its application in underwater acoustic transducer," *Applied Acoustics*, vol. 175, 2021. https://doi.org/10.1016/j.apacoust.2020.107827.
67. X. Jiang, J. Kim, and K. Kim, "Relaxor-PT single crystal piezoelectric sensors," 2014. https://doi.org/10.3390/cryst4030351.

68. X. Xu *et al.*, "Phase structure and electrical properties of 0.28PIN-0.32PZN–(0.4-x) PT-xPZ piezoelectric ceramics," *Crystals (Basel)*, vol. 13, no. 9, 2023. https://doi.org/10.3390/cryst13091362.
69. S. Ouyang, J. Zhang, F. Yu, and L. Qin, "A vibration energy harvester based on Ca3TaGa3Si2O14 piezoelectric crystal for high temperature applications," *Appl Phys Lett*, vol. 115, no. 25, 2019. https://doi.org/10.1063/1.5127658.
70. J. Topete, and T. Gómez Álvarez-Arenas, "A comparative study of air-coupled ultrasonic transducers based on piezoelectric composites and ferroelectret foams," *The 22nd International Congress on Sound and Vibration*, no. July, 2015.
71. Z. Luo *et al.*, "Energy harvesting study on single and multilayer ferroelectret foams under compressive force," *IEEE Transactions on Dielectrics and Electrical Insulation*, vol. 22, no. 3, 2015. https://doi.org/10.1109/TDEI.2015.7116323.
72. C. Ravikumar, and V. Markevicius, "Ferroelectret polypropylene foam-based piezoelectric energy harvester for different seismic mass conditions," *Actuators*, vol. 12, no. 5, 2023. https://doi.org/10.3390/act12050215.
73. C. H. Park, Q. Van Duong, Y. J. Moon, K. Ha, and S. T. Choi, "Enhanced thermo-electro-mechanical characteristics of purified P(VDF-TrFE) films for ultrasonic transducers," *Sens Actuators A Phys*, vol. 279, 2018. https://doi.org/10.1016/j.sna.2018.06.060.
74. S. E. Alavi *et al.*, "Dynamical mechanical analysis and micromechanics simulations of Spin-Crossover@Polymer particulate composites: Toward soft actuator devices," *Chemistry of Materials*, vol. 35, no. 8, 2023. https://doi.org/10.1021/acs.chemmater.3c00293.
75. T. Ikawa, H. Tabata, T. Yoshizawa, K. Utaka, O. Kubo, and M. Katayama, "Pressure-sensing properties of single-walled carbon nanotubes covered with a corona-poled piezoelectric polymer," *Appl Phys Lett*, vol. 109, no. 3, 2016. https://doi.org/10.1063/1.4959091.
76. H. Liang, G. Hao, and O. Z. Olszewski, "A review on vibration-based piezoelectric energy harvesting from the aspect of compliant mechanisms," 2021. https://doi.org/10.1016/j.sna.2021.112743.
77. M. Gholikhani, H. Roshani, S. Dessouky, and A. T. Papagiannakis, "A critical review of roadway energy harvesting technologies," 2020. https://doi.org/10.1016/j.apenergy.2019.114388.
78. A. Toprak, and O. Tigli, "Piezoelectric energy harvesting: State-of-the-art and challenges," 2014. https://doi.org/10.1063/1.4896166.
79. Z. Yang, S. Zhou, J. Zu, and D. Inman, "High-performance piezoelectric energy harvesters and their applications," 2018. https://doi.org/10.1016/j.joule.2018.03.011.
80. S. Fatimah, M. P. Kamil, D. I. Han, W. Al-Zoubi, and Y. G. Ko, "Development of anticorrosive coating on AZ31 Mg alloy subjected to plasma electrolytic oxidation at subzero temperature," *J Magnes Alloy*, vol. 10, no. 7, 2022. https://doi.org/10.1016/j.jma.2021.07.013.
81. G. Tian, W. Deng, D. Xiong, T. Yang, B. Zhang, X. Ren, B. Lan, S. Zhong, L. Jin, H. Zhang, L. Deng, "Dielectric micro-capacitance for enhancing piezoelectricity via aligning MXene sheets in composites," *Cell Rep Phy Sci*, vol. 3, no. 4, p. 100814, 2022.
82. S. Ray, A. Kumari, M. Sonkar, and P. Kumar, "Polyhydroxyalkanoate-based sensors and their applications," in *Biomaterials-Based Sensors: Recent Advances and Applications*, 2023. https://doi.org/10.1007/978-981-19-8501-0_7.
83. M. Ali, M. J. Bathaei, E. Istif, S. N. H. Karimi, and L. Beker, "Biodegradable piezoelectric polymers: Recent advancements in materials and applications," 2023. https://doi.org/10.1002/adhm.202300318.
84. M. Jafari, and G. G. Botte, "Sustainable green route for activated carbon synthesis from biomass waste for high-performance supercapacitors," *ACS Omega*, 2023. https://doi.org/10.1021/acsomega.3c09438.

85. K. H. Lee *et al.*, "Porous aerogel structures as promising materials for photocatalysis, thermal insulation textiles, and technical applications: A review," 2023. https://doi.org/10.3390/catal13091286.
86. E. D. Pinheiro, and T. Deivarajan, "A concise review encircling lead free porous piezoelectric ceramics," 2019. https://doi.org/10.12693/APhysPolA.136.555.
87. P. K. Panda, "Review: Environmental friendly lead-free piezoelectric materials," *J Mater Sci*, vol. 44, no. 19, 2009. https://doi.org/10.1007/s10853-009-3643-0.
88. V. Slabov, S. Kopyl, M. P. Soares dos Santos, and A. L. Kholkin, "Natural and eco-friendly materials for triboelectric energy harvesting," 2020. https://doi.org/10.1007/s40820-020-0373-y.
89. J. Zhang *et al.*, "Eco-friendly and recyclable all cellulose triboelectric nanogenerator and self-powered interactive interface," *Nano Energy*, vol. 89, 2021. https://doi.org/10.1016/j.nanoen.2021.106354.
90. E. A. Elvira-Hernández *et al.*, "Green energy harvesting to power electronic devices using portable triboelectric nanogenerator based on waste corn husk and recycled polystyrene," *Energy Reports*, vol. 11, 2024. https://doi.org/10.1016/j.egyr.2023.11.059.
91. S. A. Han, J. Lee, J. Lin, S. W. Kim, and J. H. Kim, "Piezo/triboelectric nanogenerators based on 2-dimensional layered structure materials," *Nano Energy*, vol. 57, 2019. https://doi.org/10.1016/j.nanoen.2018.12.081.
92. J. Wang, B. Fan, J. Fang, J. Zhao, and C. Li, "A piezoelectric energy harvester based on multi-cantilevers and magnetic force," *Energy Reports*, vol. 8, 2022. https://doi.org/10.1016/j.egyr.2022.09.005.
93. W. J. Su, and C. H. Tseng, "Design and analysis of an extended simply supported beam piezoelectric energy harvester," *Sensors*, vol. 23, no. 13, 2023. https://doi.org/10.3390/s23135895.
94. Z. Wang, L. He, Z. Zhang, Z. Zhou, J. Zhou, and G. Cheng, "Research on a piezoelectric energy harvester with rotating magnetic excitation," *J Electron Mater*, vol. 50, no. 6, 2021. https://doi.org/10.1007/s11664-021-08910-y.
95. Zeeshan, A. M. Soomro, and S. Cho, "Design and fabrication of a robust chitosan/polyvinyl alcohol-based humidity sensor energized by a piezoelectric generator," *Energies (Basel)*, vol. 15, no. 20, 2022. https://doi.org/10.3390/en15207609.
96. J. Wang, Y. Tong, C. Li, Z. Zhang, and J. Shao, "A novel vibration piezoelectric generator based on flexible piezoelectric film composed of PZT and PI layer," *Polymers (Basel)*, vol. 14, no. 14, 2022. https://doi.org/10.3390/polym14142871.
97. J. Kim, S. Yun, S. K. Mahadeva, K. Yun, S. Y. Yang, and M. Maniruzzaman, "Paper actuators made with cellulose and hybrid materials," *Sensors*, vol. 10, no. 3, 2010. https://doi.org/10.3390/s100301473.
98. H. S. Kim, Y. Li, and J. Kim, "Electro-mechanical behavior and direct piezoelectricity of cellulose electro-active paper," *Sens Actuators A Phys*, vol. 147, no. 1, 2008. https://doi.org/10.1016/j.sna.2008.05.003.
99. R. V. Chernozem *et al.*, "Diazonium chemistry surface treatment of piezoelectric polyhydroxybutyrate scaffolds for enhanced osteoblastic cell growth," *Appl Mater Today*, vol. 20, 2020. https://doi.org/10.1016/j.apmt.2020.100758.
100. C. M. Costa *et al.*, "Smart and multifunctional materials based on electroactive poly(vinylidene fluoride): Recent advances and opportunities in sensors, actuators, energy, environmental, and biomedical applications," 2023. https://doi.org/10.1021/acs.chemrev.3c00196.
101. F. Mokhtari, B. Azimi, M. Salehi, S. Hashemikia, and S. Danti, "Recent advances of polymer-based piezoelectric composites for biomedical applications," 2021. https://doi.org/10.1016/j.jmbbm.2021.104669.

102. X. Dongyu, C. Xin, S. Banerjee, and H. Shifeng, "Design, fabrication, and properties of 2-2 connectivity cement/polymer based piezoelectric composites with varied piezoelectric phase distribution," *J Appl Phys*, vol. 116, no. 24, 2014. https://doi.org/10.1063/1.4904931.
103. X. Dongyu, C. Xin, and H. Shifeng, "Investigation of inorganic fillers on properties of 2-2 connectivity cement/polymer based piezoelectric composites," *Constr Build Mater*, vol. 94, 2015. https://doi.org/10.1016/j.conbuildmat.2015.07.090.
104. X. Zhang, W. Xia, J. Liu, M. Zhao, M. Li, and J. Xing, "PVDF-based and its copolymer-based piezoelectric composites: Preparation methods and applications," 2022. https://doi.org/10.1007/s11664-022-09825-y.
105. N. Sezer, and M. Koç, "A comprehensive review on the state-of-the-art of piezoelectric energy harvesting," 2021. https://doi.org/10.1016/j.nanoen.2020.105567.
106. F. Ali, and M. Koc, "3D printed polymer piezoelectric materials: Transforming healthcare through biomedical applications," 2023. https://doi.org/10.3390/polym15234470.
107. Q. H. Chan, S. A. Alias, S. W. Quek, C. Y. Ng, and K. I. Ku Marsilla, "A review of the preparations, properties, and applications of smart biodegradable polymers," 2023. https://doi.org/10.1080/25740881.2023.2204954.
108. N. Goonoo, and A. Bhaw-Luximon, "Piezoelectric polymeric scaffold materials as biomechanical cellular stimuli to enhance tissue regeneration," 2022. https://doi.org/10.1016/j.mtcomm.2022.103491.
109. Y. Li *et al.*, "Biodegradable piezoelectric polymer for cartilage remodeling," *Matter*, vol. 7, no. 4, 2024. https://doi.org/10.1016/j.matt.2024.01.034.
110. M. Sobczyk, S. Wiesenhütter, J. R. Noennig, and T. Wallmersperger, "Smart materials in architecture for actuator and sensor applications: A review," 2022. https://doi.org/10.1177/1045389X211027954.
111. T. Marques-Almeida, V. Correia, E. Fernández Martín, A. García Díez, C. Ribeiro, and S. Lanceros-Mendez, "Piezoelectric and magnetically responsive biodegradable composites with tailored porous morphology for biotechnological applications," *ACS Appl Polym Mater*, vol. 4, no. 12, 2022. https://doi.org/10.1021/acsapm.2c01114.
112. E. S. Hosseini, L. Manjakkal, D. Shakthivel, and R. Dahiya, "Glycine-chitosan-based flexible biodegradable piezoelectric pressure sensor," *ACS Appl Mater Interfaces*, vol. 12, no. 8, 2020. https://doi.org/10.1021/acsami.9b21052.
113. A. Pryadko, M. A. Surmeneva, and R. A. Surmenev, "Review of hybrid materials based on polyhydroxyalkanoates for tissue engineering applications," 2021. https://doi.org/10.3390/polym13111738.
114. X. Zhuang, F. Wang, and X. Hu, "Biodegradable polymers: A promising solution for green energy devices," 2024. https://doi.org/10.1016/j.eurpolymj.2023.112696.
115. J. Takarada, Y. Okawa, I. Nakagawa, H. Tsuneishi, and Y. Tajitsu, "Microscopic piezoelectric response of poly (3-hydroxybutyrate-co-3-hydroxyhexanote) fiber," *Jpn J Appl Phys*, vol. 62, no. SM, 2023. https://doi.org/10.35848/1347-4065/acf586.
116. T. Dufour, "From basics to frontiers: A comprehensive review of plasma-modified and plasma-synthesized polymer films," 2023. https://doi.org/10.3390/polym15173607.
117. S. Tasoglu, and U. Demirci, "Bioprinting for stem cell research," 2013. https://doi.org/10.1016/j.tibtech.2012.10.005.
118. J. Singh, J. P. Singh, S. Kumar, and H. S. Gill, "Short review on hydroxyapatite powder coating for SS 316L," *J Electrochem Sci Eng*, vol. 13, no. 25–39, 2023.
119. R. Singh, M. Toseef, J. Kumar, J. Singh, "Benefits and challenges in additive manufacturing and its applications," In: Kaushal, S., Singh, I., Singh, S., and Gupta, A. (eds.) *Sustainable Advanced Manufacturing and Materials Processing*. pp. 137–157. CRC Press, Boca Raton, 2022. https://doi.org/10.1201/9781003269298-8

5 Bioelectronics and Biomedical Applications of Conductive Polymers

Mohamed Aboughaly

Description	Abbreviation
Conductive polymer composites	CPC
Dimethylformamide	DMF
Extracellular framework	ECM
Light radiating diode	LED
Nanocrystalline cellulose dose	NCC
Poly (D, L-lactic corrosive)	PDLLA
Poly vinylidene fluoride	PVDF
Polyanilines	PANI
Polypyrrole	PPy
Polythiophene	PTh
Polyether sulfone	PES
Polyhexamethylene biguanide hydrochloride	PHMB
Poly(3,4-ethylenedioxythiophene)	PEDOT
Triboelectric nanogenerators	TENG

5.1 INTRODUCTION

This chapter explores the use of conductive polymers and eco-friendly materials in construction and smart materials applications. This chapter summarizes the potential of biomedical applications and the utilization of conducting polymers in savvy materials, bio-actuators, hydrogels, and neural prosthetic gadgets. Conductive polymer composites are considered fabulous flexible materials due to their great natural soundness, electrical conduciveness, as well as their excellent optical and electronic properties. Moreover, the chemical properties can be tuned by joining unused functionalities to adapt to various biomedical and industrial applications. For example, conductive polymers are used in precise drug release applications using electrical

DOI: 10.1201/9781003516071-5

stimulation. Conducting polymers have the capacity to alter mechanical properties through joining, doping, and chemical adjustments.

The adaptability, lightweight, and strong vitality capacity arrangements of different conductive polymers increase their attractiveness for many applications. Conductive polymers are broadly utilized in microelectronics, batteries, photovoltaic gadgets, and light-radiating diodes (LEDs) [1–3]. Their light weight make them ideal for utilization as current carriers in electronic components [4–6]. Ordinarily, conductive polymers have similar mechanical advantages than plastic but also with the electrical conductive properties of metals. The most suggested polymers for integration with electricity are polyanilines (PANI), polypyrrole (PPy) and polythiophene (PTh) subsidiaries [4, 7–9]. This chapter centers on the component of conductivity, and on the different approaches for creating electrically conductive polymers utilized for different applications. Conductive polymers synthesized from polylactic acid thermoplastic polyester (PLA), polyaniline (PANI), and polypyrrole (PPy) are utilized as great bio-based nanocomposite films with a high scattering range. Nanocomposite films synthesized from PLA wrapped with PANI implanted particles and consistently disseminated NCC (nanocrystalline cellulose) in nanocomposite films have appeared with electrical conductivity of 2.16 Sm-1 with a 1% nanocrystalline cellulose dose (NCC). The joining of nanocrystalline suspensions has improved the consistency and viscoelasticity of the polymer composite as well as its conductivity [10].

Conductive polymers have emerged as a key component in the development of smart materials due to their advantageous mechanical, electrical, thermal, and optical properties. Conductive polymers maintain electrical conductivity and may show high electrical conductivity comparable to metals. The conductivity is tunable and can be balanced through doping, which places charge carriers into the polymer structure. Conductive polymers also upgrade mechanical adaptability and can be extended or bowed, making them flexible for different design applications [11, 12]. The extraordinary adaptability of conductive polymer composites has empowered the creation of a differing cluster of conductive composite polymers with electroactive characteristics. This envelops naturally conductive polymers, conjugated polymers with electrical conductivity, permeated conductive polymers, and electrolytic conductive hydrogels [11, 13]. These conductive polymers can be adjusted to act with electric reactions, which trigger desirable properties making them attractive for biomedical applications. Also, electroactive keen polymers within the field of tissue union and biomaterials have accomplished many successful applications [14–16].

Electroactive conductive polymers are moreover coordinated with biomaterials due to their capacity to fortify cells in built tissues. They can mimic muscle activity and convert electrical energy into mechanical energy through electromechanical pulses. Additionally, electroactive conductive polymers can convey drugs through changing internal arrangements under the impact of an electrical boost and mirror antimicrobial conduct through the conduction of power [17]. Conductive polymers are too broadly utilized in keen sensors and actuators due to their high affectability and resistivity to outside stimuli, such as weight and temperature. Also, electrochromic gadgets utilize conductive polymers to alter their color in reaction to changes in an electric field [18, 19]. Conductive polymers seem to alter their shape or volume when connected to changes in an electric field which makes them ideal for utilization in prosthetic muscles and other incitation applications. Moreover, conductive polymers may store and discharge charges when utilized in supercapacitors and batteries [20–22]. The current challenges of conductive polymers are their corruption due to oxygen and dampness presentation, whereas later progressions center on the advancement of polymer composite soundness through better polymer plans and applying protective coatings [20, 23, 24].

Also, conductive polymers are developed with exceptional thermoelectric properties that allow them to convert temperature differences into electrical energy. They can be utilized for energy harvesting applications in various wearable electronics [1, 2]. Also, conductive polymers are biocompatible which makes them suitable for usage in medical applications, such as biosensors, neural interfaces, and drug release systems. Also, dielectric elastomer actuators use conductive polymers to generate large strains and electroactive fluid pumps, insect-like robots, and artificial muscles [3, 4].

Conductive polymers may be synthesized and coordinated with hydrogels and inserted into composites or electro-spun into microfibers. They can be combined with common filaments and eco-friendly materials to make biocompatible and biodegradable composites. Their physical properties can be easily modified for various construction applications through organic bonding and functionalization. Progression in conductive polymers make them imperative materials for biosensors, neural inserts, medicate conveyance, and built tissue frameworks [28–30]. Conductive polymers permit high control capabilities of the electrical boost as well as having great electrical and optical properties. The high conductivity to mass proportion as well as their impressive biodegradable and permeable characteristics give them good advantages for being coordinated in different composite materials and tailored to the particular needs of various applications through the expansion of antibodies, chemicals, and different organic substances [31–35]. A schematic diagram of fabricating procedures and applications of conductive polymers is shown in Figure 5.1.

Integration of conductive polymers in various wound dressings and drug delivery as well as biosensors enhance the functionalization of these devices. Conductive biocompatible polymers such as polypyrrole (PPy), polythiophene, and their derivatives are recommended for their exceptional electrical properties that are used in wound-healing processes by assisting in migration, adhesion, and differentiation [5 –7]. Also, polymer composites are categorized as below from primary and secondary groups of biomaterials and categorized using various biodegradable polymeric biomaterials divided into natural and synthetic, as shown below in Figure 5.2.

5.2 VERSATILE APPLICATIONS OF CONDUCTIVE POLYMERS IN ENGINEERING APPLICATIONS

Conductive polymer composites are exceptionally fragile and troublesome to manufacture as unadulterated conducting polymer films. Subsequently, mixing conductive

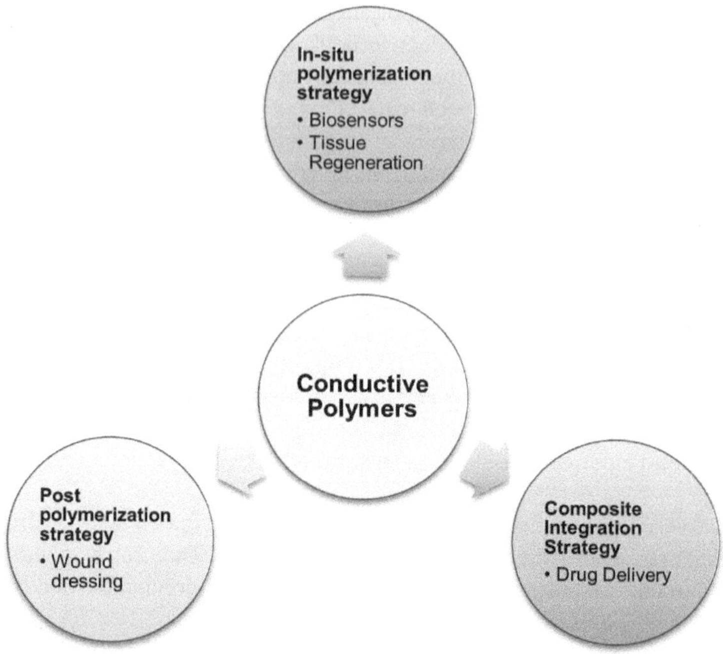

FIGURE 5.1. Different methods of conductive polymer synthesis and usage in various engineering applications.

polymers with degradable polymers is utilized to produce conductive biomaterials for the synthesis of artificial tissue applications. Conductive polymers and conductive biomaterials for tissue engineering synthesis and applications are illustrated in Figure 5.3. Manufactured and normal polymers such as PLA, PLGA, PCL and chitosan are mixed with conductive polymers. Conductive PPY and polylactic corrosive composite films are created utilizing emulsion polymerization and plunge coating strategies. Conductive PPY/poly (D, L-lactic corrosive) (PDLLA) composite films and nerve tubes and pathways were manufactured utilizing polymerization emulsion methods and dip coating strategies. The electric and micrograph analysis of PPY/PDLLA polymer applications has shown resistivity around 1.4 kΩ cm in cardiac tissue application [15, 30]. Moreover, polyhexamethylene biguanide hydrochloride (PHMB) is considered a flexible antimicrobial specialist and utilized in an assortment of applications in medication, nourishment, beauty care products, and numerous mechanical applications [56, 57]. Moreover, the integration of PHMB into chitosan layers improves the microbial resistance of different polymer composites. Moreover, the harm the cytoplasmic layers coordinate through interaction with contrarily phosphate-charged particles in bacterial cell layer phospholipids alter the structure and work of nucleic acids due to their solid interaction with them. The film shows extraordinary properties of restraining bacterial colonization. Moreover, its antimicrobial execution isn't influenced by temperature or weight making it appropriate for

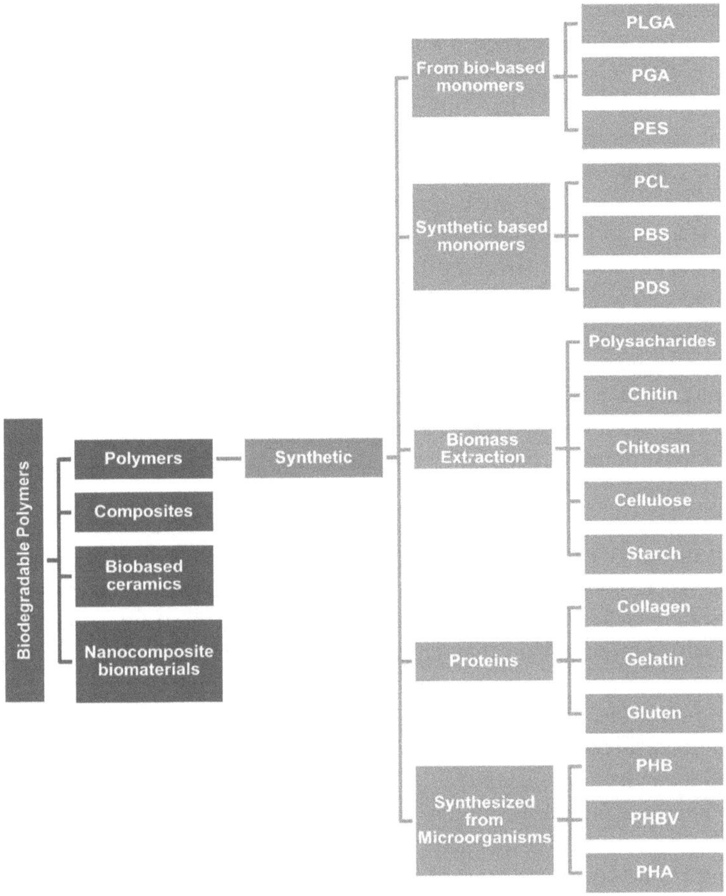

FIGURE 5.2. Primary and secondary classification of biomaterials and biodegradable polymers based on their source and processing technique.

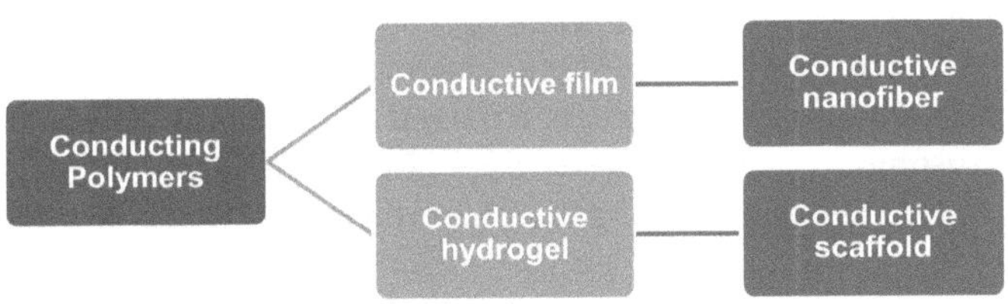

FIGURE 5.3. Conductive polymers and conductive biomaterials for tissue engineering synthesis and applications.

many mechanical applications. PHMB has additionally been presented in chitosan films, which has shown in shown a high degree of microscopic organism development for *Escherichia coli* and *Staphylococcus aureus* [57]. Additionally, incorporating PHMB into a polyether sulfone (PES) casting solution, combined with the hydrophilic anionic polyelectrolyte polymer poly-2-acrylamido-2-methylpropanesu lfonic acid, enhances the composition [57]. Synthesized polymer composites have shown excellent properties of restraint in bacterial colonization. The PES film is favored due to its tall mechanical quality, great heat, and chemical resistance. Zinc oxide nanoparticles are utilized for the alteration of layers and improving the hydrophilicity and antifouling capacity of films [56, 58]. Joining ZnO nanospheres into natural solvents, such as dimethylformamide (DMF), into the polyether sulfone casting arrangement. The synthesized compound layer appears to have high fouling resistance and water flux after the addition of ZnO nanoparticles and PHMB shapes natural and inorganic nanocomposites. This PES layer adjustment is utilized to upgrade antimicrobial properties and antifouling performance at the same time [56, 59].

The synthesized conductive polymers have the great advantage of high responsiveness to muscle tissues due to their quick reaction to electric signals, which gives their ability to conduct electric signals enormous significance for integration with various biomaterials in different tissue scaffolding applications. Conductive biomaterials with comparative conductivity to muscles are considered promising for muscle tissue synthesis and designing frameworks. Conductive biomaterials are utilized to increase the multiplication and separation conduct of electric and to synthesize responsive cells, such as muscle, bone marrow, and neuron cells [15–51]. Biomaterials can imitate the extracellular framework (ECM) structure by providing cell adhesion, and they are immune and tissue modeling. Moreover, collagen is a key protein in extracellular matrices which forms a nanofiber structure. Distinctive nano-string frameworks are created utilizing a few differential procedures such as electrospinning, stage partition, and atomic assembly.

Electrospinning is the most common method for the creation of nanofibers and nanotissues. An assortment of normal and manufactured biodegradable polymers were mixed with CPs and electro-spun into nanofibers for tissue building [52–55]. Fiber manufactured from electrospinning has a high surface region and porosity with a tunable distance of between a few nanometers and a few micrometers [56]. Conductive nanofiber composites made from gelatin and polyaniline are produced through electrospinning with a mixture of N-N-dimethylformamide and water. To stabilize the conductive nanofibers, glutaraldehyde is used as a cross-linking agent.

5.2.1 Conductive Polymers for Antimicrobial and Antibacterial Applications

Conductive polymer composites are notably has weak mechanical structure and difficult to manufacture as pure conducting polymer films. Subsequently, mixing conductive polymers with degradable polymers is utilized to produce conductive biomaterials for the synthesis of artificial tissue applications, where manufactured and normal polymers, such as PLA, PLGA, PCL and chitosan, are mixed with conductive polymers. Conductive PPY and polylactic corrosive composite films are created utilizing emulsion polymerization and plunge coating strategies. Conductive PPY/poly (D, L-lactic corrosive) (PDLLA) composite films and nerve tubes and pathways were manufactured utilizing polymerization emulsion methods and dip coating strategies. The electric and micrograph analysis of PPY/PDLLA polymer applications has shown resistivity around 1.4 kΩ cm in cardiac tissue application [15, 30]. Moreover, Polyhexamethylene biguanide hydrochloride (PHMB) is considered a flexible antimicrobial specialist and utilized in an assortment of applications in medication, nourishment, beauty care products, and numerous mechanical applications [56, 57]. Moreover, the integration of PHMB into chitosan layers improves the microbial resistance of different polymer composites. Moreover, the harm the cytoplasmic layers coordinate through interaction with contrarily phosphate-charged particles in bacterial cell layer phospholipids alters the structure and work of nucleic acids due to its solid interaction with them. The film appears extraordinary properties of restraining bacterial colonization. Moreover, its antimicrobial execution isn't influenced by temperature or weight, making it appropriate for many mechanical applications. PHMB is additionally presented in chitosan films and has appeared in tall shows of microscopic organisms development for *Escherichia coli* and *Staphylococcus aureus* [57]. Moreover, PHMB into polyether sulfone layer casting arrangements mixed with hydrophilic polyelectrolyte anionic polymer poly-2-acr ylamido-2-methylpropanesul-fonic corrosive compounds [57]. Synthesized polymer composites have shown excellent properties for the restraint of bacterial colonization. PES film is favored due to its good mechanical qualities, excellent heat, and chemical resistance. Zinc oxide nanoparticles are utilized for the alteration of layers and improving the hydrophilicity and antifouling capacity of the film [56, 58]. The incorporation of ZnO nanospheres into organic solvents like dimethylformamide (DMF) within polyether sulfone also serves as a casting solution. The synthesized compound layer appears to have high fouling resistance and water flux after the addition of ZnO nanoparticles and PHMB of natural and inorganic nanocomposites. This PES layer adjustment is utilized to upgrade both the antimicrobial properties and antifouling performance at the same time [56, 59].

5.2.2 Antimicrobial and Antifouling Polymers Based on Electrical Stimulation

Antimicrobial and antifouling polyether sulfone membranes are manufactured through development of novel PHMB (zinc oxide), ZnO, and polyhexamethylene biguanide hydrochloride nanocomposites on PES membrane surfaces embedded with polydopamine as an intermediate connection layer [8]. ZnO and PHMB nanocomposites have exhibited a high antimicrobial rate exceeding 99% with a small quantity of PHMB (wt_{ZnO}: wt_{PHMB} = 1:0.01) [8]. Another significant benefit of biodegradable polymers is their versatility and biocompatibility, which come from their adjustable properties and functions. This adaptability enables the creation of materials ranging from bioactive hydrogels to biodegradable thermoplastic polymers. Incorporating nanoparticles into the polymer matrix leads to the formation of innovative nanocomposites, broadening the spectrum of properties and functionalities of these polymer composites. Below is a flow chart for electroactive polymer composites and their properties, as shown in Figure 5.4.

Ionic polymers exhibit conductivity due to the presence of polyelectrolytes, whereas electrically conductive polymers transfer electrons through intrinsic mechanisms related to their chemical bonds or conductive particles. In biomedical applications, the biodegradability and compatibility of conductive polymers used in tissue engineering can be constrained by the doping levels required for electrical conduction, as high concentrations may lead to tissue inflammation. To increase the biocompatibility of conductive polymers, they are doped with biomolecules or ions, which improves the composite's properties by leveraging its chemical, electrical, and physical structures [9].

Developing alternative antibacterial materials with conducting polymers and carbon nanotubes offers a promising approach to addressing the growing issue of antibiotic resistance. This strategy can be further enhanced by employing electric stimulation, which bypasses the limitations associated with microbial adhesion to the polymer surface. These advancements hold potential across various industries, including medical, food, and textiles [7, 10]. Novel ZnO and PHMB nanocomposite

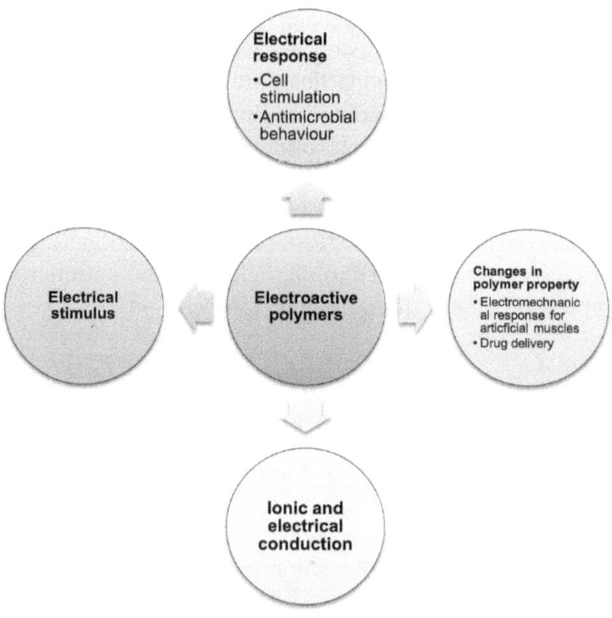

FIGURE 5.4. Overview of electroactive polymers and various mechanisms for electric stimulus.

membranes are being used that can achieve a higher than 95% bactericidal rate [11]. Also, ZnO and PHMB (Polyhexamethylene biguanide) is a cationic polymer nanocomposite with excellent antimicrobial and antiviral properties. PHMB possesses around 89% FRR after three cycles of SA filtration. ZnO and PHMB membranes achieve higher than 91% stability and long-term durability. Phosphorylcholine functionalized compounds have demonstrated the highest stability in electrochemical and antifouling properties during electrical stimulation.

Developing polymeric devices or coatings with antimicrobial or antifouling properties can aid in preventing the formation of biofilms and infections. These devices accomplish this by either preventing or destroying bacteria near the surface or by minimizing the initial attachment of proteins and bacteria. Antibacterial nanocomposites are enhanced with 0D fullerenes, carbon nanotubes, graphene, and their derivatives, including graphene oxide and reduced graphene oxide [12]. Surface modification of membranes with bacterial-based polymers is considered a promising approach that can help reduce bacterial cell activity and inhibit their proliferation on various cell membrane surfaces [11]. The antibiotic and medical industries need green materials, with sizes less than 100 nm, to be incorporated into polymer nanocomposites. These composites often include 1D carbon nanotubes, 2D-based graphene, and its derivatives, such as graphene oxide derivatives, and reduced graphene oxide or nanosheets. Conducting polymers serves as carriers for biomolecules, improving the speed, stability, and sensitivity of composite materials in biomedical devices. They can switch between conducting and semiconducting states and have the capability to modify mechanical properties through chemical changes and doping. These polymers are highly valuable in biomedical applications and neural prosthetic devices [12–14]. Antifouling polymers are also materials that may stand up and achieve better performance compared to hydrophilic and bioinert polymers.

Antifouling polymers are typically hydrophilic polymers with charged groups and polar ions that can form strong non-bonding interactions with water molecules. This property is exceptionally valuable for nanoscale applications such as nanomedicine which needs surface adjustments at the atomic level [60, 62, 63]. Antifouling polymers such as polyethylene glycol and its options act as a protecting sedate particle and restorative proteins in safe frameworks with nanoscale congregations. Integrating conductive polymers into effective delivery systems targeting tissues helps prevent the attachment of cells and particles to surfaces. These coatings are used in both marine and biomedical device applications [62, 64].

Integration of conductive polymers in nanoscale medical applications such as nanoparticles and surface adjustments at the atomic level are able to create antifouling polymers such as polyethylene glycol that play a crucial role in safeguarding therapeutic particles, proteins, and genes within a protective system, interacting with various nanostructures to facilitate efficient drug delivery to target tissues.. For coatings, antifouling polymers help in anticipating a cell's grip and connection of atoms to surfaces, and are they utilized broadly in marine and biomedical gadget applications [64]. The major polymer composites used as conductive polymers are polyoxazoline, polyzwitterions, and polyethylene glycol due to their ability to enhance biocompatibility, reduce adsorption of protein, and increase the performance of biomedical and drug delivery systems [17]. Antifouling polymers are synthesized materials that stand up to non-specific intelligence with proteins, cells, and other biomolecules. Conductive polymers can be doped either p-type or n-type, leading to the creation of positive or negative bipolarons, respectively. These charge carriers are introduced along the polymer chain, enabling electronic conductivity. In n-type doping, the negatively charged carriers are less stable compared to the positively charged ones. In the case of polypyrrole (PPy), its conductivity arises from p-type doping. [37,65]. Underneath is a schematic chart of electronic groups and chemical structures of polarons and bipolarons that are undoped of polypyrrole (PPy), as shown below in Figure 5.5.

5.2.3 Conductive Polymers for Biosensors and Bio-Actuators

Conducting polymers are utilized broadly in the manufacture of biochips and atomic gadgets and therapeutic determination and related applications. Natural polymers known as conducting polymers are progressively well known due to their excellent optical and electrical properties. Conductive polymers exhibit excellent electrical conductivity and mechanical strength comparable to metals and inorganic semiconductors, while retaining the distinctive advantages of polymers, such as flexibility and ease of blending with conventional polymers. Conductive polymers are categorized based on electrical charges, such as particles and delocalized pi electrons or conductive nanomaterials. Conductive polymers are also connected to organic tests, and maintain excellent biocompatibility, which makes them great candidates for natural and restorative applications. Conductive polymer sensor components are utilized in electrochemical biosensors, material sensors, as well as heat sensors.

Integration of polymer composites with natural fibers or materials such as a combination of ceramic-based calcium phosphate crystallites and natural fibers such as collagen filaments may shape thick, solid structures and improve the physical properties of natural and inorganic materials [14, 19, 66]. Conductive polymers act as

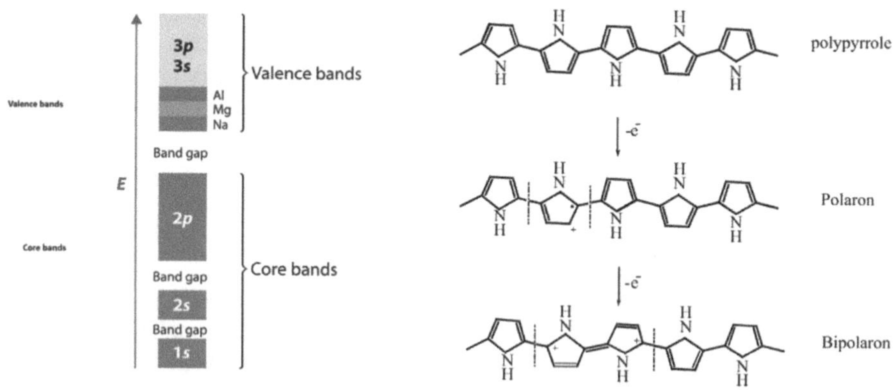

FIGURE 5.5. Electronic bands and chemical structures of undoped, polaron, and bipolarons fully doped states of polypyrrole (PPy).

a cheaper elective to different filler conductive materials such as graphite, carbon black, metal-coated glass strands, metal particles, and pieces for the arrangement of composite nanomaterials. In planning conductive polymer composites (CPC), which was developed as a fabric in bioelectronics and a few other applications. The conductive polymer composites consist of a combination of one or more non-conducting polymers and conductive biofiller materials throughout the polymer framework. The conductivity of the conductive polymer composite is determined by the permeation hypothesis, which describes the conductive phase of the CPC based on the arrangement of filler materials at a specific weight ratio [6, 15, 62]. Filler stacking has to retain a certain concentration to supply a ceaseless route for the electron transport essential for conduction and to dodge sudden drops in conductivity. Certain drawbacks of CPC incorporate heavy reliance on handling conditions, mechanical flimsiness, and reliance on handling conditions [6, 62, 67].

Conductive polymers are also integrated into piezoresistive material detection systems, which are classified into resistive metal-based and conductive polymer-based detection methods. These polymers are employed in the development of flexible arrays for resistive metal-based sensors and are incorporated into sensor systems that combine piezoresistive metals capable of detecting mechanical strain. Additionally, conductive polymers play a key role in the advancement of multimodal sensor technologies, enabling the detection of contact forces, thermal conductivity, and temperature variations, as well as applications in sensing organic skin [68, 69].

5.3 CONDUCTIVE POLYMERS IN RECENT BIOELECTRONICS APPLICATIONS

Conductive polymers are considered basic components in different therapeutic devices and are utilized for biopotential recording and electric brain incitement. Conducting polymers in different neural and molecular imprinting applications are illustrated in Figure 5.6. One of the major prerequisites for inveterate neural materials is an improvement of fabric and electrical soundness. In bioelectronic applications, conductive polymers have made colossal advances and are constantly being substituted for customary metals, such as gold (Au), platinum (Pt), or iridium (Ir), due to their high conductivity and erosion-safe properties, as well as their ability to supplant costly valuable metals and overcome mechanical, electrical, and organic impediments of designed tissues [19, 70, 71]. The expansion of conducting polymers with atomic engraving polymers is due to their simple planning by means of electrochemical polymerization that permits tall control of film thickness on the terminal surface. The conductive polymer film thickness straightforwardly influences the energy, selectivity, and biorecognition preparation, as well as the viability and downsides of the composite. The integration of conducting polymers with molecularly engraved polymers (MPs) has garnered significant mechanical interest due to the ease of fabrication of molecularly patterned conductive polymers that utilize electrochemical polymerization that permits high control of film thickness on the cathode surface. The conducting polymer film's thickness influences the energy, selectivity,

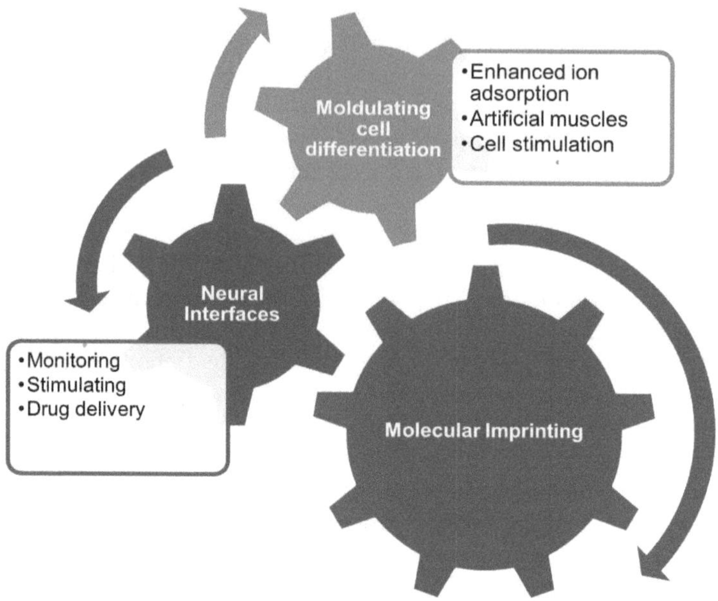

FIGURE 5.6 Conducting polymers in neural and molecular imprinting applications.

and biorecognition as well as well as the execution of detection [71–73]. Distinctive polymers such as pyrrole, phenyl diamine, aniline, amino thiophenol, amino phenyl boronic corrosive, aminophenol, and thiophene are broadly utilized as copolymers and composites. The sensor execution is exceptionally good, to the extent of small-scale molar levels. The utilization of conductive polymers has moved forward sensor execution in the case of biosensors.

Moreover, conductive polymers are utilized in an amalgamation of piezoelectric nanogenerators that change from mechanical vitality to electrical vitality. The piezoelectric nanogenerator could be an economical elective that changes mechanical vitality from its environment into electrical control. Distinctive from the working rule of TENG, by isolating negative and positive charges between piezoelectric materials beneath a remotely connected stretch [25–27].

The high piezoelectric coefficient, moo temperature fabrication, moo fetching, and mechanical adaptability of copolymers, such as poly vinylidene fluoride (PVDF), has synthesized utilitarian films for vitality gathering by encompassing mechanical vibrations. As it were the β and γ crystalline stages contribute to piezoelectricity since the course of action of exceedingly electronegative bonds between C–F molecules make a dielectric minute that produces an electric field. Several copolymers are used in energy, engineering, and storage applications, including polyvinylidene fluoride-hexafluoropropylene, polyvinylidene fluoride-chlorotrifluoroethylene, and various conductive polymer composites. The conducting polymers are utilized to progress the execution of nanostructured piezoelectric materials and are utilized to deliver electrospinning to deliver PVDF nanofibers to progress the piezoelectricity of PVDF. The piezoelectric behavior of PVDF is moved forward via the poling and mechanical extending of nanofibers.

5.4 CONCLUSION

In conclusion, this chapter delves into the pivotal role of conductive polymers and green materials used in various engineering applications and smart materials. Conductive polymers possess flexible, lightweight, and durable energy storage capabilities, which increase their utilization in diverse fields such as microelectronics, photovoltaic devices, and light-emitting diodes (LED). Their unique ability to integrate the mechanical properties of polymers with the electrical properties of metals makes them highly valuable for various engineering applications. Many commonly used polymers, such as polyaniline (PANI), polythiophene (PTh), and polypyrrole (PPy), as well as their derivatives, demonstrate exceptional electrical conductivity and versatility. Polyaniline (PANI) is known for its ease of synthesis and the ability to conduct electricity in its doped form, making it valuable for applications in sensors and electronic devices. Polythiophene (PTh), including its derivatives like poly(3,4-ethylenedioxythiophene) (PEDOT), is widely used in organic electronics due to its stability and high conductivity. Polypyrrole (PPy) and its derivatives are known for their conductivity and are used in various applications, including supercapacitors and as coatings for biomedical devices. This chapter highlights the significant advancements in conductive polymers, particularly their applications in smart materials. These polymers not only enhance electrical conductivity but also exhibit high mechanical flexibility, allowing them to be stretched or bent, which broadens their use in engineering applications. Their ability to be tailored through doping and integration with other materials results in versatile and adaptive properties suitable for a range of uses, including biomedical applications. Conductive polymers have proven vital in developing electroactive materials for tissue engineering, artificial muscles, and smart sensors. Their ability to convert electrical energy into mechanical responses and deliver drugs or mimic antimicrobial behaviors underscores their potential in advanced technologies. However, challenges such as degradation from oxygen and moisture exposure persist. Recent advancements aim to address these issues by improving polymer stability through enhanced design and protective coatings. Furthermore, conductive polymers' exceptional thermoelectric properties offer promising applications in energy harvesting and wearable electronics. Their biocompatibility further extends their utility into medical fields, such as biosensors, neural interfaces, and drug delivery systems. The integration of conductive polymers with hydrogels and natural fibers to create biodegradable and biocompatible composites exemplifies their adaptability and potential for innovation. Conductive polymers are integral to the development of piezoelectric nanogenerators which convert mechanical energy into electrical power. Piezoelectric nanogenerators use piezoelectric materials and conductive polymers that generate an electric charge through mechanical stress. This charge flow occurs due to the polarization charges created when the piezoelectric material is deformed, balancing the electrostatic potential and producing electricity.

REFERENCES

1. R. Hosaka and R. Noji, "Biomedical signal and image processing projects using matlab and labview tools," *IFMBE Proceedings 65*, vol. 17, no. 1, 2017.
2. M. Bhandari, D. P. Kaur, S. Raj, T. Yadav, M. A. S. Abourehab, and M. S. Alam, "Electrically Conducting Smart Biodegradable Polymers and Their Applications," in *Handbook of Biodegradable Materials*, 2022. https://doi.org/10.1007/978-3-030-83783-9_64-1.
3. M. Bhandari, D. P. Kaur, S. Raj, T. Yadav, M. A. S. Abourehab, and M. S. Alam, "Electrically Conducting Smart Biodegradable Polymers and Their Applications," in *Handbook of Biodegradable Materials*, 2023. https://doi.org/10.1007/978-3-031-09710-2_64.
4. I. Rendón-Enríquez *et al.*, "Thin polymer films by oxidative or reductive electropolymerization and their application in electrochromic windows and thin-film sensors," *Molecules*, vol. 28, no. 2, 2023. https://doi.org/10.3390/molecules28020883.
5. J. Alam, L. A. Dass, M. S. Alhoshan, and A. W. Mohammad, "Advances in membrane development based on electrically conducting polymers," 2013. https://doi.org/10.1002/adv.21262.
6. D. Gochnauer and T. Gilani, "Conduction mechanism in electrically conducting polymers," *Am J Undergrad Res*, vol. 14, no. 4, 2018. https://doi.org/10.33697/ajur.2018.006.
7. M. Krifa, "Electrically conductive textile materials—application in flexible sensors and antennas," 2021. https://doi.org/10.3390/textiles1020012.
8. J., Singh, J.P., Singh, S., Kumar, H.S. Gill, "Short review on hydroxyapatite powder coating for SS 316L," *J Electrochem Sci Eng,* vol. 13, 2023, 25–39. https://doi.org/10.5599/jese.1611.
9. A. M. Grancarić *et al.*, "Conductive polymers for smart textile applications," 2018. https://doi.org/10.1177/1528083717699368.
10. X. Wang, Y. Tang, X. Zhu, Y. Zhou, and X. Hong, "Preparation and characterization of polylactic acid/polyaniline/nanocrystalline cellulose nanocomposite films," *Int J Biol Macromol*, vol. 146, 2020. https://doi.org/10.1016/j.ijbiomac.2019.09.233.
11. D. Kongahage and J. Foroughi, "Actuator materials: Review on recent advances and future outlook for smart textiles," 2019. https://doi.org/10.3390/FIB7030021.
12. M. R. Repon and D. Mikučionienė, "Progress in flexible electronic textile for heating application: A critical review," 2021. https://doi.org/10.3390/ma14216540.
13. G. Kaur, R. Adhikari, P. Cass, M. Bown, and P. Gunatillake, "Electrically conductive polymers and composites for biomedical applications," 2015. https://doi.org/10.1039/c5ra01851j.
14. L. Ghasemi-Mobarakeh *et al.*, "Application of conductive polymers, scaffolds and electrical stimulation for nerve tissue engineering," 2011. https://doi.org/10.1002/term.383.
15. B. Guo and P. X. Ma, "Conducting polymers for tissue engineering," *Biomacromolecules*, vol. 19, no. 6, 2018. https://doi.org/10.1021/acs.biomac.8b00276.
16. P. Shokrollahi, Y. Omidi, L. X. Cubeddu, and H. Omidian, "Conductive polymers for cardiac tissue engineering and regeneration," 2023. https://doi.org/10.1002/jbm.b.35293.
17. J. Singh, A. W. Hashmi, S. Ahmad, and Y. Tian, "Critical review on biodegradable and biocompatibility magnesium alloys: Progress and prospects in bio-implant applications," *Inorg Chem Commun*, 2024, 113111.
18. A. Dubey, X. Tao, C. Cochrane, and V. Koncar, "Textile based three-layer robust flexible and stable electrochromic display," *IEEE Access*, vol. 8, 2020. https://doi.org/10.1109/ACCESS.2020.3028655.

19. J., Singh, S. Singh, "A review on machine learning aspect in physics and mechanics of glasses," *Mater Sci Eng B*, 284, 2022, 115858. https://doi.org/10.1016/j.mseb.2022.115858.
20. Z. Zhu, J. Di, X. Liu, J. Qin, and P. Cheng, "Coiled polymer fibers for artificial muscle and more applications," 2022. https://doi.org/10.1016/j.matt.2022.02.018.
21. X. Hu *et al.*, "Coiled polymer artificial muscles having dual-mode actuation with large stress generation," *J Bionic Eng*, vol. 20, no. 4, 2023. https://doi.org/10.1007/s42235-023-00353-x.
22. T. Mirfakhrai, J. D. W. Madden, and R. H. Baughman, "Polymer artificial muscles," *Materials Today*, vol. 10, no. 4, 2007. https://doi.org/10.1016/S1369-7021(07)70048-2.
23. Y. Lee *et al.*, "Hydrogen barrier properties of atomic layer deposited Al 2 O 3 with different oxidants for ingazno thin film transistor," *ECS Meeting Abstracts*, vol. MA2019-02, no. 24, 2019. https://doi.org/10.1149/ma2019-02/24/1129.
24. Y. Zhao, J. Kang, W. Huang, P. Kong, D. An, and G. He, "Copper nanowire/polydopamine-modified sodium alginate composite films with enhanced long-term stability and adhesion for flexible organic light-emitting diodes," *ACS Appl Mater Interfaces*, 2023. https://doi.org/10.1021/acsami.3c13443.
25. J. Li, A. B. Huckleby, and M. Zhang, "Polymer-based thermoelectric materials: A review of power factor improving strategies," 2022. https://doi.org/10.1016/j.jmat.2021.03.013.
26. K. Tada, "Toward long and strong electroactive supercoiled polymer artificial muscles: Fabrication with constant-load springs," *IEICE Transactions on Electronics*, vol. E106.C, no. 6, 2023. https://doi.org/10.1587/transele.2022OMS0006.
27. K. Tada and M. Kaku, "S-shaped nonlinearity in electrical resistance of electroactive supercoiled polymer artificial muscle," *IEICE Transactions on Electronics*, vol. E103C, no. 2, 2020. https://doi.org/10.1587/transele.2019ECS6013.
28. G. Kiran Raj *et al.*, "Conductive polymers and composites-based systems: An incipient stride in drug delivery and therapeutics realm," 2023. https://doi.org/10.1016/j.jconrel.2023.02.017.
29. T. Distler and A. R. Boccaccini, "3D printing of electrically conductive hydrogels for tissue engineering and biosensors – A review," 2020. https://doi.org/10.1016/j.actbio.2019.08.044.
30. A. Filimon, "Perspectives of conductive polymers toward smart biomaterials for tissue engineering," *Conducting Polymers*, 2016. https://doi.org/10.5772/63555.
31. I. Armentano, E. Fortunati, L. Torre, and J. M. Kenny, "Recent advances in conductive composites based on biodegradable polymers for regenerative medicine applications," *Handbook of Composites from Renewable Materials*, vol. 1–8, 2017. https://doi.org/10.1002/9781119441632.ch101.
32. D. A. A. Ruzaidi *et al.*, "Advocating electrically conductive scaffolds with low immunogenicity for biomedical applications: A review," 2021. https://doi.org/10.3390/polym13193395.
33. K. Dashtian *et al.*, "A comprehensive review on advances in polyoxometalate based materials for electrochemical water splitting," 2024. https://doi.org/10.1016/j.ccr.2023.215644.
34. M. Alagarsamy, S. R. Barkunan, N. Jayapal, A. Murugan, P. Muralikrishnan, and P. J. Ramulu, "Reinforcement of nanocellulose as green agent in the electronic applications associated with the composites of polymer matrix," *Int J Polym Sci*, vol. 2023, 2023. https://doi.org/10.1155/2023/9645190.
35. W. Serrano-Garcia, I. Bonadies, S. W. Thomas, and V. Guarino, "New insights to design electrospun fibers with tunable electrical conductive–semiconductive properties," 2023. https://doi.org/10.3390/s23031606.

36. S. S. Athukorala *et al.*, "3D printable electrically conductive hydrogel scaffolds for biomedical applications: A review," 2021. https://doi.org/10.3390/polym13030474.
37. A. Gelmi, M. K. Ljunggren, M. Rafat, and E. W. H. Jager, "Influence of conductive polymer doping on the viability of cardiac progenitor cells," *J Mater Chem B*, vol. 2, no. 24, 2014. https://doi.org/10.1039/c4tb00142g.
38. A. K. Rana, F. Scarpa, and V. K. Thakur, "Cellulose/polyaniline hybrid nanocomposites: Design, fabrication, and emerging multidimensional applications," 2022. https://doi.org/10.1016/j.indcrop.2022.115356.
39. A. U. Haq *et al.*, "Intrinsically conductive polymers for striated cardiac muscle repair," 2021. https://doi.org/10.3390/ijms22168550.
40. N. Jalilinejad *et al.*, "Electrically conductive carbon-based (bio)-nanomaterials for cardiac tissue engineering," 2023. https://doi.org/10.1002/btm2.10347.
41. A. M. Elsherbini and S. A. Sabra, "Nanoparticles-in-nanofibers composites: Emphasis on some recent biomedical applications," 2022. https://doi.org/10.1016/j.jconrel.2022.05.037.
42. L. Qiao *et al.*, "Antibacterial conductive self-healing hydrogel wound dressing with dual dynamic bonds promotes infected wound healing," *Bioact Mater*, vol. 30, 2023. https://doi.org/10.1016/j.bioactmat.2023.07.015.
43. C. Korupalli *et al.*, "Conductive materials for healing wounds: Their incorporation in electroactive wound dressings, characterization, and perspectives," 2021. https://doi.org/10.1002/adhm.202001384.
44. L. Nie *et al.*, "Fabrication and desired properties of conductive hydrogel dressings for wound healing," 2023. https://doi.org/10.1039/d2ra07195a.
45. S. Peressotti, G. E. Koehl, J. A. Goding, and R. A. Green, "Self-assembling hydrogel structures for neural tissue repair," 2021. https://doi.org/10.1021/acsbiomaterials.1c00030.
46. S. Siddiquee, M. G. J. Hong, and M. M. Rahman, *Composite Materials: Applications in Engineering, Biomedicine and Food Science*, 2020. https://doi.org/10.1007/978-3-030-45489-0.
47. A. Warhurst, "Sustainability indicators and sustainability performance management," *Sustainability Indicators and Sustainability Performance Management*, vol. 43, no. 43, 2002.
48. K. Roshanbinfar, L. Vogt, F. Ruther, J. A. Roether, A. R. Boccaccini, and F. B. Engel, "Nanofibrous composite with tailorable electrical and mechanical properties for cardiac tissue engineering," *Adv Funct Mater*, vol. 30, no. 7, 2020. https://doi.org/10.1002/adfm.201908612.
49. H. Esmaeili *et al.*, "Electroconductive biomaterials for cardiac tissue engineering," 2022. https://doi.org/10.1016/j.actbio.2021.08.031.
50. N. G. Kozan, M. Joshi, S. T. Sicherer, and J. M. Grasman, "Porous biomaterial scaffolds for skeletal muscle tissue engineering," 2023. https://doi.org/10.3389/fbioe.2023.1245897.
51. R. Dong, P. X. Ma, and B. Guo, "Conductive biomaterials for muscle tissue engineering," 2020. https://doi.org/10.1016/j.biomaterials.2019.119584.
52. Y. Zhang, C. Zhang, and Y. Wang, "Recent progress in cellulose-based electrospun nanofibers as multifunctional materials," 2021. https://doi.org/10.1039/d1na00508a.
53. L. Shen *et al.*, "Highly porous nanofiber-supported monolayer graphene membranes for ultrafast organic solvent nanofiltration," *Sci Adv*, vol. 7, no. 37, 2021. https://doi.org/10.1126/sciadv.abg6263.
54. N. Zarrin, H. Tavanai, A. Abdolmaleki, M. Bazarganipour, and F. Alihosseini, "An investigation on the fabrication of conductive polyethylene dioxythiophene (PEDOT) nanofibers through electrospinning," *Synth Met*, vol. 244, 2018. https://doi.org/10.1016/j.synthmet.2018.07.013.

55. Y. Cong, S. Liu, and H. Chen, "Fabrication of conductive polypyrrole nanofibers by electrospinning," *J Nanomater*, vol. 2013, 2013. https://doi.org/10.1155/2013/148347.
56. M. Li *et al.*, "Development of an antimicrobial and antifouling PES membrane with ZnO/poly(hexamethylene biguanide) nanocomposites incorporation," *Chemical Engineering Journal*, vol. 481, 2024. https://doi.org/10.1016/j.cej.2024.148744.
57. W. Y. Wang, H. W. Hu, J. C. Chiou, K. F. Yung, and C. W. Kan, "Poly(hexamethylene biguanide) hydrochloride (PHMB)-based materials: Synthesis, modification, properties, determination, and application," 2023. https://doi.org/10.1039/d3py01148h.
58. M. Li *et al.*, "Development of an antimicrobial and antifouling PES membrane with ZnO/poly(hexamethylene biguanide) nanocomposites incorporation," *Chemical Engineering Journal*, vol. 481, 2024. https://doi.org/10.1016/j.cej.2024.148744.
59. A. Salama *et al.*, "Nanocellulose-based materials for water treatment: Adsorption, photocatalytic degradation, disinfection, antifouling, and nanofiltration," 2021. https://doi.org/10.3390/nano11113008.
60. J. Cui *et al.*, "A comprehensive review on the inherent and enhanced antifouling mechanisms of hydrogels and their applications," 2024. https://doi.org/10.1016/j.ijbiomac.2024.130994.
61. Z. K. Zander and M. L. Becker, "Antimicrobial and antifouling strategies for polymeric medical devices," *ACS Macro Lett*, vol. 7, no. 1, 2018. https://doi.org/10.1021/acsmacrolett.7b00879.
62. L. Benny Mattam, A. Bijoy, D. Abraham Thadathil, L. George, and A. Varghese, "Conducting polymers: A versatile material for biomedical applications," 2022. https://doi.org/10.1002/slct.202201765.
63. A. M. Díez-Pascual, "State of the art in the antibacterial and antiviral applications of carbon-based polymeric nanocomposites," 2021. https://doi.org/10.3390/ijms221910511.
64. S. Ahmad, A. W. Hashmi, J. Singh, K. Arora, Y. Tian, F. Iqbal, M. Al-Dossari, M. I. Khan, "Innovations in additive manufacturing of shape memory alloys: Alloys, microstructures, treatments, applications," *J Mater Res Technol*, 2024, vol. 32, 4136–4197.
65. S. Biswas, Y. J. You, Y. Lee, J. W. Shim, and H. Kim, "Efficiency improvement of indoor organic solar cell by optimization of the doping level of the hole extraction layer," *Dyes and Pigments*, vol. 183, 2020. https://doi.org/10.1016/j.dyepig.2020.108719.
66. R. Balint, N. J. Cassidy, and S. H. Cartmell, "Conductive polymers: Towards a smart biomaterial for tissue engineering," 2014. https://doi.org/10.1016/j.actbio.2014.02.015.
67. A. K. Bakhshi and G. Bhalla, "Electrically conducting polymers: Materials of the twentyfirst century," *J Sci Ind Res (India)*, vol. 63, no. 9, 2004.
68. S. Nambiar and J. T. W. Yeow, "Conductive polymer-based sensors for biomedical applications," 2011. https://doi.org/10.1016/j.bios.2010.09.046.
69. A. D. Bendrea, L. Cianga, and I. Cianga, "Review paper: Progress in the field of conducting polymers for tissue engineering applications," 2011. https://doi.org/10.1177/0885328211402704.
70. X. Qian, M. Su, F. Li, and Y. Song, "Research progress in flexible wearable electronic sensors," 2016. https://doi.org/10.6023/A16030156.
71. J. Gamboa, S. Paulo-Mirasol, F. Estrany, and J. Torras, "Recent progress in biomedical sensors based on conducting polymer hydrogels," May 15, 2023, *American Chemical Society*. https://doi.org/10.1021/acsabm.3c00139.
72. J. I. Hahm, "Functional polymers in protein detection platforms: Optical, electrochemical, electrical, mass-sensitive, and magnetic biosensors," 2011. https://doi.org/10.3390/s110303327.
73. M. Li *et al.*, "Development of an antimicrobial and antifouling PES membrane with ZnO/poly(hexamethylene biguanide) nanocomposites incorporation," *Chemical Engineering Journal*, vol. 481, Feb. 2024. https://doi.org/10.1016/j.cej.2024.148744.

6 Photocatalytic Materials
Properties, Synthesis, Applications, and Sustainability

Hardeep Kaur, Niyati Gupta, Kritika Charak, Sukhman Kaur, Chandra Mohan, and Noushi Zaidi

Description	**Abbreviation**
Hydroxyl radicals	HO
Reactive oxygen species	ROS
Scanning electron microscopy	SEM
Transmission electron microscopy	TEM
Turnover number	TON
Ultraviolet	UV
Volatile organic compound	VOC
X-ray diffraction	XRD

6.1 INTRODUCTION

For decades, there has been immense research and development in the field of photocatalytic materials [1]. Photocatalytic materials like titanium dioxide, zinc oxide, and tungsten trioxide have extensive applications in attaining a sustainable environment [2]. Photocatalytic materials trigger chemical reactions upon exposure to ultraviolet (UV) and visible light [3, 4]. A photocatalyst is a material or substance that accelerates several chemical reactions when exposed to light radiation, resulting in the conversion of solar energy into other useful forms. These materials utilize energy from light sources like the sun, to drive reactions that would be slow or non-spontaneous otherwise. Photocatalytic materials consist of two key components which are semiconductors and the light source. Majorly photocatalytic materials are made from semiconductors, having specific band gap which allows them to absorb light and generate electron–hole pairs. Some of the most commonly exploited photocatalytic semiconductors are zinc oxide (ZnO), titanium dioxide (TiO_2), and cadmium sulfide

 DOI: 10.1201/9781003516071-6

(CdS). The two major categories of photocatalysts are organic photocatalysts and inorganic photocatalysts [5,6]. Organic photocatalysts are the substances that assist chemical reactions under light exposure, by utilizing organic materials instead of traditional inorganic matter like TiO_2 or ZnO. Organic photocatalysts are employed in various bioremediation processes like water and air purification, and the degradation of pollutants, as shown in Figure 6.1 [7].

Among all these photocatalysts, inorganic semiconductors have been widely employed due to their relatively high efficiency, low cost, good robustness, and easy availability. In order to facilitate the excitation of electrons from the semiconductor's valence band to its conduction band, an adequate amount of energy is needed to be absorbed by the photocatalyst. Inorganic photocatalysts are those materials which enable chemical reactions due to exposure to light, typically concerned with compounds like TiO_2, ZnO, and CdS [8]. Owing to their higher stability, efficiency, and robustness, inorganic photocatalysts are widely employed in environmental remediations, water and air purification, self-cleaning surfaces, and the degradation of organic pollutants, which are listed as some of the widespread applications of inorganic photocatalytic materials [9–12]. UV light is often used because its energy is higher, but visible-light-active photocatalysts are increasingly being developed [13].

Photocatalysis is a process that involves a semiconductor catalyst interacting with photons to trigger chemical reactions. When photons encounter energy equal to or higher than the catalyst's bandgap energy hitting its surface, they excite electrons from the valence band of the semiconductor to its conduction band, generating electron–hole pairs [14], as shown in Figure 6.2. On the catalyst surface, these charge carriers perform redox reactions with the adsorbed species, resulting in the generation of radicals or ions that drive the required reactions [15]. Consequential

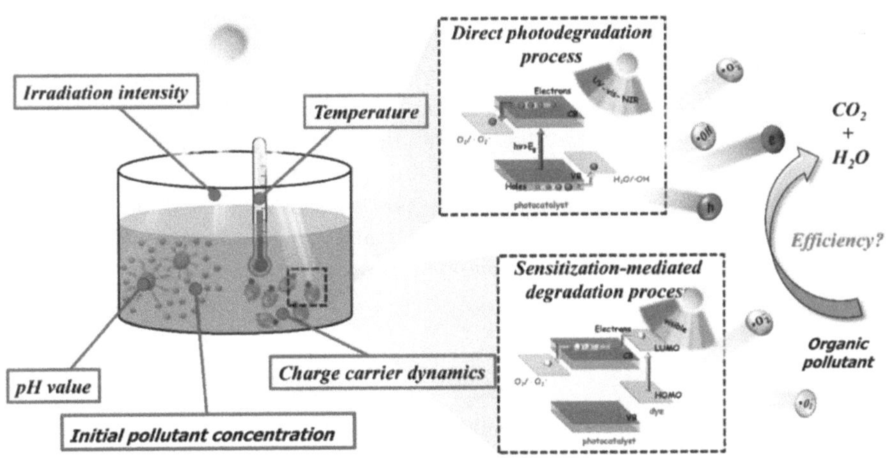

FIGURE 6.1 Schematic representation of operational parameters affecting the photocatalytic degradation of organic pollutants [7] (Credit: Creative Commons Attribution 4.0 International License).

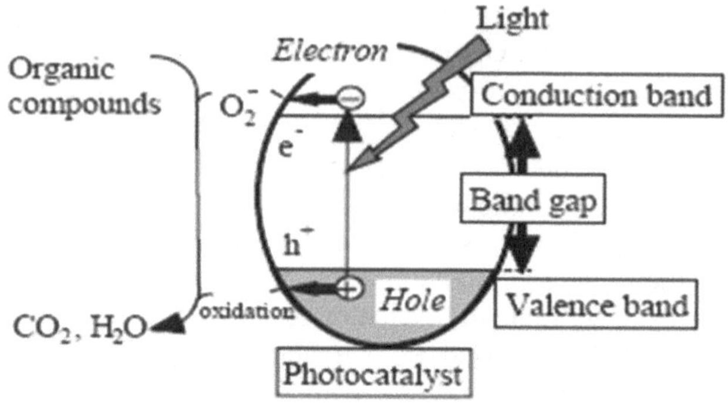

FIGURE 6.2 General mechanism of photocatalysis [22] (Credit: Creative Commons Attribution 4.0 International License).

recombination and transfer of these charge carriers determine the productivity of the process. Altogether, photocatalysis uses semiconductor's light-induction properties to accelerate chemical transformations, contributing to environment sustainability for various applications from water purification to pollutant degradation [16].

The success of this reaction depends on the capacity of a photocatalyst to generate long-lived electrons and holes, which results in the formation of strong free radicals [13]. Additionally, its fate relies on factors such as the specificity of used photocatalytic material, along with their operational variables like solution acidity, duration of irradiation, and the introduction of substances that scavenge holes. Photocatalytic materials have a large number of applications in varied areas, including environmental purification, energy production, photocatalytic coatings, self-cleansing materials, and many more [11–19]. Photocatalysis offers discreet benefits over conventional methods due to its eco-friendly nature and higher efficiency [20]. In contrast to conventional techniques, photocatalysis works under mild conditions, generally using sunlight or artificial light as the only energy source, therefore minimizing energy consumption and reducing carbon footprint. In addition to this, it accelerates the degradation of pollutants without the generation of harmful derivatives, which contributes to cleaner water and air. Moreover, this process of photocatalysis can also be employed on a wide range of substances and surfaces, offering adaptable solutions for various environmental and industrial challenges. In general, photocatalysis amounts to an effective and sustainable approach to addressing pollution and promoting green chemistry practices [21].

6.2 FUNDAMENTALS OF PHOTOCATALYSIS

Photocatalysis works on certain principles involving various interconnected processes, including light absorption, surface reactions, and charge carrier generation. All three processes are described below along with their visual representation in Figure 6.3.

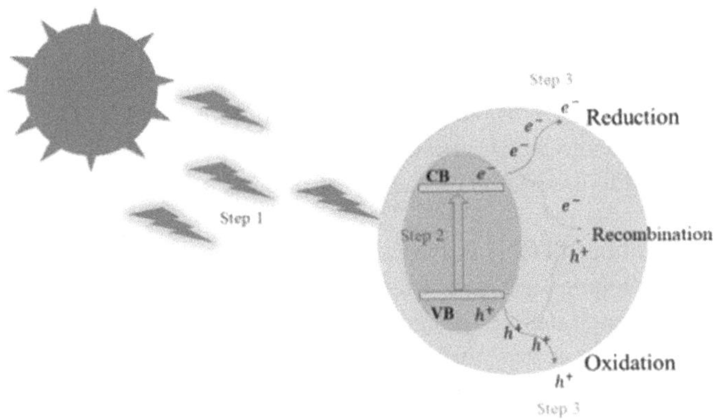

FIGURE 6.3 Visual description of three steps of photocatalysis [23] (Credit: Creative Commons Attribution 4.0 International License).

6.2.1 Light Absorption

We know that photocatalytic materials use light energy in order to direct chemical reactions, such as photocatalytic decomposition of surface stains. It is mandatory for these materials to absorb light at a suitable wavelength to obtain the desired results from the reaction. The band gap energy i.e. (energy difference between conduction and valence bonds) is essential to figure out the wavelength of light that can be absorbed [24].

6.2.2 Surface Reactions

The molecules that react with other molecules adsorb onto the surface of the photocatalyst and these adsorbed reactants undergo redox reactions facilitated with the help of charge carriers [25]. In reduction, conduction band electrons reduce an acceptor molecule, whereas in oxidation, valence band holes oxidize a donor molecule.

6.2.3 Charge Carrier Generation

The formation of charge carriers involve the effective dissociation of electron–hole pairs as this is very important to protect against heat and energy. These dissociated holes as well as electrons move to the photocatalyst's surface where they combine with adsorbed reactants [26].

There are a lot of key factors that influence this mechanism such as the presence of electron donors/acceptors, ionic strength, pH, etc. [27]. Photocatalysts are of different types as well and having specific applications that are quite efficient. Semiconductor photocatalysts include titanium dioxide (TiO_2), which is the most widely used photocatalyst because it has strong oxidative power and chemical stability. Zinc oxide (ZnO) is another photocatalyst with high electron mobility, and it is useful in antibacterial coatings. Metal oxides such as iron oxide (Fe_2O_3), bismuth vanadate ($BiVO_4$), and tungsten trioxide (WO_3) are used for specific applications like the degradation of pollutants and water splitting [28].

Photocatalytic efficiency and performance metrics are crucial for inspecting and evaluating the effectiveness of photocatalysts in numerous fields. These metrics help to compare various photocatalysts and intensify their performance. Some fundamental criteria are used to assess the efficiency of photocatalysis [29].

a) Turnover number (TON) is the standard measurement of as to how many substrate molecules are converted per photocatalyst molecule per unit of time, showing the increased catalytic activity per active site.
b) Rate constant expresses the rate of a photocatalytic reaction; if the rate constant is high, the reaction rates would be higher and vice versa.
c) Stability and durability evaluate the ability of a photocatalyst to maintain its activity even after multiple uses and are checked by rhythmic photocatalytic cycles and observing changes in them [30].

6.3 MECHANISM OF PHOTOCATALYTIC DEGRADATION

Photocatalytic degradation is a technique that uses light energy to speed up a reaction that breaks down pollutants in the environment [31]. This approach has attracted a lot of interest because of its potential uses in cleaning water and air, as well as in breaking down organic pollutants and dangerous microorganisms. It is necessary to understand the workings of photocatalytic degradation to improve the execution and effectiveness of photocatalysts. In this segment, the thorough process of photocatalytic degradation is explained, mainly focusing on the functions of different parts and the steps involved [32].

6.3.1 Absorption and Excitation

A photocatalyst absorbs the light energy whenever it is unveiled to light and further proceeds to uplift electrons from the valence band to the conduction band and then these excited electrons present in the conduction band react with oxygen molecules in order to form superoxide radicals, which are very reactive species and can degrade major pollutants [28]. This excited state can be described by the following Equation 6.1:

$$TiO_2 + h\vartheta \rightarrow e^- + h^+ \quad (6.1)$$

Here, $h\,\nu$ signifies the energy of the photon.

6.3.2 Movement of Charge Carriers

Photocatalysis largely hinges on the behavior of these charge carriers, electrons, and holes. Ideally, they should move to the surface of the photocatalyst without merging [33]. The merging of electrons and holes means a reduction in photocatalytic efficiency, as the energy is wasted as heat instead of being utilized in redox reactions. To improve the separation of charge carriers and reduce mergers, various methods are employed, like doping the semiconductor using either metals or non-metals, creating interfaces with other semiconductors (heterojunctions), or differing the properties of the photocatalyst's surface [34].

6.3.3 Reactive Oxygen Species (ROS) Formation

Once the charge carriers arrive at the photocatalyst's surface, they play a crucial role in the creation of reactive oxygen species [35]. The electrons from the photogenerated state reduce oxygen molecules that are adsorbed on the surface, forming superoxide anions, shown in Equation 6.2.

$$O_2 + e^- \rightarrow O_2^- \tag{6.2}$$

Simultaneously, holes can oxidize water molecules as shown in Equation 6.3, or hydroxide ions to generate hydroxyl radicals in Equation 6.4:

$$H_2O + h^+ \rightarrow HO^{\cdot} + H^2 \tag{6.3}$$

$$OH^- + h^+ \rightarrow HO^{\cdot} \tag{6.4}$$

These ROS are highly reactive and are essential for the breakdown of pollutants [36].

6.3.4 Pathways for Degradation

The produced ROS can attack pollutants through a variety of pathways, eventually resulting in their mineralization into water (H_2O), and carbon dioxide (CO_2), along with some other harmless substances such as organic compounds and bacteria as represented in Figure 6.4. The specific degradation pathway relies on the characteristics of the pollutant. For instance, ROS can attack the chromophore groups by breaking down organic dyes and decomposing the complex molecules into simpler, non-toxic substances [37].

Hydroxyl radicals ($HO^{\cdot}$) These radicals are among the most effective oxidizing agents [39]. They can act non-selectively on a wide array of organic pollutants by stripping away hydrogen atoms, adding to double bonds, and engaging in electron transfer reactions.

$$Pollutant + HO^{\cdot} \rightarrow Degraded\ products \tag{6.5}$$

Superoxide radicals ($O_2^{\cdot-}$) These radicals are not as reactive as hydroxyl radicals but can still play a role in degradation [40] as in Equation 6.6, often by forming hydroperoxyl radicals ($HO_2^{\cdot}$) or in conjunction with other ROS.

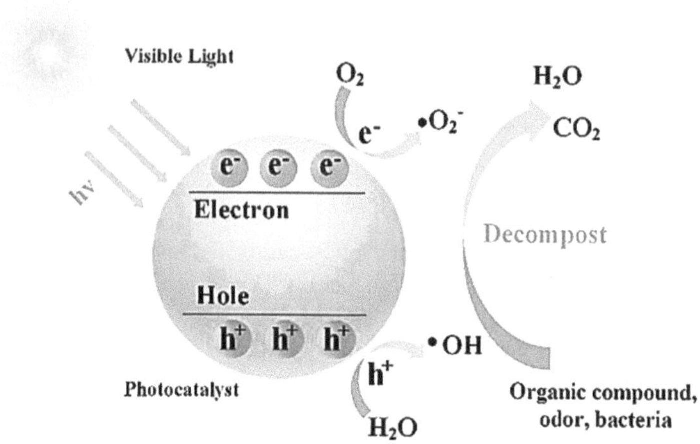

FIGURE 6.4 Pictorial representation of pathway of photocatalytic degradation [38] (Credit: Creative Commons Attribution 4.0 International License).

$$Pollutant + O_2^- \rightarrow Degraded\ products \tag{6.6}$$

6.3.5 Complete Mineralization

The ultimate goal of photocatalytic degradation is the complete mineralization of pollutants, which means breaking them down into inorganic end products like CO_2 and H_2O, as represented in Equation 6.7 [41]. This is often achieved through a series of intermediate steps, involving the continuous action of ROS on the degradation intermediates:

$$Intermediated + \frac{HO^{\cdot}}{O_2^-} \rightarrow CO_2 + H_2O \tag{6.7}$$

6.4 FACTORS AFFECTING PHOTOCATALYTIC ACTIVITY

Several factors intricately influence the efficacy of photocatalysis, a process reliant on semiconductor materials activated by light to catalyze chemical reactions.

6.4.1 Bandgap

In recent decades, wide-bandgap semiconductor photocatalysts have gained importance due to their role in environmentally friendly wastewater treatment processes. Among various oxides, ZnO and TiO_2 are the most promising materials, frequently studied for their high chemical stability, low cost, and non-toxic nature. However, their large bandgap restricts light absorption to the UV region, preventing the absorption of visible light. Additionally, the rapid recombination of photogenerated electron–hole pairs limits effective pollutant degradation as depicted in Figure 6.5 [42]. These limitations can be mitigated by broadening their light absorption range from UV to visible light through bandgap tuning, which can be achieved by introducing impurities or employing new synthesis techniques. Generally, semiconductor photocatalysts with bandgaps between 1.23 and 3 eV encompass the oxidation and reduction potential ranges of H_2O. The redox potentials for photocatalytic water splitting are $EH^+/H_2 = -4.44$ eV and $EO_2/H_2O = -5.67$ eV, with a difference of 1.23 eV, which is the minimum bandgap required for a material to be an effective photocatalyst [43].

6.4.2 Particle Size

A material's shape, size, and morphology significantly influence its photocatalytic properties. In bulk materials, the overlap of atomic orbitals produces bands with small energy gaps. In contrast, nanoscale metals have discontinuous orbitals that form discrete energy levels [45], which can be tuned by altering the nanoparticle diameter, as illustrated in Figure 6.6.

As particle size decreases, electrons become more confined, leading to higher energy levels and quantized atomic orbitals [46]. This quantum confinement effect in nanomaterials results in tunable electrical and optical properties. Consequently, The bandgaps of bulk and nanostructured materials vary, and changes in particle size affect the bandgap energy, light absorption capacity, and the average free path of photogenerated charge carriers [47]. When the size of the particle is smaller than

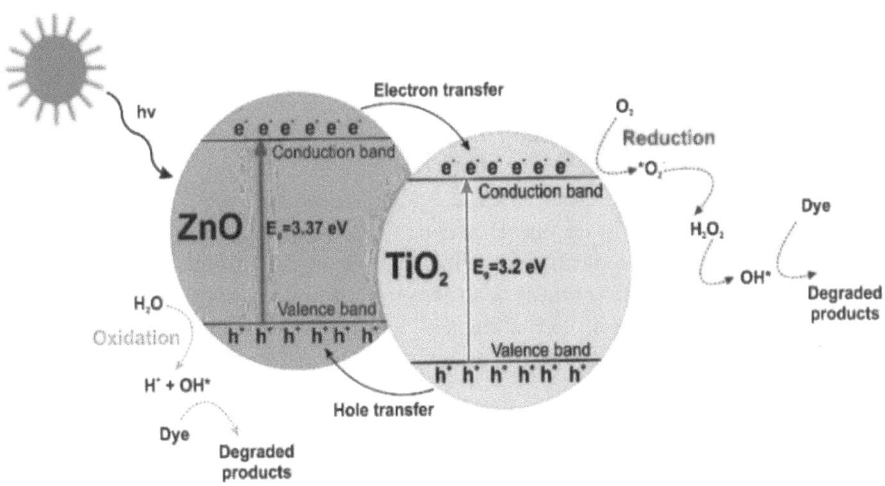

FIGURE 6.5 The p–n heterostructures shown in ZnO photocatalyst and TiO_2 photocatalyst [44] (Credit: Creative Commons Attribution 4.0 International License).

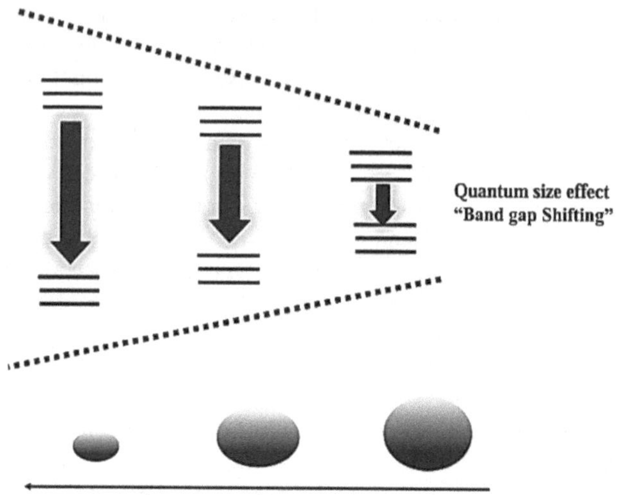

FIGURE 6.6 The effect of particle size on the bandgap (Credit: Created by authors).

the space charge layer's thickness, photogenerated charge carriers migrate from the bulk phase to the material's surface via diffusion, participating in surface redox reactions. Reducing the charge migration distance improves the separation efficiency and charge separation rate. Therefore, small particles typically have a high specific surface area and better adsorption properties, facilitating the interaction between the catalyst and reactants [48]. This results in abundant active sites and a large light absorption area, enhancing photocatalytic performance.

6.4.3 Doping

Doping metal-oxide semiconductors with different elements is a common strategy to enhance their performance. This process can influence various parameters, including morphology, particle size, bandgap, binding energy, and lattice defects [49]. For instance, doping SnO_2 with Ag reduces its average particle size from 52.3 nm to approximately 45.5 nm. However, Co doping in an Ag/SnO_2 system did not significantly decrease the particle size. Increasing the amount of dopants can impact particle size and potentially lead to a more uniform grain size distribution. The research [50] found that increasing Co doping narrowed the particle size distribution, though it also led to the formation of agglomeration clusters in pure SnO_2 due to the dopants [51].

6.5 TYPES OF PHOTOCATALYTIC MATERIALS

Photocatalytic materials are usually made up of semiconductors and metal oxides. The types of photocatalytic materials are:

6.5.1 Titanium Oxide (TiO_2)

Titanium oxide, particularly in its nanostructured forms like titanium dioxide (TiO_2), is renowned for its photocatalytic properties, making it a valuable material in various applications. Upon exposure to ultraviolet light, TiO_2 undergoes photoexcitation, generating electron–hole pairs that drive redox reactions on its surface [52]. This capability enables it to break down organic pollutants, purify water, and sanitize surfaces through photocatalysis, a process that harnesses light energy to drive chemical reactions. This property has immense potential for environmental remediation, wastewater treatment, and self-cleaning surfaces in the healthcare, construction, and automotive industries. Furthermore, TiO_2 stability, abundance, and low cost enhance its attractiveness for large-scale implementation [53]. However, challenges such as its limited absorption in visible light and the recombination of electron–hole pairs hinder its efficiency, prompting ongoing research to enhance its performance and expand its applicability in sustainable technology. However, to increase the efficiency of (TiO_2) as photocatalysts, modifications can be done to obtain the desired product, as described in Figure 6.7.

6.5.2 Zinc Oxide (ZnO)

Zinc oxide (ZnO) photocatalysts have attracted considerable interest owing to their unique properties and versatile applications in environmental restoration and solar energy conversion. As a wide-bandgap semiconductor, ZnO demonstrates excellent photocatalytic activity under ultraviolet (UV) light irradiation, enabling the degradation of organic pollutants, water disinfection, and the production of renewable fuels

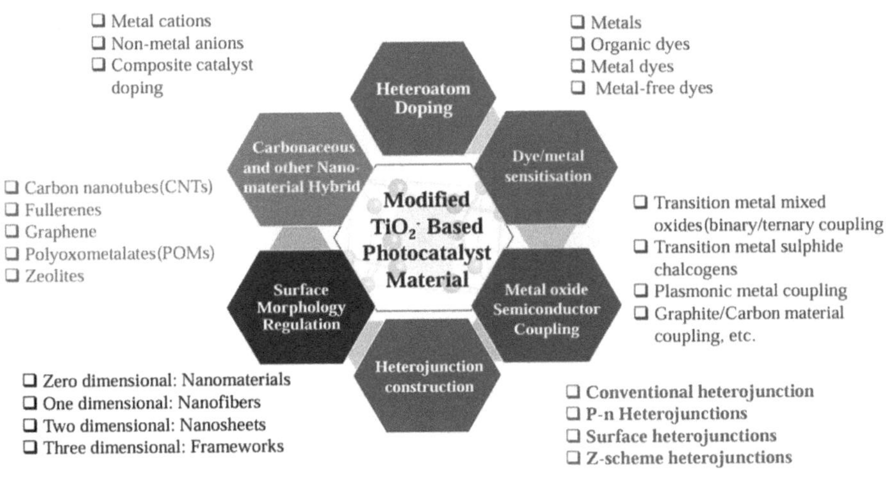

FIGURE 6.7 Modified TiO_2-based photocatalyst material [52] (Credit: Creative Commons Attribution 4.0 International License).

through water splitting and ZnO reduction. Additionally, its low toxicity, abundance, and low cost enhance its appeal for large-scale applications [54]. However, challenges such as rapid electron-hole recombination and photo corrosion under prolonged UV exposure limit the practical use of ZnO photocatalysts. Nowadays, more focus is on strategies to improve their photocatalytic efficiency and stability, including surface modification, heterojunction formation, and doping with other elements, aiming to unlock their full potential for sustainable energy and environmental technologies [54].

6.5.3 Cuprous Oxide (Cu_2O)

Cuprous oxide (Cu_2O) has attracted considerable attention as a photocatalyst because of its distinct properties and potential uses in solar energy conversion and environmental remediation. As a p-type semiconductor possesses a narrow bandgap, Cu_2O strongly absorbs light in the visible region of the electromagnetic spectrum, allowing it to harness a significant portion of solar radiation for photocatalytic reactions compared to wide-bandgap materials like titanium dioxide, as shown in Figure 6.8.

Cu_2O photocatalysts have shown effectiveness in various applications, such as pollutant degradation, splitting water to produce hydrogen, and reducing CO_2 to generate fuels or valuable chemicals. The photogenerated charge carriers in Cu_2O, comprising holes (h^+) and electrons (e^-), take part in redox reactions at the catalyst surface, driving the desired transformations [55].

However, challenges such as limited charge carrier mobility and stability under prolonged light exposure hinder the practical application of Cu_2O photocatalysts. More efforts are focused on overcoming these limitations through strategies like

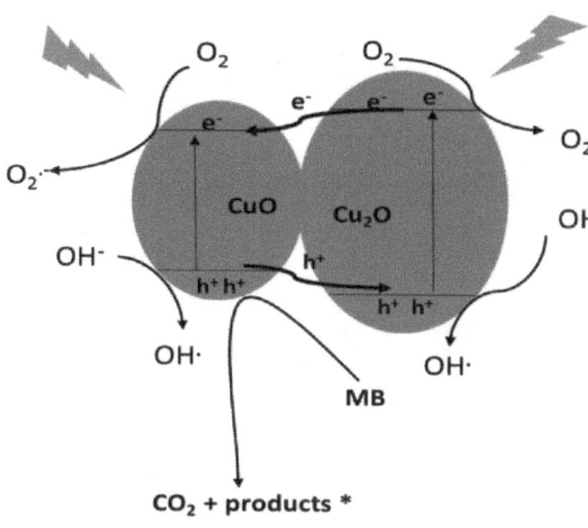

FIGURE 6.8 Band structure-related photocatalysis performed by Cu_2O Photocatalysts [56] (Credit: Creative Commons Attribution 4.0 International License).

morphology control, heterojunction formation, and surface modification to inflate the photocatalytic performance along with their stability.

6.5.4 Graphene-Based Photocatalysts

Graphene is a two-dimensional allotrope of carbon having some exceptional mechanical, electronic, and optical properties, and has emerged as a promising material for photocatalysis. Graphene-based photocatalysts combine the remarkable properties of graphene with the photocatalytic activity of other semiconducting materials, offering enhanced performance in light-driven reactions. Excellent electron mobility, high surface area, and strong light absorption in graphene make it an ideal framework for supporting photocatalytic nanoparticles and promoting charge separation [57]. By combining graphene with different semiconductors like TiO_2, ZnO, or some other metal oxides, composite photocatalysts have been developed with improved effectiveness for degrading pollutants, producing hydrogen, and the conversion of carbon dioxide into significant chemicals [58]. Graphene's conductivity and ability to entrap and transfer photogenerated electrons effectively reduce recombination losses, resulting in the enhancement of the overall photocatalytic activity. Moreover, the chemical stability and mechanical strength of the graphene contribute to the durability and recyclability of graphene-based photocatalysts [57].

6.5.5 Composite Photocatalysts

Composite photocatalysts are made up of two or more materials with supplementary properties and offer enhanced performance compared to the individual components for a diverse range of photocatalytic applications [59]. Researchers out there, aim to improve the advantages of different constituents of composite photocatalysts, like efficient charge separation, improved light absorption, and enhanced catalytic activity, while diminishing their respective limitations. It is visualized in Figure 6.9 [60].

A standard approach necessitates integrating a semiconductor photocatalyst with a co-catalyst or support material. For example, coupling a wide-bandgap semiconductor like titanium dioxide (TiO_2) with a co-catalyst like platinum (Pt) or a support material like graphene can enhance the photocatalytic activity by facilitating the transfer of charge and decreasing the recombination of electron–hole pair [58]. Furthermore, the formation of heterojunction between two semiconductors with integrative band structures results in the creation of an internal electric field that promotes charge separation and migration, leading to improved photocatalytic efficiency. Examples include coupling titanium dioxide with zinc oxide (ZnO) or bismuth oxyhalides with other metal oxides [62].

Photocatalysts are substances that speed up chemical reactions under light and have a broad range of implementations across various fields owing to their potential to harness solar energy [44]. Photocatalysts such as TiO_2, ZnO, Cu_2O, graphene-based materials, and composite photocatalysts have broad applications due to their distinctive properties that are depicted in Table 6.1.

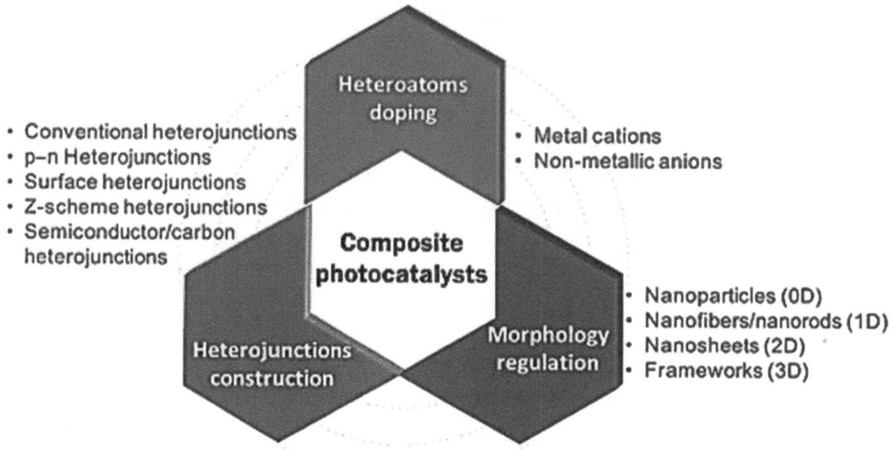

FIGURE 6.9 The image demonstrates design and synthesis strategies for composite photocatalysts [61] (Credit: Creative Commons Attribution 4.0 International License).

TABLE 6.1
Different Types of Photocatalysts with Their Applications in Various Fields

Type of Photocatalysts	Applications	Ref.
a. Titanium oxide	• In solar cell material • In the restoration of polluted air and contaminated water environments	[52, 63]
b. Zinc oxide	• Antibacterial against both gram-positive and gram-negative bacteria. • As light-emitting coating	[64,65]
c. Copper oxide	• Photoelectrochemical water splitting is a cutting-edge method for producing renewable hydrogen • In the reduction of CO_2 to methane and other one-carbon compounds	[66, 67]
d. Graphene-based photocatalysts	• Water splitting for hydrogen production • Pollutant degradation in environmental remediation	[62, 68]
e. Composite Photocatalysts	• In selective photoreduction of CO_2 to fuels using visible light	[69]

6.6 CHARACTERIZATION OF PHOTOCATALYTIC MATERIALS

Photocatalytic materials can facilitate and quicken photoreaction while absorbing light, and they are generally used to purify the environment by producing hydrogen and synthesizing organic matter [70]. Characterization of the photocatalytic materials is a necessity as it helps us to understand the mechanism, properties, and certain techniques. There are various methods used to characterize photolytic materials which are elaborated as follows.

6.6.1 Structural Characterization

Structural characterization is crucial when it comes to photocatalytic materials because it helps to understand the physical properties that affect their photocatalytic performance. Optimization of various techniques is done to figure out the structural aspects, such as morphology, surface area, crystallinity, etc. Some of the techniques used for structural characterization of photocatalytic materials are mentioned below [71]. Scanning electron microscopy (SEM) technique is used for the examination of the particle size and surface composition. This method is used to reveal the shape of the compound. In this technique, an intense electron beam is used to scan the surface of a material, resulting in its high-resolution images [72]. Transmission electron microscopy (TEM) This technique is generally used to prepare elaborate images of the internal structure of a material. This method gives out high-resolution images by transmitting electrons through the precisely prepared thin specimen. The X-ray diffraction (XRD) approach is employed in determining the size of the particle. The crystalline structure of a particle can also be examined using this technique as it analyzes the diffraction pattern of the particle.

6.6.2 Optical Properties

Understanding the ability to conduct photocatalytic reactions and absorb light characterizing optical properties is important. Here are some methods employed to characterize the optical properties of photocatalytic materials. UV-visible spectroscopy technique is used to measure the absorption of light in UV-visible regions to determine the optical band gap and assess light absorption efficiency by analyzing the wavelengths of light absorbed by the material [73]. Also, the Raman spectroscopy technique is used to get information about the electronic properties and the lattice structure. It measures the scattering of monochromatic light and identifies crystal phases and studies material changes [74].

6.6.3 Surface Area and Porosity

Photocatalytic reactions are highly affected by the surface area and porosity of photocatalysts, the greater the surface area of the catalyst, the more will the presence of sites for pollutant absorption leading to the generation of highly reactive species, whereas porosity is responsible for faster reactions as the pores give narrowed or restricted space for the absorption of light which further strengthen or increase the reaction rate [75].

6.6.4 Electron Microscopy Techniques

These techniques serve as a powerful tool to understand the morphology, composition, and structure of photocatalytic materials at high resolution. The main techniques include SEM and TEM along with their complementary techniques [76]. These techniques examine texture, particle size, crystallographic distribution, lattice structure information, etc. It is essential to analyze certain image distributions and to study certain sensitive materials and biological specimens in their native state.

6.6.5 Photocatalytic Activity Testing

Photocatalytic activity testing is important to evaluate the efficiency and performance of photocatalytic materials in different applied fields, such as environmental purification, organic synthesis, and hydrogen production, etc. Some of the methods used for photocatalytic activity testing are:

a) Hydrogen production: This involves evaluating the photocatalyst's ability to split water into hydrogen and oxygen below light irradiation.
b) CO_2 reduction: It involves the ability of the photocatalysts to reduce CO_2 to efficient products like CO, methane, etc. The products are quantified using mass spectroscopy [77].

6.7 APPLICATIONS OF PHOTOCATALYSTS

Photocatalysts find applications across various fields, leveraging their ability to harness light energy to drive chemical reactions. Some key applications include:

6.7.1 Water and Air Purification

In water purification, photocatalysts can be proved quite significant, as they can utilize light energy to degrade contaminants [78]. By considering the case of titanium dioxide and zinc oxide, which can be activated by light, produce ROS, which effectively disinfects water by eliminating hazardous microbiomes. Moreover, these compounds catalyze oxidation reactions by breaking down organic pollutants making them harmless [79]. These catalysts also help in the decomposition and removal of heavy metals (e.g., lead, mercury, arsenic, and cerium, etc.) from water. These mechanisms enabled photocatalysts to offer an efficient and sustainable way to improve the water quality against various applications, such as drinking water treatment and industrial wastewater remediation, improving the health of the environment and human well-being [80]. Furthermore, photocatalytic materials like titanium dioxide and zinc oxide can also be used to degrade aero pollutants by generating ROS on exposure to light. These photocatalytic materials decompose and oxidize harmful gases, airborne pathogens, and volatile organic compounds (VOCs) [81].

6.7.2 Degradation of Organic Pollutants

The degradation of organic pollutants is largely driven by the use of photocatalysts, that is to say, photocatalysts use light energy to start oxidation processes that decompose deleterious materials. Reactive oxygen species are produced when materials like titanium dioxide and zinc oxide are illuminated and oxidize organic pollutants into water and carbon dioxide [79]. It is an effective photocatalytic process that removes pollutants from various environmental contaminants, including industrial wastewater, agricultural runoff and urban emissions [82]. The use of photocatalysts allows human health as well as our ecosystems from the impact of organic pollutants they offer us the remediation of wastes in a sustainable and effective way [83,84].

6.7.3 Antimicrobial and Antiviral Applications

As illustrated in Figure 6.10, owing to the promising applications both in antimicrobial and antiviral fields, photocatalysis allows generating ROS on the photocatalyst surfaces through harnessing light energy. When exposed to light and thus to ROS, such as hydroxyl radicals, materials such as titanium dioxide and zinc oxide have demonstrated potent antimicrobial and antiviral properties. With the capability to effectively destroy bacterial cells, viruses, and microorganisms generally, these ROS reduce viability and prevent proliferation [85]. Methods of photocatalysis offer a cost-effective and environmentally friendly approach for disinfection with applications to the prevention of infection and control of microbial contamination in healthcare, food industry, and public spaces [86].

6.7.4 Energy-Storing Devices

Photocatalysts are necessary for energy storage devices, especially for harvesting solar energy. These allow for processes such as photocatalytic water splitting: using sunlight to break water into its components, hydrogen and oxygen and provide a clean fuel source. In addition, they help improve the efficiency of fuel cells by converting

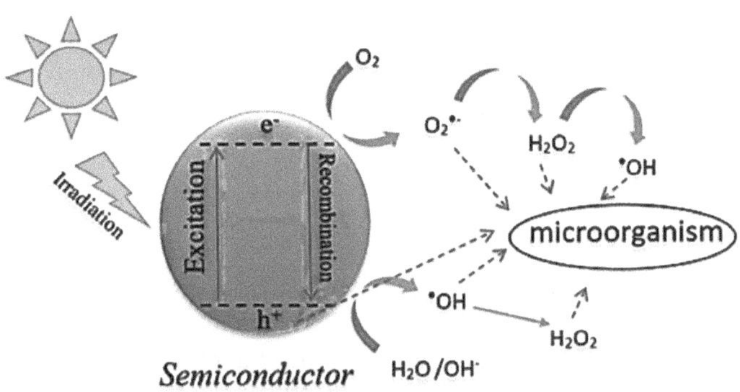

FIGURE 6.10 The schematic diagram for the photocatalytic mechanism of a semiconductor [87] (Credit: Creative Commons Attribution 4.0 International License).

sunlight to electrons to power electrochemical reactions that yield electrical power [88]. On top of that photocatalysts aid in rechargeable batteries to reduce charging times and enhance energy densities. They also help produce solar fuels like hydrogen and methanol but in a more sustainable way than fossil fuels. Superimposed upon this, the addition of photocatalytic materials to supercapacitors improves their energy storage capacity and thus provides high-performance solutions for electric vehicles and grid storage. Overall, photocatalysts have a very important role in the development of renewable energy technologies and in forming a more sustainable energy future [80, 89].

6.7.5 Biomedical Sciences

Metal and metal-oxide nanoparticles have seen recent developments in the field of nanotechnology and have been used worldwide in research institutes, bioenergy-based industries, biomedical sciences, and the fields of study on the environment. By utilizing light energy, the bacteria dissolve it to generate reactive oxygen species that can kill bacteria, viruses, and other pathogens. In addition, this technology also presents a promising route to reduce healthcare infections, fostering a safer healthcare and healthcare environment [82].

6.8 ENVIRONMENTAL AND ECONOMIC IMPACT

Photocatalysts generate certain chemical reactions when exposed to light and are known for their potential environmental and economic impacts. To assess these impacts, we need to see the life cycle of photocatalytic materials, from extraction of materials to recycling:

6.8.1 Life Cycle of Photocatalytic Materials

Extraction of Raw Materials

Raw materials such as titanium oxide extracted from rutile ores lead to the depletion of natural resources, as it is known that mining activities degrade land and destroy habitats. These are energy-consuming processes, so they require a good amount of energy mostly procured from fossil fuels, leading to greenhouse gas emissions. Further processes release a lot of pollutants in the air contributing to air pollution [90].

Processing of Raw Materials

Changing raw materials into applicable photocatalysts generally involves processes with high temperatures that are very energy-consuming, these processes contain a lot of chemicals that are harmful and need careful handling and need to be disposed of in a way that does not degrade the environment [45].

Manufacturing

Production of photocatalytic materials generally can be energy intensive; the manufacturing process produces waste materials that must be properly managed, and moving these raw materials and end products involves transportation leading to the emission of greenhouse gases.

6.8.2 Cost–Benefit Analysis

Throughout the life cycle cost–benefit analysis (CBA) involves comparing costs associated with manufacturing, implementation, and maintenance. Initial development costs involve research and development expenses, the cost of procuring raw materials, and the production of end products. Implementation cost involves expenses for the installation of these products and costs associated with infrastructure modifications. Maintenance of these products is necessary to ensure efficient performance. There are expenses related to recycling these materials and mitigating any kind of harmful environmental impacts. There are certain economic benefits like fewer energy costs from enhanced efficiency in various applications and reduced need for artificial lighting, monetary values are assigned by quantifying the cost of materials, manufacturing, and implementation [88].

6.8.3 Environmental Regulations and Standards

These standards are important in ensuring the proper use of photocatalytic materials while reducing negative environmental impacts. There are certain sets of international standards and regulations, such as European Union Regulations, United States regulations, and ISO Standards that ensure proper testing and record-keeping. Regulations ensure that the manufacturing, use, and disposal of these materials do not harm the environment to reduce health risks, proper recycling makes sure to minimize or reduce the contamination released from the manufacturing of these products. We need to consider future directions by harmonizing standards and regulations to ensure efficient performance and reduce negative environmental impacts [91].

6.8.4 Sustainable Development Goals

Photocatalytic compounds play a crucial role in advancing environmental protection and sustainability by setting several key objectives [1]. The objective is to promote limited production and consumption-based approaches that minimize resource use and waste generation. In addition, the climate mitigation approach used when developing these materials helps to control the emission of greenhouse gases hence addressing the problem of global warming. With regard to marine pollution, photocatalytic technologies aid in reducing pollution levels and thereby protect marine life [92]. They also help in the protection of biodiversity on land and in water by dynamically cleaning air and water sources. On another note, photocatalytic materials are also used to increase public awareness and teach the communities about sustainable practices, while drawing others to get involved in cutting environmental impacts. Thus, these efforts complement the importance of photocatalytic materials in promoting a healthier planet, and a future healthier planet, for current and future generations.

6.9 FUTURE PERSPECTIVE

With the pursued advancement of material science and the desire for sustainable solutions to global problems, the future of photocatalytic materials is full of promise in all fields. Photocatalytic materials are now emerging key players in grappling with critical issues such as air pollution, water scarcity and the transition to clean energy societies [93]. It is one exciting direction of the research, as one can improve the efficiency and expand the application scope of existing photocatalysts, e.g., titanium (TiO_2). These materials have already improved under scientists' active work to improve the light absorption capabilities, creating these materials as active material that responds beyond ultraviolet (UV) light into the visible and near-infrared spectra. This development could increase their effectiveness enormously for example, in tasks where light is broad spectrum such as water purification or air decontamination.

Massively there is immense potential for photocatalytic materials to be integrated into common products and infrastructure. Further, think of buildings that are coated with photocatalytic surfaces that break down air pollutants such as nitrogen oxides (NO_x) and VOCs, and thus thereby aid in the creation of cleaner urban environments [94]. Just as could self-cleaning surfaces that are treated with photocatalysts be less dependent on chemical cleaners, making maintenance a less environmentally painful process. Photocatalytic materials play a fundamental role in the design of next-generation solar technologies in the realm of photovoltaics. These materials are used in innovations in photoelectrochemical cells and in artificial photosynthesis to convert solar energy into storable fuels, such as hydrogen. Not only do they promise to decrease the need for fossil fuels but also hold the potential to get us closer to carbon-neutral energy systems.

Interdisciplinary collaboration is necessary to realize these potentials. The research and development to scale up production methods, improve material durability, and ensure the safety requirements of these devices must be done in close collaboration between researchers, engineers and policymakers. If today's challenges can be overcome, photocatalytic materials can become mainstream solutions from a positive environmental and economic standpoint. Finally, photocatalytic materials offer a transformative technology that promises to address many of the most serious environmental and energy issues of our time. We need to keep innovating in this sector and continue to invest in order to create a future where clean air, clean water and sustainable energy are available to all, and create a healthier and more resilient planet for our children and grandchildren to come.

6.10 CONCLUSION

The potential of photocatalytic materials to transform a field from the nano to minute perspective is explored, across applications spanning traditional and novel boundaries and enabling innovation across a spectrum of applications. These materials represent a sustainable approach to addressing global challenges from environmental remediation to energy production [95]. Practical solutions for purifying air and water polluted with chemicals have emerged through the use of photocatalysts, because they bend light energy to power chemical reactions, thereby improving our quality of life and protecting our ecosystems. They make contributions in the field of energy by leading the development of clean and efficient technologies including the conversion of solar energy into fuels such as hydrogen through water splitting and to the pursuit of renewable energy sources [96]. Additionally, photocatalytic materials present prospects for improving surfaces by providing self-cleaning, low maintenance, and interrelated hygienic properties within built environments. They have wider application in healthcare settings where they have a leading role in the sterilization of medical equipment and surfaces to reduce infection risk and improve patient safety [97]. There is a similar versatility for the photocatalyst in the food industry like in shelf-life extension, food preservations, and disinfection to keep it safe.

With these technologies in constant evolution, their integration into other industrial processes and products is becoming influenced by advancements in engineering and material science. The future of innovations is thus inspired by the ongoing refinement of photocatalytic materials which promises not only to meet current environmental and energy challenges but also to advance systematically such technologies. The potential for photocatalytic materials to promote a sustainable future continues to grow large and promising, and in doing so serves as an example of catalysts for the progress of the 21st century and beyond.

REFERENCES

1. Fresno F, Portela R, Suárez S, et al. Photocatalytic materials: Recent achievements and near future trends. *J Mater Chem A*. 2014;2:2863–2884.
2. Shaikh SF. Nanomaterials applied in coatings: synthesis, structures, properties, and applications. *Coatings*. 2022;12:10–12.
3. Tryba B, Miądlicki P, Rychtowski P, et al. The superiority of TiO2 supported on nickel foam over Ni-Doped TiO_2 in the photothermal decomposition of acetaldehyde. *Materials (Basel)*. 2023;16.
4. Moridon SNF, Arifin K, Mohamed MA, et al. TiO2 nanotubes decorated with Mo_2C for enhanced photoelectrochemical water-splitting properties. *Materials (Basel)*. 2023;16:1–12.
5. Chahal S, Phor L, Kumar A, et al. Enhanced photocatalytic degradation of organic dye by CeO_2/CNT/GO hybrid nanocomposites under UV light for wastewater treatment. *Environ Sci Pollut Res [Internet]*. 2023;30:124964–124975. https://www.scopus.com/inward/record.uri?eid=2-s2.0-85149222487&doi=10.1007%2Fs11356-023-26184-1&partnerID=40&md5=4a23a13fb347d42c5a1dd36b80d0f269.

6. Kumari H, Chahal S, Kumar P, et al. Photocatalytic degradation of RB dye via cerium substituted SnO_2 photocatalysts. *Mater Sci Eng B*. 2023;296. https://www.scopus.com/inward/record.uri?eid=2-s2.0-85163434421&doi=10.1016%2Fj.mseb.2023.116654&partnerID=40&md5=0bed09da935913ba980c160708246d27.
7. Wang G, Yang Y, Han D, et al. Oxygen defective metal oxides for energy conversion and storage. *Nano Today*. 2017;13:23–39.
8. Dai Z, Yang A, Bao X, et al. 26 Facile non-enzymatic electrochemical sensing. *Sensors*. 2019;19:1–13.
9. Singh J, Singh S. A review on machine learning aspect in physics and mechanics of glasses. *Mater Sci Eng B*. 2022;284:115858. https://doi.org/10.1016/j.mseb.2022.115858.
10. Singh J, Singh JP, Kumar S, et al. Short review on hydroxyapatite powder coating for SS 316L. *J Electrochem Sci Eng*. 2023;13:25–39. https://www.scopus.com/inward/record.uri?eid=2-s2.0-85148748776&doi=10.5599%2Fjese.1611&partnerID=40&md5=9dbb686d916cc26266b626a0873c5c0a.
11. Sreenivasan R, Goel A, Bourell DL. Sustainability issues in laser-based additive manufacturing. *Phys Procedia*. 2010;5:81–90. http://dx.doi.org/10.1016/j.phpro.2010.08.124.
12. Kulkarni SD, Manjunatha, Chandrasekhar U, et al. Design and optimization of polyvinyl-nitride rubber for tensile strength analysis. *Int J Interact Des Manuf*. 2024;18:5893–5908.
13. Luque Morales M, Luque Morales PA, Chinchillas Chinchillas M de J, et al. Theoretical and experimental study of the photocatalytic properties of ZnO semiconductor nanoparticles synthesized by prosopis laevigata. *Materials (Basel)*. 2023;16.
14. Malathi A, Madhavan J, Ashokkumar M, et al. A review on $BiVO_4$ photocatalyst: Activity enhancement methods for solar photocatalytic applications. *Appl Catal A Gen*. 2018;555:47–74.
15. Mirzaeifard Z, Shariatinia Z, Jourshabani M, et al. ZnO photocatalyst revisited: effective photocatalytic degradation of emerging contaminants using S-Doped ZnO nanoparticles under visible light radiation. *Ind Eng Chem Res*. 2020;59:15894–15911.
16. Wang D, Jiao S, Zhang S, et al. Growth mechanism and optical properties of phosphorus-doped ZnO nanorods synthesized by hydrothermal method. *Phys Status Solidi Appl Mater Sci*. 2017;214:1–7.
17. Mangala Nagasundari S, Muthu K, Kaviyarasu K, et al. Current trends of silver doped zinc oxide nanowires photocatalytic degradation for energy and environmental application. *Surfaces and Interfaces*. 2021;23:100931.
18. Ahmad S, Hashmi AW, Singh J, et al. Innovations in additive manufacturing of shape memory alloys: alloys, microstructures, treatments, applications. *J Mater Res Technol*. 2024;32:4136–4197.
19. Singh J, Hashmi AW, Ahmad S, et al. Critical review on biodegradable and biocompatibility magnesium alloys: Progress and prospects in bio-implant applications. *Inorg Chem Commun*. 2024;169:113111.
20. Yu W, Zhang J, Peng T. New insight into the enhanced photocatalytic activity of N-, C- and S-doped ZnO photocatalysts. *Appl Catal B Environ*. 2016;181:220–227.
21. Hitam CNC, Jalil AA, Triwahyono S, et al. Effect of carbon-interaction on structure-photoactivity of Cu doped amorphous TiO_2 catalysts for visible-light-oriented oxidative desulphurization of dibenzothiophene. *Fuel*. 2018;216:407–417.
22. Bhatti DT, Parikh SP. Recent progress in doped TiO_2Photocatalysis and hybrid advanced oxidation processes for organic pollutant removal from wastewater. *Curr World Environ*. 2022;17:146–160.
23. Mahmoud MA, Alsehli BR, Alotaibi MT, et al. A comprehensive review on the application of semiconducting materials in the degradation of effluents and water splitting. *Environ Sci Pollut Res Int*. 2024;31:3466–3494.

24. Shah AA, Bhatti MA, Tahira A, et al. Facile synthesis of copper doped ZnO nanorods for the efficient photo degradation of methylene blue and methyl orange. *Ceram Int.* 2020;46:9997–10005.
25. Rahman AFA, Jalil AA, Triwahyono S, et al. Strategies for introducing titania onto mesostructured silica nanoparticles targeting enhanced photocatalytic activity of visible-light-responsive Ti-MSN catalysts. *J Clean Prod.* 2017;143:948–959.
26. Xie W, Li R, Xu Q. Enhanced photocatalytic activity of Se-doped TiO_2 under visible light irradiation. *Sci Rep.* 2018;8:1–10.
27. Frank SN, Bard AJ. Heterogeneous photocatalytic oxidation of cyanide and sulfite in aqueous solutions at semiconductor powders. *J Phys Chem.* 1977;81:1484–1488.
28. Zhang S, Lu X. Treatment of wastewater containing reactive brilliant blue KN-R using TiO_2/BC composite as heterogeneous photocatalyst and adsorbent. *Chemosphere.* 2018;206:777–783.
29. Liu D, Wang J, Zhou J, et al. Fabricating I doped TiO_2 photoelectrode for the degradation of diclofenac: Performance and mechanism study. *Chem Eng J.* 2019;369:968–978.
30. Cao Y, Gao Q, Li Q, et al. Synthesis of 3D porous MoS_2/g-C_3N_4 heterojunction as a high efficiency photocatalyst for boosting H_2 evolution activity. *RSC Adv.* 2017;7:40727–40733.
31. Shan AY, Ghazi TIM, Rashid SA. Immobilisation of titanium dioxide onto supporting materials in heterogeneous photocatalysis: A review. *Appl Catal A Gen.* 2010;389:1–8.
32. Zhuang HL, Hennig RG. Single-layer group-III monochalcogenide photocatalysts for water splitting. *Chem Mater.* 2013;25:3232–3238.
33. Bora T, Sathe P, Laxman K, et al. Defect engineered visible light active ZnO nanorods for photocatalytic treatment of water. *Catal Today.* 2017;284:11–18.
34. Han T, Fan T, Chow SK, et al. Biogenic N-P-codoped TiO_2: Synthesis, characterization and photocatalytic properties. *Bioresour Technol.* 2010;101:6829–6835.
35. Singh NB. Green synthesis of nanomaterials. *Handb Microb Nanotechnol.* 2022;225–254.
36. Garba ZN, Zhou W, Zhang M, et al. A review on the preparation, characterization and potential application of perovskites as adsorbents for wastewater treatment. *Chemosphere.* 2020;244:125474.
37. Briffa J, Sinagra E, Blundell R. Heavy metal pollution in the environment and their toxicological effects on humans. *Heliyon.* 2020;6:e04691.
38. Wang Q, Zheng K, Yu H, et al. Laboratory experiment on the nano-TiO_2photocatalytic degradation effect of road surface oil pollution. *Nanotechnol Rev.* 2020;9:922–933.
39. Kumar A, Sanger A, Kumar A, et al. Fast response ammonia sensors based on TiO_2 and NiO nanostructured bilayer thin films. *RSC Adv.* 2016;6:77636–77643.
40. hahrezaei M, Habibzadeh S, Babaluo AA, Hosseinkhani H, Haghighi M, Hasanzadeh A, Tahmasebpour R. Study of synthesis parameters and photocatalytic activity of TiO_2 nanostructures. *J. Experiment. Nanosci.* 2017;12(1):45-61.
41. Akter S, Lee SY, Siddiqi MZ, et al. Ecofriendly synthesis of silver nanoparticles by terrabacter humi sp. nov. and their antibacterial application against antibiotic-resistant pathogens. *Int J Mol Sci.* 2020;21:1–19.
42. Hu C, Lu T, Chen F, et al. A brief review of graphene–metal oxide composites synthesis and applications in photocatalysis. *J Chinese Adv Mater Soc.* 2013;1:21–39.
43. Yoshida M, Yamakata A, Takanabe K, et al. ATR-SEIRAS investigation of the fermi level of Pt cocatalyst on a GaN photocatalyst for hydrogen evolution under irradiation. *J Am Chem Soc.* 2009;131:13218–13219.
44. Siwińska-Stefańska K, Kubiak A, Piasecki A, et al. TiO_2-ZnO binary oxide systems: Comprehensive characterization and tests of photocatalytic activity. *Materials (Basel).* 2018;11:1–19.

45. Pawade VB, Bhanvase BA. An introduction to photocatalysts and their applications. *Photocatal Energy Environ Sustain*. 2024;1-1-1–24.
46. Mittal M, Sharma M, Pandey OP. Fast and quick degradation properties of doped and capped ZnO nanoparticles under UV-Visible light radiations. *Sol Energy*. 2016;125:51–64.
47. Yakout SM. Inclusion of cobalt reinforced Ag doped SnO2 properties: electrical, dielectric constant, magnetic and photocatalytic insights. *J Mater Sci Mater Electron*. 2019;30:17053–17065.
48. Varghese Alex K, Tamil Pavai P, Rugmini R, et al. Green synthesized Ag nanoparticles for bio-sensing and photocatalytic applications. *ACS Omega*. 2020;5:13123–13129.
49. Yang B, Liu T, Guo H, et al. High-performance meta-devices based on multilayer meta-atoms: Interplay between the number of layers and phase coverage. *Sci Bull*. 2019;64:823–835.
50. Entradas T, Cabrita JF, Dalui S, et al. Synthesis of sub-5 nm Co-doped SnO_2 nanoparticles and their structural, microstructural, optical and photocatalytic properties. *Mater Chem Phys*. 2014;147:563–571.
51. Cho S, Jang JW, Kim J, et al. Three-dimensional type II ZnO/ZnSe heterostructures and their visible light photocatalytic activities. *Langmuir*. 2011;27:10243–10250.
52. Bethel C, Altin I, Bacaksiz E, et al. Titanium dioxide (TiO_2) -based photocatalyst materials activity enhancement for contaminants of emerging concern (CECs) degradation : In the light of modification strategies. *Chem Eng J Adv*. 2022;10:100262.
53. Ajmal A, Majeed I, Malik N. RSC Advances Principles and mechanisms of photocatalytic dye degradation on TiO_2 based photocatalysts: A comparative overview. *RSC Adv*. 2014;70:37003–37026.
54. Nguyen TT, Nam SN, Kim J, et al. Photocatalytic degradation of dissolved organic matter under ZnO-catalysed artificial sunlight irradiation system. *Sci Rep*. 2020;10(1):13090.
55. Masudy-Panah S, Katal R, Khiavi ND, et al. A high-performance cupric oxide photocatalyst with palladium light trapping nanostructures and a hole transporting layer for photoelectrochemical hydrogen evolution. *J Mater Chem A*. 2019;7:22332–22345.
56. Zhao W, Liu C. Mesoporous Cu-Cu_2O@TiO_2 heterojunction photocatalysts derived from metal-organic frameworks. *RSC Adv*. 2020;10:14550–14555.
57. Ahmed MA, Mohamed AA. Recent progress in semiconductor/graphene photocatalysts: Synthesis, photocatalytic applications, and challenges. *RSC Adv*. 2022;13:421–439.
58. Tu W, Zhou Y, Liu Q, et al. Robust hollow spheres consisting of alternating titania nanosheets and graphene nanosheets with high photocatalytic activity for CO_2 conversion into renewable fuels. *Adv Funct Mater*. 2012;22:1215–1221.
59. Wang J, Zhao J, Sun lu, et al. A review on the application of photocatalytic materials on textiles. *Text Res J*. 2015;85:1104–1118.
60. Guerra FD, Attia MF, Whitehead DC, et al. Nanotechnology for environmental remediation : Materials and applications. *Molecules* 2018;23(7):1760.
61. Ge J, Zhang Y, Heo YJ, et al. Advanced design and synthesis of composite photocatalysts for the remediation of wastewater: A review. *Catalysts*. 2019;9(2):122.
62. Zhang N, Zhang Y, Xu Y. Nanoscale Recent progress on graphene-based photocatalysts : Current status and future perspectives. *Nanoscale* 2012;4(19):5792–5813.
63. Du X, Men K, Xu Y, et al. Enhanced capacitance performance of Al_2O_3-TiO_2 composite thin film via sol-gel using double chelators. *J Colloid Interface Sci*. 2015;443:170–176.
64. Lee BR, Goo JS, Kim YW, et al. Highly efficient flexible organic photovoltaics using quasi-amorphous ZnO/Ag/ZnO transparent electrodes for indoor applications. *J Power Sources*. 2019;417:61–69.

65. Bharti K, Lone SA, Singh A, et al. Green synthesis of luminescent gold-zinc oxide nanocomposites: cell imaging and visible light–induced dye degradation. *Front Chem.* 2021;9:1–12.
66. Rams J, Torres B, Pulido-González N, et al. Magnesium alloys: Fundamentals and recent advances. *Encycl Mater Met Alloy.* 2021;1:2–10.
67. Habisreutinger SN, Schmidt-Mende L, Stolarczyk JK. Photocatalytic reduction of CO_2 on TiO_2 and other semiconductors. *Angew Chemie – Int Ed.* 2013;52:7372–7408.
68. Soc C, Xiang Q, Jaroniec M. Graphene-based semiconductor photocatalysts. *Chem Soc Rev.* 2012;41:782–796.
69. Yan SC, Lv SB, Li ZS, et al. Organic-inorganic composite photocatalyst of g-C3N4 and TaON with improved visible light photocatalytic activities. *Dalt Trans.* 2010;39:1488–1491.
70. Falah M, Mackenzie KJD. Photocatalytic nanocomposite materials based on inorganic polymers (Geopolymers): A review. *Catalysts.* 2020;10:1–20.
71. Linsebigler AL, Lu G, Yates JT. Photocatalysis on TiO2 surfaces: Principles, mechanisms, and selected results. *Chem Rev.* 1995;95:735–758.
72. Almayyali AOM, Muhsen HO, Merdan M, et al. Two-dimensional ZnI_2 monolayer as a photocatalyst for water splitting and improvement of its electronic and optical properties by strains. *Physica E Low Dimens Syst Nanostruct.* 2020;126:114487.
73. Ahuja R, Srivastava R. Introduction to photocatalytic materials for clean energy, renewable chemicals production, and sustainable catalysis. *Nanoscale Adv.* 2024;6(15):3695–3698.
74. Fiorentino B, Bartolomei V, Soulas R, et al. Characterization of photocatalytic paints : A relationship between the photocatalytic properties – release of nanoparticles and volatile organic compounds. *Environ Sci Nano.* 2017;4:1998–2009.
75. Alajilani H, Jamjoum A, Umar K, et al. Synthesis, Characterization, and photocatalytic activities of graphene oxide/metal oxides nanocomposites : A review. *Front Chem.* 2021;9:1–24.
76. Montini T, Gombac V, Hameed A, et al. Synthesis, characterization and photocatalytic performance of transition metal tungstates. *Chem Phys Lett.* 2010;498:113–119.
77. Xu H, Xu Y, Li H, et al. Synthesis, characterization and photocatalytic property of $AgBr/BiPO_4$ heterojunction photocatalyst. *Dalton Trans.* 2012;41:3387–3394.
78. Shivaji K, Monica ES, Devadoss A, et al. Synthesizing green photocatalyst using plant leaf extract for water pollutant treatment. In Naushad, M., Rajendran, S., Lichtfouse, E. (eds), *Green Photocatalysts. Environmental Chemistry for a Sustainable World*, vol 34., 25–46. Springer, Cham, 2020.
79. Pavel M. Photocatalytic degradation of organic and inorganic pollutants to harmless end products : Assessment of practical application potential for water and air cleaning. *Catalysts.* 2023;13(2):380.
80. Su T, Zhu X. Advanced photocatalytic materials for environmental and energy applications. *Materials (Basel).* 2023;16:2–5.
81. Zaleska A, Hanel A, Nischk M. Photocatalytic air purification. *Recent Patents on Eng.* 2014;4(3):200–216.
82. Sayed M, Danish S, Estrella LL, et al. Photocatalytic applications of metal oxides for sustainable environmental remediation. *Metals.* 2021;11:1–25.
83. Di Paola A, García-López E, Marcì G, et al. A survey of photocatalytic materials for environmental remediation. *J Hazard Mater.* 2012;211–212:3–29.
84. Kang W, Chen S, Yu H, et al. Photocatalytic ozonation of organic pollutants in wastewater using a flowing through reactor. *J Hazard Mater.* 2021;405:124277.

85. Vasseghian Y, Le VT, Joo S, et al. Spotlighting graphene-based catalysts for the mitigation of environmentally hazardous pollutants to cleaner production : A review. 2022;365:132702.
86. Gusain R, Gupta K, Joshi P, et al. Adsorptive removal and photocatalytic degradation of organic pollutants using metal oxides and their composites: A comprehensive review. *Adv Colloid Interface Sci.* 2019;272:102009.
87. Elgohary EA, Mohamed YMA, El Nazer HA, et al. A review of the use of semiconductors as catalysts in the photocatalytic inactivation of microorganisms. *Catalysts.* 2021;11:1–15.
88. Gautam S, Agrawal H, Thakur M, et al. Metal oxides and metal organic frameworks for the photocatalytic degradation: A review. *J Environ Chem Eng.* 2020;8:103726.
89. Assadi AA, Baaloudj O, Khezami L, et al. An overview of recent developments in improving the photocatalytic activity of TiO_2-based materials for the treatment of indoor air and bacterial inactivation. *Materials (Basel).* 2023;16:1–22.
90. Li R, Li T, Zhou Q. Impact of titanium dioxide (TiO_2) Modification on its application to pollution treatment — A review. *Catalysts.* 2020;10(7):804.
91. Theerthagiri J, Chandrasekaran S, Elakkiya V, et al. Recent developments of metal oxide based heterostructures for photocatalytic applications towards environmental remediation. J Solid State Chem. 2018;267:35–52.
92. Jeevanandam J, Barhoum A, Chan YS, et al. Review on nanoparticles and nanostructured materials : history, sources, toxicity and regulations. *Beilstein J Nanotechnol.* 2018;9:1050–1074.
93. Stankic S, Suman S, Haque F, et al. Pure and multi metal oxide nanoparticles : Synthesis, antibacterial and cytotoxic properties. *J Nanobiotechnology.* 2016;14:1–20.
94. Zhang R, Gao L. Preparation of nanosized titania by hydrolysis of alkoxide titanium in micelles. *Mater Res Bull.* 2002;37:1659–1666.
95. Bianchi CL, Universita M, Bianchi CL. Current Opinion in Chemical Engineering SWOT Analysis of Photocatalytic Materials Towards Large-Scale Environmental SWOT Analysis of Photocatalytic Materials Towards Large Scale Environmental.
96. Singh J, Dutta T, Kim KH, et al. 'Green' synthesis of metals and their oxide nanoparticles : Applications for environmental remediation. *J Nanobiotechnology.* 2018;16:1–24.
97. Yu JC, Yu J, Ho W, et al. Preparation of highly photocatalytic active nano-sized TiO_2 particles. *Chem Commun (Camb).* 2001;7:1942–1943.

7 Bio-Based and Recyclable Materials

Hardeep Kaur, Kritika Charak, Sukhman Kaur, Niyati Gupta, Subhendu Chakroborty, and Karunesh Tiwari

Description	Abbreviation
High-density polyethylene	HDPE
Life-cycle assessment	LCA
Polyethylene	PE
Polyethylene terephthalate	PET
Polyhydroxyalkanoates	PHA
Polylactic acid	PLA
Polypropylene	PP

7.1 INTRODUCTION

The growing industrialization and capitalism in this age and time have led to increasing demands for the effective use of standard plastics, majorly derived from petroleum-based resources that are non-replenishable. Expeditious dependency on fossil-based plastics has led to multiple environmental issues, including resource depletion, greenhouse gas emissions, and the collection of persistent overtime plastic waste in our ecosystems [6]. This chapter aims to provide a complete overview of the present state of bio-based and recyclable materials [7]. All of these environmental problems due to extreme plastic waste and the depletion of fossil fuel resources have created significant interest in sustainable materials [8]. Standard plastics, derived from non-renewable petroleum-based sources, present us with a considerable number of environmental challenges since they are environmentally persistent and also contribute to pollution. Thus, there have been a lot of exponential shifts toward the progress and consumption of bio-based and recyclable materials [9]. All of these materials are produced in such a way that they ease the environmental impact along with standard plastics, fostering a better sustainable and circular economy. Bio-based materials are obtained from renewable biological sources, such as plants, animals, and microbes as mentioned above. Bio-based materials possess multiple benefits

DOI: 10.1201/9781003516071-7

such as reducing dependency on limited fossil fuels and also helping in lowering greenhouse gas emissions [10]. Adding more to their benefits these polymers not only imitate the functional properties of prevailing plastics but are also biodegradable under suitable conditions. This property of biodegradability of these materials makes sure that they can break down into harmless components, as a result reducing their extended environmental footprint. While on the contrary, recyclable materials reinforce the reuse and recovery of existing materials to create new products as can be seen in Figure 7.1.

The upcycling of materials is very essential for the circular economy, waste we want to minimize or lessen and maximize the efficiency of resources [12]. Long-established recycling processes have been majorly applied to petroleum-based plastics, such as polyethylene terephthalate (PET) and high-density polyethylene (HDPE). At the same time, the onset of modern recycling technologies has increased the extent to a wider range of materials, including bio-based polymers. The recyclability of materials is decided by their ability to go through multiple cycles of processing without degradation of their congenital properties. This characteristic is necessary for keeping up the material's functionality and economic value over a prolonged period [13–15]. The amalgamation of bio-based and recyclable materials into various industrial applications entails a thorough understanding of their properties, performance, and environmental impacts. Bio-based polymers, such as PLA and PHA, have applications ranging from retailing and textiles to medical devices and agricultural films. These materials demonstrate mechanical properties that are equivalent to conventional plastics, accompanied by additional benefits of biodegradability and reduced carbon footprint [16]. Regardless of how, challenges such as higher cost, restricted thermal stability, and degradation conditions must

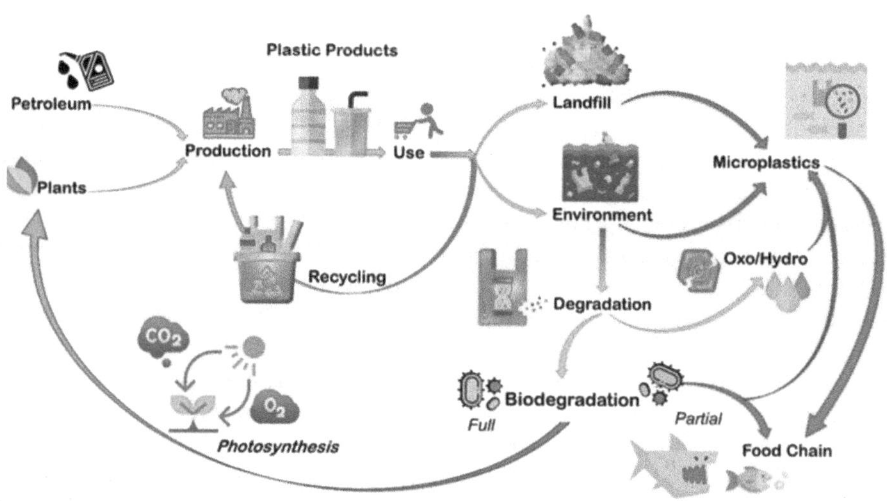

FIGURE 7.1 Evolution, usage, and fate of bio-based materials [11]. (Credit: Commons Attribution 4.0 International License)

be conveyed in such a way that enhances their enterprise and feasibility for more widespread adoption. Likewise, recyclable materials especially those produced from advanced recycling techniques such as chemical recycling and upcycling, provide the necessary potential for sustainable material management. Chemical recycling mainly involves breaking down polymers into their monomers or other basic chemicals, which can then be repolymerized into new matter [17–19]. In this way, we can address the various constraints of mechanical recycling, which most of the time results in the degradation of materials and reduced quality whereas, upcycling transforms waste materials into products that are efficient along with higher value and multiple uses. The development of structured and cost-effective recycling processes is essential for the successful application of recyclable materials in today's economy. Based on environmental sustainability, life-cycle assessment (LCA) is a very crucial tool that helps in assessing the environmental impacts altogether of bio-based and recyclable materials [20, 21]. LCA deals with the entire life-cycle of material, starting from the extraction of raw material and then moving on to production and then to usage and end-of-life disposal. By comparing the environmental performance of bio-based and recyclable materials with standard plastics, LCA offers significant insights into the possible benefits and commutations associated with their use. These evaluations help researchers, policymakers, and industry stakeholders in making well-informed decisions about the selection of materials and various management practices [22, 23]. The growing interest and increasing demand for bio-based and recyclable materials are further navigated by regulatory structure and consumer demand for sustainable products. Governments and international organizations are implementing severe regulations and motives to promote sustainable materials and minimize plastic waste. The demand for eco-friendly products has increased over time pertaining to the fact consumers are getting more and more aware of the environmental impact of their purchasing decisions. The interaction between regulation and consumer behavior is intensifying the adaptation toward a more sustainable materials economy. The production and implementation of bio-based and recyclable materials display an encouraging pathway toward addressing the various environmental threats posed by conventional plastics. These materials not only provide sustainable alternative that aligns with the proposition of the circular economy, but also give emphasis to efficiency of resources, waste reduction, and environmental protection [24]. In order to realize the full potential, various technical, economic, and regulatory and ethical challenges need to be addressed through multifaceted research and collaboration. Nonetheless, for non-biodegradable bioplastics to be adequately recycled using standard recycling infrastructure and technology, better recyclability capacity, retrieval quantity, as well as efficient separation through waste management systems are a necessity. Compostable bioplastics could be recommended over non-biodegradable plastics for single-use items such as food and beverage packaging and biomedical equipment when recycling and repurposing are not attainable.

7.2 BIO-BASED POLYMERS

Currently, the field of bio-based polymers has boosted remarkable interest pertaining to the fact that there are rapidly growing industrialized economies and their increasing demand for ingenious materials that have less harmful effects on the environment [25, 26]. Bio-based polymers or biopolymers are acquired from renewable resources and have been recognized as the key to attaining sustainability [27–29]. They are eco-friendly alternatives to conventional polymers derived from petroleum-based raw materials. Moreover, they significantly reduce fossil fuel dependency and lower carbon dioxide emissions, promoting a more sustainable and circular economy. Biopolymers are used across various industries from packaging and textiles to automotive parts and medical devices, marking a revolutionary move toward more environmentally friendly manufacturing techniques [30]. Bio-based polymers are used in both general and engineering applications due to enhanced physical durability and processability, as shown in Figure 7.2. PLA has found applications in the packaging industry. Bio-based polymers can be broadly categorized based on their source materials and synthesis processes. Physical modification and polymer processing optimization, which includes extruder screw design, processing parameter optimization, additive selection, and post-orientation for strain-induced crystallization, are the current methods used to increase reprocessibility [31].

7.2.1 Types of Bio-Based Polymers

Polyhydroxyalkanoates (PHAs)

PHAs are a type of bio-based polymer that is produced by microbial fermentation and is biodegradable in various environments. PHAs are biocompatible, which means they are suitable for various medical applications. They are derived from

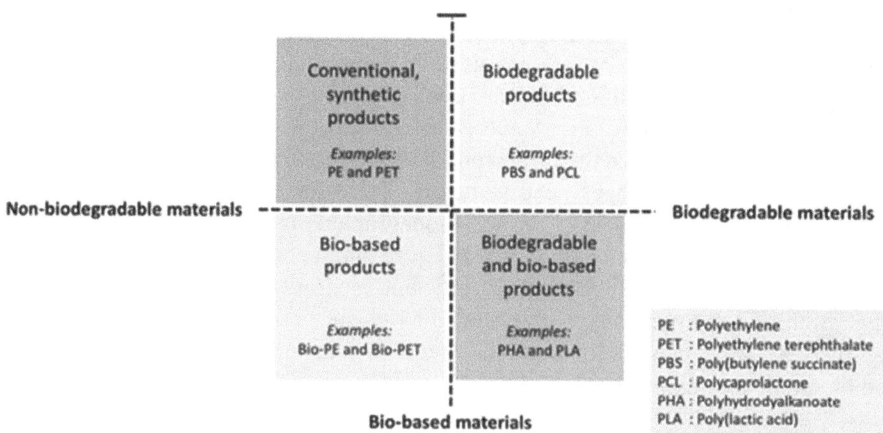

FIGURE 7.2 Examples of plastic products made from different materials [32] (Credit: Commons Attribution 4.0 International License)

renewable resources like sugar. PHAs are quite versatile have various applications and help reduce greenhouse gas emissions.

Polylactic Acid (PLA)

PLAs are derived from organic sources like plant starch that is used in fermented sugarcane or corn. They are commonly used in packaging, medical implants, and 3D printing. It is environmentally friendly and has a relatively low melting point. They offer a sustainable alternative to petroleum-based plastics.

Bio-Based Polyethylene (Bio-PE)

It is derived from ethanol procured from sugarcane and various other feedstocks and is non-biodegradable and recyclable. Bio-PE is of various types like High-Density Polyethylene and Low-Density Polyethylene [33]. They are used in plastic containers, packaging, and thin films. Bio-based polymers possess a range of properties and have innumerable applications.

Bio-Based Polyethylene Terephthalate (Bio-PET)

It is partially derived from biomass (e.g., sugarcane) along with having numerous applications in the packaging of food, in fibers and in bottles. It has properties similar to conventional PET being durable and partially bio-based.

Polybutylene Succinate (PBS)

It is created from the polymerization of bio-based succinic acid and 1,4-butanediol, which is another biodegradable polymer with applications in packaging, agricultural films, and disposable products. Natural rubber, sourced from the latex of rubber trees, is a biodegradable material known for its elasticity and durability, used in tires, footwear, and medical devices.

7.2.2 Properties of Bio-Based Polymers

Biodegradable materials are gaining traction due to their environmental benefits and potential to reduce waste. Understanding their properties is crucial for their effective application and development. This chapter delves into the mechanical, thermal, chemical, and biodegradation properties of biodegradable materials, as well as their advantages and limitations [34].

Biodegradable materials are made in a way that they can carry out their predetermined function within a biological frame without leading to any adverse reactions in the surrounding tissues or body systems. Several bio-based polymers, particularly PLA and PHAs are biocompatible, making them completely acceptable for medical applications [35]. Renewability also being a property of biopolymers is obtained from renewable sources such as microbes, plants, and algae and its main aim is to reduce dependency on fossil fuels. The mechanical properties of biodegradable materials, such as stiffness, strength, elasticity, and elongation, are essential for figuring out their compatibility for varied applications and their uses in different industries. Tensile strength also comes under mechanical properties it can be defined as the resistance or defiance of a substance to break under tension and it is measured by its tensile strength or power [36]. Biodegradable polymers like polylactic acid (PLA) and polyhydroxyalkanoates (PHA) show moderate tensile strength, which makes them acceptable for applications such as food and beverage packaging and disposable items [37]. Though, they may not be able to keep up with the tensile strength of standard plastics like polyethylene (PE) and polypropylene (PP). Elasticity and flexibility can be defined as the ability of a material to regain its original shape after going through deformation. PLA is comparatively more fragile and rigid, while materials that are cellulose derivatives and starch-based are more flexible. It is crucial to amplify the elasticity of biodegradable materials for implementation in various fields, such as agricultural film impact resistance is the capacity of a material to resist unforeseen forces without cracking. Biodegradable materials typically have lesser impact resistance compared to conventional plastics. Although, blending or strengthening them with biomass fibers can enhance their durability as well as stability making them acceptable for challenging applications [38]. Thermal properties of biodegradable materials exhibit crystallization temperature, thermal conductivity, melting temperature, glass transition temperature, and thermal stability, which are essential for refining. The melting temperature is the point at which a material transforms or adapts from solid to liquid. PHAs have a wider range of melting temperatures relying on their structural framework, whereas PLA has a comparatively low melting temperature, hence restricting its uses [39]. The Glass Transition Point is known as the temperature at which a material transforms from a fragile state to a flexible, rubber-like state. For example, PLA has a temperature of around 55–65°C, which influences its execution in different environments given the temperature is also varied [40]. There is a vital need to understand the concept of temperature as it is essential for applications like packaging, where materials must maintain their properties under different conditions. Thermal degradation can be defined as the breakdown of a material when laid bare at higher temperatures. Biodegradable materials may deteriorate at lower temperatures than traditional plastics, leading to challenges during processing. However, improvement in polymer mixture and stabilizers can make thermal stability more efficient [41]. The chemical properties of biodegradable materials are important as they help to understand their structural framework as well as reactivity such as hydrophobicity, moisture absorption, and biodegradability, sensitivity to pH. All of these properties have an impact on the material's performance.

Biodegradable materials generally have lower chemical resistance compared to traditional plastics. For instance, PLA is susceptible to hydrolysis and may degrade when exposed to moisture and certain chemicals. It can be noted that this property is essential for biodegradability but restricts its use in environments with excessive chemicals. Moisture Absorption can be defined as the material's ability to absorb moisture and it has a major effect on performance and biodegradation rate [42]. We can take the example of starch-based plastics which have high moisture absorption rates, causing swelling and reducing mechanical strength in sultry conditions. Coating with hydrophobic polymers can help in reducing this issue. The biggest advantage of biodegradable materials is their capability to break down into natural elements through microbial activity. The rate of biodegradation depends on various

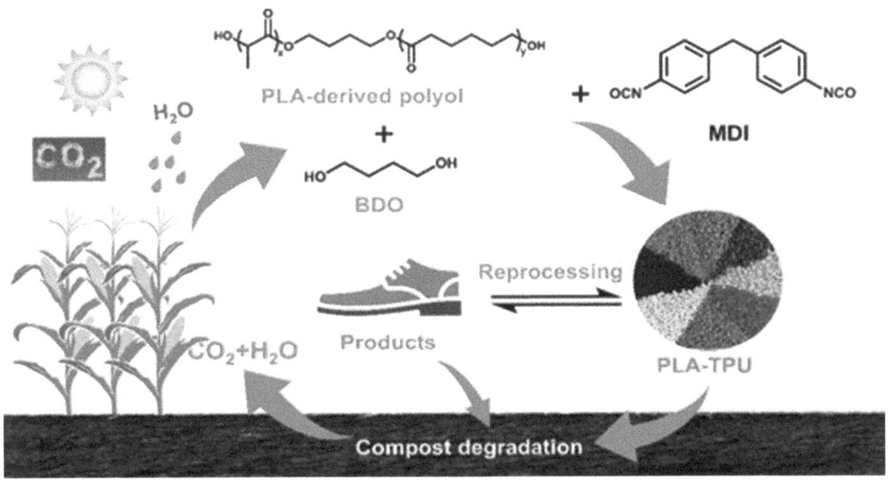

FIGURE 7.3 Schematic diagram for the full life-cycle of PLA [45] Credit: Commons Attribution 4.0 International License)

factors such as the composition of a material, polymer structure, environmental conditions, and the presence of microorganisms. PLA deteriorates effectively in industrial composting facilities but may take a longer amount of time in natural environments [43]. PHAs can decay in both aerobic and anaerobic conditions, providing resourcefulness in waste management shown in Figure 7.3. Environmental properties, including sustainability, carbon footprint, and end-of-life options, are vital for assessing the complete effect of biodegradable materials. Biodegradable materials mostly are procured from renewable resources like corn, sugarcane, and agricultural waste, reducing dependency on fossil fuels. This renewable sourcing contributes to sustainability by promoting a circular economy [44]. The production of biodegradable materials generally results in lower carbon emissions compared to conventional plastics. However, the overall carbon footprint depends on agricultural practices, transportation, and processing methods. Life-cycle assessments are essential for evaluating the true environmental impact.

7.3 APPLICATIONS OF BIOPOLYMERS

PHAs and PLA are used in medical sutures and implants due to their biocompatibility. PHAs and starch-based polymers are used in mulch films, reducing plastic waste. Food packaging uses polymers like PLA and PHAs because of their biodegradability and good barrier properties, whereas Bio-PE and cellulose-based polymers are used for packaging non-edible items [46]. Biopolymers are used in a vehicle's interior depleting the environmental impact of automotive manufacturing. Bio-based polymers have many uses, ranging from consumer goods and packaging to medical equipment and agriculture, helpful in promoting a circular economy and creating a more sustainable future [47]. However, there are certain challenges like: High production cost: Bio-based polymers are generally more costly as compared to fossil-based ones due to the high cost of raw materials and processing technologies. There is huge competition in feedstock supply as its availability can be affected by various environmental factors. The present infrastructure doesn't meet the criteria for bio-based polymers, it is built according to the needs of fossil-based polymers [48]. Awareness among consumers: People are skeptical about the performance of bio-based polymers and are hesitant to use these materials. Not all bio-based polymers meet certain characteristics like resistance to temperature, durability, etc. As time progresses there are new opportunities that are being presented to us such as Sustainability: As time progresses adaptation of sustainable bio-based polymers will increase leading to a healthy environment [49]. Technology advancement: With time there will be advances in technology leading to the production of more high-quality products. Development in feedstocks helps create more sustainable products [50]. Bio-based polymers play a major role in a sustainable future and their current development holds the potential to lessen environmental hazards leading to preservation and resource management.

7.4 RECYCLING TECHNIQUES FOR BIO-BASED MATERIAL

Recycling of bio-based products plays a major role in sustainable development, which aims to minimize adverse environmental effects, conserve resources, and reduce waste. Renewable biological resources are the source of bio-based materials, which encompass a variety of goods like bioplastics, natural fibers, and biocomposites [51]. The field of recycling bio-based materials is diverse and dynamic, and therefore essential for sustainable development shown in Figure 7.4. Hence, the

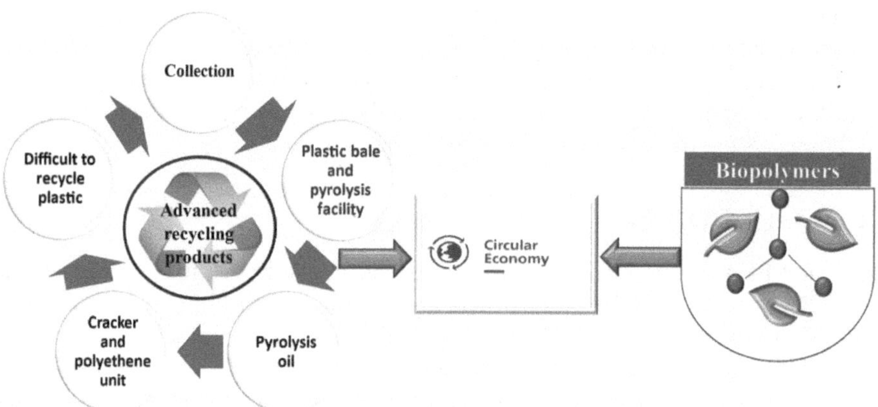

FIGURE 7.4 Representation of the effect of biopolymers on the economy (Credit statement: created by author)

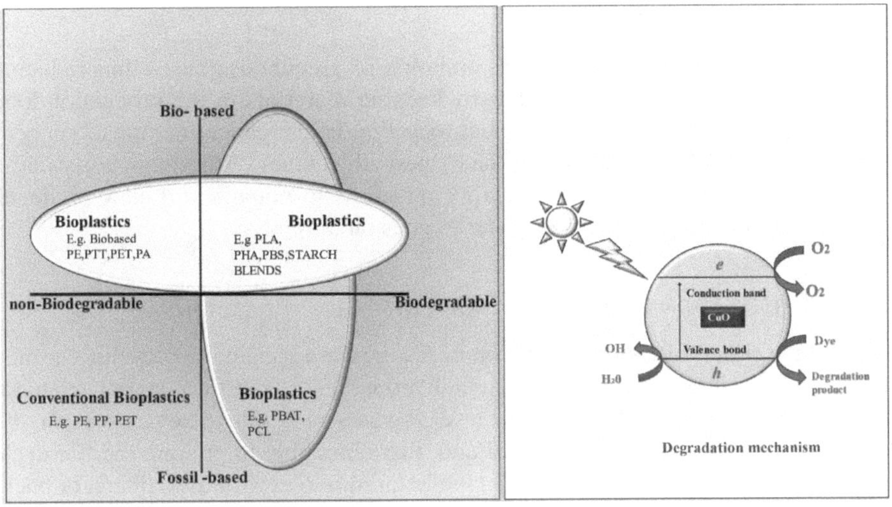

FIGURE 7.5 Visual representation of various recycling techniques (Credit statement: created by author)

variety of these materials and the available recycling methods, integrating bio-based materials into recycling systems brings both special opportunities and challenges. Recycling of bio-based materials can be done through various techniques which are mechanical recycling, chemical recycling, and biological recycling.

Mechanical recycling is a vital process that entails processing materials to retrieve plastic waste and utilize it in completely furnished products without altering their structural framework [52]. This approach most of the time is used for recycling bioplastics such as PHAs and PLA. Mechanical recycling is a step-by-step process. The first step is 'collecting' and 'sorting'. Materials are gathered and organized based on their quality, next step includes 'Shredding' and 'Washing'. The material that is gathered and arranged is shredded into smaller pieces and scoured to remove any kind of dirt or impurities [53]. These minced materials are further liquefied and reprocessed into small-scale slivers or granules which are known as 'melting' and 'reprocessing'. The last step is 'manufacturing'. The small-scale particles are then used to develop modern-day. There are possible advantages as well as disadvantages of mechanical recycling which are shown below in Figure 7.5.

7.5 ADVANTAGES OF BIOPOLYMERS

7.5.1 Energy Efficiency

Mechanical recycling usually needs less energy than producing virgin products/materials from raw sources [54]. Cost-effective: Mechanical recycling has some economic benefits as it is helpful in the growth of our economy and provides job opportunities in various recycling industries.

7.5.2 Environment Friendly

This process generally results in less emission of greenhouse gases thus reducing adverse environmental impacts. Thus by keeping materials in the production loop and out of the environment, mechanical recycling helps reduce the impact on ecosystems and wildlife. This is one of the biggest advantages of mechanical recycling as this process can retrieve good quality materials to manufacture new products, making sure this cycle of reuse of material goes on [55].

7.5.3 Preservation of Material Properties

At the same time, this is one of the main features of mechanical recycling: it may conserve the primary part of the original properties of plastic material strength, durability, and elasticity. That means recycled plastics are able to be put to use in the same applications as pristine material and, therefore, able to prolong the life-cycle of plastic. rPET will be recycled and manufactured into new bottles, fibers, or packaging material in a way that the quality is preserved and equivalent to the original product.

7.5.4 Waste Reduction

Mechanical recycling is significantly important in the recovery of plastic wastes from landfills, incineration, and the environment. As it recycles waste to make new materials, the buildup of total quantities of waste is avoided, an especially relevant aspect because of global issues brought about by plastic pollution. It minimizes waste in landfills to a great mark, and as a consequence, results in a great reduction in environmental pollution from plastics and harm to wildlife and ecosystems.

7.5.5 Economic Benefits

The mechanical recycling industry can, thus, begin an economic growth spurt via the job opportunity presented in the collection, sorting, and processing of plastic waste. Additionally, by providing a market for recycled plastics, mechanical recycling serves as a foundation to undergird sustainable materials-and-products industries that may, over time, foster new entrepreneurial ventures and inventive activity in recycling technologies and sustainable product design.

7.5.6 Scalability

Mechanical recycling is a relatively straightforward and scalable process, which means it can be implemented across various regions and industries with relative ease. The basic steps of collecting, sorting, cleaning, shredding, and reprocessing plastics can be adapted to different scales, from small community-based recycling initiatives to large industrial operations. This scalability makes mechanical recycling accessible and practical for widespread use, including in areas with limited resources.

7.6 DISADVANTAGES OF BIOPOLYMERS

7.6.1 Inefficient Sorting and Separation

Infrastructure needs not being met leads to inefficient sorting, therefore it has become important to invest in infrastructure to collect, sort, and separate. Also, there are high chances of the presence of contaminants or any impurity and it can hinder the recycling process.

7.6.2 Downcycling

Most of the time the material that is recycled is of lower quality as compared to the original so these recycled products are used for low-value applications [56]. Chemical recycling is a process that involves breaking down a material's chemical structure to obtain other chemicals that can be used to produce new materials. This technique is suitable for certain bioplastics that are difficult to recycle [57]. Certain steps need to be followed for the proper execution of this recycling process.

7.6.3 Depolymerization

Polymers are broken down chemically into smaller compounds or monomers. Monomers are refined or purified to remove any contamination called refining. Refined monomers are polymerized to form the latest polymers called repolymerization. High energy requirement: It is one of those processes that require a huge amount of energy. Therefore it is less efficient compared to mechanical recycling [19–60].

7.6.4 Scalability Matters

One of the biggest challenges when it comes to chemical recycling is scaling up to manage large amounts of waste. Biological recycling uses biological processes to break down organic materials, changing waste into valuable products like biogas, compost, etc. This technique directly contributes to waste management and resource recovery and ensures efficient decomposition of organic materials [59, 61]. It is useful in soil amendment and the production of energy while having certain industrial applications.

7.7 COMPOSTING

It involves the aerobic decomposition of organic materials by microbes in the presence of oxygen and produces compost. The steps involved are as follows:

- Collecting and sorting: Waste is collected and then sorted to remove any impurities like plastic or any kind of debris. Shredding: It allows faster decomposition of materials as they are shredded.

- Piling and aeration: The organic waste is piled into bins and then aerated to maintain oxygen levels, most importantly for microbial activity [62, 63]. Monitoring: Temperature, pH, and moisture are monitored to make sure it stays within the optimal range. Decomposition: Microbes break down organic matter into smaller and simpler compounds and it continues to decompose and stabilize.
- Screening and curing: Compost is screened and cured for several months allowing the compost to stabilize. Let us see the advantages and disadvantages of biological recycling.

7.7.1 Advantages

One of the biggest examples of composting is soil enrichment: Composting helps in the yield of compost that is packed with enriched nutrients further helping to boost the richness of soil and assisting in plant growth.

Economically viable: It not only is affordable but also helps in overall economic growth and gives job opportunities in various biogas industries.

Reduced environmental impacts: Biological recycling is one of those processes that does not affect the environment negatively and emits very small amounts of greenhouse gases [64].

7.7.2 Disadvantages

- High space requirement: Composting is one of those processes that require a notable amount of space to function.
- Time-consuming: Composting and anaerobic digestion take some weeks to months to finish its process [38].
- Management challenges: Biological recycling requires continuous maintenance to manage processes like composting and anaerobic digestion and to ensure good results. While having countless applications, recycling techniques face certain challenges like infrastructure limitations. The present recycling infrastructure is built for traditional plastics, so it requires significant advancement for new materials.
- Economic durability: Recycling bio-based materials can be less economically viable as compared to modern bio-based materials shown in Figure 7.6. The cost of these processes is certainly high.
- Awareness issues: There is very little consumer awareness of bio-based materials and their importance.
- Maintaining high quality: It is not possible to deliver high-quality materials every time because of infrastructure and cost limitations [27]. Upholding standards: Certification of bio-based materials is necessary for proper identification and is needed to facilitate better sorting. Creating a table for the recycling of bio-based materials involves outlining different types of bio-based materials and their recyclability characteristics.

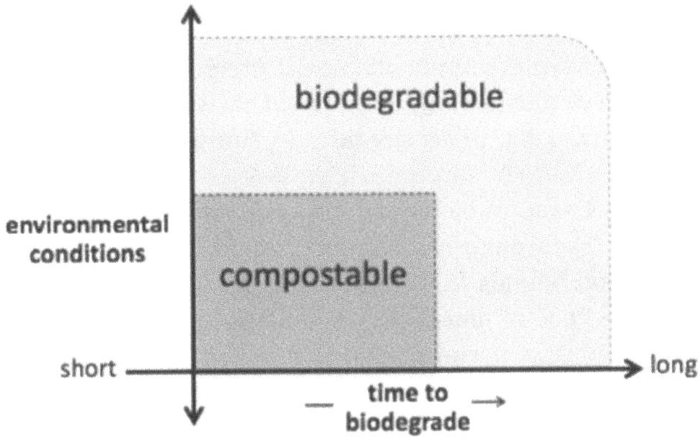

FIGURE 7.6 Graph between biodegradation time and environmental conditions [32] (Credit: Commons Attribution 4.0 International License)

Table 7.1 provides a snapshot of how different bio-based materials fare in terms of recyclability. Remember that specific materials may have variations in recyclability based on their composition, local recycling infrastructure, and technological advancements [65]. There are various recycling techniques for bio-based materials that are evolving more and more as time progresses and each of these techniques offers unique advantages, disadvantages, and challenges and is paving the way toward a more sustainable and circular economy [66].

TABLE 7.1
A Comprehensive Overview of Various Bio-Based Materials, Focusing on Different Aspects

Bio-based Material	Recyclability	Remarks
Bioplastics	Can be recycled	Depends on type (e.g., PLA)
Bio-Based Fibers	Generally recyclable	Can be recycled into new fibers or composted
Bio-Based Polymers	Varies	Some are recyclable, others may require specialized processes
Bio-Based Packaging	Recyclable in certain systems	Depends on local recycling capabilities
Bio-Based Composites	Limited recycling options	Often challenging due to mixed materials
Bio-Based Chemicals	Not typically recycled	Often used in processes rather than as end products
Bio-Based Fuels	Not applicable	Used and consumed rather than recycled

7.8 FUNDAMENTALS OF BIODEGRADABLE MATERIALS

The past few decades have gradually changed from conventional methods to more sustainable and environment-friendly products. One such innovation is biodegradable materials. Biodegradable materials refer to substances capable of undergoing decomposition by microorganisms into harmless compounds like water, carbon dioxide, and biomass. The term biodegradability denotes the capacity of a substance to break down naturally through the actions of microorganisms into simpler, environmentally benign compounds [67]. With the increased urbanization and modernization, the widespread use of non-biodegradable materials like plastic has resulted in water pollution as plastic waste is majorly dumped into oceans, resulting in the extinction of various aquatic habitats around the globe. The major threat is that these human-made polymers take millions of years to degrade in our environment, which is quite a long period. Also, some of these polymers are known to cause life-threatening conditions in our bodies. To cope with the issue, researchers discovered and developed innovative biodegradable polymers that do not cost the life of any living organism [68].

Polymers undergo biodegradation through two primary stages. The initial step, chain cleavage or depolymerization, breaks down long polymeric chains into smaller oligomeric fragments. This process is vital as large macromolecules cannot penetrate the outer membranes of living cells. Hydrolysis and oxidation are the primary biochemical reactions involved, often aided by extracellular enzymes which may act either endo (randomly cleaving internal linkages) or exo (sequentially cleaving terminal monomer units). The subsequent stage of mineralization takes place within the cell. Biomass, minerals, salts, water, and gases like methane and carbon dioxide (CO_2) are produced from small oligomeric pieces. Usually, respirometry methods—which track the evolution of CO_2—are used to determine the extent of biodegradation. Additional parameters for evaluation include the rate of molecular weight loss, changes in polymer physical properties (e.g., tensile strength per ASTM standard D3826-98), the rate of microbial culture colony growth when in contact with the material, oxygen uptake (e.g., biochemical oxygen demand), and radioactive tracer techniques utilizing 14C labeling. Factors affecting biodegradation—the process of biodegradation is affected by multiple factors, some of which are molecular weight, polymer structure, morphology, and chemical treatment.

Polymers having higher molecular weight generally degrade more slowly than lower molecular weight ones. The biodegradation rate decreases with the polymer's molecular weight increase [69]. This is because larger polymer chains require more time and energy for microorganisms and enzymes to break them down into smaller, biologically accessible fragments. As a result, polymers with lower molecular weights often exhibit faster biodegradation rates, as they offer more accessible sites for enzymatic and microbial attack.

Biodegradable polymers feature linkages susceptible to hydrolysis throughout their chain, making them prone to degradation when exposed to microorganisms and hydrolytic enzymes. Polymers with hydrophobic and hydrophilic components degrade more readily than those with solely hydrophobic or hydrophilic structures.

Enzymes target amorphous regions within polymers more readily than crystalline regions [70]. This preference stems from molecules in amorphous areas being spaced further apart, rendering them more vulnerable to degradation. Radiation and chemical treatments: The photolysis of polymers under UV and γ-ray irradiation generates radicals and/or ions, leading to cleavage and crosslinking of the polymer chains. Oxidation further complicates this process, altering the polymer's biodegradation capability.

7.9 CLASSES OF BIODEGRADABLE POLYMERS

Biodegradable polymers can be classified into several categories based on their origin and chemical composition [71]. There are three major classes of biodegradable materials which are: Natural biodegradable polymers. Biodegradable polymers can be found in all living things, either naturally occurring or altered. When natural macromolecules containing hydrolysable connections, such as proteins and polysaccharides, like cellulose and starch, are exposed to watery conditions, their hydrolytic enzymes can cause biodegradation [72]. These natural polymers' hydrophilic and hydrophobic nature significantly influences their biodegradability, performance, and durability under humid conditions. Synthetic biodegradable polymers [73]. Synthetic biodegradable polymers refer to polymeric materials created by humans, using either renewable monomers from natural sources or monomers derived from petrochemicals. The biodegradation of these synthetic polymers involves the same enzyme-catalyzed reactions in aqueous environments as natural polymers.[74] Polyesters with easily hydrolysable bonds along their chains are the primary type of polymeric products. Additionally, non-steer-based polyvinyl alcohol (PVA) is also considered biodegradable.

Modified biodegradable polymers are also natural biodegradable polymers that are modified with fillers and additives. Carbohydrates, also known as saccharides, are naturally occurring organic compounds composed of carbon, hydrogen, and oxygen which can be seen in Figure 7.7. They are highly prevalent in plants, constituting up to 80% of a plant's dry weight. Carbohydrates can be categorized into four chemical groups: monosaccharides, disaccharides, oligosaccharides, and polysaccharides [75]. Polysaccharides are biodegradable polymers comprised of simple monosaccharide units. Cellulose, a linear polysaccharide consisting of glucose units, is likely the most abundant organic compound on the planet, serving as the fundamental structural component of plant cells. Microbial enzymes (cellulases) catalyze the degradation of cellulose into glucose. Starch, the second most abundant polysaccharide, is also composed of glucose units. In plants, starch exists in the form of small insoluble granules [76]. When exposed to hot water, starch readily undergoes gelatinization, forming a paste that can be shaped into a film seen in Table 7.2. However, these films tend to become brittle upon drying due to their sensitivity to water. To mitigate brittleness, starch can be combined with other materials such as plasticizers and synthetic polymers to produce less brittle films [16, 39, 77]. Biodegradable polymers are classified into three main categories: natural polymers, synthetic polymers, and semi-synthetic polymers.

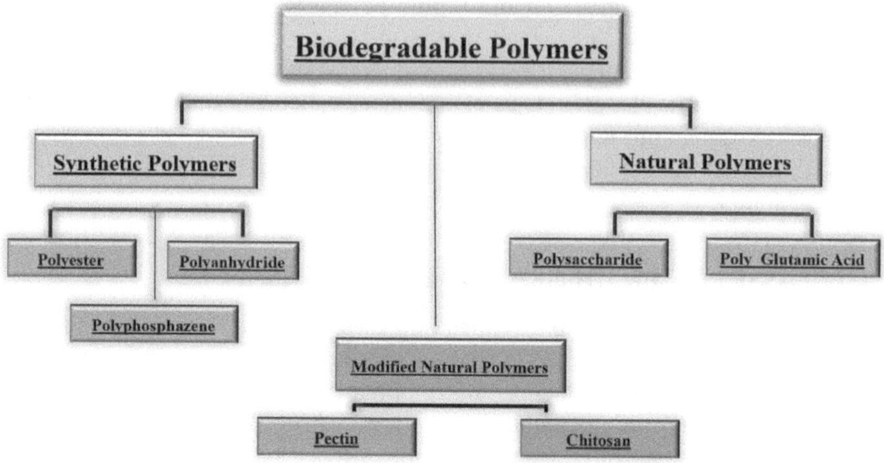

FIGURE 7.7 Classification of biodegradable polymers (Credit: created by authors)

TABLE 7.2
An Overview of the Origin and Applications of Different Biodegradable Polymers

Class	Origin	Application	References
Natural biodegradable polymers	All living organisms	Support for new tissue growth	[78]
Synthetic biodegradable polymers	Created by humans from petrochemicals or renewable sources	In packaging industries	[70]
Modified biodegradable polymers	Modification in natural biodegradable polymers		

Natural Polymers: These are derived from renewable resources and are naturally occurring. Examples include starch, cellulose, and proteins like collagen and gelatin. Polysaccharides like starch are commonly used in packaging, as they can be easily broken down by microorganisms. The natural polymer that is found in abundance is cellulose. It is a carbohydrate polymer observed in plant cell walls and has various applications in manufacturing products such as cotton fabrics, films, hydrogels paper, and rayon. Synthetic biodegradable polymers are artificial polymers formulated in such a way that they decay within certain environmental conditions. Some distinguished examples are polylactic acid (PLA), polycaprolactone (PCL), and polyhydroxyalkanoates (PHA). PLA is procured from corn starch or sugarcane and is very eco-friendly. They are broadly used in food packaging, 3D printing, and drug delivery due to their versatile properties. PCL is yet another environment-friendly polymer that is well known for its use in disposable utensils, medical implants and tissue engineering. PHAs, produced by bacterial fermentation, are completely environmentally safe and have various applications in the packaging and agriculture industries.

7.10 APPLICATIONS

Bio-based and biodegradable materials are increasingly being used across various industries due to their environmental benefits. Here are some applications.

i. Bioplastics are used for creating packaging materials such as bags, containers, and films that are biodegradable and compostable [40].
ii. Compostable Food Packaging is ideal for disposable cutlery, plates, and takeaway containers [27].
iii. Bio-composite materials are used in building materials such as insulation, panels, and bricks made from renewable resources like straw or bamboo. Green Concrete incorporates bio-based additives to improve sustainability.
iv. Bio-based plastics are used in car interiors, such as dashboards and seats, to reduce the reliance on fossil fuels [79]. Biodegradable Lubricants and Fluids provide environmentally friendly alternatives to traditional petroleum-based products.
v. Biodegradable casings for electronic devices reduce electronic waste and facilitate easier recycling. Bio-based Circuit Boards are made from renewable resources, improving the sustainability of electronic products.

7.11 FUTURE TRENDS AND PERSPECTIVE

Recent advances in bio-based material research have led to the development of sustainable, eco-friendly alternatives to traditional plastics. Innovations include biodegradable polymers derived from plants, enhanced by genetic engineering and nanotechnology, improving their strength and durability. These materials promise to reduce environmental impact and promote a circular economy. Future trends in biodegradable materials include the development of advanced biopolymers with enhanced performance, increased use of agricultural and industrial waste as raw materials, and improved composting and recycling technologies. Additionally, there will be a focus on scalability, cost reduction, and regulatory support to facilitate widespread adoption and environmental impact reduction. Integrating bio-based and recyclable materials into circular economy models enhances sustainability by reducing waste and conserving resources. These materials, derived from renewable sources, can be reused or composted, minimizing environmental impact. This integration fosters closed-loop systems, promoting resource efficiency and supporting long-term ecological balance. Biodegradable materials are increasingly becoming the focus of research and development due to their potential to address environmental concerns related to waste and pollution. Here are some key future trends and perspectives in the field of biodegradable materials:

7.11.1 Advancements in Material Science

Research is going on across the globe in order to commence new types of biopolymers with enhanced effects, such as good amount of strength, durability, and degradation rates. Revolution in this area could lead to the formation of satisfactory materials for a broad range of applications. Inculcating nanotechnology or nanomaterials such as carbon nanotubes or graphene into biodegradable materials can improve their structural framework, barrier properties, and functional characteristics hence, they become more competitive with standard plastics.

7.11.2 Sustainable Sourcing

The sourcing of substances or materials is done in a way that they are replenishable and environmentally safe. This practice of obtaining products from biofuel-rich sources such as PHA and PLA helps sustain the environment and also helps reduce carbon footprint.

7.11.3 Environmental Impact and Regulation

Anthropological activities have a severe impact on our biodiversity. Comprehensive life-cycle assessments are being run to assess the gradual increase in the negative environmental impact of biodegradable materials. This is essential to figure out areas that need improvement for improvement. Regulatory support is needed in cases like these to support the development and use of biodegradable materials through rules and regulations. This support can increase market acquisition and lead to a new wave of innovation.

7.11.4 Challenges and Research Directions

- One of the major challenges is making biodegradable materials cost-competitive with conventional plastics. Research is focused on improving production processes and economies of scale to reduce costs.
- Ensuring that biodegradable materials perform as well as or better than traditional materials in terms of durability, usability, and safety is crucial. Research is directed toward overcoming these performance trade-offs.
- Biodegradable materials require specific conditions (e.g., temperature, humidity, microbial presence) to degrade effectively. Research focuses on the materials that can degrade in a wider range of environments, including marine and landfill conditions.

7.12 CONCLUSION

Although biopolymers have properties similar to petroleum-based commodity polymers, their high cost and limited production have hindered widespread use, as they take longer to decompose and are consumed by soil and natural components. Derived from renewable resources, bio-based polymers like polylactic acid, polyhydroxyalkanoates (PHA), and starch-based plastics offer biodegradability, low toxicity, and a reduced carbon footprint, making them eco-friendly alternatives to petrochemical polymers. However, these materials encounter challenges such as limited mechanical performance, high production costs, and processing difficulties. Technological advances and increased research are critical to overcome these barriers and improve the performance and economic feasibility of bio-based polymers, and their large-scale applications in a wider range of industries. These biodegradable polymers, in general, are less stiff compared with traditional plastics and therefore potentially less suitable for certain applications. They also exhibit positive impact resistance, hence appropriate to apply in applications where the material may be under stress or impact. Biodegradable polymers are generally softer than classic plastics, which is likely to limit their suitability for specific applications. They also exhibit good impact resistance, hence appropriate to apply in applications where the material may be under stress or impact. Although they present considerable environmental benefits—reduction in plastic waste and cutting greenhouse gases, biodegradable polymers also have inherent weaknesses such as limited durability, higher production costs, and variable degradation rates. Until now, to make them more practical and cost-effective to be adopted on a wider scale, continuous innovation and upgrading would be necessary to enhance their properties.

REFERENCES

1. Canabarro N, Soares JF, Anchieta CG, et al. Thermochemical processes for biofuels production from biomass. *Sustain Chem Process*. 2013;1:22.
2. Xu L, Liu X, Sun K, et al. Corrosion behavior in magnesium-based alloys for biomedical applications. *Materials (Basel)*. 2022;15(7):2613.
3. Zhou L, Yuan T, Li R, et al. Anisotropic mechanical behavior of biomedical Ti-13Nb-13Zr alloy manufactured by selective laser melting. *J Alloys Compd [Internet]*. 2018;762:289–300. https://linkinghub.elsevier.com/retrieve/pii/S0925838818318929.
4. Moghadasi K, Mohd Isa MS, Ariffin MA, et al. A review on biomedical implant materials and the effect of friction stir based techniques on their mechanical and tribological properties. *J Mater Res Technol [Internet]*. 2022;17:1054–1121. https://linkinghub.elsevier.com/retrieve/pii/S2238785422000503.
5. Aboutalebianaraki N, Neal CJ, Seal S, et al. Biodegradable Mg-Sc-Sr alloy improves osteogenesis and angiogenesis to accelerate bone defect restoration. *J Funct Biomater*. 2022;13(4):261.
6. Wendels S, Avérous L. Biobased polyurethanes for biomedical applications. *Bioact Mater*. 2021;6:1083–1106.
7. Singh J, Singh JP, Kumar S, et al. Short review on hydroxyapatite powder coating for SS 316L. *J Electrochem Sci Eng [Internet]*. 2023;13:25–39. https://www.scopus.com/inward/record.uri?eid=2-s2.0-85148748776&doi=10.5599%2Fjese.1611&partnerID=40&md5=9dbb686d916cc26266b626a0873c5c0a.
8. Restrepo CV, Villa CC. Synthesis of silver nanoparticles, influence of capping agents, and dependence on size and shape: A review. *Environ Nanotechnology, Monit Manag*. 2021;15:100428.

9. Singh H, Singh S, Prakash C. Current trends in biomaterials and Bio-manufacturing. In: Prakash C, Singh S, Singh R, et al., editors. *Biomanufacturing*, Cham: Springer International Publishing; 2019. pp. 1–34. https://doi.org/10.1007/978-3-030-13951-3_1.
10. Singh J, Singh C. Computational analysis of convective heat transfer across a vertical tube. *FME Trans.* 2021;49:932–940.
11. Versino F, Ortega F, Monroy Y, et al. Sustainable and bio-based food packaging: A review on past and current design innovations. *Foods.* 2023;12(5):1057.
12. Singh J, Singh S. A review on Machine learning aspect in physics and mechanics of glasses. *Mater Sci Eng B [Internet].* 2022;284:115858. https://doi.org/10.1016/j.mseb.2022.115858.
13. Song J, Winkeljann B, Lieleg O. Biopolymer-based coatings: Promising strategies to improve the biocompatibility and functionality of materials used in biomedical engineering. *Adv Mater Interfaces.* 2020;7. https://doi.org/10.1002/admi.202000850.
14. Berger LM. Application of hard metals as thermal spray coatings. *Int J Refract Met Hard Mater [Internet].* 2015;49:350–364. http://dx.doi.org/10.1016/j.ijrmhm.2014.09.029.
15. Ribeiro C, Sencadas V, Correia DM, et al. Piezoelectric polymers as biomaterials for tissue engineering applications. *Colloids Surf B Biointerfaces [Internet].* 2015;136:46–55. http://dx.doi.org/10.1016/j.colsurfb.2015.08.043.
16. Ahmad S, Hashmi AW, Singh J, et al. Innovations in additive manufacturing of shape memory alloys: Alloys, microstructures, treatments, applications. *J Mater Res Technol.* 2024;32:4136–4197.
17. Hussain SA, Dey B, Bhattacharjee D, et al. Unique supramolecular assembly through Langmuir – Blodgett (LB) technique. *Heliyon [Internet].* 2018;4:e01038. https://doi.org/10.1016/j.heliyon.2018.e01038.
18. Ma S, Zhang X, Yu B, et al. Brushing up functional materials. *NPG Asia Mater.* 2019;11(1):24.
19. Du J. Challenges in molecular dynamics simulations of multicomponent oxide glasses. In: Massobrio C, Du J, Bernasconi M, et al., editors. *Molecular Dynamics Simulations of Disordered Materials, Springer Series in Materials Science*, Springer International Publishing AG Switzerland; 2015. pp. 157–180.
20. Liu YY, He J, Li YD, et al. Biobased epoxy vitrimer from epoxidized soybean oil for reprocessable and recyclable carbon fiber reinforced composite. *Compos Commun.* 2020;22:100445.
21. Sreenivasan R, Goel A, Bourell DL. Sustainability issues in laser-based additive manufacturing. *Phys Procedia [Internet].* 2010;5:81–90. http://dx.doi.org/10.1016/j.phpro.2010.08.124.
22. Chojnacka K, Moustakas K, Witek-Krowiak A. Bio-based fertilizers: A practical approach towards circular economy. *Bioresour Technol.* 2020;295:122223.
23. Thakur L, Vasudev H, Singh J, et al. *Artificial Intelligence and Machine Learning in the Thermal Spray Industry: Practices, Implementation, and Challenges.* CRC Press; 2023.
24. Mazhandu ZS, Muzenda E, Mamvura TA, et al. Integrated and consolidated review of plastic waste management and bio-based biodegradable plastics: Challenges and opportunities. *Sustainability.* 2020;12:1–57.
25. Singh J, Dutta T, Kim KH, et al. 'Green' synthesis of metals and their oxide nanoparticles: Applications for environmental remediation. *J Nanobiotechnol.* 2018;16(1):84.
26. Santosh S, Harris WBJ, Srivatsan TS. Environment-induced degradation of shape memory alloys: Role of alloying and nature of environment. *Materials (Basel).* 2023;16:5660.

27. Van De Velde K, Kiekens P. Biopolymers: Overview of several properties and consequences on their applications. *Polym Test.* 2002;21:433–442.
28. Liu J, Sun L, Xu W, et al. Current advances and future perspectives of 3D printing natural-derived biopolymers. *Carbohydr Polym [Internet].* 2019;207:297–316. https://doi.org/10.1016/j.carbpol.2018.11.077.
29. Avcu E, Baştan FE, Abdullah HZ, et al. Electrophoretic deposition of chitosan-based composite coatings for biomedical applications: A review. *Prog Mater Sci.* 2019;103:69–108.
30. Salleh A, Naomi R, Utami ND, et al. The potential of silver nanoparticles for antiviral and antibacterial applications: A mechanism of action. *Nanomaterials.* 2020;10:1–20.
31. Hazrati KZ, Sapuan SM, Zuhri MYM, et al. Extraction and characterization of potential biodegradable materials based on dioscorea hispida tubers. *Polymers (Basel).* 2021;13:1–19.
32. Azhar NNH, Ang DTC, Abdullah R, et al. Bio-based materials riding the wave of sustainability: Common misconceptions, opportunities, challenges and the way forward. *Sustainability.* 2022;14(9):5032.
33. Maraveas C. Production of sustainable and biodegradable polymers from agricultural waste. *Polymers (Basel).* 2020;12(5):1127.
34. Yaashikaa PR, Kumar PS, Saravanan A, et al. Bioconversion of municipal solid waste into bio-based products: A review on valorisation and sustainable approach for circular bioeconomy. *Sci Total Environ.* 2020;748:141312.
35. Nitsch CK, Walters LJ, Sacks JS, et al. Biodegradable material for oyster reef restoration: First-year performance and biogeochemical considerations in a coastal lagoon. *Sustainability.* 2021;13(13):7415.
36. Crisan CM, Mocan T, Manolea M, et al. Review on silver nanoparticles as a novel class of antibacterial solutions. *Appl Sci.* 2021;11:1–18.
37. Barabadi H, Mojab F, Vahidi H, et al. Green synthesis, characterization, antibacterial and biofilm inhibitory activity of silver nanoparticles compared to commercial silver nanoparticles. *Inorg Chem Commun.* 2021;129:108647.
38. Hodge I, Quille P, O'Connell S. A review of potential feed additives intended for carbon footprint reduction through methane abatement in dairy cattle. *Animals.* 2024;14(4):568.
39. Jain AS, Pawar PS, Sarkar A, et al. Bionanofactories for green synthesis of silver nanoparticles: Toward antimicrobial applications. *Int J Mol Sci.* 2021;22(21):11993.
40. Ivanović T, Hischier R, Som C. Bio-based polyester fiber substitutes: From GWP to a more comprehensive environmental analysis. *Appl Sci.* 2021;11(7):2993.
41. Naganthran A, Verasoundarapandian G, Khalid FE, et al. Synthesis, characterization and biomedical application of silver nanoparticles. *Materials (Basel).* 2022;15:1–43.
42. Pryshchepa O, Pomastowski P, Buszewski B. Silver nanoparticles: Synthesis, investigation techniques, and properties. *Adv Colloid Interface Sci.* 2020;284:87–100.
43. Gomes HIO, Martins CSM, Prior JAV. Silver nanoparticles as carriers of anticancer drugs for efficient target treatment of cancer cells. *Nanomaterials.* 2021;11(4):964.
44. Mikhailova EO. Silver nanoparticles: Mechanism of action and probable bio-application. *J Funct Biomater.* 2020;11(4):84.
45. Shou T, Hu S, Wu Y, et al. Biobased and recyclable polyurethane for room-temperature damping and three-dimensional printing. *ACS Omega.* 2021;6:30003–30011.
46. Lago A, Delgado JF, Rezzani GD, et al. Multi-component biodegradable materials based on water kefir grains and yeast biomasses: Effect of the mixing ratio on the properties of the films. *Polymers (Basel).* 2023;15(12):2594.
47. Fasake V, Dashora K. Characterization and morphology of natural dung polymer for potential industrial application as bio-based fillers. *Polymers (Basel).* 2020;12(12):3030.

48. Huq A, Rahman MM, Balusamy SR. Green synthesis and potential antibacterial applications of. *Polymers (Basel).* 2022;14(4):742.
49. Perera KY, Hopkins M, Jaiswal AK, et al. Nanoclays-containing bio-based packaging materials: Properties, applications, safety, and regulatory issues. *J Nanostructure Chem.* 2024;14:71–93.
50. Guerrero-Martin CA, Silva-Marrufo O, Ortega-Ramírez AT, et al. Development of a biodegradable material with oregano stick as a prototype of substitute for wooden agglomerate material. *Sustainability.* 2023;15:1–13.
51. Balla E, Daniilidis V, Karlioti G, et al. Poly(lactic acid): A versatile biobased polymer for the future with multifunctional properties and applications. *Polymers (Basel).* 2021;13:1822.
52. Azimi B, Maleki H, Zavagna L, et al. Bio-based electrospun fibers for wound healing. *J Funct Biomater.* 2020;11(3):67.
53. Garrison TF, Murawski A, Quirino RL. Bio-based polymers with potential for biodegradability. *Polymers (Basel).* 2016;8:1–22.
54. Cywar RM, Rorrer NA, Hoyt CB, et al. Bio-based polymers with performance-advantaged properties. *Nat Rev Mater.* 2022;7:83–103.
55. Tan YT, Chua ASM, Ngoh GC. Deep eutectic solvent for lignocellulosic biomass fractionation and the subsequent conversion to bio-based products – A review. *Bioresour Technol.* 2020;297:122522.
56. Martins VFR, Pintado ME, Morais RMSC, et al. Recent highlights in sustainable bio-based edible films and coatings for fruit and vegetable applications. *Foods.* 2024;13(2):318.
57. Buléon A, Colonna P, Planchot V, et al. Starch granules: Structure and biosynthesis. *Int J Biol Macromol.* 1998;23:85–112.
58. Bauchy M. Structural, vibrational, and thermal properties of densified silicates: Insights from molecular dynamics. *J Chem Phys.* 2012;137:15733.
59. Viallet V, Seznec V, Hayashi A, et al. *Glasses and Glass-Ceramics for Solid-State Battery Applications.* Springer Handbooks; 2019.
60. Brauer DS, Rüssel C, Kraft J. Solubility of glasses in the system P2O5-CaO-MgO-Na2O-TiO2: Experimental and modeling using artificial neural networks. *J Non Cryst Solids.* 2007;353:263–270.
61. Chen D, Heyer S, Ibbotson S, et al. Direct digital manufacturing: Definition, evolution, and sustainability implications. *J Clean Prod.* 2015;107:615–625.
62. Singh NB. Green synthesis of nanomaterials. *Handb Microb Nanotechnol.* 2022;225–254. https://doi.org/10.1016/B978-0-12-823426-6.00007-3.
63. Azimi B, Bafqi MSS, Fusco A, et al. Electrospun ZnO/Poly(Vinylidene fluoride-trifluoroethylene) scaffolds for lung tissue engineering. *Tissue Eng - Part A.* 2020;26:1312–1331.
64. Abbew A, Shao D, Champagne P. Jo ur. *Sci Total Environ.* 2020;143729.
65. Braun O, Coquery C, Kieffer J, et al. Spotlight on the life cycle of acrylamide-based polymers supporting reductions in environmental footprint: Review and recent advances. *Molecules.* 2022;27(1):42.
66. Al-Tayyar NA, Youssef AM, Al-hindi R. Antimicrobial food packaging based on sustainable Bio-based materials for reducing foodborne Pathogens: A review. *Food Chem.* 2020;310:125915.
67. Zhang Q, Song M, Xu Y, et al. Bio-based polyesters: Recent progress and future prospects. *Prog Polym Sci.* 2021;120:101430.
68. Nova Institute for Ecology and Innovation. *Bio-Based Building Blocks and Polymers Market Study.* 2014.

69. Samir A, Ashour FH, Hakim AAA, et al. Recent advances in biodegradable polymers for sustainable applications. *NPJ Mater Degrad.* 2022;6(1). https://doi.org/10.1038/s41529-022-00277-7.
70. Karthika M, Shaji N, Johnson A, et al. Biodegradation of green polymeric composites materials. *Bio Monomers Green Polym Compos Mater.* 2019;141–159. https://doi.org/10.1002/9781119301714.ch7.
71. Addis CC, Koh RS, Gordon MB. Preparation and characterization of a bio-based polymeric wood adhesive derived from linseed oil. *Int J Adhes Adhes.* 2020;102:102655.
72. Kumar A, Negi YS, Bhardwaj NK, et al. Synthesis and characterization of cellulose nanocrystals/PVA based bionanocomposite. *Adv Mater Lett.* 2013;4:626–631.
73. Nomura K, Wang X. Acyclic Diene Metathesis (ADMET) Polymerization for the synthesis of chemically recyclable bio-based aliphatic polyesters. *Catalysts.* 2024;14:1–17.
74. Schiffman JD, Schauer CL. A review: Electrospinning of biopolymer nanofibers and their applications. *Polym Rev.* 2008;48:317–352.
75. Hasan MM, Zhou Y, Mahfuz H, et al. Effect of SiO2 nanoparticle on thermal and tensile behavior of nylon-6. *Mater Sci Eng A.* 2006;429:181–188.
76. Sanyal S, Park SJ, Chelliah R, et al. Emerging trends in smart self-healing coatings: A focus on micro/nanocontainer technologies for enhanced corrosion protection. *Coatings.* 2024;14(3):324.
77. Sunitha K, Vasudev H. A short note on the various thermal spray coating processes and effect of post-treatment on Ni-based coatings. *Mater Today Proc [Internet].* 2022;50:1452–1457. https://linkinghub.elsevier.com/retrieve/pii/S2214785321058132.
78. Chesterman J, Zhang Z, Ortiz O, et al. *Biodegradable Polymers.* Principles of Tissue Engineering INC; 2020.
79. Kang W, Chen S, Yu H, et al. Photocatalytic ozonation of organic pollutants in wastewater using a flowing through reactor. *J Hazard Mater.* 2021;405:124277.

8 Energy-Efficient Thermal Barrier Coatings

A Focus on Green Technology and Smart Materials

Amrinder Mehta and Hitesh Vasudev

Description	Abbreviation
Hydroxyapatite	HAP
High-efficiency particulate air	HEPA
Solution precursor plasma spray	SPPS
Thermal barrier coatings	TBCs
Volatile organic compound	VOC
Yttrium aluminum garnet	YAG

8.1 INTRODUCTION

Thermal barrier coatings, which are important in industries such as aerospace and energy, are being transformed by green technology and smart materials. On the technological side, environmental management in TBCs explores human-invariant green technology aiming at minimizing the environmental impact of business by applying environmentally sound techniques and product inputs. This includes using renewable raw materials, conserving energy in production, and using techniques that retreat from emitting dangerous byproducts [1, 2, 3].

Various applications of micro/nanocontainer-based coatings are shown in Figure 8.1 and all these coatings are controlled by stimuli. These coatings are categorized into eight segments: pH sensitive, redox sensitive, photo sensitive, ion sensitive, magnetically sensitive, multi-sensitive, thermal sensitive, and force sensitive. Each segment provides graphical outputs and molecular structures with illustrations of how these coatings respond to their stimuli. In the middle, the segment title 'micro/

DOI: 10.1201/9781003516071-8

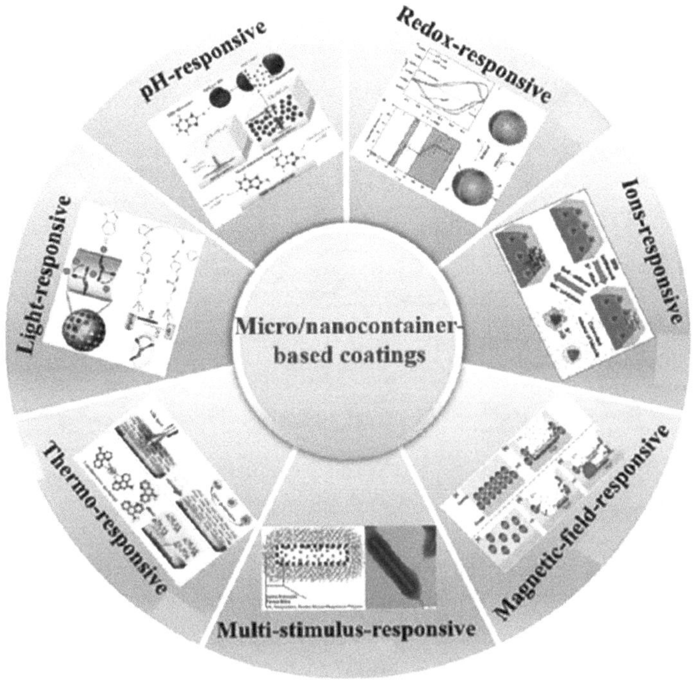

FIGURE 8.1 Stimulus-responsive micro/nanocontainer coatings respond to various applications [4]. (Credit: Creative Commons Attribution 4.0 International License).

nanocontainer-based coatings' unifies all the segments bearing in mind modern functions in material science. In turn, this diagram nicely captures a broad spectrum of uses and possibilities for these smart coatings in various sectors [4]. Intelligent materials can also optimize TBC functionality by adjusting the characteristics of their operation depending on the state of the operating conditions such as temperature range. In the TBCs, phase-change materials are used which can work as heat accumulators and protect the temperature from rising to levels which can cause damage to the satellite. Also, self-healing materials undergo processes through which they fix slight breaks or damage on the coating independently, hence increasing its durability. Integrating green technology with smart materials in TBCs not only optimizes the environmental impact but also improves performance and durability of components ideal for the enhancement of industrial uses [5, 6, 7]. Solution precursor plasma spray (SPPS) is a recently developed, environmentally friendly technique for the deposition of TBCs that offers considerable advantages over conventional powder-based thermal spray processes. Traditional techniques employ powder feed stocks, and they practiced high hot material losses, high energy utility, and poor control of micro-structural evolutions. Conversely, SPPS operate with a liquid precursor solution that fully eliminates these disadvantages [8]. The specific details of the Solution Precursor Plasma Spraying (SPPS) process are shown in Figure 8.2. It starts

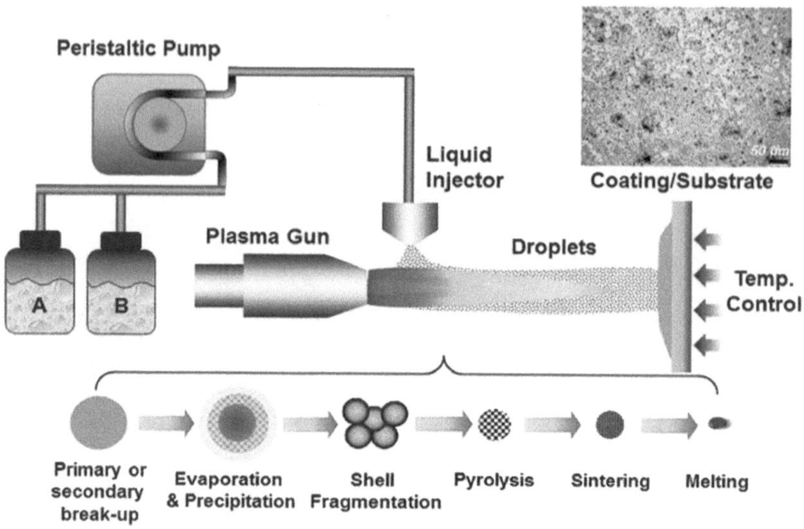

FIGURE 8.2 Physiochemical transformation of the precursor into a coating is depicted in the SPPS process schematic at the bottom [9]. (Credit: Permission of Springer Nature).

with a peristaltic pump that suck in liquids of precursors from two vials marked A and B. These precursors are then sprayed into a plasma gun through a liquid injector which propels the droplets toward the substrate. The plasma gun illustrated with orange yellow color representing heat process expels the precursors in form of blue droplets to a surface labeled coating/substrate. In a lower position of this setup, the image elaborates the different physiochemical transformations that the precursor goes through in the process of SPPS. First, a primary or secondary break up may split the precursor into smaller droplets. These droplets then evaporate, produce precipitation, and fragment the shell. The next process is pyrolysis in which the material breaks down due to heat followed by sintering in which particles combine. Last is the fusion and solidification to produce actually a coherent layer in the substrate [9].

In SPPS, the liquid phase containing the preferable aspect for the coating material is thrown into high-temperature plasma plume where the material vaporizes, decompose, and nuclease to form a fine particle size. These particles then coalesce on the substrate and give rise to a thin film possessing a nanostructured microstructure. This process also facilitates better control over the content and arrangement of pores within the TBCs and enhance the thermal characteristics of the final products such as lower thermal conduction and better resistance to sintering and thermal shock [10].

Another benefit associated with SPPS is its possibility to minimize wastage of materials. Unlike other technologies, the precursor solution is atomized and fed directly into the plasma so that almost no material is wasted through overspray. Also, SPPS is less energy-intensive compared to powder-based techniques since it does not involve onerous procedures, such as powder reconditioning [11]. In addition, the possibility to provide the necessary nanostructured coatings allows creating TBCs with the ultimate characteristics, longer service life and improved solidity. These benefits indicate that SPPS is a better technology for the production of high-performance TBCs and in line with the current push to adopt green processes in all industries [12]. Smart materials especially auto-healing ceramics are cutting-edge in the efforts to increase the life of TBCs. TBCs are used in numerous high-temperature applications such as gas turbines, and these advanced materials ensure they autonomously repair damage and greatly increase their useful life. Self-healing ceramics function based on the idea of having materials that heal at high temperatures thus filling the cracks and other pores that may be created during operation [13]. For instance, certain systems intentionally incorporate TiC or SiC particles that oxidize in high-temperature and oxygen conditions and as a result expand and close up the cracks that may have been made. The other strategies on repair involve the use of such material as $MoSi_2$ or Al_2O_3/Ni-Al particles that may be melted and rendered to flow and fill any damaged layer of the coating [14].

8.2 ENERGY EFFICIENCY

Ceramic materials of new generation or high-performance TBCs with low thermal conductivity and high-temperature resistance being incorporated into GE's gas turbines has a key role to the end users in enhancing efficiency and therefore optimizing fuel consumption and emission reduction. The development of TBC technology plays a large role in increasing economic and environmental efficiency in multiple sectors such as aerospace and power generation [15]. The principal use of TBCs is to act as a thermal barrier to protect the metallic substrates in a gas turbine from very high operating temperatures. Through designing and synthesizing TBCs with lower thermal conductivity, more heat is shielded from the metal substrate enabling the turbine to run at high temperatures. This rise in operating temperature of the gas turbine enhances the thermodynamic efficiency of the cycle [16]. Increased stability of TBCs implies their improved ability to perform in high-temperature conditions without failure or deterioration. This can be done to achieve even higher operating temperatures, which, in turn, set the bar for turbine efficacy. When the efficiency of the turbines rise, the amount of fuel needed to generate the same amount of power decreases, hence, less fuel is consumed [17]. The saving on fuel consumption is twofold. Firstly, it reduces operating expense thereby sandwiching gas turbines to be cheaper to run. Second, and perhaps more importantly from an emissions point of view, it results in a dramatic decrease in emissions. This implies that with less fuel consumed, there will be reduced output of the greenhouse gases as well as other pollutants [18]. New generations of TBC materials and designs such as nanostructure ceramics, rare earth oxides and multi lay coating systems have also revealed potential toward these objectives. The yttria-stabilized zirconia (YSZ) with controlled porosity and new thin film deposition technique such as solution precursor plasma spray have shown significant increase in thermal insulation [19]. Advanced thermal barrier coating systems incorporating nanostructured and multi-layered architectures illustrate state-of-art coatings for achieving improved insulation effectiveness for high-temperature applications in aircraft turbines and aerospace engines. All these designs are new concepts intended to address some of the challenges of TBC materials and structures [20]. Nanocomposite TBCs have fine-scale structures or heterogeneous particles present in the TBC matrix. It can also down the thermal conductivity through enhancing the phonon scattering at grain boundaries and interface. Nanostructured coatings also provide higher mechanical properties and are resistant to sintering at increased temperatures that promote low thermal conductivity for longer periods of usage [21].

8.3 SUSTAINABLE MANUFACTURING

The integrated feedback systems and *in-situ* diagnostic techniques are producing new dimensions in thermal spray processes related to TBC [22]. Higher end technologies are being implemented to perfect spray applications, minimizing wastage of materials as well as greatly enhancing the quality of the coatings. *In-situ* monitoring is a technique of observing and analyzing the spray process at real-time using different instrumentations and diagnostics [23, 24]. Such equipment may include high-speed cameras, pyrometers, spectrometer, and acoustic sensors. These devices measure the particle temperature, particle velocity distribution, particle size distribution, and substrate temperature while coating deposition. The closed-loop structure of the feedback system utilizes this real time data to make instant changes to the spray parameters [25, 26]. For instance, if the particle temperature is sensed to be low, the process can be regulated by increasing the plasma power or reducing the feed rate of powder [27, 28]. Due to this dynamic control, it is easy to achieve comparable coating thickness without significant disparities in environmental condition or feedstock material. GN has adopted AI and machine learning to enhance thermal spray in the reduction of emission and overall consumption of the material [29].

Figure 8.3 is an illustration of a simplified cross-sectional view of typical thermal barrier coating systems employed on turbine blades in jet engines. This layer is quite helpful while cooling the turbine blade from the internal side; a ceramic layer is applied which is of 200–250 μm thin and can withstand temperature up to 1300°C. These advanced technologies are being integrated to optimize spray processes, reduce material waste, and significantly improve coating quality. *In-situ* monitoring involves real-time observation and measurement of the spray process using various sensors and diagnostic tools. These may include high-speed cameras, pyrometers, spectrometers, and acoustic sensors. These devices collect data on parameters such as particle temperature, velocity, size distribution, and substrate temperature during the coating deposition [30, 31]. The closed-loop feedback system takes this real-time data and uses it to make immediate adjustments to the spray parameters. If the particle temperature is detected to be too low, the system can automatically increase the plasma power or adjust the powder feed rate. This dynamic control allows for consistent coating quality even when faced with variations in environmental conditions or

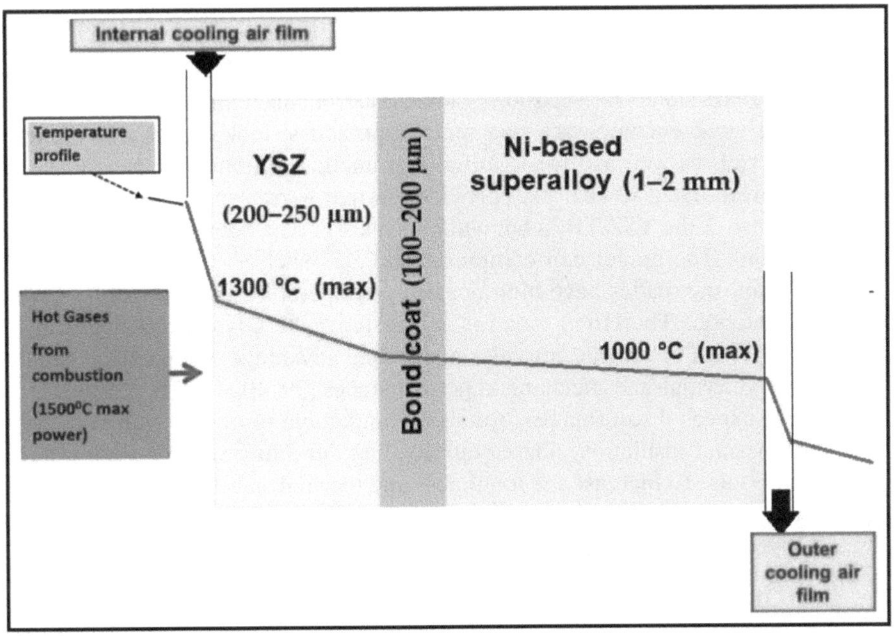

FIGURE 8.3 Schematic representation of the thermal barrier coating (TBC) component within a thermal gradient and air-cooling system [30]. (Credit: Creative Commons Attribution 4.0 International License).

feedstock materials. AI and machine learning are being applied to thermal spray to reduce emissions and material consumption [32, 33].

Recently, LCA has been extended to TBCs to examine their full life performance in terms of sustainability. This approach considers each phase in the life-cycle of a TBC from the mining of raw materials, production, utilization, and disposal/recycling. They can be applied to evaluate and reduce the impact of TBC technologies in the environment [34, 35]. At the same time these assessments' attempts are being made to come up with long-lasting approaches in application of TBC. One emerging segment is the effort to seek raw materials that can be recycled or converted to reusable rather than using the most common polymer mask type. Most conventional mask materials are polymers that are non-recyclable, and their improper disposal will lead to generation of wastes and pollution [36]. To some extent, the common scientific and technological theme in these initiatives is the LCA approach to comprehensively assess the environmental impact of TBC technologies and use sustainable materials for creating masks. With this respect, it is important to work through the whole of the life-cycle and focusing on areas of waste generation as the norm for sustainable TBC industry that is consistent with these environmental objectives of the aerospace and energy sectors [37].

8.4 ALTERNATIVE MATERIALS

Some of the reasons that have motivated research into better TBCs are reduction in the use of rare earth materials, erosion of toxic material, and enhanced performance. This ongoing work encompasses two main approaches: looking for and pioneering new materials as well as investigating biomimetic solutions [38]. Scientists are developing novel materials that may provide thermal barrier properties equal to or superior to those of the YSZ TBCs but with much lesser dependence on rare elements such as yttrium. The reader can compare stabilizers such as CaO or MgO, Ta_2O to yttria. Composite oxides have high operating temperature and low coefficient of thermal conduction. Therefore, creating multi-step materials can be practical by using the properties of various ceramics and taking advantage of nanoscale geometry to improve thermal-and-mechanical performances [39, 40, 41]. These alternative materials are expected to offer performance comparable to or superior to conventional YSZ thermal insulation, phase stability, and durability while utilizing fewer strategic materials. To increase the toughness and thermal stability of the composite in a manner analogous to the multi-scale structure of nacre and bone, introducing aspects derived from living tissue healing and applying them to increase the coating durability. To lessen thermal stresses, it is possible to establish gradual changes of the composition or structure such as those of natural interfaces. Adapting geometries observed in heat dissipation patterns in nature such as shark skin or lotus leaves to optimize thermal dissipation [42, 43, 44]. These approaches are based on the concepts derived from biological systems to promote better performance of TBCs, such as crack resistance, thermal cycling durability, and efficiency in heat dissipation. The integration of the state-of-the-art materials and biomimetic concepts present hope for not only the enhanced performance next generation TBC's but also for sustainability and durability [45]. This research is applicable not only to environmental and resource issues but also to unlocking greater possibilities in high-temperature applications to achieve better and more effective turbine engines in the future.

The field of thermal barrier coatings is advancing rapidly to meet the increasing demands for high-temperature protection in the aerospace and energy sectors. Research has highlighted the need for TBCs with enhanced performance and minimal environmental impact. Advanced green technologies, particularly the development of SPPS yttrium aluminum garnet (YAG) TBCs, have shown promising results in improving thermal insulation and operational temperatures beyond conventional TBCs [46, 47, 48]. Studies investigating the erosion-corrosion behavior of processed composites emphasize the importance of protecting materials in harsh environments. The analysis of the performance of hard surfacing alloy coatings under slurry erosion conditions reflects the relevance of robust materials to prolong service life [49]. The role of surface topology in enhancing the durability of coatings is also crucial, as seen in examinations of composite coatings that improve mechanical and thermal stability. The utilization of Ni- and Co-alloy coatings has demonstrated effectiveness in reducing maintenance through improved erosion resistance [50]. Comprehensive reviews on TBC development via thermal spray techniques showcase advancements in material properties and process optimizations. Additionally, the incorporation of self-healing ceramics as smart materials suggests an ongoing trend toward sustainable, adaptive TBC solutions [51]. This integration of innovative materials and processes supports the goal of extending component life and reducing maintenance needs. Reviews on oxidation and wear behavior of advanced alloy coatings further demonstrate how new approaches are enhancing the durability of TBCs. Collectively, these studies underscore the synergy between advanced green technologies and high-performance TBCs, aligning with the goal of enhancing safety and reducing costs in high-temperature applications.

8.5 FUTURE SCOPE

8.5.1 Integration with Other Technologies

The use of TBCs with several AM technologies is a major innovation in component manufacturing and could be beneficial in energy efficiency and materials saving. These strategies address the strengths of the two technologies to design enhanced manufacturing systems that are efficient and more sustainable. 'Freeform fabrication' as it is also called means components are built by additive processes and this makes it possible for difficult to manufacture shapes to be created out of stock materials without much wastage [52, 53]. This means for TBC production is the ability to deposit specific coating material at the point of contact, without wastefully spraying across a broad surface area, as is often the case with conventional thermal spray techniques [54]. The integration allows for producing parts with complex internal cooling, as well as internally cooled TBCs of variable thickness, which are not manufacturable by other means [55]. This may result in better thermal regulation and, possibly, the utilization of fewer materials in creating components and boosting their effectiveness. Moreover, the use of TBCs together with AM means fewer processing steps can be achieved since a notable process – coating application – can be incorporated into the component manufacturing process [56]. These methods of coating can also bring about a drastic reduction of energy consumption particularly from several processes of coating as well as from transportation between the various stages of manufacture [57, 58]. Integration of TBCs with AM also enables demonstrations of new coating designs at a faster rate and pace, which could lead to enhancements the thermal protection systems in the future [59, 60]. However, some problems still lie in the control of coating uniformity as well as the release of residual stress, but this integration points to a brighter future in more efficient production of high-temperature components with better material utilization.

8.5.2 Smart Functionalities

Thermal barrier coatings that include sensor and real-time monitoring, and self-diagnostic and self-healing properties provide the latest advancement in high-temperature materials technology. Some of these new TBCs targets for improving performance, durability and reliability of components in hostile environments, such as gas turbines, aerospace applications, etc. [61]. Thin film thermocouples and piezoelectric materials are incorporated into TBC layers to continuously monitor temperature on coated parts and strain, respectively [62]. Such close observation allows the identification of coating wear and tear in their most incipient stage, the enhancement of the engine operating mode, the support of the predictive maintenance process, and the generation of essential data for thermal and mechanical simulations [63]. Intelligent sensors for self-diagnosing TBC systems assimilate the sensor data to diagnose and prognosticate the coating's remaining fatigue life. This capability makes it possible for interventions and maintenance to be done in good time, thus eliminating disastrous failures [64]. Smart coating using self-repair mechanism of TBCs is the next generation of coating systems frontier. These systems are designed to provide the ability of self-healing to repair damage and increase coating life and performance. Techniques consist of filling healing agents in micro spheres, in-situ phase transformation healing, and healing through oxidation [65]. Although, there exists a lot of promise in these technologies that enhance component life, cut costs of maintenance and improve safety, risks remain in durability of sensors, effect of the sensors on the performance of coating, and in creating effective self-healing strategies that are effective in extreme environments [66]. Future work involves developing solutions to these issues and incorporating these smart functionalities into realistic TBC systems.

8.5.3 Environmental Protection

Innovative filtration and exhaust treatment systems in thermal spray booths have been introduced for pollution control. High-efficiency particulate air (HEPA) filters to trap small dust particles and other particulate matter. Electrostatic precipitators to get rid of charged particles. Carbon filtration to used to eliminate VOCs and chlorobenzene and other hazardous chemicals [67]. There are two advanced technologies like controlled oxidization units that can effectively transform toxic gases into less noxious compounds from the ecosystems and gas absorption for treatment of water-soluble gases and particles by wet scrubbers. These technologies operate in combination to achieve high reductions of HAP emissions and help enhance the quality of indoor and outdoor air [68]. It plays an important role in helping thermal spray facilities meet the strict standards of environmental permits, while at the same time cutting the risk of exposure to dangerous substances by the workers. Moreover, some of the systems may allow the recovery of those captured materials and possibly recycle thus enhancing sustainability in thermal spray processes. Major trends in the field include development of eco-friendly, highly efficient, and smart TBC systems by materials and processes, and incorporation of information control technology [69]. These advancements are focused on increasing the efficiency, as well as on decreasing the eco-inefficiency of TBC-coated components throughout their life-cycle.

8.6 CONCLUSION

Current advances in green technology and smart material have brought changes to TBCs in aerospace and energy markets. The use of SPPS has enhanced the control of composition and structure hence better thermal treatments and less utilization of resources. TBC damages are tackling by introducing smart materials of which the self-healing ceramics are currently being developed to self-repair and enhance the useful years of TBC. There are new designs of TiO_2-based TBCs that are starting to incorporate nanoscale structures onto the layers and multiple layering of the TBC designs to further improve the insulation properties. Feedback systems along with in-situ monitoring are enhancing the spray processes, minimizing the use of materials and enhancing the quality of the coatings. Reducing environmental impact includes life-cycle assessments and recyclable masking materials. New materials and biomimetic approaches are under consideration to decrease various materials' use of REEs and enhance performance. Adoption of TBC technology with additive manufacturing and combining it with additional features, such as incorporated sensors for constant monitoring of the system or self-identification of the problem and its subsequent rectification, is one future development of TBC technology. Efficient filters and exhaust management is being used to reduce emissions and enhance the quality of air in thermal sprays.

REFERENCES

1. Bahamirian M. A comparative study on the phase stability of ZrO2-8 wt.% Y2O3: Nano-and micro-particles. *Advanced Ceramics Progress.* 2022;8:53–60.
2. Dai M, Song X, Lin C, Liu Z, Zheng W, Zeng Y. Investigation of microstructure changes in Al 2 O 3-YSZ coatings and YSZ coatings and their effect on thermal cycle life. *Journal of Advanced Ceramics.* 2022;11:345–353.
3. Fang H, Wang W, Yang Z, Yang T, Wang Y, Huang J, Ye D. Phase stability, thermal shock behavior and CMAS corrosion resistance of Yb2O3-Y2O3 co-stabilized zirconia thermal barrier coatings prepared by atmospheric plasma spraying. *Surface Coatings Technology.* 2021;427:127864.
4. Sanyal S, Park S, Chelliah R, Yeon S-J, Barathikannan K, Vijayalakshmi S, Jeong Y-J, Rubab M, Oh DH. Emerging trends in smart self-healing coatings: A focus on micro/nanocontainer technologies for enhanced corrosion protection. *Coatings.* 2024;14:324.
5. Li Q, Yin R, Zhang D, Liu H, Chen X, Zheng Y, Guo Z, Liu C, Shen C. Flexible conductive MXene/cellulose nanocrystal coated nonwoven fabrics for tunable wearable strain/pressure sensors. *Journal of Materials Chemistry A.* 2020;8:21131–21141.
6. Manimaran S, Dhas AAG, Joy N, Raja K, Saikia K, Kumar S. *Analysis of Surface Texture of Ni-based Super Alloy With Thermal Barrier Coating System (TBC) for Gas Turbine blade.* AIP Conference Proceedings; 2020.
7. Mrdak M, Lačnjevac Č, Bajić D. Characterization of zirconium oxidecoating stabilized with cerium and yttrium oxide deposited on the bonding coating nickel chromium aluminum cobalt yttrium oxide. *Materials Protection.* 2023;64:5–12.
8. Nehe P, Sivakumar G, Kumar S. Solution precursor plasma spray (SPPS) technique of catalyst coating for hydrogen production in a single channel with cavities plate type methanol based microreformer. *Chemical Engineering Journal.* 2015;277:168–175.
9. Kumar R, Jiang C, Cottom B, Gell M, Jordan EH. Development of solution precursor plasma spray (SPPS) yttrium aluminum garnet (YAG) Coatings for engine components using a high enthalpy cascaded arc gun: Part I. *Journal of Thermal Spray Technology.* 2023;32:1482–1504.

10. Joulia A, Bolelli G, Gualtieri E, Lusvarghi L, Valeri S, Vardelle M, Rossignol S, Vardelle A. Comparing the deposition mechanisms in suspension plasma spray (SPS) and solution precursor plasma spray (SPPS) deposition of yttria-stabilised zirconia (YSZ). *Journal of the European Ceramic Society.* 2014;34:3925–3940.
11. Fan W, Bai Y. Review of suspension and solution precursor plasma sprayed thermal barrier coatings. *Ceramics International.* 2016;42:14299–14312.
12. Gell M, Jordan EH, Teicholz M, Cetegen BM, Padture NP, Xie L, Chen D, Ma X, Roth J. Thermal barrier coatings made by the solution precursor plasma spray process. *Journal of Thermal Spray Technology.* 2008;17:124–135.
13. Omran K, Ahmed MS, Elmeniawi M, Sallam H, Naga S. A review on thermal barrier coatings. *The Egyptian International Journal of Engineering Sciences Technology.* 2021;36:1–6.
14. Sankar J, Kumar S. Synthesis of rare earth based pyrochlore structured (A2B2O7) materials for thermal barrier coatings (TBCs)-A review. *Current Applied Science Technology.* 2021;21(3):601–617.
15. Almeida E. Surface treatments and coatings for metals. A general overview. 1. Surface treatments, surface preparation, and the nature of coatings. *Industrial Engineering Chemistry Research.* 2001;40:3–14.
16. Ataie A, Mostaghimi J, Pershin L, Xu P. Fabrication of nanostructured cobalt ferrite coatings using suspension plasma spraying (SPS) technique. *Surface Coatings Technology.* 2017;328:451–461.
17. Bakan E, Vaßen R. Crack healing mechanisms in atmospheric plasma sprayed Yb-silicate coatings during post-process heat treatment. *Journal of the European Ceramic Society.* 2023;43:3684–3693.
18. Bakhsheshi-Rad H, Hamzah E, Ismail A, Daroonparvar M, Kasiri-Asgarani M, Jabbarzare S, Medraj M. Microstructural, mechanical properties and corrosion behavior of plasma sprayed NiCrAlY/nano-YSZ duplex coating on Mg–1.2 Ca–3Zn alloy. *Ceramics International.* 2015;41:15272–15277.
19. Bobzin K, Öte M, Knoch M. Surface pre-treatment for thermally sprayed ZnAl15 coatings. *Journal of Thermal Spray Technology.* 2017;26:464–472.
20. Bobzin K, Öte M, Linke T, Sommer J, Liao X. Influence of process parameter on grit blasting as a pretreatment process for thermal spraying. *Journal of Thermal Spray Technology.* 2016;25:3–11.
21. Brossard S, Munroe P, Tran A, Hyland M. Study of the splat formation for plasma sprayed NiCr on aluminum substrate as a function of substrate condition. *Surface Coatings Technology.* 2010;204:2647–2656.
22. Singh J, Singh R, Singh S, Vasudev H, Kumar S. Reducing scrap due to missed operations and machining defects in 90PS pistons. *International Journal on Interactive Design Manufacturing.* 2023;17:2527–2539.
23. Buchwalder A, Zenker R. Pre-and post-surface treatments using electron beam technology for load-related application of thermochemical and PVD hard coatings on soft substrate materials. *Surface Coatings Technology.* 2019;375:920–932.
24. Singh J, Singh S. Support vector machine learning on slurry erosion characteristics analysis of Ni-and Co-alloy coatings. *Surface Review Letters.* 2023:2340006. https://doi.org/10.1142/s0218625x23400061.
25. Chen H, Fan M, Zhu W, Norton A. High temperature oxidation behaviour of combustion flame sprayed CoNiCrAlY coatings. *Surface Coatings Technology.* 2020;385:125431.
26. Singh R, Toseef M, Kumar J, Singh J. Benefits and challenges in additive manufacturing and its applications. *Sustainable Advanced Manufacturing and Materials Processing.* CRC Press; 2022. pp. 137–157.

27. Singh J, Mohapatra S, Kumar S. Performance analysis of pump materials employed in bottom ash slurry erosion conditions. *Jurnal Tribologi.* 2021;30:73–89.
28. Sunitha K, Vasudev H. A short note on the various thermal spray coating processes and effect of post-treatment on Ni-based coatings. *Materials Today: Proceedings.* 2022;50:1452–1457.
29. Cinca N, Lima CRC, Guilemany JM. An overview of intermetallics research and application: Status of thermal spray coatings. *Journal of Materials Research Technology.* 2013;2:75–86.
30. Mehta A, Vasudev H, Singh S, Prakash C, Saxena KK, Linul E, Buddhi D, Xu J. Processing and advancements in the development of thermal barrier coatings: A review. *Coatings.* 2022;12:1318.
31. Yedida VS, Vasudev H. A review on the development of thermal barrier coatings by using thermal spray techniques. *Materials Today: Proceedings.* 2022;50:1458–1464.
32. Lin J, Stinnett TC. Development of thermal barrier coatings using reactive pulsed dc magnetron sputtering for thermal protection of titanium alloys. *Surface Coatings Technolog.* 2020;403:126377.
33. Prashar G, Vasudev H, Mehta A. Application of Bimodal stuructured thermal sprayed coatings in power plant components. *Advanced Ceramic Coatings for Energy Applications.* 2024:273–287. https://doi.org/10.1016/B978-0-323-99620-4.00012-9.
34. Daroonparvar M, Bakhsheshi-Rad HR, Saberi A, Razzaghi M, Kasar AK, Ramakrishna S, Menezes PL, Misra M, Ismail AF, Sharif S. Surface modification of magnesium alloys using thermal and solid-state cold spray processes: Challenges and latest progresses. *Journal of Magnesium Alloys.* 2022;10:2025–2061.
35. Yedida VS, Mehta A, Vasudev H, Singh S. Role of numerical modeling in predicting the oxidation behavior of thermal barrier coatings. *International Journal on Interactive Design Manufacturing.* 2024;18:5653–5662.
36. Di Girolamo G, Blasi C, Brentari A, Schioppa M. Microstructural, mechanical and thermal characteristics of zirconia-based thermal barrier coatings deposited by plasma spraying. *Ceramics International.* 2015;41:11776–11785.
37. Di Girolamo G, Blasi C, Pagnotta L, Schioppa M. Phase evolution and thermophysical properties of plasma sprayed thick zirconia coatings after annealing. *Ceramics International.* 2010;36:2273–2280.
38. Di Girolamo G, Blasi C, Schioppa M, Tapfer L. Structure and thermal properties of heat treated plasma sprayed ceria–yttria co-stabilized zirconia coatings. *Ceramics International.* 2010;36:961–968.
39. Fan X, Zou B, Gu L, Wang C, Wang Y, Huang W, Zhu L, Cao X. Investigation of the bond coats for thermal barrier coatings on Mg alloy. *Applied Surface Science.* 2013;265:264–273.
40. Fanicchia F, Axinte D, Kell J, McIntyre R, Brewster G, Norton A. Combustion flame spray of CoNiCrAlY & YSZ coatings. *Surface Coatings Technology.* 2017;315:546–557.
41. Fanicchia F, Axinte DA. Transient three-dimensional geometrical/thermal modelling of thermal spray: Normal-impinging jet and single straight deposits. *International Journal of Heat Mass Transfer.* 2018;122:1327–1342.
42. Fanicchia F, Maeder X, Ast J, Taylor A, Guo Y, Polyakov M, Michler J, Axinte D. Residual stress and adhesion of thermal spray coatings: Microscopic view by solidification and crystallisation analysis in the epitaxial CoNiCrAlY single splat. *Materials Design.* 2018;153:36–46.
43. Fauchais PL, Heberlein JV, Boulos MI, Fauchais PL, Heberlein JV, Boulos MI. Overview of thermal spray. *Thermal Spray Fundamentals: From Powder to Part.* 2014:17–72.

44. Fauchais PL, Heberlein JV, Boulos MI, Fauchais PL, Heberlein JV, Boulos MI. Industrial applications of thermal spraying technology. *Thermal Spray Fundamentals: From Powder to Part*. 2014:1401–1566.
45. Gan JA, Berndt CC. Nanocomposite coatings: Thermal spray processing, microstructure and performance. *International Materials Reviews*. 2015;60:195–244.
46. Dutta V, Thakur L, Singh B, Vasudev H. A study of erosion–corrosion behaviour of friction stir-processed chromium-reinforced NiAl bronze composite. *Materials*. 2022;15:5401.
47. Kattimani MA, Venkatesh P, Masum H, Math MM, Bahadurdesai VN, Mustafkhadri S, Prasad CD, Vasudev H. Design and numerical analysis of tensile deformation and fracture properties of induction hardened inconel 718 superalloy for gas turbine applications. *International Journal on Interactive Design Manufacturing*. 2024;18:5939–5949.
48. Kulkarni SD, Manjunatha, Chandrasekhar U, Manjunath K, Prasad CD, Vasudev H. Design and optimization of polyvinyl-nitride rubber for tensile strength analysis. *International Journal on Interactive Design Manufacturing*. 2024;18:5893–5908.
49. Majji BGR, Vasudev H, Bansal A. A review on the oxidation and wear behavior of the thermally sprayed high-entropy alloys. *Materials Today: Proceedings*. 2022;50:1447–1451.
50. Poojari M, Hanumanthappa H, Prasad CD, Jathanna HM, Ksheerasagar AR, Shetty P, Shanmugam BK, Vasudev H. Computational modelling for the manufacturing of solar-powered multifunctional agricultural robot. *International Journal on Interactive Design Manufacturing*. 2023;18:1–12.
51. Prashar G, Vasudev H, letters. A review on the influence of process parameters and heat treatment on the corrosion performance of Ni-based thermal spray coatings. *Surface Review Letters*. 2022;29:2230001.
52. Garcia-Alonso D, Serres N, Demian C, Costil S, Langlade C, Coddet C. Pre-/during-/post-laser processes to enhance the adhesion and mechanical properties of thermal-sprayed coatings with a reduced environmental impact. *Journal of Thermal Spray Technology*. 2011;20:719–735.
53. Ghadami F, Aghdam ASR. Improvement of high velocity oxy-fuel spray coatings by thermal post-treatments: A critical review. *Thin Solid Films*. 2019;678:42–52.
54. Singh J, Kumar S, Mohapatra S. Study on solid particle erosion of pump materials by fly ash slurry using Taguchi's orthogonal array. *Tribologia-Finnish Journal of Tribology*. 2021;38:31–38.
55. Gisario A, Puopolo M, Venettacci S, Veniali FJIJoRM, Materials H. Improvement of thermally sprayed WC–Co/NiCr coatings by surface laser processing. *International Journal of Refractory Metals Hard Materials*. 2015;52:123–130.
56. Góral M, Kotowski S, Dychtoń K, Drajewicz M, Kubaszek T. Influence of low pressure plasma spraying parameters on MCrAlY bond coat and its microstructure. *Key Engineering Materials*. 2014;592:421–424.
57. Prashar G, Vasudev H. Surface topology analysis of plasma sprayed Inconel625-Al2O3 composite coating. *Materials Today: Proceedings*. 2022;50:607–611.
58. Sharanabasava H, Prasad CD, Ramesh M. Effect of Mo-and SiC-reinforced NiCr microwave cladding on microstructure, mechanical and wear properties. *Journal of The Institution of Engineers: Series D*. 2023;104:539–551.
59. Govande AR, Chandak A, Sunil BR, Dumpala R. Carbide-based thermal spray coatings: A review on performance characteristics and post-treatment. *International Journal of Refractory Metals Hard Materials*. 2022;103:105772.

60. Saxena A, Mehta A, Vasudev H, Prashar G, Jasim NY. A short review on machine learning for the purpose of optimizing and predicting the properties of polymeric nanocomposites. *Materials Today: Proceedings.* 2023. https://doi.org/10.1016/j.matpr.2023.07.347.
61. Gupta M, Li X-H, Peng R, Ottosson A, Gillberg P, Girgulis J. Understanding the effect of bondcoat surface treatment on enhanced lifetime of suspension plasma sprayed thermal barrier coatings. *Surface Coatings Technology.* 2024;482:130716.
62. Singh J, Gill HS, Vasudev H. Computational fluid dynamics analysis on role of particulate shape and size in erosion of pipe bends. *International Journal on Interactive Design Manufacturing.* 2023;17:2631–2646.
63. Hejrani E, Sebold D, Nowak W, Mauer G, Naumenko D, Vaßen R, Quadakkers W. Isothermal and cyclic oxidation behavior of free standing MCrAlY coatings manufactured by high-velocity atmospheric plasma spraying. *Surface Coatings Technology.* 2017;313:191–201.
64. Hospach A, Mauer G, Vaßen R, Stöver D. Columnar-structured thermal barrier coatings (TBCs) by thin film low-pressure plasma spraying (LPPS-TF). *Journal of Thermal Spray Technology.* 2011;20:116–120.
65. Hospach A, Mauer G, Vaßen R, Stöver D. Characteristics of ceramic coatings made by thin film low pressure plasma spraying (LPPS-TF). *Journal of Thermal Spray Technology.* 2012;21:435–440.
66. Ito K, Shima T, Fujioka M, Arai M. Improvement of oxidation resistance and adhesion strength of thermal barrier coating by grinding and grit-blasting treatments. *Journal of Thermal Spray Technology.* 2020;29:1728–1740.
67. Jamali H, Mozafarinia R, Razavi RS, Ahmadi-Pidani R. Comparison of thermal shock resistances of plasma-sprayed nanostructured and conventional yttria stabilized zirconia thermal barrier coatings. *Ceramics International.* 2012;38:6705–6712.
68. Koivuluoto H. A review of thermally sprayed polymer coatings. *Journal of Thermal Spray Technology.* 2022;31:1750–1764.
69. Kromer R, Costil S, Verdy C, Gojon S, Liao H. Laser surface texturing to enhance adhesion bond strength of spray coatings–Cold spraying, wire-arc spraying, and atmospheric plasma spraying. *Surface Coatings Technology.* 2018;352:642–653.

9 An Art of Nanostructures in Energy-Efficient Solar Cell Applications

M. Abila Jeba Queen, Y. Julit, K.C. Bright, and Arish Dasan

9.1 INTRODUCTION

Solar energy represents the most abundant source of energy that has been produced and utilized for billions of years. Energy from renewable sources is completely pollution-free and almost clean. The energy that reaches Earth is approximately 4×10^{15} MW, which is 200 times more than the current global need. The usage of solar power has grown rapidly and it now it accounts for 2.1% of global energy [1]. Global awareness about the necessity of renewable energy pushes the evolution of solar energy technologies producing a green transition by changing its design and composition. The availability of solar resources is still limitless around the world but fabricating solar cells with minimum cost and good efficiency is a challenge. From a research point of view, researchers are incorporating nanotechnology into solar energy technology, thereby many frameworks have been formulated with different configurations in order to enhance efficiency [2]. Currently, there are three generations of solar cells that have been categorized according to their dimensions: the first generation is composed of crystalline silicon, the second generation is made of thin film solar cells, and the third generation is made up of dye-sensitized solar cells (DSSC), polymer solar cells (PSC), and perovskite solar cells [3].

Solar cells made in the form of crystalline silicon are considered as major part of the world's solar cell market due to their properties of non-toxicity, tuneable band gap, and availability of silicon. The crystalline silicon solar cell reaches 27% practical efficiency at an energy gap between 1.35 and 1.5 eV [4]. Next generation thin film solar cells have been introduced with certain advantages, such as tuneable band gap, low operational cost, high open circuit voltage, and high scalability. A gallium arsenide film solar cell produces the highest output efficiency of about 29% and works at a low-temperature coefficient and has minimal external conditions effect. Third-generation solar cells include DSSC, PSC, hybrid multiple junction solar cells, organic solar cells, and perovskite solar cells [5]. Because of the specific properties of DSSCs, they are simple in structure and low cost, possibly replacing silicon solar cells, but they have the limitation of only using infrared light. Initially, researchers

 DOI: 10.1201/9781003516071-9

were able to attain an efficiency of 7%, but due to their continuous efforts it can now reach up to 12% [6]. With the benefit of the simplest structure and processing, cost-effectiveness, and usage of large-area applications, organic solar cells have reached 15% efficiency in halogenated solvent [7]. Perovskite solar cells have a high absorption coefficient, carrier mobility, and tuneable band gap, and achieve 23% efficiency at a low cost. An interesting way to improve the efficiency of the solar cell is by fabricating multi-junction cells; the literature reported that the maximum attainable efficiency is about 47% in the case of six multi-layered junctions [8].

Quantum dots are a special type of semiconductor nanocrystal, with the benefits of tuneable band gap and intermediate bands, it can be easily processed, and easily molded into sheets and arrays. As the quantum dot size reaches the material exciton Bohr radius size, quantum confinement effects play a major role thus electron energy levels are considered as a discrete one. The tuneable band gap of quantum dots made a significant contribution to the construction of nanostructure solar cells, which harvest more solar energy from the sun. Due to the intrinsic dipole moments that emit a maximum of three photons because of multiple exciton generations, this lead to a drastic increase in the solar power efficiency of 20% to 65% [9].

In order to harvest sunlight, as early as 1980 a wide band of material was sensitized with dye that generates charge carriers; this was developed commercially as dye-sensitized solar cells [10]. The light harvesting efficiency of the semiconductor material electrode with loaded dye is very low, this in turn produces very low solar cell efficiency. In silicon material both sensitized cells decouple the charge transport and production happens in the same material. But in sensitized devices, charge generation is created inside the dye and photo-generated charges are injected into a transport media; this media only allows the transportation of charge carriers and it limits the recombination of charge carriers, as a result the fabrication cost of dye-sensitized solar cells is greatly reduced [11]. When the light irradiates on a dye-sensitized device, quantum dots or nanomaterials absorb the sun's energy, thus the electrons are excited from the valence band to the conduction band and generate electron–hole pairs. Charge separation is attained by the variation of the conduction band gap in the quantum dot and metal oxide; the conduction band electrons of the quantum dot are trapped in the TiO_2 metal oxide conduction band. The carrier electrons diffuse from the porous titanium oxide to the metal oxide substrate and reach the counter electrode; the other side of the quantum dot is regenerated by the redox couple and then regeneration is significantly reduced by the catalysis followed by the recombination mechanism, which deteriorates cell performance [12].

Nanofluids are different from conventional fluids as they experience better thermal properties. They are utilized in different fields such as heat exchangers, automotive industries, solar power generation, etc. The dispersion and suspension of the nanoparticles in the host fluid affords the highest thermal properties. Heat transfer is a most important property in solar collectors and plays an important role in energy-saving applications with varying operational temperatures. Nanofluids contain particles of less than 100 nm [13, 14]. These nanofluids are mainly used for heat transfer in different applications, such as fuel cells, vehicle thermal management, and boiler flue gas temperature. The good thermos-physical properties of nanofluids increase the efficiency of the solar cell.

In this chapter, the potential roles of nanostructures for solar cell applications are analyzed in detail. From the research, it was found that nanomaterials with semiconductor oxides, carbon nanotubes, and quantum dots have been investigated thoroughly for enhancing conversion efficiency. The core of the chapter deals with solar cells based on nanomaterials, quantum dots-based solar cells, and nanofluid-based solar collectors. Overall, a comprehensive outline of improving solar cell adsorption efficiency of nanostructures is discussed.

9.2 SOLAR ENERGY TECHNOLOGIES

Solar energy is a sustainable and renewable energy resource available on an enormous scale and also used as an alternative source of power. Using advanced technology one can harvest electricity from the sun's energy through photovoltaic and solar thermal collectors [1].

French physicist Edmond Becquerel in 1839 discovered that a solid material generates an electric current when exposed to sunlight. Due to the photovoltaic effect, certain materials convert the sun's energy into electrical energy. This was initially studied by Adam and Day in the year 1876, and later they were able to construct solar cells from selenium material with a maximum efficiency of 1 to 2%. Albert Einstein in the year 1904 explained the photovoltaic effect based on photon theory [15] (as shown in Figure 9.1).

In 1916 a Polish scientist Jan Czochralski invented the process of producing pure crystalline solids which was a big breakthrough in modern electronics (Figure 9.2). Initially, prepared silicon solar cells showed about 6% efficiency [16] due to the continuous effort by researchers in the field of solar technology to develop solar cells with higher efficiency, but this came with the disadvantage of higher fabrication costs. This disadvantage can be overcome with the usage of green- and nanotechnology.

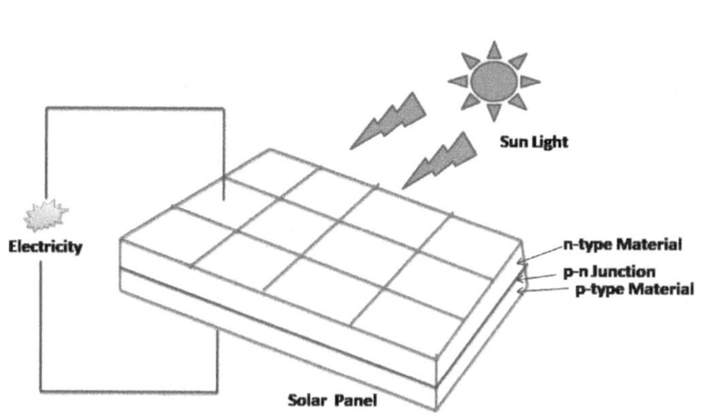

FIGURE 9.1 Schematic diagram of the photoelectric effect (Credit statement: created by author).

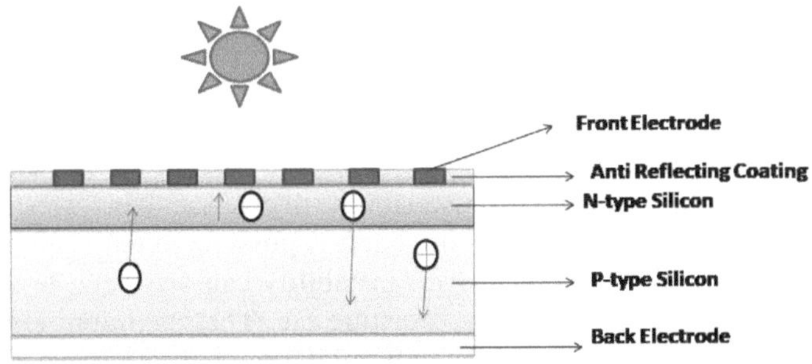

FIGURE 9.2 Photoelectric effect in solar cells (Credit statement: created by author).

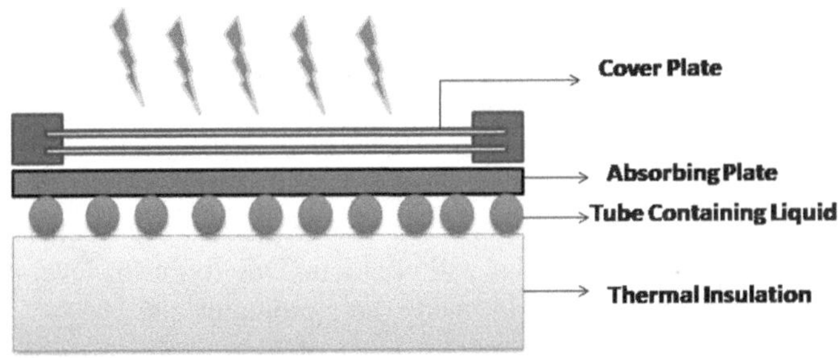

FIGURE 9.3 Schematic representation of solar thermal collector (Credit statement: created by author).

A solar thermal collector consists of heat transfer liquids, such as water or non-freezing liquids (Figure 9.3). A flat plate solar collector, which is utilized in solar water heaters, produces higher efficiency than a direct solar energy conversion system [17]. Solar collectors in the form of flat plates are used mostly in domestic applications like space heating, water-heating, solar dryers, and low-temperature appliances. Due to the advancement of solar energy technology, recent solar collectors are made by concentrating solar power systems with mirrors, and lenses that capture large quantities of sunlight. Wide ranges of solar collectors with different shapes are also available in the market: parabolic reflectors, Fresnel reflectors, sterling dish, and tower collectors. Nowadays higher power efficiency is attained by combining conventional technology with nanotechnology and developing innovative nanofluid-based solar collectors [1].

9.3 NANOMATERIALS-BASED SOLAR CELLS

At the beginning of 2009 perovskite solar cells were developed with the perovskite structure ABX_3, where A represents a cation consisting of organic and inorganic materials, and B and X are a metal cation and halide anion, respectively. The most effective way to increase efficiency can be attained because of the photon's absorption energy; the energy is greater than the band gap of the semiconductor; moreover, the excess energy is released in the form of thermal energy. Therefore to improve efficiency, various nano-absorbers with different band gaps can be combined as tandem solar cells, which reduces thermal loss [18]. Another important factor that determines the efficiency of the perovskite solar cell is its fabrication, the purity of the material, and the method used for the deposition. It involves energy cost, raw material selection, and payback time, but it is possible to fabricate using a low-temperature solution technique. However, instability can arise due to atmospheric conditions like temperature, humidity, moisture etc. The maximum efficiency one can harvest through the perovskite solar cell is 25% by incorporating titanium oxide and the efficiency improved by 4% [19].

The schematic representation of solar cells made up of nanostructures is given in Figure 9.4, consisting of nanomaterials, contact junctions to provide the electrical contact, and substrates. The substrate used in solar cell fabrication is made up of glass and the top contact is a transparent conductor that allows absorption of radiation. Nanomaterials appear in various forms such as nanorods, nanotubes, dots, or any form incorporated in between the contact surface. The photons from the sun's radiation are absorbed by the nanomaterials effectively due to the increased density of states of the system. The electric field separates the electron–hole pairs that are collected by the bottom solar cell contacts. Due to Fermi-level pinning and nanoscale, the electric field is built-in inside the nanomaterials.

The solar cell consists of a transparent layer of indium tin oxide that acts as a top electrode due to its conductivity and optical properties. However indium is a rare metal with crystal defects, low melting temperature, high cost, and less flexibility; researchers are attracted to solar cells free of indium [20]. Carbon nanotubes are excellent materials that have been used as solar collectors because of their exceptional electrical, thermal, photonic, and mechanical properties. Luo et al.[21] replaced the indium tin oxide with titanium oxide with carbon nanotubes. They found that single-wall carbon nanotube were synthesized cheaply with a solution or deposition technique [21]. In order to increase efficiency, carbon nanoparticles were incorporated into titanium oxide, an electron transporting layer, but the doping concentration had to be controlled precisely; more dopants decrease its efficiency [22].

FIGURE 9.4 Schematic representation of the nanostructure-based solar cell. (Credit statement: created by author).

9.3.1 Quantum Dot-Based Solar Cells

Quantum dot-based solar cells have been considered as having a large potential because of their maximum thermodynamic conversion efficiency, reported as a maximum of 66% when increasing conversion efficiency either by increasing photocurrent or photovoltage (Figure 9.5). The photoelectrodes with quantum dot arrays are made up of 3D arrays with small quantum dots leading to long-range electron transport by the formation of electronic coupling and mini bands. The delocalized 3D mini bands slow down the cooling and transport the carriers to produce higher photo potential. Ionization may also be expected in quantum dot arrays, increasing photocurrent. However, the above two effects cannot occur simultaneously [23].

Quantum dot-based solar cells are fabricated by sandwiching metal-semiconductor junction quantum dots between a metallic electrode and a counter. Lead selenide (PbSe) quantum dot solar cells are prepared by the process of dip coating, in which PbSe quantum dot films of a thickness of 140 nm have been reported with a 2.2% efficiency [24]. With 150 nm layered metal-semiconductor junctions, a sandwich of ZnO and PbS dots produce an efficiency of 1.53 %. Photoelectrons are generated and pass through the metallic electrode with higher probability but the disadvantage that arises in this model is a lower reduction of hole–current injection at metallic contact [25]. By introducing a transition metal oxide in between the metal electrode and quantum dot, the hole–current injections are highly reduced. In a PbS quantum dot, the addition of transition metals oxides molybdenum and vanadium reduces hole extraction and increases the 4.4% conversion efficiency, thereby enhancing binding and making the cell efficient for hole extraction [26].

9.3.2 Quantum Dot Sensitized with Solar Cells

Quantum dots sensitized with solar cells are the next promising new area of solar cells based on dye-sensitized nanoparticles with a quantum yield greater than 1.0 due to impact ionization [23]. The basic principle and idea behind the quantum dot sensitized with the solar cells and the dye-sensitized cells are almost similar, the only difference is quantum dots are the current injection source (Figure 9.6). In order to improve the efficiency of the quantum dot sensitized with solar cells, the following parameters have to be under consideration they are conducting electrodes, counter

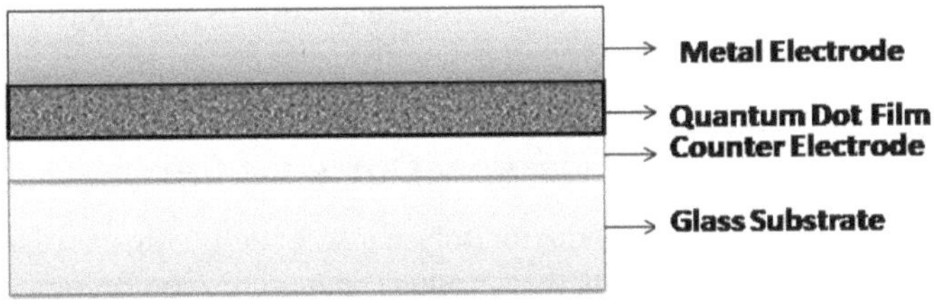

FIGURE 9.5 Schematic diagram of the quantum dot-based solar cell (Credit statement: created by author).

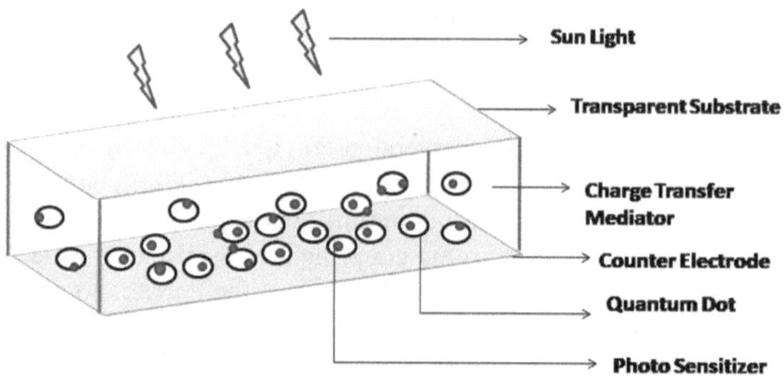

FIGURE 9.6 Schematic representation of a quantum dot sensitized with solar cells (Credit statement: created by author).

electrodes, TiO_2 nanostructure layer, energy levels of quantum dot and the electrolyte used in the cell.

Furthermore, the incorporation of CdSe, CdS and CdTe quantum dot sensitizer with proper energy gap matching improves the efficiency [27]. Investigation on the effect of dot size in the solar cell was studied by Kongkanand et al. [28]. and their result revealed that size confinement of cadmium selenide quantum dots results in higher charge injection rate and IPCE whereas larger particles produce well visible region absorption.

The addition of dye to the semiconductor quantum dots has been under research interest in recent years because natural or artificial dyes enhance the power conversion efficiency. Some of them reported the dyes extracted from the fruits also. Zaban et al [29]. Quantum dot incorporated dye bi-layer sensitized solar, made up of TiO_2 nanocrystal, CdS quantum dot and N719 dye. The configuration increased the charge separation efficiency which brings in a 25 per cent increase in cell efficiency compared with monolayer quantum dot cells. The same research group synthesized colloidal quantum dots in the form of core, shell and absorption is tuned by SQ02 dye molecules. Higher energy transfer efficiency and improved solar cell stability properties are attained when the quantum dots in the form of colloids are incorporated into TiO_2 photoelectrode. The efficiency can be further enhanced by incorporating quantum dots with some other dye and used as a sensitizer. The dye-sensitized quantum dots cells performance has further increased by utilizing different nanostructures, optimizing the preparation technique, using alternate wide band gap semiconductors instead of TiO_2 and using different electrolytes.

9.3.3 Polymer Incorporated Quantum Dot Solar Cells

Quantum dot with organic semiconductor polymer solar cells consists of quantum dot junctions with polymeric materials that produce many electron-hole pairs in which the quantum dots are dispersed into the semiconductor organic mixture shown in Figure 9.7 [23]. In 2002 hybrid nanorod polymer-added solar cells were fabricated

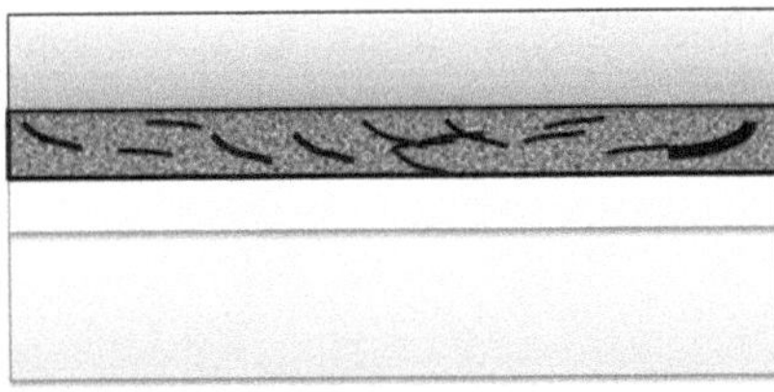

FIGURE 9.7 Schematic diagram quantum dot polymer solar cells (Credit statement: created by author).

by sandwiching the quantum dot in a polymer with a power conversion efficiency of 1.7%. The polymer material MEH-PPV is dispersed in CdSe quantum dots, when exposing light the photo-generated holes are traced into the polymer matrix and attain 1.8% power conversion efficiency [30].

9.4 NANOFLUID-BASED SOLAR COLLECTORS

Nanofluids are a solid liquid mixture, developed by dispersing metallic or non-metallic solid nanoparticles into liquids. The dispersing nanomaterials may be nanoparticles, nanofibers, nanowires and nanorods with sizes lesser than 100 nm. When the nanomaterials are mixed with liquids, there is molecular scale interaction and conducts heat effectively. Choi in 1995 developed the concept of heat transfer at the Argonne National Laboratory, USA [31]. Due to the combined effect of heat transfer phenomena such as conduction and convection and the effect of collision in which energy transfer through γ-particles, nanofluids bring excellent thermal conductivity, diffusivity, heat transfer, viscosity etc. In recent years, research on nanofluids has focused on the fields of solar thermal storage solar collectors [32, 33].

Nanofluids are classified as metallic nanofluids and non-metallic nanofluids into two broad categories (Figure 9.8). Metallic nanofluid contains Cu, Zn, Si, Al, Fe, Ti, Ni, Ag and Au metallic nanoparticles. Non-metallic nanofluid contains aluminum oxide, zinc oxide, titanium oxide, copper oxide, silicon carbide, carbon tubes and composites. The enhancement in the thermochemical properties depends on the quantity, size and shape of nanoparticles [34]. Based on the composition, the nanofluids can be further classified as mono nanofluid and hybrid nanofluid. The mono nanofluids are metals, metal oxides, carbides, nitride and carbon with base fluid but the hybrid nanofluids are composed of a mixture of nanomaterials and composites and base fluid. [35]. Tracing the thermochemical properties importantly the heat transfer phenomena of nanofluid one can apply nanofluid for the solar collector application. This technology enhances the efficiency and performance of the system [36]. Nanofluid is composed of base fluid and nanoparticles as illustrated in Figure 9.9 and a schematic representation is given in Figure 9.10.

Using nanofluid, the efficiency of the solar collector was potentially increased by 5–10% with graphite/therminol VP-1 and aluminum/therminol VP-1base of 0.05% volume concentration. [37, 38]. Solar collector's efficiency improved by changing the

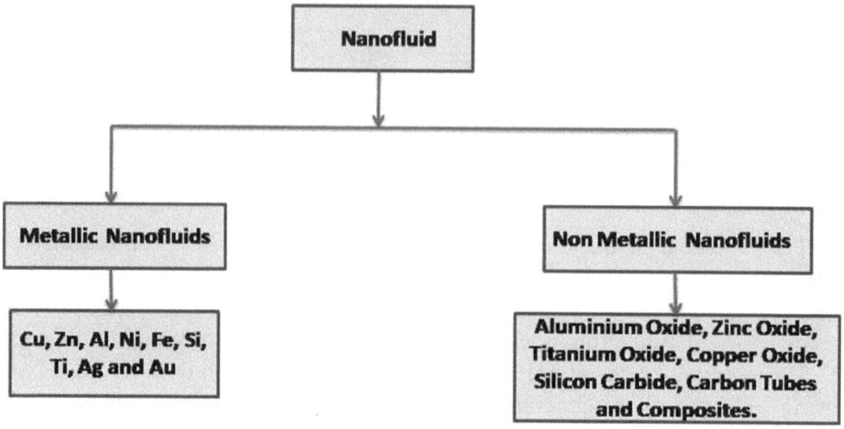

FIGURE 9.8 Nanofluids classification (Credit statement: created by author).

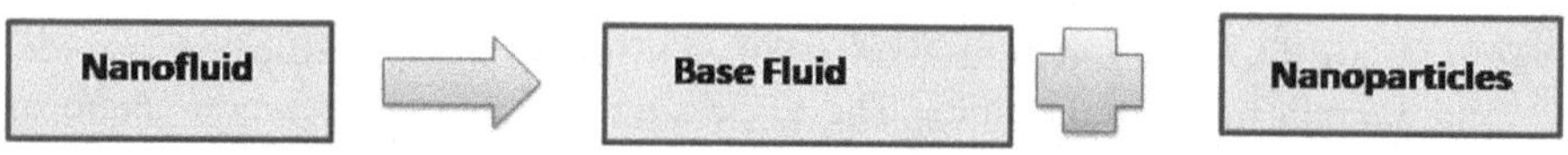

FIGURE 9.9 Composition of nanofluid (Credit statement: created by author).

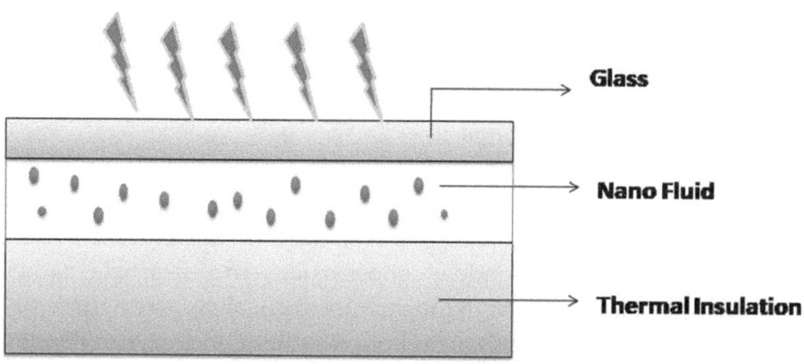

FIGURE 9.10 Schematic representation of nanofluid (Credit statement: created by author).

particle volume fraction, collector height and glass cover transmissivity. This phenomenon was examined by aluminum oxide nanoparticles/water base of different volume fractions 0.8% to 1.6% increasing the efficiency by 8% [39].

About 95% of the incoming sunlight was observed by the nanofluids aluminum oxide nanoparticles/water, zinc oxide nanoparticles/water and magnesium oxide nanoparticle/water nanofluids lesser than 10 ppm volume fraction [40]. The effect of the surfactants and pH of the isoelectric point on the nanofluid enhances solar collector efficiency by 28% proven with multiwall carbon nanotubes/water nanofluid and surfactant Triton X-100 [41].

The effect of nanofluids as the absorption medium enhances its efficiency. The absorption medium of graphite nanoparticles with particle size 30 nm leads to a 3% increase in efficiency, a 5% increase arises by silver nanoparticles of particle size varies between 20–40 nm and efficiency rises when the particle size reduces [42].

Carbon nanotubes/water nanofluid experiences excellent light heat conversion characteristics compared to titanium oxide/ water nanofluid. Using multi-walled carbon nanotubes about 37% solar collector size reductions are possible [43]. The production and usage of solar collector using nanoparticle-enhanced Ionic Liquids are reported and their finding reveals that enhanced thermal conductivity of 5% based on the type of base fluid and its ionic concentration. The heat capacity of aluminum oxide nanofluids was enhanced by 23% whereas 26% for silica nanofluids [44]. Nanofluid made up of Fe_3O_4/water with varying magnetic field also increases its efficiency, without field its efficiency rises 24% with the application of 50 Hz frequency the efficiency increase as 28% [45]. Al_2O_3 nanoparticles/water of 0.4 wt% attains a maximum of 46.7% efficiency whereas copper nanoparticles/water improves about 76.8% [46].

Base fluid with a mixture of ionic liquid and water composed of Ti_3C_2 nanoparticles solar cells have more thermal stability and show the highest efficiency 81.2% at 0.2 wt% concentration [47]. Higher thermal stability of the solar cell has been achieved with the Ti_3C_2 nanoparticles and soybean oil as a base fluid achieving a higher efficiency of 84.3% by lowering the thermal surface temperature compared to the water base [48]. An optical filtration channel made with Ag/water nanofluid operates at the temperature of 45° C reaches an efficiency of 67%, thereby the nanofluid flows over the solar cell [49]. Table 9.1 shows the increase in the efficiency of nanofluids.

TABLE 9.1
Nanofluid with Increased Efficiency

Nanoparticles	Base Fluid	Increase in Efficiency (%)	References
Aluminium	Graphite / therminol VP-1	5–10	[37, 38]
Aluminium oxide	Water	8	[39]
Multi-wall carbon nanotubes	Water	28	[41]
Carbon nanotubes	Water	37	[43]
Fe_3O_4	Water	28	[45]
Al_2O_3	Water	76.8	[46]
Ti_3C_2	Water	81.2	[47]
Ti_3C_2	Soybean oil	84.3	[48]

Stability is a major factor which influences the nanoparticles and the fluid motion. Brownian motion leads to which the small-sized nanoparticles are well distributed in the fluid and the nanoparticles are settled at the bottom due to gravity; moreover, Vander Waals attraction forces and the electric repulsive force also influence the nanofluid [50]. The instability generates a loss of nanofluid properties and sedimentation. According to the research, there are four effective ways to improve this stability: continuous agitation and stirring make good dispersion, the usage of surfactants and they are compatible with base fluid, pH can be modified by the addition of non-reactive alkaline or acid solution and functionalization.

Even though the nanofluid increases the thermal properties thereby the collector's efficiency increases, the usage of nanofluids as solar collectors faces many challenges. It is different to fabricate nanofluids with a uniform distribution of nanoparticles. The fabrication cost of nanofluid is much higher. A long-time usage of nanofluids causes erosion and corrosion. Nanofluids raise the viscosity and pumping power [51].

9.5 CONCLUSION

Global population rises and modern industrialization leads to scarcity of resources, climatic change and pollution. In order to sustain life and save the environment, we must develop an innovative approach for the massive scaling of renewable energy resources. Major research has to be developed in the field of energy transport mechanisms and harvesting the maximum amount of green energy. Solar thermal engineering industries have been working with the objective of increasing heat transfer rates. The biggest challenge identified in the solar industry is the production of cost-effective solar cells. In conventional solar cells photons with energies greater than band gap are produced but the energy is wasted in the form of heat energy. This can be overcome by means of nanostructure materials in which the band gap can be tuned to produce higher conversion efficiency. In quantum dots, due to the size variation effect and intermediate bands are processed with organic polymers and sensitized dyes making the solar cells efficient. For solar thermal applications, nanofluid plays a pivotal role due to its higher transfer coefficients, high boiling point and moderate pressure. Moreover, ideal nanofluids for solar collector applications are not available yet.

This chapter provided an overview of the art of nanostructures in energy-efficient solar cell applications. A brief introduction to the fundamentals of solar cells, technology integrated with the nanomaterials, and different generations of solar cells was discussed. Solar technologies discussed the key factor in which the energy harnessing takes place were briefly elaborated in this chapter. Further, the application of nanomaterials for solar cells, quantum dots and dye-sensitized materials with their energy conversion efficiencies and their challenges were discussed. Later the advanced technology of nanofluids and its applications were discussed. The following conclusion could be made: Nanomaterials in solar cells improve efficiency but toxicity and cost-effectiveness depend on the selection of the materials; non-silicon solar cells are also possible but limited with service and stability; carbon nanotubes also enhance efficiency; nanofluid is an advanced technology with limitations of sedimentation, stability, and high production cost. Possible future work may be the development of different architectures, the combination of different nanoparticles with varying particle sizes, and eco-friendly nanoparticle fabrication in solar cells. Making use of solar energy with nanostructured material is probable in the future with the global focus of achieving clean and green energy.

REFERENCES

1. Nagarajan PK, Subramani J, Suyambazhahan S, Ravishankar Sathyamurthy. Nanofluids for solar collector applications: A review. *Energy Procedia*, 61, **2014**, 2416–2434.
2. Lina Wang, Mavd PR, Teles Ahmad Arabkoohsar, Haoshui Yu, Kamal AR, Ismail, Omid Mahian, Somchai Wongwises. A holistic and state-of-the-art review of nanotechnology in solar cells. *Sustain Energy Technol Assessments*, 54, **2022,** 102864.
3. Yang M, Wu J, Lan Z, Lin J, Huang M, Fan L. Hotspots, frontiers, and emerging trends of tandem solar cell research: A comprehensive review. *Int J Energy Res*, 46, **2022,**104–123. https://doi.org/10.1002/er.7074.
4. Andreani LC, Bozzola A, Kowalczewski P, Liscidini M, Redorici L. Silicon solar cells: Toward the efficiency limits. *Adv Phys X,* 4, **2019**, 12. https://doi.org/10.1080/23746149.2018.1548305.
5. Neukom M, Züfle S, Jenatsch S, Ruhstaller B. Opto-electronic characterization of third-generation solar cells. *Sci Technol Adv Mater*, 19, **2018**, 291–316. https://doi.org/10.1080/14686996.2018.1442091.
6. Nandan Arka G, Bhushan Prasad S, Singh S. Comprehensive study on dye sensitized solar cell in subsystem level to excel performance potential: A review. *Sol Energy*, 226, **2021**, 192–213. https://doi.org/10.1016/j.solener.2021.08.037.
7. Xue R, Zhang J, Li Y, Li Y. Organic solar cell materials toward commercialization. *Small,* 14, **2018**, 1–24. https://doi.org/10.1002/ smll.201801793.
8. Green M, Dunlop E, Hohl-Ebinger J, Yoshita M, Kopidakis N, Hao X. Solar cell efficiency tables. *Prog Photovoltaics Res Appl,* 29, **2021**, 3–15. https://doi.org/10.1002/pip.3371.
9. Khalil Ebrahim Jasim. Quantum dots solar cells. In Kosyachenk, Leonid A. **2015,** 303–331. IntechOpen.
10. Gerischer H, Michel-Beyerle ME, Rebentrost F, Tributsch H. Sensitization of charge injection into semiconductors with large band gap. *Electrochim Acta*, 13, **1968**, 1509–1515.
11. Bisquert J. *The Physics of Solar Cells*. **2017**, Boca Raton: CRC Press.
12. Zhao K, Pan Z, Zhong X. Capping ligand-induced self-assembly for quantum dot sensitized solar cells. *J Phys Chem Lett*, 7, **2016**, 406–417.
13. Wen D, Ding Y. Experimental investigation into convective heat transfer of nanofluids at the entrance region under laminar flow conditions. *Int J Heat Mass Tr*, 47, **2004**, 5181–5188.
14. Wen D, Ding Y. Effective thermal conductivity of aqueous suspensions of carbon nanotubes (CNT Nanofluids). *J Thermophys Heat Tr*,18, **2004**, 481–485.
15. Luque A, Hegedus S. *Handbook of Photovoltaic Science and Engineering*. **2003,** England: Wiley.
16. Chow TT. A review on photovoltaic /thermal hybrid solar technology. *Appl Energy*, 87, **2010**, 2: 365–379.

17. Jaisankar S, Ananth J, Thulasi S, Jayasuthakar ST, Sheeba KN. A comprehensive review on solar water heaters. *Renew Sustain Energy Rev*, 15, **2011,** 6: 3045–3050.
18. Wang R, Huang T, Xue J, Tong J, Zhu K, Yang Y. Prospects for metal halide perovskite-based tandem solar cells. *Nat Photonics*, 15, **2021,** 411–425. https://doi.org/10.1038/s41566-021-00809-8.
19. Yoo JJ, Seo G, Chua MR, Park TG, Lu Y, Rotermund F. Efficient perovskite solar cells via improved carrier management. *Nature*, 590, **2021**, 587–593. https://doi.org/10.1038/s41586-021-03285-w.
20. Rajanna PM, Lund PD, Nasibulin AG. Hybrid heterojunction solar cells based on single-walled carbon nanotubes and amorphous silicon thin films. *Wiley Interdiscip Rev Energy Environ,* **2021,** 1–16. https://doi.org/10.1002/wene.402.
21. Luo Q, Wu R, Ma L, Wang C, Liu H, Lin H. Recent advances in carbon nanotube utilizations in perovskite solar cells. *Adv Funct Mater*, 31, **2021**, 1–30. https://doi.org/10.1002/adfm.202004765.
22. Batmunkh M, Macdonald TJ, Shearer CJ, Bat-Erdene M, Wang Y, Biggs MJ. Carbon nanotubes in TiO_2 nanofiber photoelectrodes for high-performance perovskite solar cells. *Adv Sci,* 4, **2017,** 1–11. https://doi.org/10.1002/ advs.201600504.
23. Nozik AJ. Spectroscopy and hot electron relaxation dynamics in semiconductor quantum wells and quantum dots. *Annu Rev Phys Chem*, 52, **2001**, 193.
24. Luther M, Gao J, Lloyd MT, Semonin OE, Beard MC, Nozik AJ. Stability assessment on a 3% bilayer PbS/ZnO quantum dot heterojunction solar cell. *Adv Mater*, 22, **2010,** 33, 3704–3707.
25. Pattantyus Abraham AG, Kramer IJ, Barkhouse AR, Wang X, Konstantatos G, Debnath R, Levina L, Raabe I, Nazeeruddin MK, Gratzel M, Sargent EH. De-pleted-heterojunction colloidal quantum dot solar cells. *ACSNano*, 4, **2010**, 6, 3374–3380.
26. Gao J, Perkins CL, Luther JM, Hanna MC, Chen H-Yu, Semonin OE, Nozik AJ, Ellingson RJ, Beard MC. n-Type transition metal oxide as a hole extraction layer in PbS quantum dot solar cells. *Nano Lett*, 11, **2011**, 3263.
27. Peng A, Peng X. Formation of high-quality CdTe, CdSe, and CdS nanocrystals using CdO as precursor. *J Am Chem Soc,* 123, **2001,** 1, 183–184.
28. Kongkanand A, Tvrdy K, Takechi K, Kuno M, Kamat PV. Quantum dot solar cells. Tuning photoresponse through size and shape control of CdSe-TiO2 architecture. *J Am Chem Soc*, 130, **2008**, 12, 4007–4015.
29. Shalom M, Albero J, Tachan Z, Martínez Ferrero E, Zaban A, Palomares E. Quantum dot-dye bilayer-sensitized solar cells: Breaking the limits imposed by the low absorbance of dye monolayers. *J Phys Chem Lett*, 1, **2010**, 1134–1138.
30. Sun B, Marx E, Greenham NC. Photovoltaic devices using blends of branched CdSe nanoparticles and conjugated polymers. *Nano Lett,* 3, **2003,** 7, 961–963.
31. Choi SUS. Developments and applications of Non-newtonian flows. *ASME FED*, 66, **1995,** 99–105.
32. Das SK, Putra N, Thiesen P, Roetzel W. Temperature dependence of thermal conductivity enhancement for nanofluids. *ASME J Heat Tr*, 125, **2003**, 567–574.
33. Patel HS, Das SK, Sundarajan T, Nair AS, George B, Predeep T. Thermal conductivity of naked and monolayer protected metal nanoparticles based nanofluids: Manifestation of anomalous enhancement and chemical effects. *Appl Phys Lett*, 83, **2003,** 931–933.
34. Hong TK, Yang HS, Choi CJ. Study of the enhanced thermal conductivity of Fe nanofluids. *J Appl Phys*, 97, **2005**, 064311.
35. Bhanvase B, Barai D. *Introduction to Nanofluids. Nanofluids Heat Mass Transfer.* **2021**, Academic Press, 3–42. doi:10.1016/B978-0-12-821955-3.00011-X.

36. Omid Mahian, Ali Kianifar, Soteris, Kalogirou A, Loan Pop, Somchai Wongwises. A review of the applications of nanofluids in solar energy. *Int J Heat Mass Tr*, 57, **2013**, 582–594.
37. Taylor RA, Phelan PE, Otanicar TP, Walker CA, Nguyen M, Trimble S, Prasher R. Applicability of nanofluids in high flux solar collectors. *J Renew Sustain Energy*, 3, **2011,** 023104.
38. Khullar V, Tyagi H, Phelan PE, Otanicar TP, Singh H, Taylor RA. Solar energy harvesting using a nanofluids-based concentrating solar collector. *Proceedings of MNHMT2012 3rd Micro/Nanoscale Heat & Mass Transfer International Conference*, Atlanta, USA, **2012**.
39. Tyagi H, Phelan P, Prasher R. Predicted efficiency of a low-temperature nanofluid based direct absorption solar collector. *J Sol Energy Eng*, 131, **2009**, 041004.
40. Li Y, Xie H, Yu W, Li J. Investigation on heat transfer performances of nanofluids in solar collector. *Mater Sci Forum*, 694, **2011,** 33–36.
41. Yousefi T, Shojaeizadeh E, Veysi F, Zinadini S. An experimental investigation on the effect of pH variation of MWCNT-H2O nanofluid on the efficiency of flat-plate solar collectors. *Sol Energy*, 86, **2012,** 771–779.
42. Otanicar TP, Phelan PE, Prasher RS, Rosengarten G, Taylor RA. Nanofluid based direct absorption solar collector. *J Renew Sustain Energy*, 2, **2010,** 033102.
43. Faizal M, Saidur R, Mekhilef S. Potential of size reduction of flat-plate solar collectors when applying MWCNT nanofluid. *4th International Conference on Energy and Environment (ICEE 2013), Conference Series: Earth and Environmental Science*, 16, **2013,** 012004.
44. Titan Paul C, Morshed AKM M, Jamil A.Khan. Nanoparticle enhanced ionic liquids (NEILS) as working fluid for the next generation solar collector. *Procedia Engineering, 5th BSME International Conference on Thermal Engineering*, 56, **2013,** 631–636.
45. Rahmanian S, Rahmanian-Koushkaki H, Omidvar P, Shahsavar A. Nanofluid-PCM heat sink for building integrated concentrated photovoltaic with thermal energy storage and recovery capability. *Sustain Energy Technol Assess*, 46, **2021,** 101223. https://doi.org/10.1016/j.seta.2021.101223.
46. Rejeb O, Sardarabadi M, Menezo C, Passandideh-Fard M, Dhaou MH, Jemni A. Numerical and model validation of uncovered nanofluid sheet and tube type photovoltaic thermal solar system. *Energy Convers Manag,* 110, **2016**, 367–377. https://doi.org/10.1016/j.enconman.2015.11.063.
47. Das L, Habib K, Saidur R, Aslfattahi N, Yahya SM, Rubbi F. Improved thermophysical properties and energy efficiency of aqueous ionic liquid/mxene nanofluid in a hybrid pv/t solar system. *Nanomaterials*, 10, **2020**,1–26. https://doi.org/10.3390/nano10071372.
48. Rubbi F, Habib K, Saidur R, Aslfattahi N, Yahya SM, Das L. Performance optimization of a hybrid PV/T solar system using Soybean oil/MXene nanofluids as A new class of heat transfer fluids. *Sol Energy*, 208, **2020,** 124–138. https://doi.org/10.1016/j.solener.2020.07.060.
49. Abdelrazik AS, Saidur R, Al-Sulaiman FA. Investigation of the performance of a hybrid PV/thermal system using water/silver nanofluid-based optical filter. *Energy*, 215, **2021**, 119172. https://doi.org/10.1016/j.energy.2020.119172.
50. Angayarkanni SA, Philip J. Review on thermal properties of nanofluids: Recent developments. *Adv Colloid Interface Sci*, 225, **2015,** 146–176. https://doi.org/ 10.1016/j.cis.2015.08.014.
51. Mangesh Gupta, Ram Bilas Prasad. Potential use of nanofluids in solar collectors: A review. *S-JPSET*, 12, **2020**, 1.

10 Nanomaterials for Engineering Applications

Nigar K. Mujawar, Babaso V. Udugade, Ganesh D. Mote, Shivaji B. Sadale, and Sawanta S. Mali

Description	Abbreviation
Circular economy strategies	CES
Environmental Protection Agency	EPA
European Chemicals Agency	ECHA
Global Positioning System	GPS
Greenhouse gas	GHG
Heating, ventilation, and air conditioning	HVAC
Metal–organic frameworks	MOFs

10.1 INTRODUCTION OF GREEN NANOTECHNOLOGY

Technology usually refers to the application of various techniques, skills, methods, and processes for any and all practical purposes or to achieve specific objectives such as scientific investigation or research. "A technology that is environmentally friendly in its production, supply chain, or usage" is called green technology or green tech. Green tech/green technology is "an umbrella term that covers a wide range of products, systems, and equipment that are less harmful to the natural environment and their resources limit the negative effect on human activities" [1]. Green nanotechnology is a mixture of challenges, environmentally friendly, creative products or services, sustainable solutions, and sustainable technology, as seen in Figure 10.1.

Green nanotechnology is using nanotechnology to improve the environmental sustainability of processes that produce negative externalities. Green technology, also called clean or sustainable technology, deals with creating products or services that mitigate the negative impact on the environment. It involves processing green nano-products and making products to promote sustainability [2]. Research on nanomaterials is helping to boost the efficiency of solar panels, hydrogen fuel cells, and emissions batteries. New computational techniques are being employed in areas of nanoscience to refocus energy-based nanotechnology, including energy storage,

DOI: 10.1201/9781003516071-10

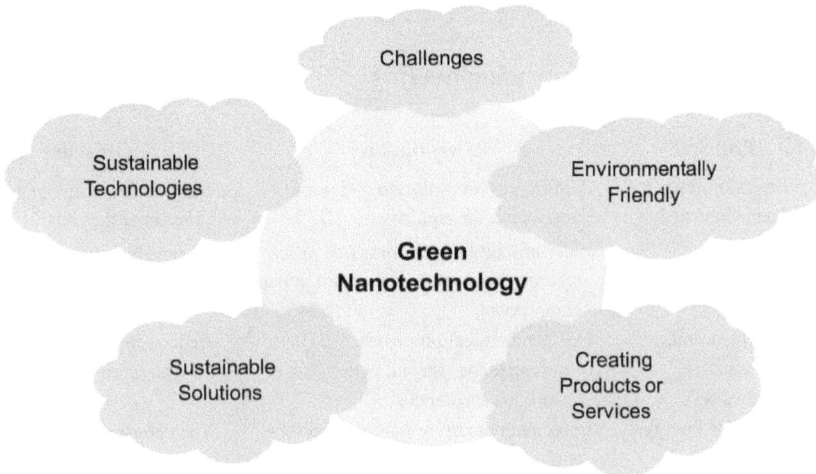

FIGURE 10.1 Various aspects of green nanotechnology (Credit statement: created by author).

conversion, and production processes. These innovations are instead designed to reduce material consumption and speed of processes, enhance energy intensity, and progressively increase renewable resources usage while reducing the usage of non-renewable resources. Green nanotechnology has shown much potential for water treatment because it effectively removes recalcitrant pollutants and reduces reliance on finite resources. The change toward green technologies enables business organizations to embrace green energy, including solar, wind, biofuels, and geothermal power, thus reducing pollution and costs in the development process.

Sustainability in engineering practice is important in creating sustainable solutions that will help minimize the effects of our actions on the environment and utilize the available resources. Green technologies such as nanomaterials aim to develop sustainable technologies that reduce wastage, utilize natural resources, and conserve energy. Ultra-fine particles and large molecules such as polymer batteries are nanomaterials with better mechanical, electrical, and thermal characteristics, which help to improve performance across industries. New nanomaterials provide adaptability properties that enhance a material's durability and performance, thus contributing to sustainability by solving environmental problems and using resources efficiently [1–2].

10.2 PRINCIPLE OBJECTIVES OF GREEN ENGINEERING

Green engineering is concerned with developing clean, efficient processes, products, and systems. A brief of its primary objectives and principles is presented in Table 10.1.

TABLE 10.1
Principle Objectives of Green Engineering

Sr. No.	Principle	Description	Objective
1	Prevention Rather Than Control	Aim to prevent pollution and waste generation through design.	Minimize environmental impact at the source.
2	Atom Economy	Fully utilizing all products in a process whereby they are incorporated into an end product.	Diminish waste and enhance efficiency.
3	Less Hazardous Chemical Syntheses	Design chemical processes to minimize the use and generation of hazardous substances.	Reduce risks to health and the environment.
4	Design for Energy Efficiency	Implement energy-efficient processes and technologies.	They consume less energy to run the appliances and thus have a less negative effect on the environment.
5	Use of Renewable Feedstocks	Use renewable, sustainable, or recyclable materials instead of non-renewable resources.	Promote resource sustainability.
6	Design for Degradation	Design products to degrade into non-toxic substances after their useful life.	Reduce long-term environmental impact.
7	Real-time Analysis for Pollution Prevention	Utilize real-time monitoring to prevent pollution and manage environmental impact proactively.	Enable immediate adjustments to minimize emissions.
8	Accident Prevention through Inherently Safer Chemistry	Design processes to be inherently safer, reducing the likelihood of accidents.	Enhance safety and minimize risks.
9	Green Chemistry	Optimize chemical processes using principles of green chemistry to reduce the ecological footprint.	Integrate environmental considerations into chemistry.
10	Sustainable Engineering	Focus on engineering solutions that are economically, socially, and environmentally sustainable.	Ensure long-term viability and balance.
11	Lifecycle Thinking	One should consider not only the production of products but also the use of materials, purification/extraction, and disposal.	Identify and address environmental impacts throughout the product's life.
12	Resource Efficiency	Ensure the efficient use of resources such as energy, water, and materials in the construction project.	Reduce consumption and waste.
13	Pollution Prevention	Implement practices that avoid the creation of pollutants.	Protect the environment and human health [3–4].

10.3 EXAMINATION OF GLOBAL ENVIRONMENTAL ISSUES

An examination of global environmental issues is illustrated in Figure 10.2.

10.3.1 Climate Changes

Climate alterations are mainly attributed to the increase in levels of GHG, including CO_2, CH_2, and N_2O, within the atmosphere through the burning of fossil fuels, deforestation, industrial emissions, etc. Climate change increases global temperatures, heat waves, storms, droughts, loss of habitats, endangered species, melting of ice caps, and rising sea levels [5].

10.3.2 Pollution

Pollutions may be categorized as air pollution, water pollution, and soil pollution through industrial activity emissions, vehicles burning fossil fuels, methane emission in agriculture activities, SO_2, NO_x, PM, and VOCs. Among these are respiratory ailments, smog and acid rain. Some of the dangerous effects are respiratory ailment, smog and acid rain [6]. Water pollution is the act of polluting water through industrial discharges, agricultural leakage, and improper waste disposal. This will lead to a degradation of the aquatic habitats and the occurrence of algal blooms, yielding risks to human health. Where there is soil pollution in the form of contaminants, including heavy metals, pesticides, and industrial chemicals, this will influence the fertility of the soils and, consequently, the productivity of crops, and this will be responsible for what can enter the food chain [7].

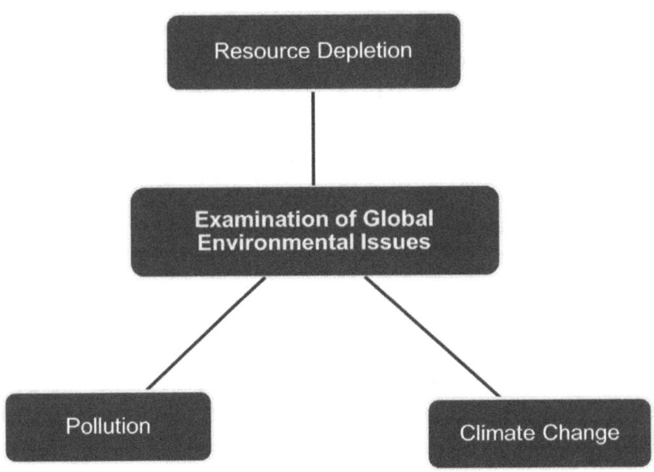

FIGURE 10.2 Examination of global environmental issues (Credit statement: created by author).

10.3.3 Resource Depletion

Fossil energy resources, which include coal, oil, natural gas, minerals, and metals, are exploited and consumed in a manner that results in the exhaustion of the resources and pollution of the environment. Deriving these various resources causes adverse effects on the environment, such as the destruction of habitats, pollution, and long-term negative impacts on the ecosystem. They are scarce resources; hence, their exploitation is a potential danger to the future supply and the environment. On the other hand, renewable materials like water, forests, and fish can be replenished in the natural environment after some time. Nonetheless, such resources may cause severe environmental problems if they are used up to exceed the power of regeneration. Excessive use of water can cause dry up of the water sources, cutting down trees disrupts the ecological balance, and overfishing affects the fish population. This kind of overuse affects the stability of ecosystems and the capacity of these resources to reproduce them, hence the issue of sustainability. And the regulation of this consumption with natural replacement is the key to sustaining ecological integrity and resource stock [8].

10.4 CONTRIBUTION OF GREEN TECHNOLOGY TO MITIGATING ENVIRONMENTAL IMPACT

The contribution of green technology to mitigating environmental impact is also represented in Figure 10.3.

10.4.1 Renewable Energy

Photovoltaic cells or solar thermal systems transform solar energy and reduce the amount of fossil energy, thus reducing greenhouse gas (GHG) emissions. Photovoltaic cells are designed in a manner that they can collect the sun's rays and turn them into electricity. On the same note, solar thermal systems tap on these rays to heat a particular fluid that, in turn, rotates a turbine to produce power. Wind power uses wind turbines, which transform wind into electricity, making wind power a renewable energy source for generating electricity with insignificant emissions of air pollutants and GHGs. Wind technology employs wind flows to produce electricity without burning fuel, utilizing fossil energy. Hydropower is the generation of electric power through a system that controls water flow or through river resources such as dams or run-of-river plants. Compared to fossil fuels, it is helpful in the reduction of emissions. Nonetheless, it needs to be well managed in a way that would not disrupt the functioning of ecosystems in water systems that serve as dwelling places for several living organisms. These renewable energy sources are essential, as they reduce the harmful impact of other sources on the environment and nature [9].

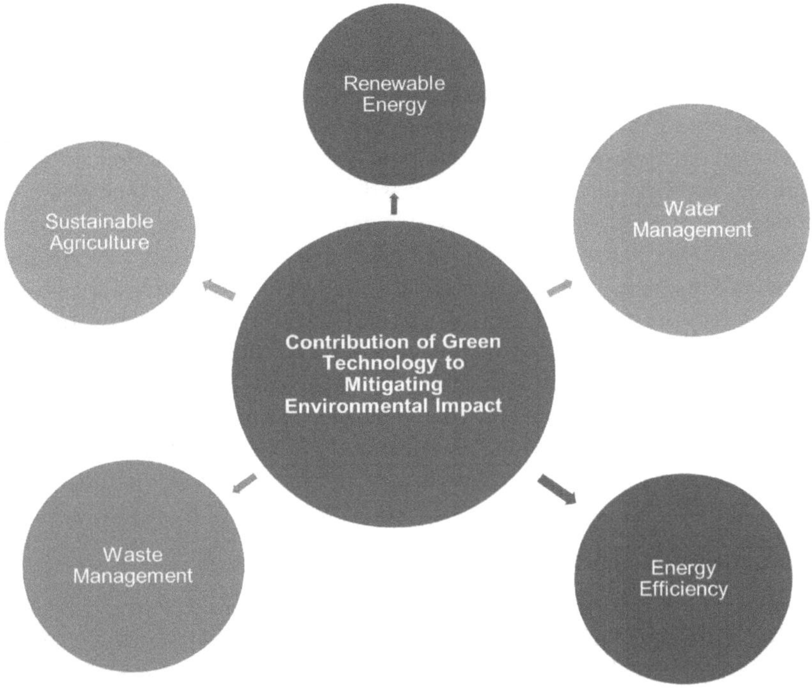

FIGURE 10.3 Contribution of green technology to mitigating environmental impact (Credit statement: created by author).

10.4.2 Energy Efficiency

Green Building Design also involves the application of energy-efficient materials and technological practices to reduce energy use and, at the same time, reduce GHG emissions. This includes various lighting technologies such as light-emitting diodes and LEDs that record lower energy use and durability than other types of lighting. Further, HVAC systems are required to efficiently control climate control for proper and efficient building operations for energy efficiency. Smart grids are a new kind of electrical grid that has been designed to enhance the distribution of energy through the application of advanced technologies. They aid in generating renewable power, such as wind and solar, which are often intermittent and affect the energy supply. They provide better control of utility consumption and utilization with data and communication technology, thus making intelligent grids more efficient and effective [10].

10.4.3 Waste Management

Recycling and upcycling are techniques used in reshaping materials to reduce the waste of resources. Recycling is taking used materials from other people, such as paper, glass, and plastic, and transforming them into new products, reducing raw materials and energy used and the use of landfills. Upcycling is a step ahead of this concept and involves the reuse of waste and junk in a manner that would create something of increased worth or use. These practices help attain the organizational goal of optimizing the utilization of resources and or preventing adverse impacts on the environment as wastes are eradicated and natural resources are preserved. Waste-to-energy technologies are systems utilized to capture energy from waste items, by burning or digestion. Incineration involves burning waste at high temperatures to produce heat useful for generating electricity. Specifically, the biological process occurs without oxygen, resulting in biogas, a form of fuel. These are utilized in waste management and provide renewable energy; hence, they help manage waste and energy sources [11].

10.4.4 Sustainable Agriculture

Precision Farming incorporates Global Positioning System (GPS) and other techniques into farming to improve various farming methods. GPS enables farmers to locate and observe fields and their features more accurately; sensors record information about the soil, the crops, and the climate. This reduces the wastage of water, fertilizers, and pesticides, thus setting minimal standards for environmental degradation. It also reduces losses and pollution and improves the amount and quality of crops produced on a farm. Organic farming is mainly said to be a form that does not use synthetic chemicals and applies natural processes when cultivating crops. It encourages practices like making compost, intercropping, and using biocontrol methods. It conserves soil fertility and structure, minimizes chemical pollution in the water, and preserves the earth's diverse plant and animal life. Organic farming does not involve chemicals, and its methodologies are inclined toward natural resources; therefore, it helps to pave a healthier environment and better agricultural practices [12].

10.4.5 Water Management

Water recycling entails taking wastewater and processing it to reuse it in processes such as agricultural and industrial use. It decreases the need for new water and avoids polluting it by treating the used water and using it again. The technologies assist in making this recycled water safe and fit for the specific purposes for which it will be employed, hence saving scarce water and reducing environmental effects. Desalination is the process by which seawater is turned into drinking water. Water scarcity is solved here by providing a new source of fresh water in places that can barely afford the classic approaches to water supply. Nevertheless, desalination consumes much energy and has environmental issues connected with concentrated brine disposal and its effect on sea life. Thus, Green Engineering may be defined as the systematic creation and deployment of engineering solutions with ambient characteristics for their usage and disposal. In that respect and prospect, green engineering embraces and anticipates the future in solving as many environmental problems as possible through technology such as water recycling and desalination to forge a sustainable future or world [13].

10.5 NANOMATERIALS: APPLICATIONS AND PROPERTIES

Nanotechnology represents "a ground-breaking field focused on manipulating matter at the nanoscale typically between 1 and 100 nanometres". This technology involves designing, producing, and applying materials and devices at this minute scale, where unique physical, chemical, and biological properties emerge [14]. Nanotechnology significantly impacts engineering since it enables the production of new materials with increased properties and the upgrading of existing technologies. Nanotechnology drives advancements in materials science, electronics, biology, and environmental engineering, affecting various industries such as medicine, energy, and manufacturing. Nanotechnology's precision and control drive breakthroughs in performance and sustainability, making it a critical field of research and application [15]. Applications of nanomaterials in green technology are briefly discussed in Table 10.2. Various properties of nanomaterials are mentioned in Figure 10.4.

Figure 10.5 represents the different classes of nanomaterials. Nanomaterials are materials with structural components sized between 1 and 100 nanometers. Because of their nanoscale size, these materials have distinct properties that differ significantly from their bulk counterparts. The size of nanomaterials is one of their most important aspects since it determines their properties and behavior. Materials at the nanoscale exhibit quantum effects, and have a higher surface-to-volume ratio. These characteristics can alter electrical, optical, and magnetic properties and increase reactivity and surface energy [16]. Nanomaterials can also assume several shapes, including nanoparticles (spherical or irregular), nanotubes, nanowires, and nanosheets. Depending on the material and the production method, their structure might be amorphous, crystalline, or a mixture [17]. There is also a brief discussion on introducing different types of nanomaterials, refer to Table 10.3.

Size is another attribute among the characteristics defining nanomaterials, where most of their properties are size-dependent. The nanoscale of material can induce quantum mechanics, and the material's color, light absorption, and other properties also differ significantly from the properties of the same material at the macroscale. This kind of behavior often results in what is known as large property not present in large-scale material. Notably, the ratio of nanoparticle surface area to volume is indeed high, which is also entirely consistent with its peculiarities. Due to this increased surface area, they have enhanced behaviors toward their surroundings, which is why nanoparticles have extensive applications in catalysts, drug delivery, and sensors [18]. In nanomaterials, greater surface exposure brings desirable

TABLE 10.2
Applications of Nanomaterials in Green Technology

Sr. No.	Application	Role of Nanomaterials	Benefits
1	Solar Cells	a) Enhanced Efficiency: Using quantum dots nanowires for better light absorption and electron transport. b) Flexible and Lightweight Panels: Incorporation of nanostructured materials like nanotubes and graphene for flexible solar panels. c) Cost Reduction: Utilization of nanomaterials for reduced material use and improved panel durability.	• Increased energy conversion rates. • Integration into various surfaces (e.g., clothing, building materials). • Lower production costs and improved panel longevity [24].
2	Fuel Cells	a) Catalyst Enhancement: Nanocatalysts (e.g., platinum nanoparticles) for improved reaction efficiency and reduced platinum usage. b) Improved Durability: Using nanocomposites to enhance fuel cell components' lifespan and resistance. c) Better Energy Storage: Nanomaterials for more efficient and stable energy management.	• Higher efficiency and lower catalyst costs. • Increased component durability and reliability. • Enhanced energy storage and system performance [25].
3	Water Purification	a) Nanoparticles: Some examples include metallic nanoparticles, such as silver and iron oxide, and carbon-based nanoparticles, such as carbon nanotubes and graphene oxide, to remove contaminants. b) Nanocomposites: Materials combining nanoparticles with other substances to enhance pollutant removal. c) Nanofiltration Membranes: Nanomaterial-based membranes for improved filtration efficiency and removal of microorganisms.	• High removal efficiency for contaminants and pathogens. • Improved membrane performance and lifespan. • Enhanced ability to purify water from a wide range of pollutants.

(Continued)

TABLE 10.2 (CONTINUED)
Applications of Nanomaterials in Green Technology

Sr. No.	Application	Role of Nanomaterials	Benefits
4	Air Filtration	a) Nanofibers: Electrospun nanofibers for creating highly efficient air filters. b) Nanocatalysts: Photocatalytic nanoparticles (e.g., titanium dioxide) for degrading air pollutants under UV light. c) Nanostructured Materials: Nanomaterial coatings on filters to improve pollutant capture and reduce filter clogging.	• Enhanced air quality by capturing fine particles and pollutants. • Effective degradation of gaseous pollutants. • Longer filter lifespan and reduced maintenance [26].
5	Lightweight Materials	a) Nanocomposites: Incorporation of nanotubes, graphene, and nanoclayst to reduce weight without compromising strength. b) Nanostructured Polymers: Nano-fillers create lightweight and durable polymers.	• Reduced material weight, improving fuel efficiency in transportation. • Enhanced mechanical properties and durability. • Lower energy consumption and emissions in manufacturing and usage.
6	High-Strength Composites	a) Nanotubes and Nanofibers: Carbon nanotubes and nanofibers reinforce composites, improving tensile strength and flexibility. b) Nanostructured Coatings: Application of nanocoatings to enhance surface strength and resistance to wear and corrosion.	• Improved mechanical strength and impact resistance. • Enhanced performance and longevity of materials. • Reduced need for additional materials and resources [27].

characteristic alterations in the mechanical, thermal, and electrical areas by improving hardness, flexibility, and insulating properties.

Since nanoparticles possess a small size and relatively large surface area, they have some peculiar properties. These features contribute to the progress and advancements in diverse engineering and science disciplines, illustrating how nanotechnology can revolutionize future research and implementation [19].

FIGURE 10.4 Various properties of nanomaterials (Credit statement: created by author).

10.6 INTRODUCTION TO SMART NANOMATERIALS

Smart nanomaterials are state-of-the-art materials devised to exhibit sensitive responses concerning various stimuli, including temperature, pH, light, magnetic field, or chemical conditions. Smart nanomaterials, on the other hand, may actively change their characteristics or functional requirements in response to environmental changes and, as such, are highly versatile for numerous applications. Its importance lies in its ability to enhance productivity and support facilitating special functionalities in multiple fields such as healthcare, environmental technology, electronics, or power industry. Applications of smart nanomaterials in renewable energy technologies, construction, and infrastructure development are discussed in brief in Table 10.4. The classification of smart nanomaterials is briefly discussed in Table 10.5.

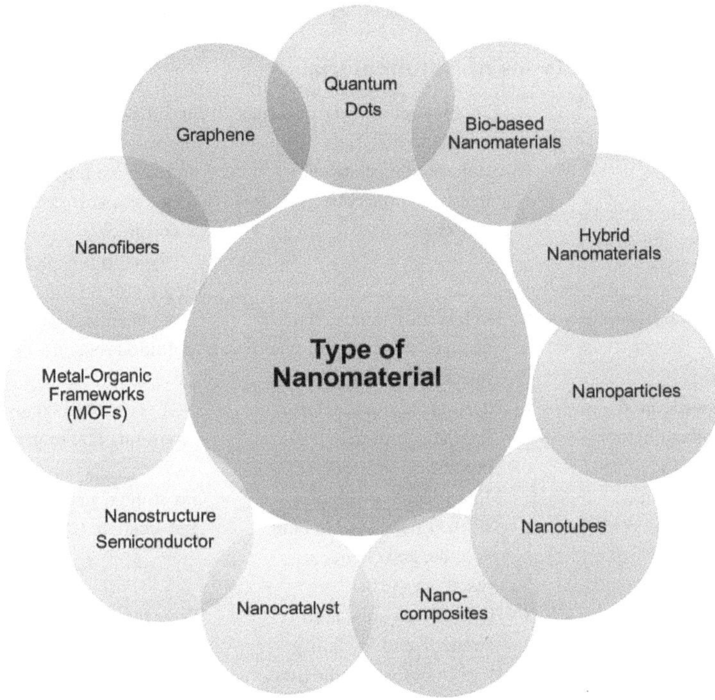

FIGURE 10.5 Types of nanomaterials (Credit statement: created by author).

TABLE 10.3
Brief Introduction of Types of Nanomaterials

Sr. No.	Type of Nanomaterial	Properties	Applications
1	Nanoparticles	Size 1–100 nm. High surface area. Some of the size-dependent properties include nanomaterials' optical, electronic and magnetic properties.	Drug delivery. Imaging and diagnostics catalysis. Electronics and semiconductors. Environmental remediation.
2	Nanotubes	Cylindrical structures with diameters in the nanometer range. High tensile strength. Excellent electrical and thermal conductivity.	Electronics (e.g., transistors, conductors). Reinforcement in composites. Energy storage (e.g., supercapacitors). Drug delivery & sensors.

(Continued)

TABLE 10.3 (CONTINUED)
Brief Introduction of Types of Nanomaterials

3	Nanocomposites	Combination of nanomaterials with bulk materials. Enhanced mechanical, thermal, and electrical properties.	Structural reinforcement (e.g., in aerospace and automotive industries). Improved coatings (e.g., anti-corrosion). Advanced electronics. Catalysis. Biomedical applications [20].
4	Nanocatalysts	High surface area, enhanced catalytic activity, and reusability.	Green chemical processes, pollution reduction, energy production.
5	Nanostructured Semiconductors	High surface area, efficient light absorption, tunable properties.	Solar cells, photocatalysis, environmental sensing.
6	Metal-Organic Frameworks (MOFs)	More specifically, high surface area, tunable pore volume, and chemical reactivity—exist.	Gas storage and separation, water purification, CO_2 capture [21].
7	Nanofibers	High surface area, high strength, and flexibility.	Water filtration, air purification, reinforcement materials.
8	Graphene	High electrical conductivity, mechanical strength, and flexibility.	Energy storage (batteries and supercapacitors), sensors, conductive materials.
9	Quantum Dots	Size-dependent optical properties, high stability.	Environmental monitoring, solar energy, water treatment [22].
10	Bio-based Nanomaterials	Biodegradable, derived from renewable resources.	Green composites, biodegradable packaging, environmental remediation.
11	Hybrid Nanomaterials	Combination of properties from different nanomaterials, such as enhanced functionality and stability.	Integrating multiple sensing capabilities. Improving performance in solar cells and batteries. Combining properties for efficient pollutant breakdown [23].

The definition of smart nanomaterials encompasses a broad range of materials that can alter their physical or chemical properties in a controlled manner. This capability is obtained through the merging of nanotechnology and materials science, which enables the design of nanoscale materials to generate precise responses. For example, thermo-responsive nanomaterials can alter their structure or solubility in reaction to temperature changes. Similarly, pH-sensitive nanomaterials can release pharmaceuticals or change their structure when exposed to varied pH levels, making them particularly helpful in targeted drug delivery systems [28]. Figure 10.6 shows the properties and functionalities of smart nanomaterials. The detailed discussion on these properties is elaborated as follows.

TABLE 10.4
Applications of Smart Nanomaterials

Sr. No	Field	Type of Smart Nanomaterial	Applications
1	Renewable Energy Technologies	a) Photovoltaic Nanomaterials (e.g., quantum dots, nanowires)	• Solar Cells: Enhanced efficiency and light absorption in solar panels. • Flexible Solar Panels: Integration into flexible substrates for versatile applications.
		b) Nanocatalysts (e.g., platinum nanoparticles)	• Fuel Cells: Improved catalyst efficiency and reduced material costs in hydrogen fuel cells. • Hydrogen Production: Enhanced performance in water splitting for hydrogen generation [35].
2	Construction	a) Self-Healing Concrete (e.g., microcapsule-based polymers)	• Concrete Repair: Autonomous repair of cracks and damage in concrete structures. • Durability Improvement: Enhanced longevity and reduced maintenance of infrastructures.
		b) Nanocoatings (e.g., nanostructured coatings)	• Surface Protection: Improved resistance to environmental degradation in building materials. • Energy Efficiency: Enhanced thermal insulation and energy savings in buildings [36].
3	Infrastructure Development	a) Smart Materials (for example, the alloys based on metals like nickel and titanium)	• Seismic Protection: Adaptive structures and components that respond to seismic activity. • Structural Adjustments: Automated adjustment of infrastructure components in response to environmental changes [37].
		b) Piezoelectric Materials (e.g., piezoelectric ceramics, nanowires)	• Vibration Control: Active control and damping of vibrations in bridges and buildings. • Energy Harvesting: Conversion of environmental vibrations into electrical energy for infrastructure sensors [38].

TABLE 10.5
Classification of Smart Nanomaterials

S. No.	Type	Description	Applications
1	Shape Memory Alloys	When exposed to specific stimuli (e.g., temperature), materials can return to their original shape after deformation. Typically composed of metals such as nickel-titanium.	• Actuators: Provide accurate movement in mechanical systems and robotics. • Medical devices: Used in surgical instruments and stents. • Aerospace: Used in components and deployable structures that change shape in reaction to temperature changes [32].
2	Piezoelectric Materials	Piezoelectric materials produce an electric charge in proportion to the mechanical stress. Such as ceramic, crystal, and nanowire parts, which have piezoelectric characteristic.	• Sensors: They are used in pressure, accelerometers, and vibration sensors. • Energy Harvesting: Couples mechanical energy from vibration or motion, into electrical energy. • Actuators: Applied in precision positioning and control systems [33].
3	Self-Healing Polymers	Polymers that can autonomously repair damage or restore their original properties after being compromised. Includes microcapsule-based, vascular, and dynamic bond systems.	• Coatings: Applied in protective coatings and paints that self-repair scratches and damage. • Structural Materials: Used in aerospace and automotive components to enhance durability. • Electronics: Incorporated in flexible and wearable electronics for improved longevity [34].

10.6.1 Responsive Behavior

The capacity of smart nanomaterials to display responsive behavior is one of their most remarkable characteristics. This involves alterations in physical characteristics in response to outside stimuli, such as size, shape, or visual qualities. For instance, photo-responsive materials can change their mechanical strength or optical characteristics when exposed to light. At the same time, thermo-responsive hydrogels can expand or contract in response to temperature.

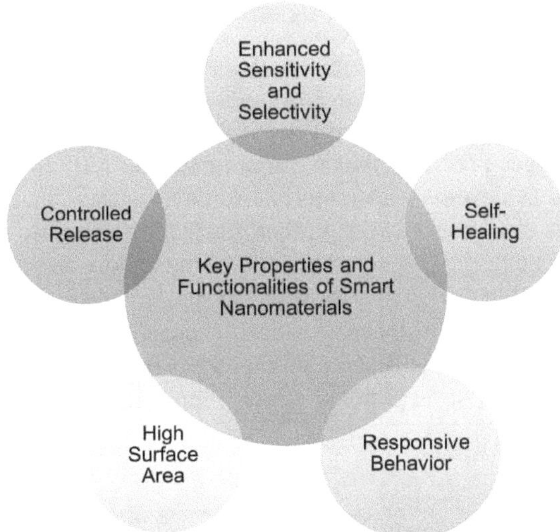

FIGURE 10.6 Key properties and functionalities of smart nanomaterials (Credit statement: created by author).

10.6.2 High Surface Area

Smart nanomaterials usually possess a large surface area-to-volume ratio, increasing their reactivity and ability to respond to stimuli from the external environment. As a result, the material can sense, catalyze, and transport drugs faster because it has more area of interaction with the surrounding environment.

10.6.3 Controlled Release

Smart nanomaterials can achieve controlled release of therapeutic substances in applications relating to drug delivery. For example, it is possible to build nanocapsules or nanocarriers to release their contents in reaction to particular triggers, like pH changes or the presence of particular proteins. By releasing medications precisely where and when needed, this focused method minimizes adverse effects and increases therapeutic efficacy [29].

10.6.4 Self-Healing

Specific smart nanomaterials can recover after damage thanks to their self-healing characteristics. The material's ability to self-heal or its dynamic bonding mechanisms allows for this. Self-healing coatings, for instance, can automatically fix cracks or scratches, preserving the material's functioning and increasing its lifespan.

10.6.5 Enhanced Sensitivity and Selectivity

Smart nanomaterials can display heightened sensitivity and selectivity toward particular stimuli or target molecules. This makes them useful in applications like biosensors and sensors, where their ability to detect and react to even the slightest environmental changes precisely makes them beneficial [30]. Smart nanomaterials are an outstanding achievement in materials science as they have numerous sensitive and changeable characteristics that other materials can hardly provide. They are invaluable for a whole host of applications, including energy storage, the monitoring of the environment, diagnostics of diseases, and sustainable manufacturing because of their ability to change their characteristics in response to stimuli from their surroundings. The progress of intelligent nanomaterials is allowing the development of new opportunities across several fields while improving the performance of these fields [31].

10.7 Environmental Benefits and Impact Assessment

10.7.1 Contribution of Smart Nanomaterials to Energy Efficiency and Waste Reduction

Energy Efficiency

Thermal Insulation using aerogels and phase-change material because aerogel is endowed with low thermal conductivity, which works like an insulator trapping air in a matrix form, thus preventing the heat prior and, in turn, reducing it. Functioning as a phase-change material that silently stores and releases thermal energy through its changing states, it provides quiet operation longitudinally in every cycle of heating or cooling across seasons. Nanomaterials improve solar cell efficiency in photovoltaics. Nanomaterials such as quantum dots and nanowires help us increase the absorption of light rays, which leads to a better conversion from sunlight into electricity by solar panels. As for this development, renewable energy sources are promoted, and the effectiveness of the systems using solar energy is enhanced. Nanomaterials in batteries and supercapacitors improve energy storage. By increasing surface area and conductivity, these materials enhance energy storage capacities, resulting in more effective energy utilization and longer-lasting power sources. Energy storage technology advances thanks to nanotechnology, enabling better energy densities and quicker charging periods [39].

Waste Reduction

Catalysis gets the help of nanocatalysts, increasing the chemical reactions' efficacy. Due to nanocatalysts' exclusive properties, such as large surface area and volume ratio, reactions get selective and complete, reducing industrial waste production. They also reduce the chances of forming unwanted products that may affect the source and efficiency of raw materials, hence improving the overall sustainability of the process. In water treatment, nanotechnology is used in water purification using nanomaterials. Nanomaterials efficiently remove impurities from water; they can eliminate bacteria, viruses, and other pollutants. This enhanced method of filtering pollutants frequently eradicates the usage of chemicals to boost the safety and quality of treated water. The recycling systems where smart nanoparticles are integrated can also help to extract metal from e-waste. It becomes easier to recycle and economically profitable because these nanoparticles can selectively and preferentially teach valuable metals, including gold and silver, from disused electronic equipment more particularly. This strategy helps reduce the expenditure of technology waste and regains valuable resources [40].

10.7.2 Environmental Impact Assessment of Smart Nanomaterials in Engineering Applications

Life Cycle Assessment (LCA)

LCA of smart nanomaterials on a material level, which includes the extraction of raw material, manufacturing process, and energy used in the production stage, is conducted. This implies evaluating the amount of emission, waste, and resources that go into extracting raw forms and their transformation into nanomaterial. It is crucial to understand these effects to minimize the environmental impact associated with nanomaterials processing. For nanomaterials to be used, there is a need to assess their performance in their intended uses, noting benefits such as reduction in energy consumption or enhancement of efficiency. This includes assessing these products' effectiveness in enhancing the products' characteristics in areas such as environmental technology, electronics and reduced resources to achieve the desired results. It includes the questions of destruction, recycling, or disposal of nanoparticles, which are discussed in end-of-life questions. This entails analysis of what benefits might be tied to their end-of-life stage, including the emission of hazardous substances on degradation or the effectiveness of recycling in reclaiming valuable materials and minimizing waste [41].

Ecotoxicity and Human Health

Toxicological tests are used to investigate material's effects on the health of people and animals, the toxicity. As elaborated earlier, such research endeavors focus on trying to understand how exactly nanomaterials may enter into and affect or harm biological systems. The various ways by which man is exposed to mercury and the health consequences of such exposure are discussed; the toxicity effects of mercury on the cells and organs of the human body, as well as the long-term impacts such exposures may have on health, are explored. Evaluating their safety helps avoid situations in which nanoparticles cause a severe threat to living organisms. Exposure risks assess all the risks associated with exposure to nanomaterials throughout the life cycle. This is part of evaluating risks associated with consumption, breathing in the vapors and being affected by the compound through touch. Risk assessments of exposure consider how deep into the body the nanomaterials can penetrate, how they can accumulate, and any effects they might have. These considerations must be considered when designing safety procedures and legal frameworks as they expose health dangers relating to nanomaterials [42].

10.7.3 Regulatory and Safety Considerations

Control ensuring compliance with current environmental and safety requirements pertains to using smart nanomaterials. This includes compliance with guidelines set by these regulatory bodies, such as the European Chemicals Agency (ECHA) in Europe, and the Environmental Protection Agency (EPA) in the United States. Compliance helps reduce environmental and public health threats by ensuring responsible usage of nanomaterials. It entails following guidelines for safe production, use, and disposal. Assessing whether the usage of nanomaterials is in line with more general sustainability objectives is necessary for sustainability. This entails evaluating their effect on lowering the total environmental footprint and endorsing resource reuse, recycling, and recovery as essential tenets of the circular economy. The objective is to ensure that nanomaterials improve the environmental performance of goods and processes, reduce waste, and conserve resources to promote sustainable development [43].

10.8 CHALLENGES AND CONSIDERATIONS IN GREEN ENGINEERING WITH NANOMATERIALS

Challenges and considerations in green engineering with nanomaterials are highlighted in Figure 10.7. This section includes various environmental and health impacts of nanomaterials and their safety, regulatory, and ethical considerations associated with green nanotechnology.

10.8.1 Environmental and Health Impacts of Nanomaterials

Environmental Impacts

Nanomaterials' mobility and persistence can cause their accumulation in the environment. Nanomaterials may endure in ecosystems and show unexpected mobility, possibly traveling through soil and water systems because of their small size and unique qualities. Unintentional accumulation and consequences on environmental quality, including soil structure disturbance and groundwater contamination, may arise after this. Because nanomaterials can be hazardous to plants, soil organisms, and aquatic life, venomousness to ecosystems is problematic. Compared to bulk materials, their small size makes it easier for them to enter biological systems, which could intensify their toxicity and bioavailability. This can trouble the ecological equilibrium and affect the well-being and functionality of different organisms. It is undistinguishable how enduring the use of nanomaterials will disturb them, especially in light of their tendency to degrade and yield potentially dangerous by-products. To assess the long-lasting environmental impact, research must determine how nanomaterials degrade over time and whether they release toxic compounds that might impend ecosystems and public health [44].

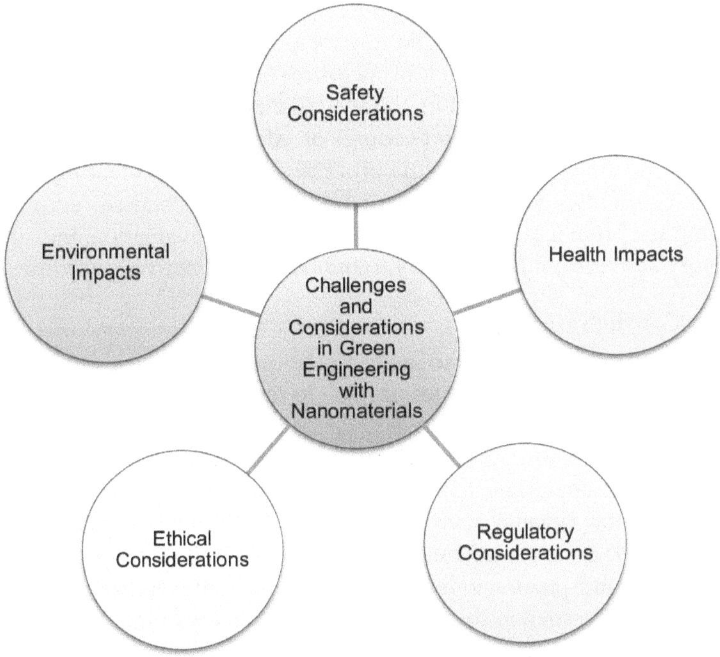

FIGURE 10.7 Challenges and considerations in green engineering with nanomaterials (Credit statement: created by author).

Health Impacts

Breathing into the air that contains nanoparticles of such size is dangerous since they can profoundly penetrate through the lungs. This possibility of entering the lower respiratory tract enhances the risks of contracting respiratory diseases and other diseases, inclusive of inflammation or fibrosis, besides decreased lung functioning. Nanomaterials have shown the capability of penetrating the epidermis and reaching circulation or other tissues when ingested or absorbed. If the characteristics of the nanomaterials are unfavorable or inappropriately incorporated into biological systems, this can be hazardous to your health. It means that the chronic effects of nanoparticles within the body may cause diseases that may be hard to detect at first sight. To this end, there is a need for constant surveillance and research to establish any latent dangers and muster adequate measures to prevent them. Chronic use of exercise may have adverse health effects on the human body bagi Triennial (employee) which was revealed during in the study conducted by Papp et al. [45].

10.8.2 Safety, Regulatory, and Ethical Considerations Associated with Green Nanotechnology

Safety Considerations

Risk evaluation is important in evaluating the safety of nanoparticles in different applications. This necessitates effective evaluations to understand risks where they are and develop tolerance levels and the best ways of handling these. Risk assessment enables the definition of possible threats to human health and the atmosphere for nanomaterials and the definition of measures for reducing such threats. This is because there are no standardized methods for defining the safety of nanomaterials. There might be differences in the procedures of safety assessment, as well as regulations applied. This may complicate the processes involved in creating regularity in safety measures, applying the law, and corroborating study findings across research. In enhancing the efficiency and reliability of the safety evaluation and supervising the regulatory activities, it is necessary to indicate the effective procedures [46].

Regulatory Considerations

Often, the current regulations fail to interpret the unique properties and risks associated with nanoparticles. These issues are due to the particle size of these materials and their reactivity and inclination toward certain peculiar behaviors. Current regulations may not sufficiently capture these problems. To mitigate risks early, guidance on the design of nanomaterials, exposure routes, disposal methods and safe use triggers the need for new rules to cover the whole lifetime of nanotechnology. Specific laws are required to ensure wide-ranging protection, effective risk management, and protection of the atmosphere and human health. This will help bridge gaps and help people become accustomed to the dynamic nanotechnology environment [47].

Ethical Considerations

A key component of sustainable development is ensuring that nanomaterials are used to respect moral standards and don't adversely impact environments or populations already at risk. Public perception and engagement are essential to resolve problems because they foster responsible innovation and a culture of transparency among stakeholders. Clear labeling and communication about potential dangers ensure that people are well-informed. Informed consent is essential for applications requiring direct human exposure, such as consumer products. Green nanotechnology has great potential to improve environmental sustainability, but its responsible and advantageous application depends on overcoming these moral and practical issues [48].

10.9 FUTURE TRENDS AND INNOVATIONS IN GREEN ENGINEERING WITH NANOMATERIALS

10.9.1 Emerging Trends and Future Directions

Advanced Functional Nanomaterials

Self-repair materials represent a novel advancement in nanotechnology, as these materials can self-heal in the event of harm. The materials' capacity to mend themselves increases their longevity and lowers the need for replacements. When subjected to abrasion, deterioration, or other forms of damage, self-healing materials can return to their former characteristics and functions, increasing their longevity and lowering maintenance expenses. The purpose of responsive nanomaterials is to respond to alterations in their surroundings, including shifts in pH or temperature. These materials are more effective and versatile in various applications because they can perform particular tasks by reacting to outside inputs. Sensitive nanomaterials, for instance, are helpful in smart coatings, sensors, and medication delivery systems because they can change their characteristics or release active substances in response to changes in their surroundings. This versatility enhances performance in various applications and permits more exact control [49].

Integration with Digital Technologies

Nanomaterial-focused sensors incorporate nanomaterials to perceive the environment in real time. These new generations of sensors are highly sensitive and accurate in measuring changes in environmental parameters ranging from pollutants to gases or temperature. Due to the assimilation of nanomaterials, these sensors can function more effectively, have optimum resource utilization, properly manage pollution and take necessary action to change the atmosphere. Smart coatings and films apply nanoparticles to coatings for more complex application performance. Such coatings can possess properties such as enhanced mechanical characteristics that cause materials to last longer, resist corrosion that causes the surfaces to acquire a rusty and deteriorated look, and self-cleaning, where the coating surface allows dirt and other deposits to slide off. These smart coatings that contain nanoparticles offer heightened performance and stability, qualifying them for building uses in the auto and space industries [50].

Sustainable Nanomaterial Production

Green synthesis methods are the development of sustainable methods for producing nanomaterials. The focal goals of these technologies are to use renewable resources and reduce the invention of hazardous by-products throughout the synthesis. For instance, green synthesis can use less hazardous chemicals or plant-based resources, which lessens the impact of nanomaterial creation on the environment. These techniques seek to preserve the nanomaterials' functionality and performance while increasing the production process's sustainability. The circular economy strategies (CES) for recycling and reusing materials throughout their lives are part of approaches in nanomaterials. This entails creating nanomaterials with end-of-life concerns in mind, such as recoverability and simplicity of disassembly, to guarantee effective recycling or repurposing. The advantages of circular economy methods are that they aim to make waste unnecessary and ensure proper resource utilization. These tactics, however, are amicable to each other in that it is possible to diminish the impact of nanomaterials on the environment while, at the same time, boosting the effectiveness of resources [51].

Enhanced Energy Applications

The advancement of solar cells made from nanomaterials remains the foundation for creating new solar technologies in the future. Such innovations as perovskite and quantum dot solar cells that require nanomaterials for improved efficiency and reduced costs are among the developments. Perovskite cells might be more effective than silicon-based cells because they utilize a specific material already famous for its high efficiency in capturing sunlight and relatively low production cost. Composed of semiconductor particles with a nanometer degree of dimensionality, quantum dot solar cells utilize the restrictions of quantum mechanics to attain a higher photo-electric conversion coefficient through the broadened absorbance of solar radiation. These changes should enhance solar energy costs and make it more available to more people. The primary target of advanced energy storage is to fabricate supercapacitors and batteries derived from nanomaterials to work more effectively than the existing ones. Since broader nanomaterials elevate energy densities, supercapacitors and batteries can store more energy with reduced bulkiness. They also improve energy storage system reliability and effectiveness by accelerating charging time and extending battery life. The rise in demand for reliable energy storage equipment that delivers higher performances, especially in renewable energy applications, makes innovations in this science more pertinent [52].

10.9.2 Potential Advancements and Research Areas in Sustainable Engineering Technologies

Sustainable Water Management

Nanomaterial-Oriented Investigating nanomaterials that greatly enhance water filtering and purification procedures is the focus of filtration research. The advanced materials enhance the removal of pathogens and the formation of pollutants from water supply sources. Because of their boundless surface areas, carbon nanotubes, graphene, and nanosized metal oxides can provide reactive sites capable of enhancing the filtering systems. Additionally, they can precisely target certain impurities, improving the effectiveness and dependability of water purification. Technological advancements in nanomaterials have improved seawater desalination's economic viability and efficiency, which benefits desalination technologies. By boosting water permeability and salt rejection rates, nanomaterials such as nanofiltration membranes and nanoparticle-coated filters—improve the efficiency of desalination processes. Through these developments, desalination will become a more practical option for resolving the world's water scarcity problems by lowering energy requirements and operating costs [53].

Green Construction Materials

Assisted with nanomaterials, the construction materials are blended or reinforced with nanoparticles to increase thermal conductivity, mechanical load and resistance to weather conditions. Aerogels and coatings were two special forms of nanomaterials that improved insulating properties and reduced energy's heating and cooling demands [54]. To this end, nanomaterials improve such aspects of construction by making the building materials stronger and more enduring, thus reducing the amount of construction material wasted by having to replace worn-out or damaged products. These developments are for more sustainable building designs and eco-friendly construction. Studies in carbon capture and Storage technologies strive to design nanomaterials that effectively capture and sequester carbon dioxide (CO_2) emitted from industries. The nanomaterials have high surface areas and reactivity, thus enhancing the performance of devices used in collecting CO_2 in general, nanoparticles and nanofibers in particular. These technologies help a lot in the fight against the effects of climate change because they lock up CO_2 away somewhere and then release it as it comes out into the atmosphere [55].

Biodegradable and Eco-Friendly Nanomaterials

Biodegradable polymers are nanomaterials derived from renewable resources, and these materials disintegrate safely without affecting the environment. Some of these polymers form conventional plastics that are ecologically unfriendly because they endure in the environment by being non-biodegradable. In contrast, others have the added advantage of complying with the natural tendency of being biodegradable through natural biological systems. Examples are plant-derived polymers like polylactic acid (PLA), which break down into innocuous by-products. Creating these biodegradable nanoparticles helps manage materials sustainably and tackles the growing plastic waste problem. Non-toxic nanomaterials aim to reduce hazards to ecosystems and human health by developing safe substitutes for conventional nanoparticles. Research aims to find substances that are not hazardously poisonous when consumed, inhaled, or exposed to the environment. Because they are safer to employ in consumer goods, industrial applications, and environmental technology, non-toxic nanomaterials reduce adverse outcomes and encourage safer innovation processes [56, 57].

Smart Agriculture

The use of pesticides and fertilizers has significantly advanced agricultural technology. These nanomaterials improve the accuracy and efficiency of agricultural inputs like insecticides and fertilizers. These products use fewer chemicals because they can more efficiently deliver nutrients or active agents to specific locations thanks to nanoparticles. This method increases crop quality and yield while lowering the adverse effects on the environment and using fewer toxic chemicals. Nano fertilizers, for instance, can deliver nutrients in a controlled release way to improve nutrient uptake and decrease runoff. Soil remediation is a process that involves developing nanomaterials that will be used to clean up the soil. Such contaminants as heavy metals and organic poisons can be treated through nanomaterials, for instance, by concentrating them and eliminating them [58, 59]. Some of these methods include the adsorption of the pollutants by the nanoparticles or the catalytic degradation of the pollutants. This method facilitates the recovery of contaminated soils and supports sustainable agriculture by enhancing soil health and restoring fertility [60, 61].

Enhanced Recycling and Waste Management

Using nanoparticles in sorting and recycling processes can improve the recovery and separation of valuable components from waste streams. This is known as nanomaterials for waste sorting. Nanomaterials with enhanced sensitivity and selectivity for various waste kinds, including nanofibers or nanocomposites, can boost sorting system efficiency. These cutting-edge materials can aid in more precisely separating diverse materials, including glass, metals, and plastics, improving recycling procedures and increasing the recovery of priceless resources. The advanced use of nanomaterials and waste-to-energy enhances waste conversion to electricity optimization. The study intends to minimize environmental effects and enhance production efficiency through applying nanomaterials within, for instance, anaerobic digestion and incineration procedures. Waste conversion catalysts, energy recovery efficiency and emissions can be enhanced through the use of nanomaterials. Essentially, nanomaterials make trash and waste disposal a renewable source of energy and also advanced waste management techniques [62]. Using nanomaterials in green engineering holds great potential for the future, as it will spur creativity, efficiency, and sustainability in various sectors. Research and development in these areas will be essential to address global issues and accomplish sustainable development goals [63].

10.10 CONCLUSION

Techniques and fresh materials help make green engineering less adverse; thus, it is more sustainable. Due to the unique characteristics of nanomaterials, they become vital for enhancing environment remediation, disposal methods and energy utilization. New Thermal Management, purer water, and using superior catalysis for waste management are some of the paradigms they bring about. However, the advantages of smart nanomaterials include the following drawbacks regarding human health and the environment. A complete understanding of these risks necessitates risk management techniques and environmental impact evaluations to maintain safety and regulatory compliance. The problems include dealing with the possible toxicity of nanomaterials, their persistence in the environment, and the requirement for new regulatory frameworks. Future directions for green engineering using nanoparticles include creating responsive and self-healing materials, fusing them with digital technologies for improved functioning, and implementing environmentally friendly manufacturing techniques. Future developments could result in biodegradable materials, improved water treatment techniques, and more effective energy storage, all of which have the potential to increase sustainability significantly. Attaining sustainable engineering solutions necessitates combining green technology with smart and nanomaterials. This integration promotes creative responses to environmental problems, increases productivity, and facilitates efficient risk management. We can significantly advance toward a sustainable future by carefully utilizing these technologies, striking a balance between environmental stewardship and technological innovation, and enhancing our standard of living.

REFERENCES

1. Qamar, M.Z., Ali, W., Qamar, M.O. and Noor, M., 2021. Green technology and its implications worldwide. *The Inquisitive Meridian*, *3*, pp.1–11.

2. Sridhar, C., Thaskeen, F., Harshitha, M., Varsha, J.R., Deepika, T., Devi and Pareek, P.K., 2022. Green technology and sustainable renewable energy analysis. In *Innovations in Computer Science and Engineering: Proceedings of the Ninth ICICSE, 2021* (pp. 617–625). Singapore: Springer Singapore.
3. Asfaw, N., Chebude, Y., Ejigu, A., Hurisso, B.B., Licence, P., Smith, R.L., Tang, S.L. and Poliakoff, M., 2011. The 13 principles of green chemistry and engineering for a greener Africa. *Green Chemistry, 13*(5), pp.1059–1060.
4. Glavič, P., 2022. Updated principles of sustainable engineering. *Processes, 10*(5), p.870.
5. Kumar, A., Nagar, S. and Anand, S., 2021. Climate change and existential threats. In *Global climate change* (pp. 1–31). Elsevier.
6. Isiah, G., 2024. Oil Exploitation and environmental sustainability in the Middle East: Issues and solutions. *Influence: International Journal of Science Review, 6*(2), pp.90–98.
7. Singh, J., Kumar, S. and Mohapatra, S.K., 2018. Optimization of erosion wear influencing parameters of HVOF sprayed pumping material for coal-water slurry. *Materials Today Proceedings, 5*, pp.23789–23795.
8. Ali, A., Audi, M. and Roussel, Y., 2021. Natural resources depletion, renewable energy consumption and environmental degradation: A comparative analysis of developed and developing world. *International Journal of Energy Economics and Policy, 11*(3), pp.251–260.
9. Al-Shetwi, A.Q., 2022. Sustainable development of renewable energy integrated power sector: Trends, environmental impacts, and recent challenges. *Science of The Total Environment, 822*, p.153645.
10. Heidari, M., Thangavel, S., Al Naamani, E. and Khashehchi, M., 2024. Emerging trends in smart green building technologies. *Sustainable Technologies for Energy Efficient Buildings*, p.313.
11. Behera, D.K., 2023. Promoting sustainable development through environmental policy, green technologies, and effective waste management: A comprehensive review. *Journal of Multidisciplinary Science: MIKAILALSYS, 1*(2), pp.179–198.
12. Gawande, V., Saikanth, D.R.K., Sumithra, B.S., Aravind, S.A., Swamy, G.N., Chowdhury, M. and Singh, B.V., 2023. Potential of precision farming technologies for eco-friendly agriculture. *International Journal of Plant & Soil Science, 35*(19), pp.101–112.
13. Subramaniam, S., Raju, N., Ganesan, A., Rajavel, N., Chenniappan, M., Prakash, C., Pramanik, A., Basak, A.K. and Dixit, S., 2022. Artificial intelligence technologies for forecasting air pollution and human health: A narrative review. *Sustainability, 14*. https://doi.org/10.3390/su14169951.
14. Logothetidis, S., 2011. Nanotechnology: Principles and applications. In *Nanostructured Materials and Their Applications* (pp. 1–22). Berlin, Heidelberg: Springer Berlin Heidelberg.
15. Sev, A. and Ezel, M., 2014. Nanotechnology innovations for the sustainable buildings of the future. *World Academy of Science, Engineering and Technology International Journal of Civil, Environmental, Structural, Construction and Architectural Engineering, 8*(8), pp.886–896.
16. Asha, A.B. and Narain, R., 2020. Nanomaterials properties. In *Polymer Science and Nanotechnology* (pp. 343–359). Elsevier.
17. Rizwan, M., Shoukat, A., Ayub, A., Razzaq, B. and Tahir, M.B., 2021. Types and classification of nanomaterials. In *Nanomaterials: Synthesis, Characterization, Hazards and Safety* (pp. 31–54). Elsevier.

18. Biju, V., 2014. Chemical modifications and bioconjugate reactions of nanomaterials for sensing, imaging, drug delivery and therapy. *Chemical Society Reviews*, *43*(3), pp.744–764.
19. Ogbonna, V.E., Popoola, A.P.I., Popoola, O.M. and Adeosun, S.O., 2020. A review on polyimide reinforced nanocomposites for mechanical, thermal, and electrical insulation application: Challenges and recommendations for future improvement. *Polymer Bulletin*, pp.1–33.
20. Njuguna, J., Ansari, F., Sachse, S., Rodriguez, V.M., Siqqique, S. and Zhu, H., 2021. Nanomaterials, nanofillers, and nanocomposites: Types and properties. In *Health and Environmental Safety of Nanomaterials* (pp. 3–37). Woodhead Publishing.
21. Gupta, R., Adaval, A. and Kumar, S., 2021. An approach for development of materials for green chemical catalytic processes: Green catalysis. In *Green Polymer Chemistry and Composites* (pp. 113–138). Apple Academic Press.
22. Gaur, M., Misra, C., Yadav, A.B., Swaroop, S., Maolmhuaidh, F.Ó., Bechelany, M. and Barhoum, A., 2021. Biomedical applications of carbon nanomaterials: Fullerenes, quantum dots, nanotubes, nanofibers, and graphene. *Materials*, *14*(20), p.5978.
23. Atmakuri, A., Palevicius, A., Vilkauskas, A. and Janusas, G., 2020. Review of hybrid fiber based composites with nano particles—material properties and applications. *Polymers*, *12*(9), p.2088.
24. Javed, H.M.A., Que, W., Ahmad, M.R., Ali, K., Irfan Ahmad, M., Haq, A.U. and Sharma, S.K., 2020. Perspective of nanomaterials in the performance of solar cells. In *Solar Cells: From Materials to Device Technology* (S. K. Sharma and Khuram Ali, editors). Springer. pp.25–54.
25. Guo, K.W., 2012. Green nanotechnology of trends in future energy: A review. *International Journal of Energy Research*, *36*(1), pp.1–17.
26. Taran, M., Safaei, M., Karimi, N. and Almasi, A., 2021. Benefits and application of nanotechnology in environmental science: An overview. *Biointerface Research in Applied Chemistry*, *11*(1), pp.7860–7870.
27. Slepičková Kasálková, N., Slepička, P. and Švorčík, V., 2021. Carbon nanostructures, nanolayers, and their composites. *Nanomaterials*, *11*(9), p.2368.
28. Behera, A., Nayak, A.K., Mohapatra, R.K. and Rabaan, A.A., 2024. Smart nanocarriers. In *Smart Micro—and Nanomaterials for Drug Delivery* (A. Behera, A. K. Nayak, R. K. Mohapatra, and A. A. Rabaan, editors). CRC Press: Boca Raton, FL, p. 23.
29. Aflori, M., 2021. Smart nanomaterials for biomedical applications—a review. *Nanomaterials*, *11*(2), p.396.
30. Khatib, M., Zohar, O. and Haick, H., 2021. Self-healing soft sensors: From material design to implementation. *Advanced Materials*, *33*(11), p.2004190.
31. Muhammad, A., Jatoi, A.S., Mazari, S.A., Abro, R., Mubarak, N.M., Ahmed, S., Shah, A., Memon, A.Q., Akhter, F. and Wahocho, S.A., 2021. Recent advances and developments in advanced green porous nanomaterial for sustainable energy storage application. *Journal of Porous Materials*, *28*(6), pp.1945–1960.
32. Nair, S., Sai, P.H., Pranav, R. and Prashanth, M., 2023, December. Smart material: Nitinol; shape memory alloy–A review. In *AIP Conference Proceedings* (Vol. 2943, No. 1). AIP Publishing.
33. Zaarour, B., Zhu, L., Huang, C., Jin, X., Alghafari, H., Fang, J. and Lin, T., 2021. A review on piezoelectric fibers and nanowires for energy harvesting. *Journal of Industrial Textiles*, *51*(2), pp.297–340.
34. Idumah, C.I., 2021. Recent advancements in self-healing polymers, polymer blends, and nanocomposites. *Polymers and Polymer Composites*, *29*(4), pp.246–258.
35. Serrano, E., Rus, G. and Garcia-Martinez, J., 2009. Nanotechnology for sustainable energy. *Renewable and Sustainable Energy Reviews*, *13*(9), pp.2373–2384.

36. Bouiti, K., Labjar, N., Benmessaoud, M., Chraka, A., Omari, M., Jebbari, S., Nasrellah, H. and El Hajjaji, S., 2024. Nano-hybrid smart coatings: Advancements in self-healing and responsive functionalities. In *Nano-hybrid Smart Coatings: Advancements in Industrial Efficiency and Corrosion Resistance* (pp. 279–302). American Chemical Society.
37. Chowdhury, M.A. and Rahman, M.M. eds., 2024. *Shape Memory Alloys-New Advances: New Advances*. BoD–Books on Demand.
38. Li, T. and Lee, P.S., 2022. Piezoelectric energy harvesting technology: From materials, structures, to applications. *Small Structures*, *3*(3), p.2100128.
39. Casini, M., 2016. *Smart Buildings: Advanced Materials and Nanotechnology to Improve Energy-Efficiency and Environmental Performance*. Woodhead Publishing.
40. Baruah, A., Chaudhary, V., Malik, R. and Tomer, V.K., 2019. Nanotechnology based solutions for wastewater treatment. In *Nanotechnology in Water and Wastewater Treatment* (pp. 337–368). Elsevier.
41. Bauer, C., Buchgeister, J., Hischier, R., Poganietz, W.R., Schebek, L. and Warsen, J., 2008. Towards a framework for life cycle thinking in the assessment of nanotechnology. *Journal of Cleaner Production*, *16*(8–9), pp.910–926.
42. Thakur, A. and Kumar, A., 2023, August. Ecotoxicity analysis and risk assessment of nanomaterials for the environmental remediation. In *Macromolecular Symposia* (Vol. 410, No. 1, p. 2100438).
43. Biswas, P. and Yadav, A., 2018. Regulations for safety assessment of nanomaterials. In *Environmental Toxicity of Nanomaterials* (pp. 447–496). CRC Press.
44. Tran, T.K., Nguyen, M.K., Lin, C., Hoang, T.D., Nguyen, T.C., Lone, A.M., Khedulkar, A.P., Gaballah, M.S., Singh, J., Chung, W.J. and Nguyen, D.D., 2023. Review on fate, transport, toxicity and health risk of nanoparticles in natural ecosystems: Emerging challenges in the modern age and solutions toward a sustainable environment. *Science of The Total Environment*, p.169331.
45. Papp, T., Schiffmann, D., Weiss, D., Castranova, V., Vallyathan, V. and Rahman, Q., 2008. Human health implications of nanomaterial exposure. *Nanotoxicology*, *2*(1), pp.9–27.
46. Kica, E. and Bowman, D.M., 2012. Regulation by means of standardization: Key legitimacy issues of health and safety nanotechnology standards. *Jurimetrics*, pp.11–56.
47. Mehta, N., Shetty, S., Prajapati, B.G. and Shetty, S., 2024. Regulatory and ethical concerns in the use of nanomaterials. In *Alzheimer's Disease and Advanced Drug Delivery Strategies* (pp. 197–212). Academic Press.
48. Sandler, R., 2009. *Nanotechnology: The Social and Ethical Issues*. Washington: Woodrow Wilson International Center for Scholars.
49. Song, T., Jiang, B., Li, Y., Ji, Z., Zhou, H., Jiang, D., Seok, I., Murugadoss, V., Wen, N. and Colorado, H., 2021. Self-healing materials: A review of recent developments. *ES Materials & Manufacturing*, *14*, pp.1–19.
50. Li, J., Xin, M., Ma, Z., Shi, Y. and Pan, L., 2021. Nanomaterials and their applications on bio-inspired wearable electronics. *Nanotechnology*, *32*(47), p.472002.
51. Omran, B.A. and Baek, K.H., 2022. Valorization of agro-industrial biowaste to green nanomaterials for wastewater treatment: Approaching green chemistry and circular economy principles. *Journal of Environmental Management*, *311*, p.114806.
52. Gohar, O., Khan, M.Z., Bibi, I., Bashir, N., Tariq, U., Bakhtiar, M., Karim, M.R.A., Ali, F., Hanif, M.B. and Motola, M., 2024. Nanomaterials for advanced energy applications: Recent advancements and future trends. *Materials & Design*, *241*, p.112930.
53. Khraisheh, M., Elhenawy, S., AlMomani, F., Al-Ghouti, M., Hassan, M.K. and Hameed, B.H., 2021. Recent progress on nanomaterial-based membranes for water treatment. *Membranes*, *11*(12), p.995.

54. Singh, J., 2021. Slurry erosion performance analysis and characterization of high-velocity oxy-fuel sprayed Ni and Co hardsurfacing alloy coatings. *Journal of King Saud University-Engineering Science*, 1–14. https://doi.org/10.1016/j.jksues.2021.06.009.
55. Alaskar, A., Mahmood, M.S., Zaid, O., Althoey, F. and Arbili, M.M., 2023. Systematic review on geopolymer composites modified with nanomaterials and thin films: Enhancing performance and sustainability in construction. *Construction and Building Materials*, *409*, p.133888.
56. Bellanthudawa, B.K.A., Nawalage, N.M.S.K., Handapangoda, H.M.A.K., Suvendran, S., Wijayasenarathne, K.A.S.H., Rathnasuriya, M.L.D., Wickramasinghe, P.G.M.U., Aberathna, A.A.A.U., Tennakoon, A. and Perera, I.J.J.U.N., 2023. A perspective on biodegradable and non-biodegradable nanoparticles in industrial sectors: Applications, challenges, and future prospects. *Nanotechnology for Environmental Engineering*, *8*(4), pp.975–1013.
57. Prashar, G. and Vasudev, H., 2022. Surface topology analysis of plasma sprayed Inconel625-Al_2O_3 composite coating. *Materials Today Proceedings*, *50*, pp.607–611.
58. Sarkar, M.R., Rashid, M.H.O., Rahman, A., Kafi, M.A., Hosen, M.I., Rahman, M.S. and Khan, M.N., 2022. Recent advances in nanomaterials based sustainable agriculture: An overview. *Environmental Nanotechnology, Monitoring & Management*, *18*, p.100687.
59. Kumar, S., Singh, J. and Mohapatra, S.K., 2018. Role of particle size in assessment of physico-chemical properties and trace elements of Indian fly ash. *Waste Management Research*, *36*, 1016–1022.
60. Singh, J., Gill, H.S. and Vasudev, H., 2022. Computational fluid dynamics analysis on role of particulate shape and size in erosion of pipe bends. *International Journal of Interactive Design and Manufacturing*, 1–16.
61. Singh, J., Mohapatra, S.K. and Kumar, S., 2021. Performance analysis of pump materials employed in bottom ash slurry erosion conditions. *Jurnal Tribologi*, *30*, 73–89.
62. Singh, J., Kumar, S. and Mohapatra, S.K., 2021. Study on solid particle erosion of pump materials by fly ash slurry using Taguchi's orthogonal array. *Tribologia - Finnish Journal of Tribology*, *38*, 31–38.
63. Singh, K., Kumar, N., Bharti, A., Thakur, P. and Kumar, V., 2024. Waste to energy conversion: Key elements for sustainable waste management. In *Integrated Waste Management: A Sustainable Approach from Waste to Wealth* (pp. 91–117). Singapore: Springer Nature Singapore.

11 Biomimetic Materials

Properties, Applications, and Recent Developments

Gaurav Prashar and Hitesh Vasudev

Description	**Abbreviation**
Hydroxyapatite	HAP
High-efficiency particulate air	HEPA
Solution precursor plasma spray	SPPS
Thermal barrier coatings	TBCs
Volatile organic compound	VOC
Yttrium aluminum garnet	YAG

11.1 INTRODUCTION

The term "biomimetics" is derived from Latin, with "bios" meaning life and "mimesis" meaning to imitate or replicate. Creating materials that imitate nature is a way to solve intricate human issues by drawing inspiration from natural elements, systems, and models. These materials exemplify the principles of waste minimization and efficiency that are ingrained in nature. Studying and imitating the shapes, functions, and ecosystems found in nature to create more sustainable designs is called biomimicry, or biomimetics. It has been referred to by various names, such as "intellectual structure" in Japan and "smart material" in the US [1]. The application of biomimetics has been ingeniously demonstrated in well-known examples, like the Eiffel Tower's design, which draws inspiration from the trabecular structure of bone.

The history of biomimetics dates back to the first and second centuries, when evidence of primitive dental implants was discovered in the cultures of ancient Rome and pre-Columbian civilizations in Central and South America. In 659 AD, Chinese literature documented the initial utilization of dental amalgam. Leonardo da Vinci's (1452–1519) work exemplifies the concept of biomimicry. He created a "flying machine" that took inspiration from the graceful movements of a bird [2]. In the Far East, General Yi Sun-sin constructed the turtleship, a formidable warship designed to combat Japanese raiders during invasions [3]. Inspired by the wings of eagles, the

DOI: 10.1201/9781003516071-11

Wright brothers were able to create a powered airplane that achieved the remarkable feat of human flight in 1903. Throughout the following century, advancements in aviation led to increased speed, improved stability, and enhanced aerodynamics of airplanes [4–5].

In the past ten years, automobile design has impacted not only the exterior aesthetics of vehicles but also their functional capabilities. The principles of economization and energy efficiency inherent in biomimetics have been integrated into automotive design, as exemplified by Daimler Chrysler's initial prototype bionic idea car (Figure 11.1). Heart pacemakers and hip and knee joints have been in use since the early and mid-twentieth century. In 1950, Otto Schmitt introduced the concept of biomimetics, which involves drawing inspiration from biology. Later, in 1960, Jack Steele named the field of studying and imitating biology as bionics. The term biomimetic was added to the Webster's Dictionary in 1974. Even though biomimetic history dates back to the first century, it only gained popularity among scientists and researchers after the publication of Janine Benyus' revolutionary book in 1997, titled *Biomimicry: Innovation Inspired by Nature* [6]. Figure 11.2 illustrates the research of biomimetics within the discipline of biomimicry. Janine Benyus along with others took the initiative to establish a social enterprise known as Biomimicry 3.8.

FIGURE 11.1 Daimler Chrysler has developed a prototype bionic car (D) that draws inspiration from the box fish (A), utilizes the skeletal structure of the blowfish for vehicle design (C), and incorporates tree growth patterns (B) [1] (Credit: Creative Commons Attribution 3.0 International License).

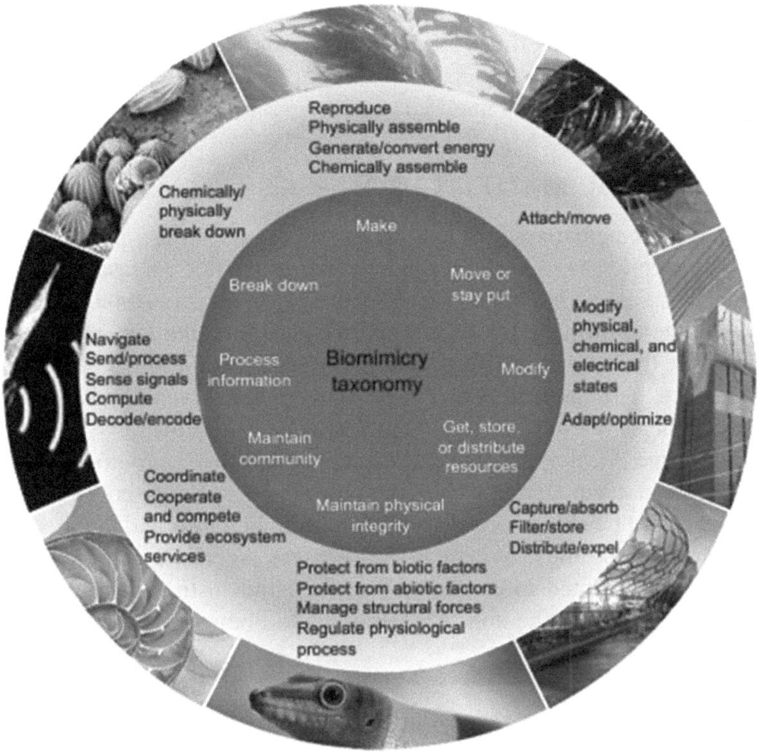

FIGURE 11.2 Categorizing the research pursuits of biomimetics within the discipline of biomimicry [1] (Credit: Creative Commons Attribution 3.0 International License).

Their goal was to foster the exchange of ideas and concepts related to biomimicry and biomimetics, while also facilitating connections among researchers, scientists, engineers, and stakeholders from various disciplines. The fundamental research approach for biomimetics consists of six distinct steps that can be employed to incorporate biomimetics into various areas such as design, product development, services, and agriculture [7].

Understanding the applications and limitations of newly discovered materials can be greatly enhanced through biomimetics research. Understanding the morphological and functional uses of the novel materials is crucial, as well as evaluating the advantages and disadvantages of biomimetics. Unraveling the outcomes of combining these elements is essential. Research is actively being conducted in these areas, but it is challenging to make significant progress. By studying the structure and function of the biological material, new avenues for innovation in material development can be explored, potentially leading to connections with other materials. Through tests and assessments, the structure and function of known materials undergo transformations and advancements, leading to the development of new materials. Through the integration of cutting-edge advancements in medicine, chemistry, and nanotechnology, new applications may emerge that have the potential to greatly enhance human life. This book chapter explores the properties, applications of biomimetics, and recent developments in the field which has expanded to encompass a range of interconnected objectives in innovative materials and processes.

11.2 BIOMIMETIC MATERIAL PROPERTIES

Within various biological systems, one can observe mechanical properties that are challenging to replicate precisely in synthetic materials [8]. It's fascinating to think about how biological architectures are composed of polymers containing basic elements like carbon, nitrogen, hydrogen, and oxygen. Just like a materials engineer, one can analyze the mechanical properties of tissues by examining how they respond to applied forces and deformations. The elastic modulus, also known as Young's modulus, is a measure of a material's stiffness. It is determined by the slope of the stress-strain curve [9]. Some materials, such as bones, are less deformable because they have a higher elastic modulus. On the other hand, softer biological tissues like skin, cartilage, and heart valves have a lower elastic modulus and are more deformable [10]. Most importantly, human tissues possess an elastic nature. This unique characteristic is attributed to elastin, a protein found in the extracellular matrix. Elastin provides not only the ability to stretch and bounce back but also a certain level of durability.

11.2.1 Tissue Engineering

The elastomeric properties of various tissues are different, including heart muscle, heart valves, and blood vessels. There has been an ongoing endeavor in the field of tissue engineering to engineer these tissues [11–15]. As a means of engineering these tissues with the desired elastomeric properties, numerous elastomeric materials have been studied. Some examples of common synthetic polymers include poly(ε-caprolactone) (PCL) and polyurethanes (PU), along with elastic materials generated from the natural ECM [15]. PCL, a type of semicrystalline polymer, is extremely elastic at room or body temperature due to its low glass-transition temperature of approximately −62°C. Physiological conditions cause PCL to degrade via a number of mechanisms [16–17], however, the degradation rate is excessively slow, reducing its appeal. Elastomeric characteristics can be achieved through structural changes in segmented PU [18]. Toxic precursors, like toluene diisocyanates, are used in the manufacture of PU, which is a big drawback for biomedical uses. Biodegradable polyurethane (PU) and urethane-based polymers utilizing less hazardous diisocyanates have been developed [19]. Researchers have looked at these polymers for potential use in vascular and other types of tissue engineering [20–22].

Aside from utilizing current polymer technologies, biomimetic methods can be employed to create elastomeric materials that mimic the desired elasticity and pliability of different tissues. The extensibility and elastic recoil qualities are imparted to numerous bodily tissues by elastin, an insoluble polymeric ECM protein. Tropoelastin, a soluble precursor to insoluble elastin, is produced during elastin synthesis (~70 kDa) [23]. The repetitive domain structure of tropoelastin alternates between sections that are hydrophobic and those that are crosslinking. At high enough temperatures, tropoelastin and repetitive peptides produced from elastin can self-assemble (coacervate) to create aggregates that are insoluble. Dan Urry was an early trailblazer in studying elastin-like polypeptides (ELPs) using recombinant technologies. The pentapeptide repeat Val-Pro-Gly-Xaa-Gly (VPGXG) in tropoelastin's hydrophobic domain is the starting point for the artificial polypeptides [24–25]. To create thermo-reversible hydrogels, researchers have engineered elastin-mimetic triblock copolymers that are hydrophobic, hydrophylic, and hydrophobic [26]. For specific tissue engineering uses, ELPs show promise as scaffolding materials [27]. The natural elastomeric proteins served as inspiration for the recent engineering of a small non-mechanical protein, GB1 (the streptococcal B1 immunoglobulin-binding domain of protein G), into a polyprotein (GB1)8. This engineered material exhibited excellent elastomeric properties [28], suggesting it could be used as a scaffold in tissue engineering.

The strength and tensile modulus of connective tissues are crucial, especially tendons and ligaments. The natural fibrous protein substance known as silk has been utilized for ages in textile production and surgical sutures due to its exceptional tensile mechanical qualities. Silk is spun from the cocoon of silkworms. It shows promise as a scaffolding material for the engineering of ligaments and tendons [29]. Natural silk consists of two parts: silk fibroin, a protein that forms the core of the filament, and a coating of sericin proteins that acts like glue. The potential toxicity of natural silk has raised some concerns [30]. One drawback of using natural silk materials for tissue engineering is the presence of sericin, which has been shown to trigger negative immunological responses [31]. The biocompatibility of regenerated silk fibroin is enhanced after sericin removal [32]. Natural block copolymers of hydrophobic and hydrophilic amino acid blocks are what make up silk fibroins. As a result of hydrogen bonding and hydrophobic interactions, the hydrophobic blocks in silk fibroins often form β-sheets or crystals, which gives them their tensile strength [32]. Polymers mimicking spider silk fibroin have been produced through genetic engineering; these have enhanced cell adhesion capability [33] and great mechanical characteristics [34].

Creating cellular materials that mimic the mechanical properties, porosity, and elastic modulus of natural bone is a constant goal in the realm of bone implants. Creating and producing an implant that meets all of these requirements is quite a difficult undertaking. The study conducted by [reveals that the developed biomimetic structural material exhibits remarkable controllability over its mechanical properties]. The rod plate-like (RPT) trabecular biomimetic structural material demonstrated remarkable mechanical load-bearing capabilities, surpassing those of natural bone trabeculae, while preserving the same total porosity (83.1%) and elastic modulus (798.1 MPa) as natural bone. This biomimetic structural material achieves a remarkable balance between strength and elasticity, making it an ideal design template for future medical implants. This material shows immense promise for use in clinical applications, particularly for bone repair. Additionally, its mechanical properties can be easily adjusted, making it a highly attractive option for tissue engineering [35].

11.2.2 Musculoskeletal Tissue Engineering

In order to more accurately replicate a flawed tissue in musculoskeletal tissue engineering, it is important to utilize materials that possess comparable chemical and physical properties to the desired tissue. There are three main types of biomaterials used for the musculoskeletal system, each with its own unique biophysical properties. These include elastic, soft, and hard biomaterials [36].

11.3 APPLICATION AREA OF BIOMIMETICS

Humans have had a significant impact on nature due to industrialization and resource extraction. Fortunately, biomimetics offers a way to break free from this destructive pattern. Biomimetics extends beyond the utilization of natural properties to drive the development of novel products. These products have a wide range of applications across various industries, including general industry, chemistry, architecture, engineering, and medicine (Figure 11.3). A harmonious relationship between humans and nature is crucial for our coexistence, and its potential applications are limitless. It is crucial to have a thorough understanding of these areas and to have specific examples for each of them.

Here we discuss only architecture, interior design, and dentistry. One of the most fascinating examples of natural architecture is the towering termite nests that can be discovered in the African savanna, reaching an impressive height of 6 meters. These nests are constructed using a combination of earth, tree bark, sand, and termite saliva, resulting in a remarkable strength that surpasses that of concrete. Termites are highly responsive to heat, and when the outside temperature reaches 40°C, their nests are able to sustain an internal temperature of 30°C. The cooling system utilized

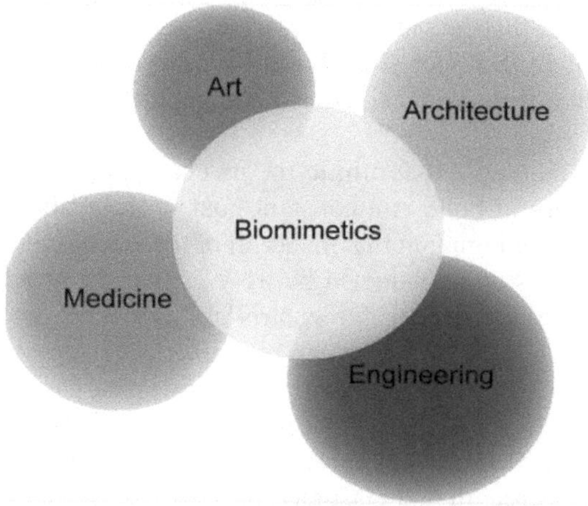

FIGURE 11.3 Application areas in biomimetics [1] (Credit: Creative Commons Attribution 3.0 International License).

in these structures consists of a series of underground canals that establish a well-ventilated environment, ensuring a consistent temperature is maintained.

This system caught the attention of Mick Pearce, a South African architect who was working with the engineering firm Arup in Harare, the capital of Zimbabwe. Utilizing a unique characteristic found in termite nests, the Eastgate Center was constructed as a multi-purpose building, making it the world's first structure to incorporate a fully natural cooling system.

This building incorporates openings on the roof and lower floors to facilitate natural ventilation, mimicking the efficient airflow found in termite nests. As hot air escapes through the roof, cool air is drawn in from below, creating a comfortable and energy-efficient environment. There are no cooling systems installed in the building.

Due to this factor, the energy usage of this building is significantly lower, by less than 10%, in comparison to other buildings with similar dimensions. This is achieved by maintaining an internal temperature of only 24°C, even when the external temperature soars to 38°C.

This is just one of the numerous instances where biomimetics can assist us in developing practical and effective designs.

When considering interior design, it's fascinating to observe how biomimetics is continually advancing with the help of cutting-edge technologies such as 3D Printing. MIT professor and designer Neri Oxman has utilized biomimetics to design innovative products, employing 3D Stratasys printers equipped with triple-jetting technology [37].

In dentistry, no biomaterial can replicate the exact mechanical, physical, and optical properties of tooth structure (such as dentin, enamel, and cementum) while also possessing the same physiological characteristics as natural teeth during function (inset Figure 11.4). Through the utilization of biomimetic therapeutic approaches, dental professionals have the ability to enhance their practices and align more closely with the intricacies of natural biological structures and their functionality [38–41].

11.4 RECENT DEVELOPMENTS IN BIOMIMETICS

There have been significant advancements in various fields thanks to biomimetics. The intricate structural architectures found in nature surpass the capabilities of conventional design and fabrication methods, impeding the advancement of biomimetic research and its application in engineering systems. Due to the advancement in additive manufacturing (3D printing), there are now exciting possibilities for manipulating and imitating the intricate structures found in nature, which are inherently multiscale, multi-material, and multifunctional [42–48]. This technique is completely transforming the way we modify the surfaces of scaffolds, allowing us to finely and precisely adjust their properties such as geometry and roughness on a sub-micrometric scale [49, 50–53]. 3D printed scaffolds offer numerous advantages over previous techniques, allowing for the design of specific geometries with improved resolution and the ability to customize production in a patient-tailored manner [54]. Utilizing 3D printing technology in the healthcare industry offers numerous benefits, including faster production and reduced material waste. Additionally, it allows for

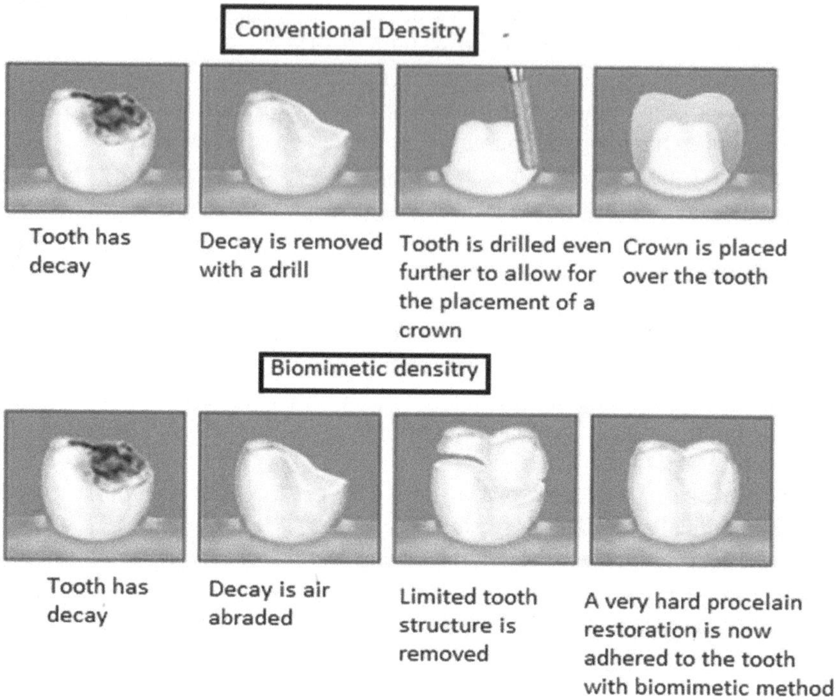

FIGURE 11.4 Comparison between two approaches to dentistry [38]. (Credit: Creative Commons Attribution 3.0 International License).

the utilization of surface modification techniques that have been previously established. As an expert in materials engineering, Shi et al. [55] introduced an innovative method to enhance the surface bioactivity of PEEK implants through the utilization of 3D printing technology. PEEK is commonly used as an orthopedic implant material because of its excellent biocompatibility and mechanical strength. However, it does have the drawback of being biologically inert. Through a combination of fused deposition modeling and sulfonation treatment, the researchers were able to enhance the development of micropores on the surface of PEEK implants. Tests on stem cells from human exfoliated deciduous teeth revealed that the final material demonstrated excellent biocompatibility and displayed significantly stronger osteogenic activity in the phalanges compared to the control groups [56].

Emerging technologies are now being developed to apply biomimetic intelligence and humanoid robots in various complex scenarios, including living systems and home environments, to replicate biological functions [57]. Researchers are finding biomimetic AI to be incredibly exciting, as it has the potential to greatly impact our lives. This field is centered around the advancements of algorithms that draw inspiration from biological working mechanisms. Examples of these include the genetic algorithm and ant colony optimization. In recent times, there have been significant breakthroughs in biomimetic intelligence. Biomimetic sensors and sensing techniques, along with evolution-inspired algorithms, have shown remarkable results. These advancements have allowed scientists to develop an intuition similar to that used in their research. The rapid advancement of science thanks to AI is constantly evolving, but the practical implementation of bionics and biomimicry from AI into human applications is still a work in progress. So far, there have been significant challenges in accurately replicating data on a large scale. In the field of biomaterial design, researchers face numerous challenges, including conducting countless experiments, developing models, making predictions, and encountering failures. This is because they need to recreate complex and dynamic environments. Artificial intelligence could potentially be a valuable tool in this process [58–61]. The combination of machine learning and AI needs to be fine-tuned, as machine learning places limitations on AI [62–64].

Robotic biomimetic technology is being applied across multiple domains, including design, chemistry, medical profession, agriculture, and material science. This concept has been integrated into robotics research, leading to the emergence of a new domain known as biomimetic robotics [65]. The motor mechanisms observed in animals and insects are currently being replicated, resulting in a notable enhancement in machine motility, which was previously constrained by the use of wheels. Altendorfer et al. [65] conducted research on the mechanisms of cockroach legs, leading to the development of RHex (Figure 11.5), a robotic system featuring legs and feet designed to surpass the limitations associated with wheeled locomotion. Wright et al. [66] are engaged in the development of robotic arms that utilize the structural mechanisms of a snake's bones and joints, as well as those of an elephant's trunk (Figure 11.6). Additionally, Full and Fearing engineered an iRobot that is sensitive to pressure, enabling it to achieve effective dry adhesion. This innovation was inspired by the micro hairs found on the soles of a gecko.

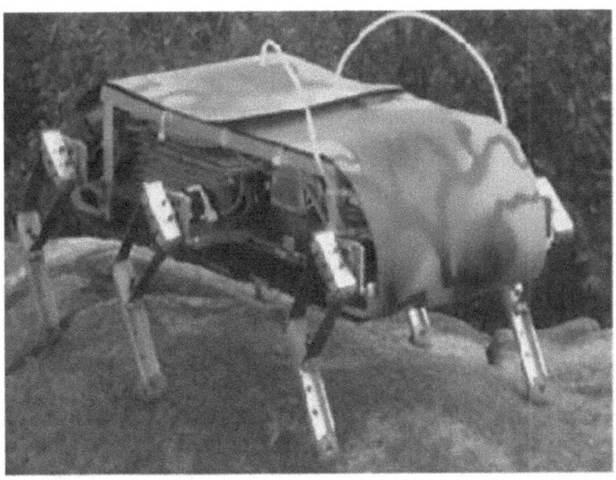

FIGURE 11.5 Image of RHex [1] (Credit: Creative Commons Attribution 3.0 International License).

FIGURE 11.6 Photograph of a snake robot [1] (Credit: Creative Commons Attribution 3.0 International License).

11.5 CONCLUSION

In this chapter, we explored the latest advancements in designing and developing materials that mimic nature and discussed their various applications. We also tried to touch on the latest breakthroughs in biomimetics and had a conversation about the intricate process of creating artificial living systems from scratch. We focused on the exciting applications of 3D bioprinting. Fast-paced 3D printing and artificial intelligence, however, face a challenge when it comes to creating biomimetic materials that can be reproduced at a real scale, matching the complexity of living systems. How challenging is it to achieve this? In today's world, the field of science is actively seeking bioengineering technologies that can be applied in medicine and clinics, drawing inspiration from nature and mimicking its processes.

REFERENCES

1. Hwang, J., Jeong, Y., Park, J.M., Lee, K.H., Hong, J.W. and Choi, J., 2015. Biomimetics: Forecasting the future of science, engineering, and medicine. *International Journal of Nanomedicine*, *10*(1), 5701–5713.
2. Vincent, J.F., Bogatyreva, O.A., Bogatyrev, N.R., Bowyer, A. and Pahl, A.K., 2006. Biomimetics: Its practice and theory. *Journal of the Royal Society Interface*, *3*(9), 471–482.
3. Pemsel, H., 1977. *A History of War at Sea: An Atlas and Chronology of Conflict at Sea from Earliest Times to the Present*. Naval Institute Press.
4. Freedman, R., 1991. *The Wright Brothers: How They Invented the Airplane*. New York: Holiday House.
5. Jakab, P.L., 1990. *Visions of a Flying Machine: The Wright Brothers and the Process of Invention*. Smithsonian Institution Press.
6. Benyus, J.M., 1997. *Biomimicry: Innovation Inspired by Nature*. Boston, MA: Mariner Books.

7. Masatsugu, S., 2010. New trends in next-generation biomimetic material technology: Learning from biodiversity. *Science & Technology Trends, Quarterly Review, 37*, 53–75.
8. Vincent, J.F.V., 2012. *Structural Biomaterials*. Princeton: Princeton University Press. https://press.princeton.edu/books/paperback/9780691154008/structural-biomaterials (accessed on 29 July 2012).
9. Muiznieks, L.D. and Keeley, F.W., 2013. Molecular assembly and mechanical properties of the extracellular matrix: A fibrous protein perspective. *Biochimica et Biophysica Acta (BBA)-Molecular Basis of Disease, 1832*(7), 866–875.
10. Coenen, A.M., Bernaerts, K.V., Harings, J.A., Jockenhoevel, S. and Ghazanfari, S., 2018. Elastic materials for tissue engineering applications: Natural, synthetic, and hybrid polymers. *Acta biomaterialia, 79*, 60–82.
11. Shinoka, T., Shum-Tim, D., Ma, P.X., Tanel, R.E., Langer, R., Vacanti, J.P. and Mayer Jr, J.E., 1997. Tissue-engineered heart valve leaflets: does cell origin affect outcome? *Circulation, 96*(9 Suppl), 2–102.
12. Shinoka, T., Shum-Tim, D., Ma, P.X., Tanel, R.E., Isogai, N., Langer, R., Vacanti, J.P. and Mayer Jr, J.E., 1998. Creation of viable pulmonary artery autografts through tissue engineering. *The Journal of Thoracic and Cardiovascular Surgery, 115*(3), 536–546.
13. Berglund, J.D., Nerem, R.M. and Sambanis, A., 2004. Incorporation of intact elastin scaffolds in tissue-engineered collagen-based vascular grafts. *Tissue Engineering, 10*(9–10), 1526–1535.
14. Stephan, S., Ball, S.G., Williamson, M., Bax, D.V., Lomas, A., Shuttleworth, C.A. and Kielty, C.M., 2006. Cell-matrix biology in vascular tissue engineering. *Journal of Anatomy, 209*(4), 495–502.
15. Isenberg, B.C., Williams, C. and Tranquillo, R.T., 2006. Small-diameter artificial arteries engineered in vitro. *Circulation Research, 98*(1), 25–35.
16. Pitt, G.G., Gratzl, M.M., Kimmel, G.L., Surles, J. and Sohindler, A., 1981. Aliphatic polyesters II. The degradation of poly (DL-lactide), poly (ε-caprolactone), and their copolymers in vivo. *Biomaterials, 2*(4), 215–220.
17. Woodward, S.C., Brewer, P.S., Moatamed, F., Schindler, A. and Pitt, C.G., 1985. The intracellular degradation of poly (ε -caprolactone). *Journal of Biomedical Materials Research, 19*(4), 437–444.
18. Lin, H.B., Sun, W., Mosher, D.F., García-Echeverría, C., Schaufelberger, K., Lelkes, P.I. and Cooper, S.L., 1994. Synthesis, surface, and cell-adhesion properties of polyurethanes containing covalently grafted RGD-peptides. *Journal of Biomedical Materials Research, 28*(3), 329–342.
19. Skarja, G.A. and Woodhouse, K.A., 2001. In vitro degradation and erosion of degradable, segmented polyurethanes containing an amino acid-based chain extender. *Journal of Biomaterials Science, Polymer Edition, 12*(8), 851–873.
20. Sunitha, K. and Vasudev, H., 2022. A short note on the various thermal spray coating processes and effect of post-treatment on Ni-based coatings. *Materials Today: Proceedings, 50*, 1452–1457.
21. Zhang, J.Y., Beckman, E.J., Piesco, N.P. and Agarwal, S., 2000. A new peptide-based urethane polymer: Synthesis, biodegradation, and potential to support cell growth in vitro. *Biomaterials, 21*(12), 1247–1258.
22. Yedida, V.V.S. and Vasudev, H., 2022. A review on the development of thermal barrier coatings by using thermal spray techniques. *Materials Today: Proceedings, 50*, 1458–1464.

23. Bellingham, C.M., Lillie, M.A., Gosline, J.M., Wright, G.M., Starcher, B.C., Bailey, A.J., Woodhouse, K.A. and Keeley, F.W., 2003. Recombinant human elastin polypeptides self-assemble into biomaterials with elastin-like properties. *Biopolymers: Original Research on Biomolecules*, *70*(4), 445–455.
24. Urry, D.W., 1997. Physical chemistry of biological free energy transduction as demonstrated by elastic protein-based polymers. *The Journal of Physical Chemistry B*, *101*(51), 11007–11028.
25. Chilkoti, A., Christensen, T. and MacKay, J.A., 2006. Stimulus responsive elastin biopolymers: Applications in medicine and biotechnology. *Current Opinion in Chemical Biology*, *10*(6), 652–657.
26. Wright, E.R. and Conticello, V.P., 2002. Self-assembly of block copolymers derived from elastin-mimetic polypeptide sequences. *Advanced Drug Delivery Reviews*, *54*(8), 1057–1073.
27. Betre, H., Ong, S.R., Guilak, F., Chilkoti, A., Fermor, B. and Setton, L.A., 2006. Chondrocytic differentiation of human adipose-derived adult stem cells in elastin-like polypeptide. *Biomaterials*, *27*(1), 91–99.
28. Cao, Y. and Li, H., 2007. Polyprotein of GB1 is an ideal artificial elastomeric protein. *Nature Materials*, *6*(2), 109–114.
29. Altman, G.H., Horan, R.L., Lu, H.H., Moreau, J., Martin, I., Richmond, J.C. and Kaplan, D.L., 2002. Silk matrix for tissue engineered anterior cruciate ligaments. *Biomaterials*, *23*(20), 4131–4141.
30. Kourie, J.I. and Shorthouse, A.A., 2000. Properties of cytotoxic peptide-formed ion channels. *American Journal of Physiology-Cell Physiology*, *278*(6), pp.C1063–C1087.
31. Altman, G.H., Diaz, F., Jakuba, C., Calabro, T., Horan, R.L., Chen, J., Lu, H., Richmond, J. and Kaplan, D.L., 2003. Silk-based biomaterials. *Biomaterials*, *24*(3), 401–416.
32. Santin, M., Motta, A., Freddi, G. and Cannas, M., 1999. In vitro evaluation of the inflammatory potential of the silk fibroin. *Journal of Biomedical Materials Research: An Official Journal of The Society for Biomaterials, The Japanese Society for Biomaterials, and The Australian Society for Biomaterials and the Korean Society for Biomaterials*, *46*(3), 382–389.
33. Bini, E., Foo, C.W.P., Huang, J., Karageorgiou, V., Kitchel, B. and Kaplan, D.L., 2006. RGD-functionalized bioengineered spider dragline silk biomaterial. *Biomacromolecules*, *7*(11), 3139–3145.
34. Guerette, P.A., Ginzinger, D.G., Weber, B.H. and Gosline, J.M., 1996. Silk properties determined by gland-specific expression of a spider fibroin gene family. *Science*, *272*(5258), 112–115.
35. Kong, D., Wang, Q., Huang, J., Zhang, Z., Wang, X., Han, Q. and Shi, Y., 2024. A biomimetic structural material with adjustable mechanical property for bone tissue engineering. *Advanced Functional Materials*, *34*(8), 2305412.
36. Del Bakhshayesh, A.R., Asadi, N., Alihemmati, A., Tayefi Nasrabadi, H., Montaseri, A., Davaran, S., Saghati, S., Akbarzadeh, A. and Abedelahi, A., 2019. An overview of advanced biocompatible and biomimetic materials for creation of replacement structures in the musculoskeletal systems: Focusing on cartilage tissue engineering. *Journal of Biological Engineering*, *13*, 1–21.
37. Singh, J., Hashmi, A. W., Ahmad, S. and Tian, Y., 2024. Critical review on biodegradable and biocompatibility magnesium alloys: Progress and prospects in bio-implant applications. *Inorganic Chemistry Communications*, *169*, 113111.
38. Singer, L., Fouda, A. and Bourauel, C., 2023. Biomimetic approaches and materials in restorative and regenerative dentistry. *BMC Oral Health*, *23*(1), 105.

39. Singh, R., Toseef, M., Kumar, J. and Singh, J. 2022. Benefits and challenges in additive manufacturing and its applications. In *Sustainable Advanced Manufacturing and Materials Processing*, eds. Kaushal, S., Singh, I., Singh, S. and Gupta, A., 137–157. CRC Press. https://doi.org/10.1201/9781003269298-8.
40. Dutta, V., Thakur, L., Singh, B. and Vasudev, H., 2022. A study of erosion–corrosion behaviour of friction stir-processed chromium-reinforced NiAl bronze composite. *Materials, 15*, 5401.
41. Kulkarni, S.D., Manjunatha, C.U., Manjunath, K.V, Prasad, C.D. and Vasudev, H., 2024. Design and optimization of polyvinyl-nitride rubber for tensile strength analysis. *International Journal on Interactive Design and Manufacturing, 18*, 5893–5908.
42. Poojari, M., Hanumanthappa, H., Prasad, C.D., Jathanna, H.M., Ksheerasagar, A.R., Shetty, P., Shanmugam, B.K. and Vasudev, H., 2023. Computational modelling for the manufacturing of solar-powered multifunctional agricultural robot. *International Journal on Interactive Design and Manufacturing (IJIDeM), 18*, 1–12.
43. Kumar, P., Singh, J. and Singh, S., 2022. Neural network supported flow characteristics analysis of heavy sour crude oil emulsified by ecofriendly bio-surfactant utilized as a replacement of sweet crude oil. *Chemical Engineering Journal Advances, 11*, 100342.
44. Singh, J., 2021. Slurry erosion performance analysis and characterization of high-velocity oxy-fuel sprayed Ni and Co hardsurfacing alloy coatings. *Journal of King Saud University - Engineering Sciences*. https://doi.org/10.1016/j.jksues.2021.06.009.
45. Prashar, G. and Vasudev, H., 2022a. Surface topology analysis of plasma sprayed Inconel625-Al_2O_3 composite coating. *Materials Today: Proceedings, 50*, 607–611.
46. Prashar, G. and Vasudev, H., 2022b. A review on the influence of process parameters and heat treatment on the corrosion performance of Ni-based thermal spray coatings. *Surface Review and Letters, 29*(1), 2230001.
47. Sharanabasava, H., Raviprakash, M., Prasad, C.D., Ramesh, M.R., Phanibhushana, M. V, Vasudev, H. and Kumar, S., 2023. Microstructure, mechanical and wear properties of SiC and Mo reinforced NiCr microwave cladding. *Advances in Materials and Processing Technologies*. https://doi.org/10.1080/2374068X.2023.2257937.
48. Singh, J., Mohapatra, S.K. and Kumar, S., 2021. Performance analysis of pump materials employed in bottom ash slurry erosion conditions. *Jurnal Tribologi, 30*, 73–89.
49. Singh, J., Singh, R., Singh, S., Vasudev, H. and Kumar, S., 2022. Reducing scrap due to missed operations and machining defects in 90PS pistons. *International Journal on Interactive Design and Manufacturing (IJIDeM), 17*(5), 1–13.
50. Chakraborty, R., Anoop, A.G., Thakur, A., Mohanta, G.C. and Kumar, P., 2023. Strategies to modify the surface and bulk properties of 3D-printed solid scaffolds for tissue engineering applications. *ACS Omega, 8*(6), 5139–5156.
51. Shopova, D., Yaneva, A., Bakova, D., Mihaylova, A., Kasnakova, P., Hristozova, M., Sbirkov, Y., Sarafian, V. and Semerdzhieva, M., 2023. (Bio) printing in personalized medicine—opportunities and potential benefits. *Bioengineering, 10*(3), 287.
52. Majji, B.G.R., Vasudev, H. and Bansal, A., 2022. A review on the oxidation and wear behavior of the thermally sprayed high-entropy alloys. *Materials Today: Proceedings, 50*, 1447–1451.
53. Singh, J., Gill, H.S. and Vasudev, H., 2022. Computational fluid dynamics analysis on role of particulate shape and size in erosion of pipe bends. *International Journal on Interactive Design and Manufacturing (IJIDeM), 17*(5), 1–16.
54. Ahmad, S., Hashmi, A.W., Singh, J., Arora, K., Tian, Y., Iqbal, F., Al-Dossari, M. and Khan, M.I., 2024. Innovations in additive manufacturing of shape memory alloys: Alloys, microstructures, treatments, applications. *Journal of Materials Research and Technology, 32*. https://doi.org/10.1016/j.jmrt.2024.08.213.

55. Shi, Y., Deng, T., Peng, Y., Qin, Z., Ramalingam, M., Pan, Y., Chen, C., Zhao, F., Cheng, L. and Liu, J., 2023. Effect of surface modification of peek artificial phalanx by 3D printing on its biological activity. *Coatings*, *13*(2), 400.
56. Wang, J., Chen, W., Xiao, X., Xu, Y., Li, C., Jia, X. and Meng, M.Q.H., 2021. A survey of the development of biomimetic intelligence and robotics. *Biomimetic Intelligence and Robotics*, *1*, 100001.
57. Raabe, D., Mianroodi, J.R. and Neugebauer, J., 2023. Accelerating the design of compositionally complex materials via physics-informed artificial intelligence. *Nature Computational Science*, *3*(3), 198–209.
58. Witte, H., Hoffmann, H., Hackert, R., Schilling, C., Fischer, M.S. and Preuschoft, H., 2004. Biomimetic robotics should be based on functional morphology. *Journal of Anatomy*, *204*(5), 331–342.
59. Singh, J., Kumar, S. and Mohapatra, S.K., 2018. Optimization of erosion wear influencing parameters of HVOF sprayed pumping material for coal-water slurry. *Materials Today: Proceedings*, *5*(11), 23789–23795.
60. Thakur, L., Vasudev, H., Singh, J. and Prashar, G., 2023. *Artificial Intelligence and Machine Learning in the Thermal Spray Industry: Practices, Implementation, and Challenges*. CRC Press.
61. Boxerbaum, A.S., Shaw, K.M., Chiel, H.J. and Quinn, R.D., 2012. Continuous wave peristaltic motion in a robot. *The International Journal of Robotics Research*, *31*(3), 302–318.
62. Singh, J., Kumar, S. and Mohapatra, S.K., 2021. Study on solid particle erosion of pump materials by fly ash slurry using Taguchi's orthogonal array. *Tribologia - Finnish Journal of Tribology*, *38*(3–4), 31–38.
63. Singh, J. and Singh, S., 2023. Support vector machine learning on slurry erosion characteristics analysis of Ni- and Co-alloy coatings. *Surface Review and Letters*. https://doi.org/10.1142/S0218625X23400061.
64. Kattimani, M.A., Venkatesh, P.R., Masum, H., Math, M.M., Bahadurdesai, V.N., Mustafkhadri, S., Prasad, C.D. and Vasudev, H., 2023. Design and numerical analysis of tensile deformation and fracture properties of induction hardened inconel 718 superalloy for gas turbine applications. *International Journal on Interactive Design and Manufacturing (IJIDeM)*, *18*, 1–11.
65. Altendorfer, R., Moore, N., Komsuoglu, H., Buehler, M., Brown, H.B., McMordie, D., Saranli, U., Full, R. and Koditschek, D.E., 2001. Rhex: A biologically inspired hexapod runner. *Autonomous Robots*, *11*, 207–213.
66. Wright, C., Johnson, A., Peck, A., McCord, Z., Naaktgeboren, A., Gianfortoni, P., Gonzalez-Rivero, M., Hatton, R. and Choset, H., 2007, October. Design of a modular snake robot. In *2007 IEEE/RSJ International Conference on Intelligent Robots and Systems*, 2609–2614. IEEE.

12 Biomimetic Materials and Their Applications

A Holistic Perspective

Anish Kumar, Abdullah Al Noor, Karuna Khajuria, and Ashish Kumar Agrahari

Description	Abbreviation
Biomimetic natural biomaterials	BNBM
Deoxyribonucleic acid	DNA
Extracellular matrix	ECM
Hydroxyapatite	HA
Metal-organic framework	MOF
Molecularly imprinted polymers	MIPs
Peptide nucleic acids	PNAs
Polyhydroxyalkanoates	PHAs
Polylactic acid	PLAs

12.1 INTRODUCTION

Recently developed biomimetic materials based on natural sources are a promising field for the development of new materials because of their multiple potential uses, including tissue engineering, regenerative medicine, and impact uses [1]. These materials are commonly known as "smart" biostructures and they mimic the conformation and characteristics of an *in vivo* ECM, offering cells a variety of biological, physical, and chemical signals [2–3]. The term "biomimetic" is derived from the Greek language. It has been derived from *bios*, which means life, and *mimetikos*, which means good imitator [4]. It was first introduced by Otto Schmitt in the 1960s [5], and it means the design and manufacturing process of materials, structures, and systems modeled on biological ones [6]. This notion encompasses processes that mimic biological systems; and the creation of materials at the nanoscale characterized by the physical, chemical, mechanical, and biological features of biogenic materials at the nano- or macro-scale. As with other such materials, biomimetic material technology includes concepts such as self-assembly and interfacial molecular

DOI: 10.1201/9781003516071-12

recognition. Other promising directions that include biopolymers and biocomposites are operational in the biomedical and related industries.

Biomimetic materials refer to those that have been designed from information obtained from natural systems, and such materials have the potential of mimicking natural tissue in terms of structure and function [7]. They focus on creating materials with enhanced mechanical performance and biocompatibility of the developed composites to mimic natural hard and soft tissues of the oral cavity [8].

Biomimetics, despite being a relatively new science, has been applied over a vast period of time. Also called intellectual structure in Japan and smart material in the United States, biomimetics was established on the understanding that nature is the best model for delivering higher levels of productivity and functionality, and therefore achieves major advancements in these areas [9]. Additionally, the elimination of waste and the consequent curtailment of research costs has provided a body of practical gains [10, 11]. Minor advances in the fabrication of BNBMs have led to the development of mechanically flexible networks, microstructure integration, and self-osteoinductivity. These properties make these materials ideal for use in the fabrication of living implants for specific applications in tissue engineering and tissue regeneration. In addition, synthesis of biomimetic materials including Se-modified carbon nitride nanosheets, small intestinal submucosa, magnesium-strontium hydroxyapatite, dimethylglyoxime-urethane polyurethane, polydimethylsiloxane, Ag/Ag@AgCl/ZnO and PDTC$(COOH)_4$/HA have exhibited promising applications in numerous areas. This chapter postulates to provide an overview of new developments in the area of biomimetic materials with a focus on the enhancement of their preparation, functionality, potential uses, and future challenges. Further, it will explore the use of biomimetic designs for impact applications in a relatively new and progressing field compared to conventional engineering structures. In conclusion, biomimetic materials prove that technology might be developed based on natural inspiration, and new materials with superior characteristics might be designed. When integrated with innovations in health, chemistry, and nanotechnology, these materials may unlock astounding uses that will enrich the lives of people all over the world (Figure 12.1).

12.2 PROPERTIES OF BIOMIMETIC MATERIALS

Biomimetic materials, derived from the aspects of design seen in living organisms, have attracted significant interest in a number of fields, including biomedicine, material science, and engineering [11]. These materials are designed to mimic the mechanical, structural, and functional characteristics of biological tissues for certain purposes, including promoting tissue repair, fashioning complex drug delivery systems, and designing new biomaterials for use in medical devices [12]. It is exactly the properties of a material that determine its possible applications when considering biomimetic materials.

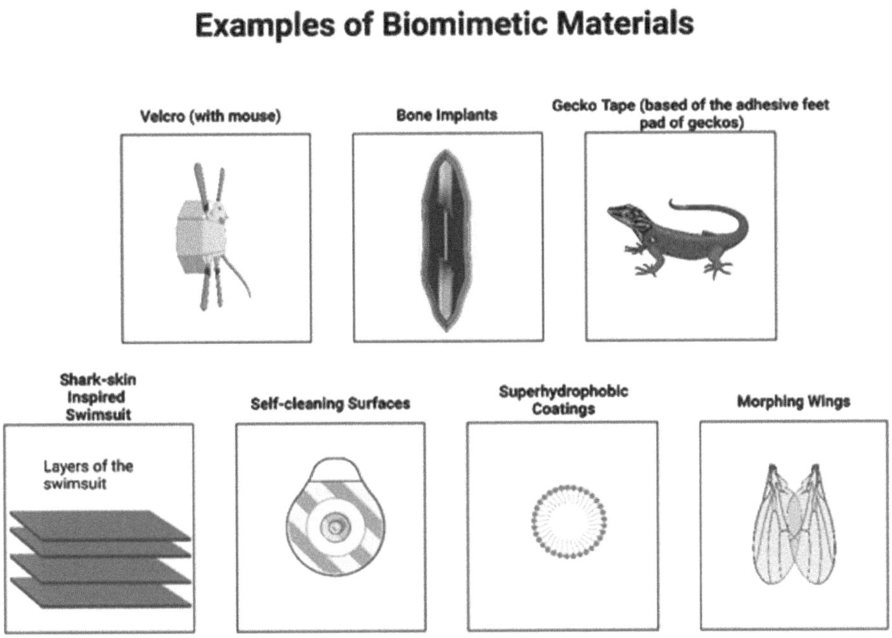

FIGURE 12.1 Examples of biomimetic materials (Credit: Created by authors).

12.2.1 Microstructure Interconnectivity

Self-assembling materials copy the porous open structure in natural tissues like bone or cartilage. These microstructures with networks of pores and channels have a central significance in the process of delivering substantial components, such as nutrients, oxygen, and the removal of waste products, which are vital for the cellular activity and sustainability of tissues [13]. Processes like 3D printing, electrospinning, and self-assembly processes are used to incorporate connected porosity into biomimetic scaffolds and structures. Most importantly, it has been possible for researchers to develop man-made scaffolds that possess hierarchical interconnected pore structures similar to that of the bone tissue so as to allow enhanced cell penetration and tissue incorporation (Figure 12.2) [14].

12.2.2 Mechanical Adaptability

Biomimetic materials must incorporate one fundamental property, namely mechanical adaptability. This implies the idea of a capacity to change their mechanical properties related to changes in the environment or mechanical stress. This behavior is realized either through the incorporation of stimuli-responsive elements or by adapting structural structures whose mechanical behavior depends on external conditions [14]. For instance, hydrogels formed from thermo-responsive polymers can have

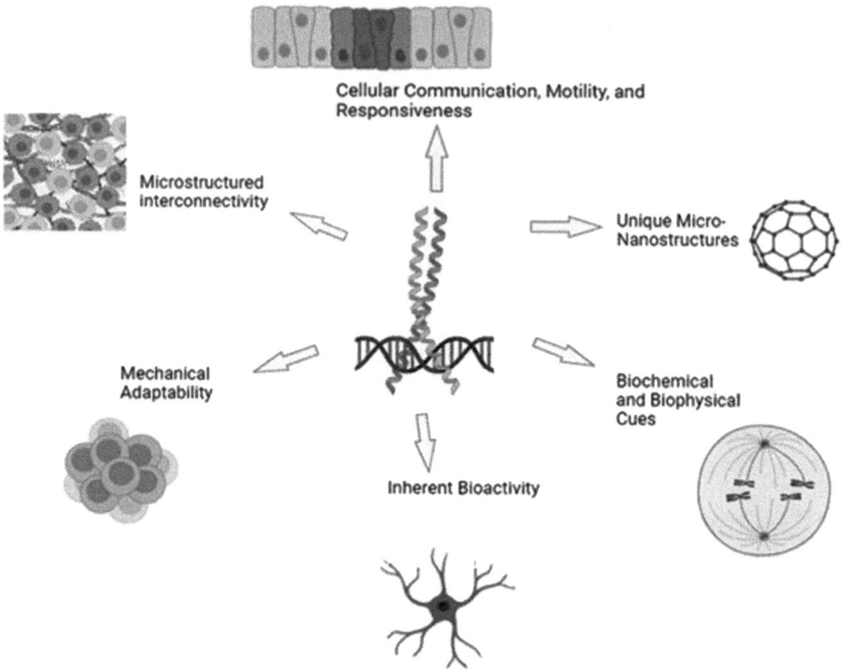

FIGURE 12.2 Properties of biomimetic material (Credit: Created by authors).

reversible changes in mechanical properties in response to changes in temperature which, in effect, mimics the dynamic mechanical property of soft tissue [15].

12.2.3 Inherent Bioactivity

It remains one of the key properties of biomimetic materials that establish the way the constructed material interacts with living systems [16]. These materials are capable of stimulating cell attachment, growth, differentiation, and tissue repair by displaying bioactive molecules or functional groups related to the natural ECM [17]. For instance, the reorientation of bioactive peptides obtained from ECM proteins like fibronectin and collagen in biomimetic matrixes may considerably improve cell–scaffold connectivity and hence the tissue repair process [18, 19].

12.2.4 Biochemical and Biophysical Cues

All these features play a crucial role in directing cellular response within biomimetic scaffolds. They contain features of the presentation of certain ligands, growth factors, or mechanical stimuli by which cell adhesion, migration, and differentiation processes are controlled [20]. Employing the exact signaling patterns found in biological organisms, biomimetic materials can effectively control cell differentiation and tissue formation [21]. For instance, surface coating with cell adhesive peptides or growth factors to enhance cell attachment and proliferation onto biomimetic scaffolds is useful in tissue engineering applications [22].

12.2.5 Unique Micro-Nanostructures

Biomimetic materials typically contain unique structures on a macroscopic and/or microscopic level imitating the structures of tissues or organisms. These structures affect several characteristics of the material, including mechanical properties, surface characteristics, and cellularity [22]. Material scientists have found that fabrication methods and material selections enable researchers to design and develop micro-nanostructures with desirable functions [22]. For example, the nanofiber-based scaffold described can mimic ECM topographical signals that influence cell alignment and tissue formation [20].

12.2.6 Cellular Communication, Motility, and Responsiveness

In addition, biomimetic materials are presumed to play a significant role in enabling cell signaling, migration, and sensitivity problems by offering an extracellular matrix that is nearly equivalent to the native tissue architecture [23]. This involves the presentation of the biochemical and biophysical signals that regulate cell activity like matrix elasticity or rigidity, ligand availability and localization [24]. In this regard, biomimetic materials' highly accurate recreation of certain cellular interactions in an *in vivo* environment may contribute to tissue morphogenesis and regeneration. For instance, such structures as the former applying the biochemical gradients and cell–cell communications typically inherent to the developing tissues might help to guide stem cells to adopt a desired form and to form corresponding tissue [25].

12.3 CLASSIFICATION OF VARIOUS BIOMIMETIC MATERIALS

Self-induced patterning techniques are recognized as complex and sophisticated. These are intentionally manufactured and employed in the forms as well as have various categorizations including natural analogs, the origin of the materials, the transduction mechanisms used, the functions empirical, or essential, and molecular or atomic structures.

12.3.1 Based on Natural Analogs

Based on source, the biomimetic materials can be categorized as bone mimics or shell mimics or mimics of spider silk. Imitating biological materials, these focus on structure, composition, or function with the aim of achieving specific properties. This includes biomimetic structure, production, or formulation and use of biogenic elements, which are identified as major components of living organisms that are known to affect material properties. For instance, authors have synthesized bone substitute materials [26]. For instance, carbon nitride nanosheets, small intestinal submucosa, magnesium-strontium hydroxyapatite, dimethylglyoxime-urethanes polyurethane, polydimethylsiloxane alloy, and Ag/Ag@AgCl/ZnO and PDTC (COOH) 4/HA are developed by mimicking natural tissue architecture and function.

12.3.2 Based on Origin

Another distinction is based on whether the biomimetic materials developed are biodegraded byproducts of bio-based materials (bio-derived) or are newly synthesized in laboratories (bioinspired). Bio-based materials are bio-proteins or polysaccharides, whereas biologically inspired materials are chemically designed to replicate biological processes [27]. Biological materials are biopolyesters of PLAs, PHAs, and their derivatives, polysaccharides of $(C_{14}H_{21}NO_{11})_nC_6H_{10}O_7$, $(C_6H_{10}O_5)_n$ cellulose, and chitosan, and proteins and poly-amino acids of collagen, gelatin, fibroin, γ-PGA, and AMPs [28–34]. These materials are derived from renewable resources, like microbes, plants, and animals, and exhibit a combination of polymer types, an intricate structure at a fine scale, and other performance characteristics.

12.3.3 Based on Transduction Mechanism

This classification of biomimetic materials classifies them according to the manner in which they convert stimuli into a response. Some of the substances may change their structure or properties in reaction to a certain cue, such as light or temperature, or when subjected to mechanical pressure, duplicating the phototransduction or mechanotransduction of cells [35].

12.3.4 Based on Function

The biomimetic materials discussed above can also be categorized using the application of the function needed: structural, self-healing, and sensing. For instance, there are self-healing polymers that work using the biological method of repair and which have the ability to heal any damage that may be on the polymer, thus helping to increase its useful life [36].

12.3.5 Based on Structure and Composition

Classification of materials may be made according to their structural and/or chemical similarity to the biological systems. This, of course, categorizes the materials based on features such as a system's organizational structure, architecture at the nanoscale, and molecular constituents [37]. For instance, synthetic hydrogels with complementing structures mimicking the ECM give higher affinity to cells and tissue remodeling [38].

12.4 DIFFERENT METHODS OF PREPARATION OF BIOMIMETIC MATERIALS

The process of biomimetic material manufacturing involves the application of numerous techniques for stable material production depending on the particular requirements related to a certain type of biomimetic material. This section looks into several typical strategies used in the fabrication of biomimetic materials and the difficulties often experienced in the synthesis as well as possible ways of overcoming challenges.

12.4.1 Template Method

Based on knowledge of the template method, a biomimetic approach for the formation of materials and structures using biomimetic principles of self-assembly and structural hierarchy has been previously claimed in literature. It entails using a sacrificial template: regardless of whether this is physical, chemical, and/or biological, to guide the construction of articulated structures from precursors or other constituent components [39]. This method enables the fabrication of biologically inspired structures at a range of micro- and nanoscales with fine control over the resulting material [39]. One of the best applications of the template method is in the synthesis of mesoporous materials, for example, mesoporous silica [40]. The materials at hand possess periodic nanopores, which are similar to pores observed in biological organisms. This comprises the formation of specific structures in a solution via the ability of block copolymers to form micelles that are used to build templates for the deposition of silica precursors. In addition, after the silica is deposited on the internal surface of the carbon nanotube the organic template is then gradually eliminated. For mesoporous silica materials with well-defined pore structures, the formation is achieved. This method mimics the arrangement structure of other biogenic materials such as bone and diatomaceous earth, which has also orderly nanopores [40].

One good example is the mimicking of biomolecules such as DNA or proteins to produce inorganic materials. DNA can be programmed to bind with particular kinds of metal ions and create a scaffold for the assembly and growth of inorganic nanoparticles at particular dimensions. This approach to DNA-templated synthesis allows the construction of site-specific metallic nanoparticles with characteristics appropriate for encouraging photocatalytic, mechanical, and electronic applications [1]. The spatial relationship of nanoparticles is similar to the one found in biological systems like enzymes stabilized in protein shells [1, 41]. Furthermore, the template method has been used to fabricate biomimetic tissue engineering scaffolds. When these natural polymers or hydrogels are used as templates for scaffold fabrication, the developed 3D scaffolds can exhibit specific mechanical properties that mimic natural tissues [41]. These scaffolds are able to recreate the micro-environment of cells and their attachments for growth and specialization, hence promoting tissue regeneration and repair [1]. The template method discussed in this section can be helpful for tunable and reproducible fabrication of biomimetic materials. This method replicates self-assembly processes and the structural hierarchy evident in nature, which can be used to develop materials useful in biomedical applications, nanotechnology, and environmental clean-up [41, 1].

12.4.2 Hydrothermal Method

In a high-pressure and high-temperature process, the hydrothermal method is used to synthesize biomimetic components since it forms components with structures as well as strength similar to natural materials. It copies processes that occur in nature, for instance, mineralization in hydrothermal vents or formation of biomimetic minerals in. Evaluation offers researchers the ability to controllably alter the variables related to reactions such as pressure, temperature, pH, and concentration of precursors to alter the chemical form and properties of the final materials [42].

The hydrothermal process is another biomimetic material preparation method, for example, hydroxyapatite (HA), which a calcium phosphate that is close to the bone and tooth mineral [43]. In the one-pot method, calcium and phosphate precursors are directly treated under hydrothermal conditions, and consequently nanocrystalline HA particles with high purity and crystallinity are obtained. Among these HA nanoparticles, they are applied in bone tissue engineering due to their biomimetic feature for cell adhesion, proliferation, and mineralization [43].

The machinery of hydrothermal synthesis is shown in Figure 12.3. The hydrothermal synthesis procedure is applied for the preparation of metal–organic frameworks

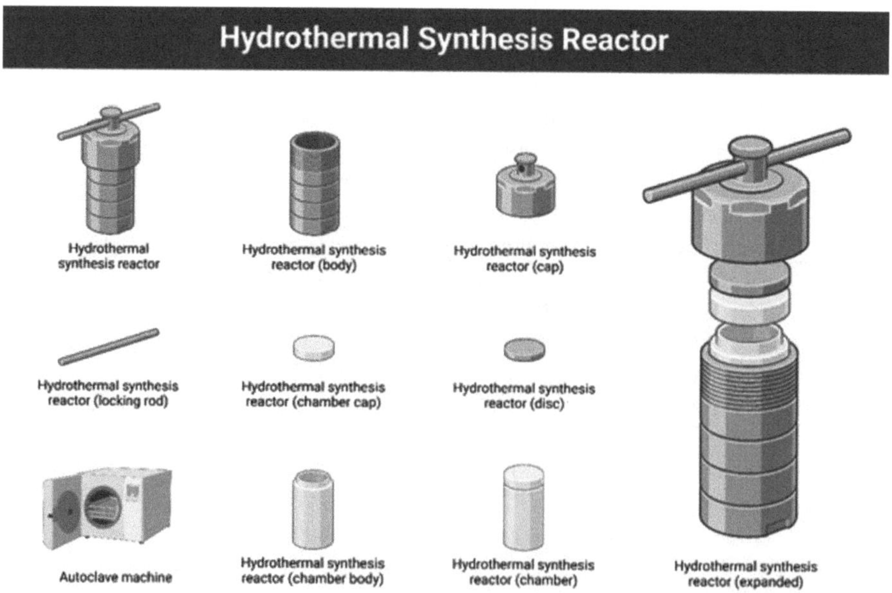

FIGURE 12.3 Machinery of hydrothermal synthesis (Credit: Created by authors).

(MOFs), and these are open-framework materials made of metal ions or clusters connected with organic ligands [44]. Because the crystallization of MOFs involves reaction steps, it is possible to achieve precise control over the size, morphology, and pore structure of the resultant crystals, thereby providing a biologically inspired structural mimic. This accurate control allows various uses of MOFs in gaseous storage, catalysis and as drug delivery systems due to their adjustable pore size and tremendously high surface area [44]. Evaluation offers researchers the ability to controllably alter the variables related to reactions such as pressure, temperature, pH, and concentration of precursors to alter the chemical form and properties of the final materials [42, 45, 46].

12.4.3 Lithography

The lithography method is very suitable for obtaining biomimetic materials since it allows the fabrication of ideal shapes and sizes of materials in the micro and nanoscale range. The DNA replication process or protein arrangement on surfaces, and lithography are the controlled formation or eradication of material on a substrate which are achieved through templates or masks [47]. This, in turn, promotes the synthesis of biomimetic materials with architectonically appropriate structures and tailor-made attributes for numerous end uses in biomedicine, nanotechnology and surface engineering.

One important usage of lithography is in the production of microfluidic systems for the culture of cells and tissue engineering. Small channels and compartments successfully embedded into the substrates using photolithography or soft lithographic processes allow for effective management of the fluid dynamic and cellular response. These microfluidic platforms mimic structural conditions present in living cells; they are useful for investigating cell communications, tissue development, and responses to medications [48].

Another use of lithography in biomimetic materials is in the production of biomolecular patterns for cell attachment and movement research. When the extracellular matrix proteins or peptides are immobilized in a 2D manner, by light or with the use of an electron microscope, researchers can control accessibility to cells and their behavior during tissue engineering [49]. These biomimetic patterns replicate the spatial guide in native tissues with regard to cell receptor affinity, cell migration, and cellular differentiation. In addition, lithography techniques have been employed to produce biomimetic structures, with hierarchical structures for applications in antifouling, low-reflective coatings and water collection [50]. For instance, hierarchical structures in the form of lotus-leaf-like or butterfly-like wings can be transferred onto substrates by techniques such as photolithography or nanoimprint lithography that give the substrate properties such as superhydrophobicity, self-cleaning, and superior light trapping [51].

Suitable lithography types provide patterned biomimetic materials with desired structural attributes and potential uses in biomedical, nanotechnological, and surficial technologies and applications. The general applicability of lithography results from the possibility to copy processes in nature and to control the deposition of

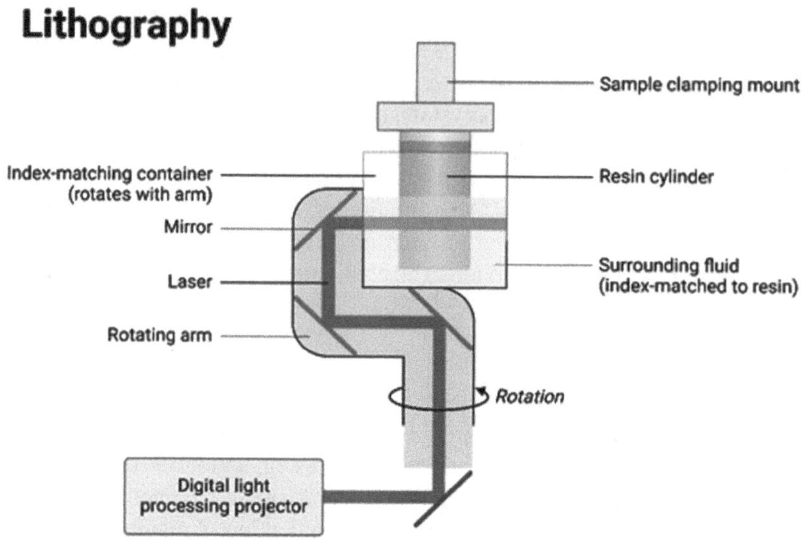

FIGURE 12.4 Lithography (Credit: Created by authors).

material at the micro- and nanoscale, which enables the fabrication of multifunctional materials with various uses (Figure 12.4).

12.4.4 Self-Assembly

The application studies executed based on self-assembly methods form the basis of biomimetic materials in which the molecules of a material self-assemble into an organized structure [52]. This technique is based on the self-assembly of functional subcomponents which are held together by mainly hydrogen bonding, hydrophobic interactions and other electrostatic forces to form structures at several hierarchal levels. The self-assembly technique has some appreciable features such as it is straightforward, feasible, and capable of providing the products with a structural hierarchy similar to biological systems [53]. An example of biomimetic material constructed via a self-assembly process is the lipid bilayer, an analog to the cell membrane present in living organisms [54]. Lipids self-assemble into the bilayer structures in aqueous conditions, due to the hydrophobic effect between the lipid tails and the polar lipid head groups. These lipid bilayers can be used to model membrane biophysics and exemplify drug delivery systems that can help to understand other cellular processes such as membrane fusion and protein–lipid interaction [55].

An example of biomimetic application of self-assembly is the formation of peptide nanostructures. Lower molecular assemblies include peptides, which are amino acid building blocks that can spontaneously form a range of nanostructures including nanofibers, nanotubes, and nanoparticles [56]. The hydrogen bonding, π-π stacking, and electrostatic interactions between charged amino acid residues drive this self-assembly process to advance. The above-designed peptide nanostructures possess structural and functional characteristics comparable to natural proteins; therefore, they are useful in tissue engineering, drug delivery, and the development of biomaterials [56]. Furthermore, self-assembly techniques have been used in supramolecular processes, where materials are created by the formation of large structures through non-covalent interactions, including host–guest interactions, metal-ligand coordination, and π-π stacking. For instance, porphyrin-based molecules can undergo self-organization through π-π stacking interactions in a well-ordered manner, which has led to the formation of functional material such as light harvesting, photocatalytic, and sensors [57]. To sum up, the present study demonstrates that self-assembly strategies are universal and effective for developing smart biomimicking materials with designed architecture and performance. By bonding building blocks with inherent self-organization interactions researchers can synthesize materials that perform as natural systems with potential uses in diagnostics/therapy, nanotechnology, and materials engineering [58].

12.4.5 3D Bio-printing

Three-dimensional bio-printing is well-defined as an intricate method of manufacturing biomimetic materials and tissues by successively layering bio-inks to form a 3D structure [59]. Due to its potential of biologically inspired tissue formation processes, 3D bio-printing has more promising prospects in stem cell research, biological developments, and pharmaceutical research owing to the direct and accurate control of cell and biological molecule positioning [60]. One of the more prominent areas that concern 3D bio-printing is the generation of tissue constructs for regenerative medicine. Research has employed bio-inks with cells encapsulated within carriers like hydrogels or decellularized extracellular matrix molecules to develop tissue-like constructs with specified architectures and functionality [59, 61].

For example, a culture of bio-printed constructs has been used in the development of vascularized tissues including skin, cartilage and liver tissue through the precise placement of cells and biomaterials in adherence to native tissue architecture [60]. And these bio-printed tissues are yet to be used in the area of organ transplantation and tissue remodeling.

Moreover, on a broader front 3D bio-printing is revolutionizing drug discovery through the construction of *in vitro* models for diseases and testing of drugs. Instead of traditional cell or tissue culture plates, researchers can generate organ-on-a-chip devices or 3D tissue models through bio-printing that mimics the niches of certain organs and tissues. This makes it easier to categorize and differentiate drug responses and that of diseases. For instance, 3D bio-printed liver models incorporating hepatocytes and stellate cells have been employed in carrying out the functionality, metabolism, and toxicity of drugs and the advancement of liver diseases [62]. These models are useful in testing drugs, which have not yet been tested on animals, and in the development of the concept of treatment tailor-made to the patient's DNA.

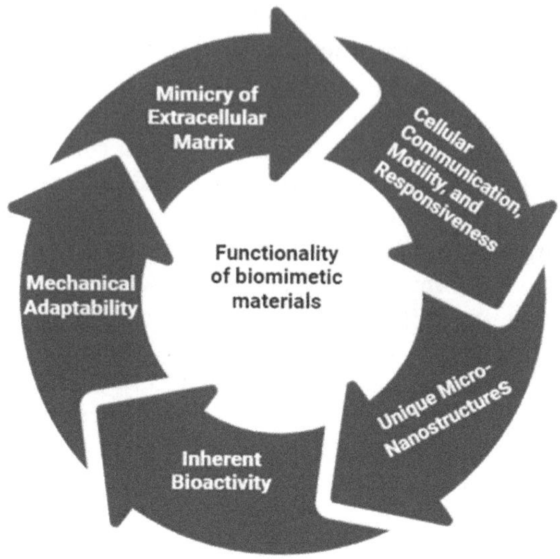

FIGURE 12.5 Functionality of biomimetic materials (Credit: Created by authors).

Among other essential uses of 3D bio-printing, there is the development of biomimetic structures for tissue handling. Such scaffolds provide an appropriate environment for cell attachment, growth, and differentiation, which are important to tissue healing and remodeling [54]. The geometry, porosity, and mechanical properties of the scaffolds can be fine-tuned based on the requirement at the tissue level and the desired application. For example, bio-printed scaffolds made from biodegradable polymers or decellularized ECM components have been used in various preclinical research studies for bone, cartilage, and cardiac tissue regeneration (Figure 12.5). These scaffolds recreate the organization of native tissue and regulate tissue regeneration in vivo [63].

Finally, 3D bio-printing is a powerful versatile technique for developing biomimetic constructs and tissues for applications in regenerative medicine, drug discovery and tissue engineering. With complete control of the scaffold made of cells and biological materials, three-dimensional bio-printing brings opportunities for generating practical tissues and organs for transplanting, disease simulations, and evaluating novel treatments [64]. With the development of various technologies, three-dimensional bio-printing has great possibilities in the field of healthcare and advancement of patient's condition (Figure 12.6).

12.5 APPLICATION OF BIOMIMETIC MATERIALS

Beginning with biomimetic materials, derived from natural processes, one can identify a wide range of possible usage scenarios in a number of fields. The following outlines several major applications of biomimetic materials (Figure 12.7).

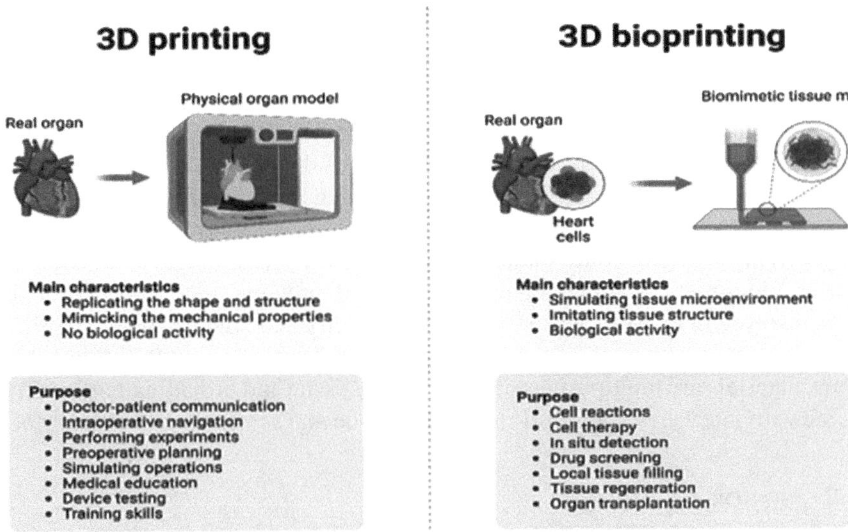

FIGURE 12.6 Advantages of 3D bio-printing over normal three-dimensional printing (Credit: Created by authors).

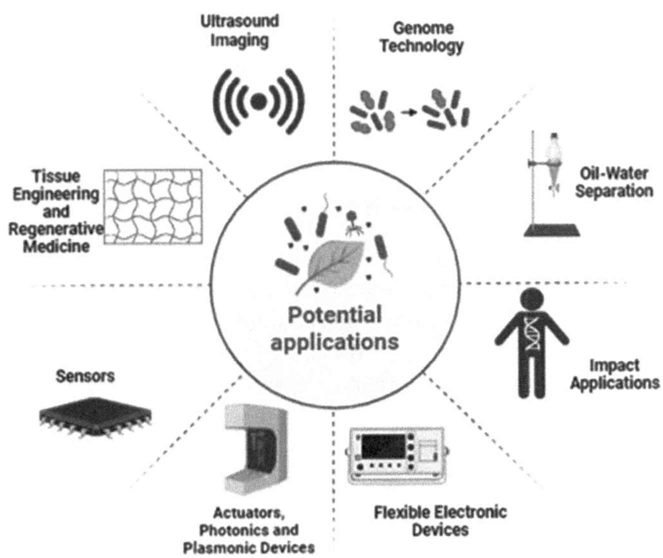

FIGURE 12.7 Potential applications of biomimetic materials (Credit: Created by authors).

12.5.1 Tissue Engineering and Regenerative Medicine

Biomimetic materials have been successful in the field of TE and RM in helping to develop specific applications for tissue regeneration and repair. These materials are intentionally engineered to mimic the stroma and functional attributes of native tissues and serve as support structures and coverings that facilitate cell attachment, growth, and specialization [65]. Hydrogels mimicking the natural or synthetic polymers resemble the ECM, for example, biomimetic hydrogels have been used as tissue engineering scaffolds to facilitate tissue repair of skin, cartilage, and bone [66]. Moreover, biomimetic coatings composed of ECM proteins or a sequence of the proteins involved in ECM enhance the biocompatibility and effectiveness of medical implants increasing the rates of tissue integration and regeneration [67]. Thus, biomimetic materials are multipurpose in engineering tissue and providing regenerative medicine with interfaces that mimic the native tissue and scaffold cell behavior [68].

12.5.2 Sensor Technology

Biomimetic materials imitating biological tissues and organs display high efficacy in the sensor field due to their ability to present novel approaches for identifying various chemical substances with superior sensitivity and selectivity [69]. For example, molecularly imprinted polymers (MIPs) mimic the binding sites made by natural receptors to allow accurate identification of target molecules in samples containing many other types of molecules [70]. Moreover, biomimetic sensors which are regarded as new techniques in sensing devices including the electroreceptors of sharks and olfactory receptors of dogs offer new horizons in sensing environmental pollutants, explosives, and different biomarkers [71].

12.5.3 Flexible Electronic Devices

Biomimetic materials extend distinctly favorable prospects in the domains of flexible electronic appliances which include the manufacture of light, stretchable as well as bio-compatible electronics with improved reliability and versatility [72–76]. These materials can replicate the function of biological tissues in terms of both electrical and mechanical properties, which enhances the compatibility of integrated devices with body tissue and improves device performance [72]. Conductive polymers and elastomers are used in biomimetic wearable sensors, implantable electronics, and electronic skin [77]. Specifically, these materials have several advantages such as high flexibility and stretchability as well as biocompatibility; thus, these materials can be applied to healthcare, robotics, and human-machine interface applications [78], schematically shown in Figure 12.8. More versatile materials like functional, fragile, and oxidation or corrosion-resistant materials can be used in future [79–88].

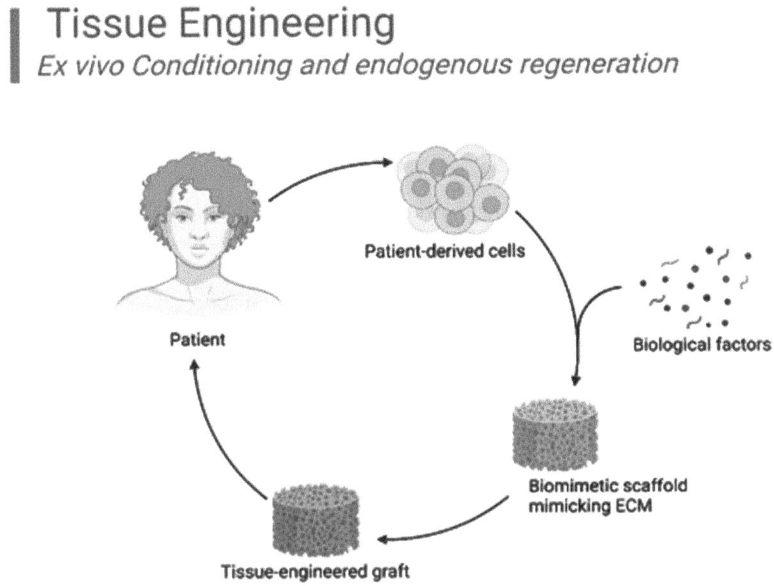

FIGURE 12.8 Tissue engineering (Credit: Created by authors).

12.5.4 Impact Applications

Biomimetic materials have the potential to significantly transform various industries and influence applications across multiple sectors. These multifaceted materials offer innovative solutions for addressing intricate challenges in healthcare, energy, environmental remediation, and other fields [89]. In the healthcare sector, biomimetic materials are propelling advancements in regenerative medicine, drug delivery, and medical diagnostics, delivering personalized treatments and enhancing patient outcomes [90]. Within the energy domain, biomimetic materials are being investigated for integration into solar cells, batteries, and fuel cells, leveraging the efficiency and sustainability observed in nature [91–95].

12.5.5 Genome Technology

Biomimetic materials exhibit significant promise in genome technology, presenting innovative approaches for DNA sequencing, gene editing, and synthetic biology by drawing inspiration from nature's intricate genetic processes [96]. These materials can mimic the structure and function of DNA, enzymes, and other biomolecules, paving the way for the development of advanced genomic tools and techniques [96]. For example, biomimetic nanomaterials such as DNA origami and peptide nucleic acids (PNAs) have been employed for precise DNA manipulation and sequencing, showcasing high specificity and sensitivity. Additionally, biomimetic systems inspired by CRISPR-Cas9, RNA interference, and other natural mechanisms have been developed for targeted gene editing and regulation, revolutionizing genetic engineering and biotechnology [97].

12.6 FUTURE CHALLENGES FOR BIOMIMETIC MATERIALS

The field of biomimetic materials presents considerable potential for transformation across various industries, including healthcare, energy, and materials science. However, realizing this potential and translating biomimetic materials into practical applications necessitates addressing several challenges [98]. These challenges encompass fundamental scientific understanding, technological development, and commercialization [99]. Below are some key future challenges within the field of biomimetic materials:

- Complexity of Biological Systems: This factor presents significant challenges in creating biomimetic materials that mimic their intricate hierarchical structures and dynamic functionalities. Fabrication techniques, material design, and scalability all pose considerable challenges in achieving precise control over structure-function relationships and dynamic behavior, making it a fundamental challenge [100].
- Bioinspired Synthesis: The emulation of natural structures and processes to create biomimetic materials demands innovative synthesis strategies. A fine balance must be struck in biomimetic synthesis methods to attain precision, efficiency, and sustainability [101]. A significant challenge lies in the development of scalable and cost-effective synthesis routes that faithfully replicate natural processes.
- Multifunctionality: Biological systems often demonstrate multifunctionality, in which a single structure or material carries out multiple tasks at the same time. In order to create biomimetic materials with multifunctional capabilities like sensing, actuation, and self-healing, it is essential to have a comprehensive grasp of structure-property correlations and to employ innovative material design methodologies [102].
- Biocompatibility and Biodegradability: It is imperative for biomimetic materials designed for biomedical applications to exhibit biocompatibility and biodegradability to ensure compatibility with living systems while minimizing adverse effects [103–105].

REFERENCES

1. Liu, S., Yu, J. M., Gan, Y. C., Qiu, X. Z., Gao, Z. C., Wang, H.,... Hou, H. H. (2023). Biomimetic natural biomaterials for tissue engineering and regenerative medicine: new biosynthesis methods, recent advances, and emerging applications. *Military Medical Research*, 10(1), 16–21.
2. Suresh Kumar, N., Padma Suvarna, R., Chandra Babu Naidu, K., Banerjee, P., Ratnamala, A., & Manjunatha, H. (2020). A review on biological and biomimetic materials and their applications. *Applied Physics A*, 126(6), 445.

3. Dharani Kumar, S., Ravi Prasath, S., Magarajan, U., & SibiNandhan, S. (2022, August). Biomimetic design approaches for impact applications: a review. In *International Symposium on Plasticity and Impact Mechanics* (pp. 219–232). Singapore: Springer Nature Singapore.
4. Hornyak, G. L., Tibbals, H. F., Dutta, J., & Moore, J. J. (2008). Biomimetics. In *Introduction to Nanoscience and Nanotechnology* (pp. 1321–1404). Boca Raton, FL: CRC Press.
5. Rawlings, A. E., Bramble, J. P., & Staniland, S. S. (2012). Innovation through imitation: biomimetic, bioinspired and biokleptic research. *Soft Matter*, 8(25), 6675–6679.
6. Du Plessis, A., Broeckhoven, C., Yadroitsava, I., Yadroitsev, I., Hands, C. H., Kunju, R., & Bhate, D. (2019). Beautiful and functional: a review of biomimetic design in additive manufacturing. *Additive Manufacturing*, 27, 408–427.
7. Gareev, K. G., Grouzdev, D. S., Koziaeva, V. V., Sitkov, N. O., Gao, H., Zimina, T. M., & Shevtsov, M. (2022). Biomimetic nanomaterials: diversity, technology, and biomedical applications. *Nanomaterials (Basel, Switzerland)*, 12(14), 2485.
8. Luo, X., Niu, J., Su, G., Zhou, L., Zhang, X., Liu, Y.,... Sun, N. (2023). Research progress of biomimetic materials in oral medicine. *Journal of Biological Engineering*, 17(1), 72.
9. Katiyar, N. K., Goel, G., Hawi, S., & Goel, S. (2021). Nature-inspired materials: emerging trends and prospects. *NPG Asia Materials*, 13(1). doi:10.1038/s41427-021-00322-y.
10. Bhushan, B. (2009). Biomimetics: lessons from nature–an overview. *Philosophical Transactions. Series A, Mathematical, Physical, and Engineering Sciences*, 367(1893), 1445–1486.
11. Sarikaya, M., Tamerler, C., Jen, A. K.-Y., Schulten, K., & Baneyx, F. (2003). Molecular biomimetics: nanotechnology through biology. *Nature Materials*, 2(9), 577–585.
12. Mann, S. (1993). Molecular tectonics in biomineralization and biomimetic materials chemistry. *Nature*, 365(6446), 499–505.
13. Abdelhamid, M. A., Ki, M. R., & Pack, S. P. (2024). Biominerals and bioinspired materials in biosensing: recent advancements and applications. *International Journal of Molecular Sciences*, 25(9), 4678.
14. Liu, X., & Ma, P. X. (2004). Polymeric scaffolds for bone tissue engineering. *Annals of Biomedical Engineering*, 32(3), 477–486.
15. Ling, X., Osotsi, M. I., Zhang, W., Wu, Y., Jin, Q., & Zhang, D. (2023). Bioinspired materials: from distinct dimensional architecture to thermal regulation properties. *Journal of Bionic Engineering*, 20(3), 873–899.
16. Annabi, N., Nichol, J. W., Zhong, X., Ji, C., Koshy, S., Khademhosseini, A., & Dehghani, F. (2010). Controlling the porosity and microarchitecture of hydrogels for tissue engineering. *Tissue Engineering. Part B, Reviews*, 16(4), 371–383.
17. Li, L.-L., An, H.-W., Peng, B., Zheng, R., & Wang, H. (2019). Self-assembled nanomaterials: design principles, the nanostructural effect, and their functional mechanisms as antimicrobial or detection agents. *Materials Horizons*, 6(9), 1794–1811.
18. Zhang, S. (2002). Emerging biological materials through molecular self-assembly. *Biotechnology Advances*, 20(5–6), 321–339.
19. Perlin, L., MacNeil, S., & Rimmer, S. (2008). Production and performance of biomaterials containing RGD peptides. *Soft Matter*, 4(12), 2331.
20. Baker, B. M., Trappmann, B., Wang, W. Y., Sakar, M. S., Kim, I. L., Shenoy, V. B., Burdick, J. A., & Chen, C. S. (2015). Cell-mediated fibre recruitment drives extracellular matrix mechanosensing in engineered fibrillar microenvironments. *Nature Materials*, 14(12), 1262–1268.
21. Hersel, U., Dahmen, C., & Kessler, H. (2003). RGD modified polymers: biomaterials for stimulated cell adhesion and beyond. *Biomaterials*, 24(24), 4385–4415.

22. Ladoux, B., Mège, R.-M., & Trepat, X. (2016). Front–rear polarization by mechanical cues: from single cells to tissues. *Trends in Cell Biology*, 26(6), 420–433.
23. Smandri, A., Al-Masawa, M. E., Hwei, N. M., & Fauzi, M. B. (2024). ECM-derived biomaterials for regulating tissue multicellularity and maturation. *iScience*, 27(3), 109141.
24. Speck, O., & Speck, T. (2019). An overview of bioinspired and biomimetic self-repairing materials. *Biomimetics (Basel, Switzerland)*, 4(1), 26.
25. Zhao, Z.-L., Shu, T., & Feng, X.-Q. (2016). Study of biomechanical, anatomical, and physiological properties of scorpion stingers for developing biomimetic materials. *Materials Science & Engineering. C, Materials for Biological Applications*, 58, 1112–1121.
26. Weiner, S., & Wagner, H. D. (1998). THE MATERIAL BONE: structure-mechanical function relations. *Annual Review of Materials Science*, 28(1), 271–298.
27. Gorb, S. N., & Gorb, E. V., Eds. (2017). *Functional Surfaces in Biology III*. Springer.
28. Cao, X.-N., Lian, S., Tong, Y., Lin, W., Jia, L., Fang, Y., & Wang, X. (2020). Fluorescent Se-modified carbon nitride nanosheets as biomimetic catalases for free-radical scavenging. *Chemical Communications (Cambridge, England)*, 56(6), 916–919.
29. Wang, J., Li, K., Xu, J., Liu, M., Li, P., Li, X., & Fan, Y. (2021). A biomimetic hierarchical small intestinal submucosa–chitosan sponge/chitosan hydrogel scaffold with a micro/nano structure for dural repair. *Journal of Materials Chemistry. B, Materials for Biology and Medicine*, 9(37), 7821–7834.
30. Scalera, F., Palazzo, B., Barca, A., & Gervaso, F. (2020). Sintering of magnesium-strontium doped hydroxyapatite nanocrystals: towards the production of 3D biomimetic bone scaffolds. *Journal of Biomedical Materials Research. Part A*, 108(3), 633–644.
31. Liu, Z., Zhang, L., Guan, Q., Guo, Y., Lou, J., Lei, D., Wang, S., Chen, S., Sun, L., Xuan, H., Jeffries, E. M., He, C., Qing, F.-L., & You, Z. (2019). Biomimetic materials with multiple protective functionalities. *Advanced Functional Materials*, 29(28). doi:10.1002/adfm.201901058.
32. Sun, Y., Wang, Y., Sui, X., Liang, W., He, L., Wang, F., & Yang, B. (2021). Biomimetic multiwalled carbon nanotube/polydimethylsiloxanenanocomposites with temperature-controlled, hydrophobic, and icephobic properties. *ACS Applied Nano Materials*, 4(10), 10852–10863.
33. Zare, M., Namratha, K., Alghamdi, S., Mohammad, Y. H. E., Hezam, A., Zare, M., Drmosh, Q. A., Byrappa, K., Chandrashekar, B. N., Ramakrishna, S., & Zhang, X. (2019). Novel green biomimetic approach for synthesis of ZnO-Ag nanocomposite; Antimicrobial activity against food-borne pathogen, biocompatibility and solar photocatalysis. *Scientific Reports*, 9(1). doi:10.1038/s41598-019-44309.
34. Zhang, Y., Li, J., Mouser, V. H. M., Roumans, N., Moroni, L., & Habibovic, P. (2021). Biomimetic mechanically strong one-dimensional hydroxyapatite/poly(d,l-lactide) composite inducing formation of anisotropic collagen matrix. *ACS Nano*, 15(11), 17480–17498.
35. Palleau, E., Morales, D., Dickey, M. D., & Velev, O. D. (2013). Reversible patterning and actuation of hydrogels by electrically assisted ionoprinting. *Nature Communications*, 4(1), 2257.
36. White, S. R., Sottos, N. R., Geubelle, P. H., Moore, J. S., Kessler, M. R., Sriram, S. R., Brown, E. N., & Viswanathan, S. (2001). Autonomic healing of polymer composites. *Nature*, 409(6822), 794–797.
37. Annabi, N., Tamayol, A., Uquillas, J. A., Akbari, M., Bertassoni, L. E., Cha, C., Camci-Unal, G., Dokmeci, M. R., Peppas, N. A., & Khademhosseini, A. (2014). 25th anniversary article: rational design and applications of hydrogels in regenerative medicine. *Advanced Materials (Deerfield Beach, Fla.)*, 26(1), 85–124.

38. Ciulla, M. G., Massironi, A., Sugni, M., Ensign, M. A., Marzorati, S., & Forouharshad, M. (2023). Recent advances in the development of biomimetic materials. *Gels (Basel, Switzerland)*, 9(10), 833.
39. Chen, W., Gan, L., & Huang, J. (2023). Design, manufacturing and functions of pore-structured materials: from biomimetics to artificial. *Biomimetics (Basel, Switzerland)*, 8(2), 140.
40. Zhao, D., Feng, J., Huo, Q., Melosh, N., Fredrickson, G. H., Chmelka, B. F., & Stucky, G. D. (1998). Triblock copolymer syntheses of mesoporous silica with periodic 50 to 300 angstrom pores. *Science (New York, N.Y.)*, 279(5350), 548–552.
41. Shiohara, A., Prieto-Simon, B., & Voelcker, N. H. (2021). Porous polymeric membranes: fabrication techniques and biomedical applications. *Journal of Materials Chemistry.B, Materials for Biology and Medicine*, 9(9), 2129–2154.
42. Wang, S., & Yu, S.-H. (2024). Advances in biomimetic materials. *Small Methods*, 8(4): e2301487.
43. Dorozhkin, S. V. (2009). Calcium orthophosphate-based biocomposites and hybrid biomaterials. *Journal of Materials Science*, 44(9), 2343–2387.
44. Furukawa, H., Cordova, K. E., O'Keeffe, M., & Yaghi, O. M. (2013). The chemistry and applications of metal-organic frameworks. *Science (New York, N.Y.)*, 341(6149). doi:10.1126/science.1230444.
45. Li, L., Zhang, Y., Lei, J., He, J., Lv, R., Li, N., & Pan, F. (2014). Water-only hydrothermal method: a generalized route for environmentally-benign and cost-effective construction of superhydrophilic surfaces with biomimetic micronanostructures on metals and alloys. *Chemical Communications (Cambridge, England)*, 50(56), 7416.
46. Shi, W., Song, S., & Zhang, H. (2013). Hydrothermal synthetic strategies of inorganic semiconducting nanostructures. *Chemical Society Reviews*, 42(13), 5714.
47. Baker, M. (2011). A living system on a chip. *Nature*, 471(7340), 661–665.
48. El-Ali, J., Sorger, P. K., & Jensen, K. F. (2006). Cells on chips. *Nature*, 442(7101), 403–411.
49. Bettinger, C. J., Langer, R., & Borenstein, J. T. (2009). Engineering substrate topography at the micro- and nanoscale to control cell function. *Angewandte Chemie (International Ed. in English)*, 48(30), 5406–5415.
50. Dong, J., Chen, J.-F., Smalley, M., Zhao, M., Ke, Z., Zhu, Y., & Tseng, H.-R. (2020). Nanostructured substrates for detection and characterization of circulating rare cells: from materials research to clinical applications. *Advanced Materials (Deerfield Beach, Fla.)*, 32(1). doi:10.1002/adma.201903663.
51. Retsch, M., Zhou, Z., Rivera, S., Kappl, M., Zhao, X. S., Jonas, U., & Li, Q. (2009). Fabrication of large-area, transferable colloidal monolayers utilizing self-assembly at the air/water interface. *Macromolecular Chemistry and Physics*, 210(3–4), 230–241.
52. Levin, A., Hakala, T. A., Schnaider, L., Bernardes, G. J. L., Gazit, E., & Knowles, T. P. J. (2020). Biomimetic peptide self-assembly for functional materials. *Nature Reviews Chemistry*, 4(11), 615–634.
53. Zhang, X., Li, J., Ma, C., Zhang, H., & Liu, K. (2023). Biomimetic structural proteins: modular assembly and high mechanical performance. *Accounts of Chemical Research*, 56(19), 2664–2675.
54. Chung, J. J., Im, H., Kim, S. H., Park, J. W., & Jung, Y. (2020). Toward biomimetic scaffolds for tissue engineering: 3D printing techniques in regenerative medicine. *Frontiers in Bioengineering and Biotechnology*, 8. doi:10.3389/fbioe.2020.586406.
55. Alberts, B., Johnson, A., Lewis, J., Raff, M., Roberts, K., & Walter, P. (2007). *Molecular Biology of the Cell*. W.W. Norton & Company.
56. Zhang, S. (2003). Fabrication of novel biomaterials through molecular self-assembly. *Nature Biotechnology*, 21(10), 1171–1178.

57. Otsuki, J. (2018). Supramolecular approach towards light-harvesting materials based on porphyrins and chlorophylls. *Journal of Materials Chemistry. A, Materials for Energy and Sustainability*, 6(16), 6710–6753.
58. Aliprandi, A., Mauro, M., & De Cola, L. (2016). Controlling and imaging biomimetic self-assembly. *Nature Chemistry*, 8(1), 10–15.
59. Vijayavenkataraman, S., Yan, W.-C., Lu, W. F., Wang, C.-H., & Fuh, J. Y. H. (2018). 3D bioprinting of tissues and organs for regenerative medicine. *Advanced Drug Delivery Reviews*, 132, 296–332.
60. Herzog, J., Franke, L., Lai, Y., Gomez Rossi, P., Sachtleben, J., & Weuster-Botz, D. (2024). 3D bioprinting of microorganisms: principles and applications. *Bioprocess and Biosystems Engineering*, 47(4), 443–461.
61. Murphy, S. V., & Atala, A. (2014). 3D bioprinting of tissues and organs. *Nature Biotechnology*, 32(8), 773–785.
62. Skardal, A., & Atala, A. (2015). Biomaterials for integration with 3-D bioprinting. *Annals of Biomedical Engineering*, 43(3), 730–746.
63. Chimene, D., Lennox, K. K., Kaunas, R. R., & Gaharwar, A. K. (2016). Advanced bioinks for 3D printing: a materials science perspective. *Annals of Biomedical Engineering*, 44(6), 2090–2102.
64. Dey, M., & Ozbolat, I. T. (2020). 3D bioprinting of cells, tissues and organs. *Scientific Reports*, 10(1), 14023.
65. Jain, T., Sujani, S., Singh, N., Al-Alawachi, S. F. A., Asha & Meheta, A. (2024). Biomimetic materials for regenerative medicine: design and applications. *E3S Web of Conferences*, 505, 04002.
66. Hussey, G. S., Dziki, J. L., & Badylak, S. F. (2018). Extracellular matrix-based materials for regenerative medicine. *Nature Reviews Materials*, 3(7), 159–173.
67. Ratner, B. D., & Bryant, S. J. (2004). Biomaterials: where we have been and where we are going. *Annual Review of Biomedical Engineering*, 6(1), 41–75.
68. Gaharwar, A. K., Singh, I., & Khademhosseini, A. (2020). Engineered biomaterials for in situ tissue regeneration. *Nature Reviews Materials*, 5(9), 686–705.
69. Shen, G. (2021). Recent advances of flexible sensors for biomedical applications. *Progress in Natural Science: Communication of State Key Laboratories of China*, 31(6), 872–882.
70. Blanco-López, M. C., Lobo-Castañón, M. J., Miranda-Ordieres, A. J., & Tuñón-Blanco, P. (2004). Electrochemical sensors based on molecularly imprinted polymers. *Trends in Analytical Chemistry: TRAC*, 23(1), 36–48.
71. Das, P. P., Nagraik, R., Sharma, A., Upadhyay, T. K., Lalhlenmawia, H., Balram, D., Lian, K.-Y., Singh, J., & Kumar, D. (2024). Recent update on biomimetic sensor technology for cancer diagnosis. *Talanta Open*, 9(100276), 100276.
72. Suresh Kumar, N., Padma Suvarna, R., Chandra Babu Naidu, K., Banerjee, P., Ratnamala, A., & Manjunatha, H. (2020). A review on biological and biomimetic materials and their applications. *Applied Physics A, Materials Science & Processing*, 126(6). doi:10.1007/s00339-020-03633-z.
73. Ahmad, S., Hashmi, A. W., Singh, J., Arora, K., Tian, Y., Iqbal, F., Al-Dossari, M., & Khan, M. I. (2024). Innovations in additive manufacturing of shape memory alloys: alloys, microstructures, treatments, applications. *Journal of Materials Research and Technology*, 32, doi:10.1016/j.jmrt.2024.08.213.
74. Kumar, P., Singh, J., & Singh, S. (2022). Neural network supported flow characteristics analysis of heavy sour crude oil emulsified by ecofriendly bio-surfactant utilized as a replacement of sweet crude oil. *Chemical Engineering Journal Advances*, 11, 100342.

75. Singh, J., Hashmi, A. W., Ahmad, S., & Tian, Y. (2024). Critical review on biodegradable and biocompatibility magnesium alloys: progress and prospects in bio-implant applications. *Inorganic Chemistry Communications,* 169, 113111.
76. Kumar, S., Singh, J., & Mohapatra, S. K. (2018). Role of particle size in assessment of physico-chemical properties and trace elements of Indian fly ash. *Waste Management and Research,* 36(11), 1016–1022.
77. Someya, T., Bao, Z., & Malliaras, G. G. (2016). The rise of plastic bioelectronics. *Nature*, 540(7633), 379–385.
78. Zhang, Z., Liu, G., Li, Z., Zhang, W., & Meng, Q. (2023). Flexible tactile sensors with biomimetic microstructures: mechanisms, fabrication, and applications. *Advances in Colloid and Interface Science*, 320(102988), 102988.
79. Dutta, V., Thakur, L., Singh, B., & Vasudev, H. (2022). A study of erosion–corrosion behaviour of friction stir-processed chromium-reinforced NiAl bronze composite. *Materials,* 15(15), 5401.
80. Kattimani, M. A., Venkatesh, P. R., Masum, H., Math, M. M., Bahadurdesai, V. N., Mustafkhadri, S., Prasad, C. D., & Vasudev, H. (2024). Design and numerical analysis of tensile deformation and fracture properties of induction hardened inconel 718 superalloy for gas turbine applications. *International Journal on Interactive Design and Manufacturing*, 18, 5939–5949.
81. Kulkarni, S. D., Manjunatha, Chandrasekhar, U., Manjunath, K. V, Prasad, C. D., & Vasudev, H. (2024). Design and optimization of polyvinyl-nitride rubber for tensile strength analysis. *International Journal on Interactive Design and Manufacturing*, 18(8), 5893–5908.
82. Majji, B. G. R., Vasudev, H., & Bansal, A. (2022). A review on the oxidation and wear behavior of the thermally sprayed high-entropy alloys. *Materials Today: Proceedings,* 50, 1447–1451.
83. Poojari, M., Hanumanthappa, H., Prasad, C. D., Jathanna, H. M., Ksheerasagar, A. R., Shetty, P., Shanmugam, B. K., & Vasudev, H. (2024). Computational modelling for the manufacturing of solar-powered multifunctional agricultural robot. *International Journal on Interactive Design and Manufacturing,* 18, 5939–5949.
84. Prashar, G., & Vasudev, H. (2022a). Surface topology analysis of plasma sprayed Inconel625-Al_2O_3 composite coating. *Materials Today: Proceedings,* 50, 607–611.
85. Prashar, G., & Vasudev, H. (2022b). A review on the influence of process parameters and heat treatment on the corrosion performance of Ni-based thermal spray coatings. *Surface Review and letters,* 29(1), 2230001.
86. Sharanabasava, H., Raviprakash, M., Prasad, C. D., Ramesh, M. R., Phanibhushana, M. V, Vasudev, H., & Kumar, S. (2023). Microstructure, mechanical and wear properties of SiC and Mo reinforced NiCr microwave cladding. *Advanced Materials Processing Technology*, 10(4). doi:10.1080/2374068X.2023.2257937.
87. Sunitha, K., & Vasudev, H. (2022). A short note on the various thermal spray coating processes and effect of post-treatment on Ni-based coatings. *Materials Today: Proceedings,* 50, 1452–1457.
88. Yedida, V. V. S., & Vasudev, H. (2022). A review on the development of thermal barrier coatings by using thermal spray techniques. *Materials Today: Proceedings,* 50, 1458–1464.
89. Antony, S., Anju, T. R., & Thomas, B. (2022). Nature-inspired biomimetic polymeric materials and their applications. In *Handbook of Biopolymers* (pp. 1–31). Singapore: Springer Nature.
90. Peppas, N. A., & Langer, R. (1994). New challenges in biomaterials. *Science (New York, N.Y.)*, 263(5154), 1715–1720.

91. Nosonovsky, M., & Bhushan, B., Eds. (2012). *Green Tribology: Biomimetics, Energy Conservation and Sustainability*. Springer Berlin Heidelberg.
92. Singh, J., Gill, H. S., & Vasudev, H. (2022). Computational fluid dynamics analysis on role of particulate shape and size in erosion of pipe bends. *International Journal on Interactive Design and Manufacturing*, 17(5), 1–16.
93. Singh, J., Kumar, S., & Mohapatra, S. K. (2021). Study on solid particle erosion of pump materials by fly ash slurry using Taguchi's orthogonal array. *Tribologia - Finnish Journal of Tribology*, 38(3–4), 31–38.
94. Singh, J., Mohapatra, S. K., & Kumar, S. (2021). Performance analysis of pump materials employed in bottom ash slurry erosion conditions. *Jurnal Tribologi*, 30, 73–89.
95. Thakur, L., Vasudev, H., Singh, J., & Prashar, G. (2023). *Artificial Intelligence and Machine Learning in the Thermal Spray Industry: Practices, Implementation, and Challenges*. Boca Raton, FL: CRC Press.
96. Yang, K., Qian, J., Zhang, C., Wang, Z., Huang, Q., Shi, G., Zhang, Z., Yang, Y., & Han, X. (2023). Biogenic materials for CRISPR delivery and therapeutics. *Biomaterials Science*, 11(9), 3016–3033.
97. Doudna, J. A., & Charpentier, E. (2014). The new frontier of genome engineering with CRISPR-Cas9. *Science (New York, N.Y.)*, 346(6213). doi:10.1126/science.125809.
98. Stuart-Fox, D., Ng, L., Barner, L., Bennett, A. T. D., Blamires, S. J., Elgar, M. A., Evans, A. R., Franklin, A. M., Hölttä-Otto, K., Hutchison, J. A., Jativa, F., Jessop, A.-L., Kelley, J., McGaw, J., Mei, J., Mirkhalaf, M., Musameh, M., Neto, C., O'Connor, A. J., & Wong, W. W. H. (2023). Challenges and opportunities for innovation in bioinformed sustainable materials. *Communications Materials*, 4(1). doi:10.1038/s43246-023-00405-z.
99. Meyers, M. A., Chen, P.-Y., Lin, A. Y.-M., & Seki, Y. (2008). Biological materials: structure and mechanical properties. *Progress in Materials Science*, 53(1), 1–206.
100. Mann, S. (2009). Self-assembly and transformation of hybrid nano-objects and nanostructures under equilibrium and non-equilibrium conditions. *Nature Materials*, 8(10), 781–792.
101. Kamat, P. V. (2008). Quantum dot solar cells. Semiconductor nanocrystals as light harvesters. *The Journal of Physical Chemistry. C, Nanomaterials and Interfaces*, 112(48), 18737–18753.
102. Place, E. S., Evans, N. D., & Stevens, M. M. (2009). Complexity in biomaterials for tissue engineering. *Nature Materials*, 8(6), 457–470.
103. Appel, E. A., del Barrio, J., Loh, X. J., & Scherman, O. A. (2012). Supramolecular polymeric hydrogels. *Chemical Society Reviews*, 41(18), 6195.
104. Malda, J., Visser, J., Melchels, F. P., Jüngst, T., Hennink, W. E., Dhert, W. J. A., Groll, J., & Hutmacher, D. W. (2013). 25th anniversary article: engineering hydrogels for biofabrication. *Advanced Materials (Deerfield Beach, Fla.)*, 25(36), 5011–5028.
105. Derby, B. (2012). Printing and prototyping of tissues and scaffolds. *Science (New York, N.Y.)*, 338(6109), 921–926.

13 Self-Healing Materials
Mechanisms, Classifications and Applications

J.M Yelwa, H. Musa, Opeoluwa O. Fasanya, B.M. Aliyu, Magaji Ladan, Shuaibu Abdullahi, J.M Nasir, A.J Aliyu, Omolade Ojo, and Alhagie Drammeh

13.1 INTRODUCTION

The advent of self-healing materials marks a revolutionary stride in materials science and engineering. These materials can repair damage on their own, much like biological systems, extending the lifespan and reliability of engineering applications (White *et al.*, 2021). By reducing wear and tear, they promote durability and lessen environmental impact (Trask *et al.*, 2007). Inspired by nature's self-healing mechanisms, self-healing materials are capable of independently detecting damage and thereafter producing a healing action (Toohey *et al.*, 2007). This is greatly in line with green technology principles of reducing waste and also the need for resource-intensive repairs (Qin & Lu, 2023). They are thus very important in fields like aerospace, automotive, and civil engineering where material failure could lead to disastrous results (Scheiner *et al.*, 2016).

Nanotechnology imbues self-healing materials; nanomaterials improve healing efficiency and responsiveness (Zhao *et al.*, 2022a). Smart materials that react to the environment further enhance this technology (Zhu *et al.*, 2020). Healing efficiency, long-term stability, and economic feasibility remain the bottlenecks toward large-scale applications (Yedida & Vasudev, 2022). In this regard, research is focused on how these materials could be optimized for practical use to meet diverse engineering demands (Yedida & Vasudev, 2022).

Self-healing materials are a relatively young domain of sustainability and durability, with inspiration from nature and state-of-the-art technology. It is predisposed to offer a sustainable solution to the degradation of various materials to green technologies toward achieving a sustainable future (Irzhak *et al.*, 2022). The history of self-healing materials began with polymers and composites at the end of the 20th century and evolved through microencapsulation techniques and vascular networks in the early 2000s (Hager *et al.*, 2010; Toohey *et al.*, 2007; Kulkarni *et al.*, 2024), and lately in this first decade of the 21st century—expanded their capabilities to metals, ceramics, and concrete (Tay *et al.*, 2023). The application of self-healing materials

DOI: 10.1201/9781003516071-13

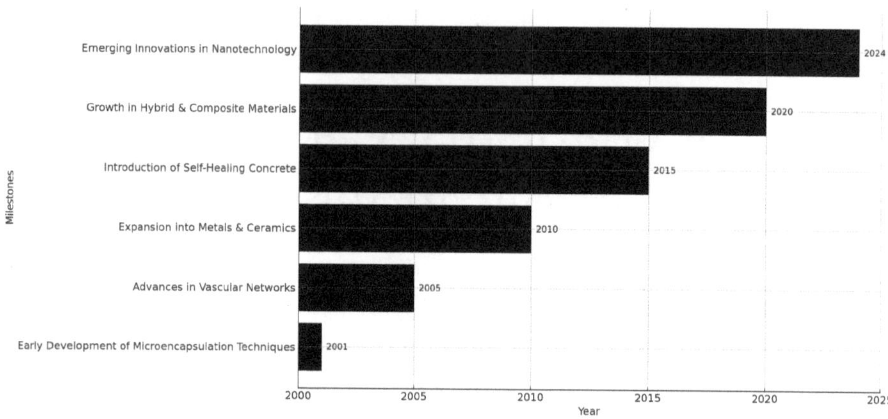

FIGURE 13.1 Evolution of self-healing materials timeline (Credit: Authors of this chapter).

to industries, including aerospace, automotive, civil infrastructure, and electronics, improves durability, safety, and sustainability (Chopra *et al.*, 2023). These materials help bring down maintenance costs by enhancing reliability and service life and will thus have a very important role in engineering design and maintenance in the future (Neal *et al.*, 2015; Kanu et al., 2019).

This chapter aims to underline the transformational potential of self-healing materials in pushing engineering and materials science further. Research into the development, applications, and challenges relating to these materials brings forward some of the real significance associated with enhancing sustainability, durability, and efficiency within different industries. Definitely, one major motivation behind this chapter is to spur further research and innovation and, finally, to help create a more resilient, safer, long-term future for engineering.

Figure 13.1 illustrates the timeline of the evolution of self-healing materials, starting with the development of polymers and composites in the late 20th century, progressing through microencapsulation techniques and vascular networks in the early 2000s, and extending to metals, ceramics, and concrete in recent years.

13.2 SCIENCE BEHIND SELF-HEALING MATERIALS

13.2.1 Core Mechanisms of Self-Healing

Self-healing materials are engineered to restore their functionality after damage. This section explores the core mechanisms enabling self-repair, focusing on intrinsic versus extrinsic healing, microencapsulation, vascular approaches, and reversible polymers and supramolecular systems.

Intrinsic versus Extrinsic Healing

The traditional intrinsic healing methods, on the one hand, rely on the intrinsic properties of the material to make good the damage. For example, reversible bonds in the polymer matrix that break and then reform (Yang *et al.*, 2022). Extrinsic healing, on the other hand, relies on external agents such as encapsulated healing chemical agents that release in case there is damage (Toohey *et al.*, 2007). In a polymer matrix, a micro-encapsulated healing agent heals scratches on the surface of an automotive application. The two classic philosophies of self-healing in materials are shown in Figure 13.2.

Intrinsic healing is the ability of the material to be self-repaired owing to the chemical or physical properties of the material itself, while extrinsic healing is the incorporation of extra elements or mechanisms in the material that would lead to self-repair at the occurrence of damage. Both of these approaches to self-healing are considered important, not only in setting forth self-healing materials and their different bestowed benefits but also in the specific avenues of application, noting that benefits may vary depending on the type of materials used intended. As shown in Figure 13.2, intrinsic healing relies on the material's inherent properties, such as reversible bonds in a polymer matrix, while extrinsic healing involves external agents like encapsulated chemicals that are released upon damage. The process starts with a material in its original, undamaged state. Over time, stress or external factors may cause damage. In intrinsic healing, the material repairs itself through dynamic bonds within its molecular structure, enabling it to self-heal without external assistance. This occurs as the molecules in the damaged area reorganize and reconnect, restoring the material to its initial condition (Ahmed *et al.*, 2024). The result is a fully restored site, thanks to the material's inherent self-repairing capabilities. In extrinsic healing, however, the material relies on external agents (such as microcapsules or fibers) embedded within it. When damage occurs, these agents are released to fill and repair the affected area. While the outcome is also a repaired site, the healing

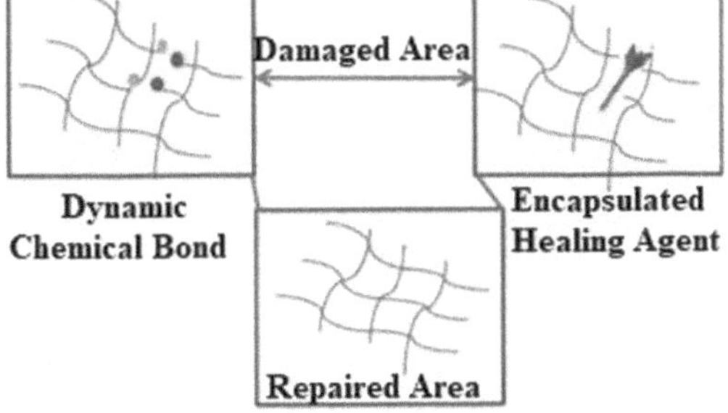

FIGURE 13.2 Mechanism of intrinsic and extrinsic self-healing (Yelwa and Musa, 2024). (Credit: Creative commons Attribution 4.0).

process here depends on the activation of these external agents rather than the material's intrinsic properties.

Microencapsulation

Microencapsulation involves the inclusion of tiny capsules within the material that contain the healing agent. In the event of a crack, the aforementioned capsules rupture, and the healing agent flows into the crack and solidifies, hence restoring the integrity of the material (Wen *et al.*, 2021).

Vascular Approaches

Recently, researchers have developed vascular self-healing materials that consist of a channel-like structure containing healing agents, which, under the induction of damage, are released into the crack to initiate healing processes similar to blood clotting in wounds (Norris *et al.*, 2013).

Figure 13.3 illustrates the vascular self-healing process which demonstrates the architecture of embedded networks influences the flow of healing agents and the repair outcome. In Figure 13.3(a), a random vascular network is shown, with irregularly woven channels that allow healing agents to flow freely in various directions upon damage. In Figure 13.3(b), an aligned network is depicted, with parallel channels that guide the agents along structured paths for more controlled healing. The corresponding microscopic images on the right capture the healing response in action. Figure 13.3(c) is associated with the random network and shows agents (in red and green) dispersing in a chaotic pattern, covering a broader area but with less precision. Figure 13.3(d) is aligned with the ordered network and demonstrates the agents flow neatly along the linear paths, producing focused and efficient repair along the damage site. This figure highlights the trade-off between broad, dispersed healing with random networks and precise, targeted restoration with aligned networks, emphasizing the role of vascular design in optimizing self-healing materials.

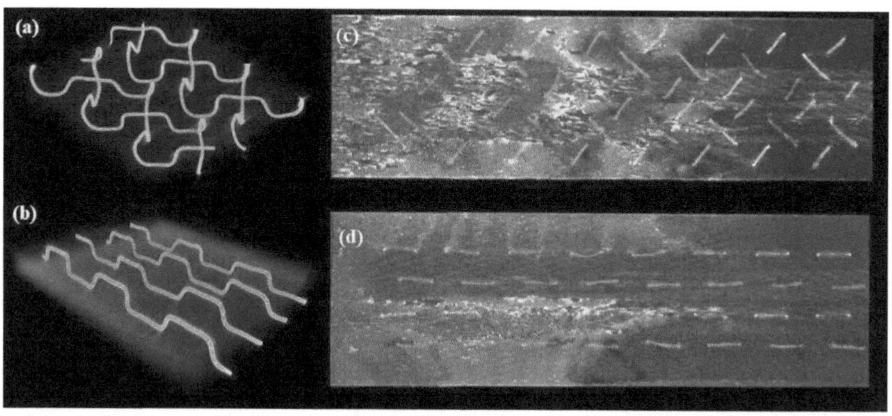

FIGURE 13.3 Schematic of vascular self-healing (Credit: Authors of this chapter).

Reversible Polymers and Supramolecular Systems

Reversible polymers are those containing dynamic bonds that can break and, under specific conditions, reform, thus allowing the self-healing of the material. Supramolecular systems use non-covalent interactions like hydrogen bonding or metal–ligand coordination to enable reversible processes for healing (Mohamadhoseini and Mohamadnia, 2021).

13.2.2 Molecular and Chemical Fundamentals

The molecular and chemical fundamentals of self-healing materials are centered on the ability to design and manipulate chemical bonds and interactions that can respond to damage in a controlled and reversible manner. These principles enable the creation of materials that can autonomously repair themselves, leading to longer-lasting, more sustainable materials with wide-ranging applications in industries such as aerospace, automotive, and construction. The molecular and chemical basics of self-healing materials encompass understanding monomers, polymers, cross-linking, catalysts, and healing agents.

Monomers, Polymers, and Cross-linking Functions

Monomers and polymers often constitute the heart of self-healing materials. In any case, their structure and nature qualify a material for self-healing or not. Cross-linking agents enhance structure and allow healing through reversible or permanent networks formed across the polymer matrix (Snyder *et al.*, 2018). Figure 13.4 depicts the chemical structure of self-healing polyurea elastomer. The arrangement and interaction of these molecules determine the material's ability to self-repair. Cross-linking functions, which connect polymer chains, play a crucial role in enhancing the material's mechanical properties and its capacity to heal after damage. This

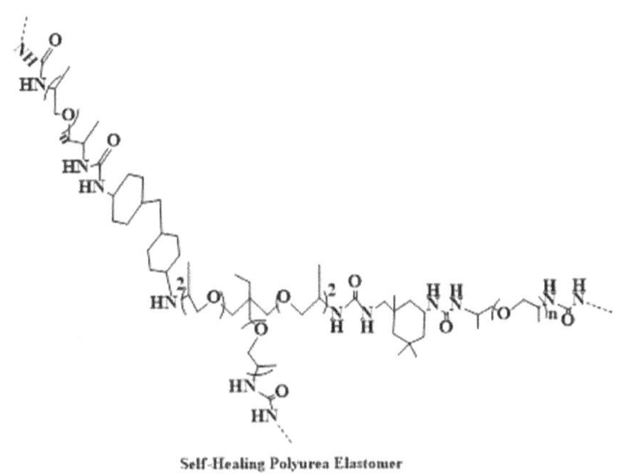

FIGURE 13.4 Chemical structure of self-healing polyurea elastomer (Credit: Authors of this chapter).

chemical structure is essential for optimizing the performance of self-healing polyurea elastomers in various engineering applications.

Catalysts and Healing Agents

The catalysts lower the activation energy needed to form bonds, hence enhancing self-healing reactions. More often than not, healing agents are encapsulated or embedded in the material and participate in the healing process by offering the required chemicals to mend the damage (Li *et al.*, 2019).

13.2.3 Physical and Mechanical Aspects

The physical and mechanical problems of self-healing materials are basic with respect to their functionality. It comprises the themes of load/stress and crack detection, initiation of healing processes, and recovery of structural integrity.

Stress and Crack Detection

A large number of self-healing materials have, in most cases, been engineered with sensors or some kind of stress-responsive elements. Embedded detection mechanisms may trigger the release of healing agents or turn on self-healing upon sensing damage (Yelwa and Musa, 2024).

Activation of Healing Mechanisms

These mechanisms of healing can be triggered by various stimuli, which include temperature changes, light, or mechanical stress. For example, heating triggers self-healing in thermally responsive polymers, while in photo-responsive materials, light is used to activate self-repair (Wu *et al*, 2008).

Restoration of Structural Integrity

Finally, the ultimate restoration of the structure to its wholesome integrity and functionality is the long-term goal of self-healing materials. This means that when cracks are filled and bonded, restoration of mechanical properties of the repaired material should be maintained to ensure still performing per-predetermined performance standards (Sharma *et al.*, 2023).

13.3 TYPES OF SELF-HEALING MATERIALS

13.3.1 Self-Healing Metals and Ceramics

Self-healing metals and ceramics are the future of materials science, giving extra durability and elongating life under extreme operating conditions. Principally, self-healing in metals is approached by using either reversible phase transformations or embedded healing agents that are selectively activated upon damage. For instance, shape-memory alloys in metals return their original shape in case of deformation, healing cracks and dents in the process (Zhu *et al.*, 2021). In ceramics, this is realized by the incorporation of self-healing agents such as silicon carbide or alumina, which react with oxygen at high temperatures to fill up the cracks (Zhang *et al.*, 2023c).

In Figure 13.5, the concept of self-healing metals is illustrated. The image shows a metal structure with a visible crack or defect. Over time, under external stress (represented by red arrows), the metal is able to heal the crack by activating mechanisms within its material composition. This process typically involves the rearrangement of atoms or diffusion of material into the damaged area, restoring the metal's structural integrity. Self-healing metals are crucial for improving durability and reliability in applications like aerospace, automotive, and structural engineering, where metals are subjected to high stress and potential damage.

13.3.2 Self-Healing Polymers and Elastomers

Polymers and elastomers are the most investigated self-healing materials due to their versatility in applications and relatively easy chemistry. Damage in self-healing polymers is repaired through different mechanisms, including hydrogen bonding, ionic interactions, or dynamic covalent bonds. Most elastomers have flexibility and usually adopt thermally reversible Diels–Alder reactions or supramolecular interactions to endow self-healing properties (Zhao *et al.*, 2021a). These materials will find applications mostly in things that involve flexibility and durability—aerospace components, and car parts.

Figure 13.6 demonstrates the self-healing mechanism of polymers, highlighting both the chemical reaction and the microscopic repair process. In Figure 13.6(a), two reactive polymer components undergo a reversible reaction when exposed to heat between 90–120°C, forming new covalent bonds. This heat-activated process enables the material to reconnect broken molecular chains, allowing the polymer to self-heal. In Figure 13.6(b), the microscopic self-healing behavior is illustrated. On the left, a crack disrupts the polymer matrix, breaking the connections between

FIGURE 13.5 Self-healing metals (Credit: Authors of this chapter).

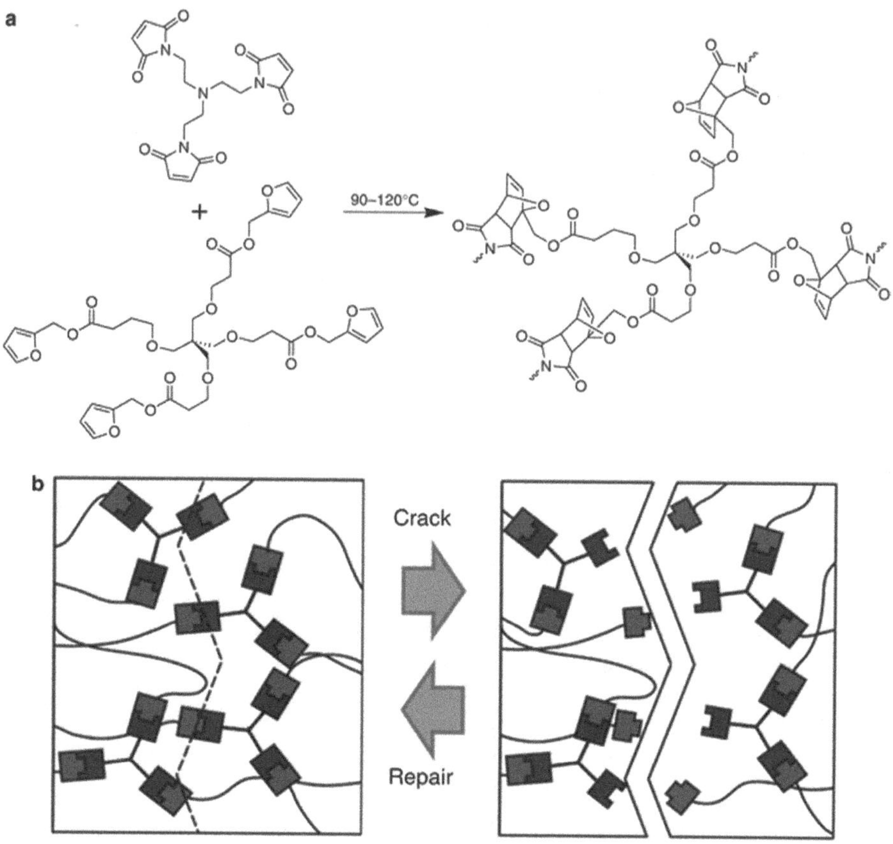

FIGURE 13.6 Self-healing mechanisms in polymers (Credit: Authors of this chapter).

components. The right side shows the polymer after repair, where the segments realign and bond back together, restoring the material's structure and functionality. This figure captures how reversible chemistry enables polymers to heal themselves, providing a durable solution to mechanical damage by reconnecting broken bonds and maintaining performance over time. Polymers and elastomers are widely used in applications requiring flexibility and resilience, such as automotive components and aerospace parts.

13.3.3 Self-Healing Coatings and Surfaces

Anti-Corrosion Coatings

Functional coatings functionality are very vital in protecting metallic surfaces against corrosion (Zhao *et al.*, 2022b, Zhao *et al.*, 2021b). Normally, healants or correlation inhibitors involved in such coatings are housed in microcapsules, breaking open whenever there is a scratch on the surface and flowing into that created space by damage, thereby blocking further propagation of the corrosion process (Zhang *et al.*, 2023a; Yedida and Vasudev, 2022). According to Panahi *et al.* (2024), this incorporation of self-healing features into coatings prolongs the life of metal structures, reducing maintenance costs and bettering the safety conditions.

Self-Cleaning Surfaces

Self-cleaning surfaces are made with photocatalytic materials or superhydrophobic coatings, which clean themselves and thus do not require manual cleaning. Besides, such surfaces can self-heal from small scratches or damage without losing their functionality. For instance, titanium dioxide-based self-healing coating uses photocatalytic reactions that degrade the organic contaminants to regain hydrophobicity on the surface (Hot *et al.*, 2023).

Figure 13.7 illustrates the self-cleaning behavior of a cotton textile, showing both macroscopic water-repellent action and the underlying microscopic structure that enables it. Figure 13.7(a) shows water droplets sitting on the fabric's surface without being absorbed, demonstrating the textile's superhydrophobic properties. This allows water and dirt particles to roll off, keeping the fabric clean with minimal effort. Figure 13.7(b) provides a microscopic view of the cotton fibers. The close-up reveals a textured fiber surface, likely treated with hydrophobic coatings or nanoscale roughness, which enhances water repellency by preventing liquid from spreading across the material. Figure 13.7 highlights the nanoscale surface modifications that convert traditional cotton into a self-cleaning textile, enabling it to repel water and dirt effectively for enhanced durability and low maintenance.

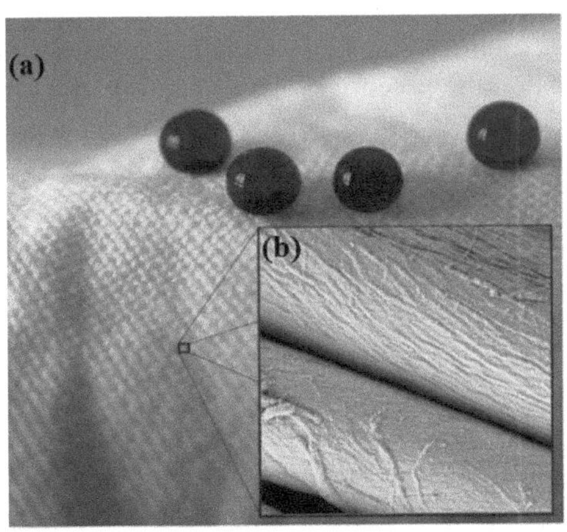

FIGURE 13.7 Self-cleaning surfaces (NIC, 2022) (Credit: Creative Commons Attribution 4.0).

13.3.4 Hybrid and Composite Self-Healing Materials

Hybrid and composite materials combine various types of mechanisms to self-heal with superior performance. Such materials include polymers, ceramics, or metals, with various nanomaterials or other strengthening agents to enhance mechanical strength and efficiency in self-healing. For example, carbon fiber-reinforced polymers that incorporate microcapsules of healing agents may be able to permit the autonomous repair of cracks while maintaining high strength and stiffness (Forintos and Czigany, 2019).

Figure 13.8 illustrates the self-healing mechanism in hybrid and composite materials, highlighting the structure and integration of healing networks across different scales. Figure 13.8(a) shows the macroscopic weave pattern of the material, with blue channels embedded within the black fiber matrix. These channels are strategically arranged to distribute healing agents across the material, ensuring widespread coverage in the event of damage. Figure 13.8(b) provides a closer view of the network, revealing the interwoven healing channels within the fiber structure. This layout ensures that healing agents can be delivered precisely to areas where damage occurs, enabling targeted and efficient repair. Figure 13.8(c) offers a microscopic close-up of individual channels embedded between structural fibers. When cracks form, the blue channels release their healing agents, sealing the damage and restoring both strength and functionality to the material. This figure highlights how embedded healing networks within composite materials allow for precise, efficient self-repair, enhancing durability and ensuring the material maintains its performance over time.

13.3.5 Emerging Materials and Innovative Composites

The self-healing materials field has been developing through the development of emerging materials and their innovative composites. Some recent developments include nanomaterials such as graphene, carbon nanotubes, and nanocellulose to enhance traditional materials' potential for self-healing and mechanical properties.

FIGURE 13.8 Hybrid and composite self-healing materials (Credit: Authors of this chapter).

In particular, innovative composites are bio-inspired materials that copy the ability of self-healing features found in natural systems, which is attracting great interest for several industrial applications (Jilani *et al.*, 2020).

13.4 ENVIRONMENTAL CONSIDERATIONS AND SUSTAINABILITY

13.4.1 Eco-Friendly Materials Used in Self-Healing Systems

Eco-friendly materials form a significant part of self-healing strategies. There has been growing interest in the application of renewable and biodegradable resources in the production of sustainable self-healing materials. For instance, biopolymers from plants and algae have been utilized, which have proven biocompatibility, and their environmental impact is low upon degradation (Urban, 2021). These include chitosan, cellulose, and lignin, among others (Guadagno *et al.*, 2021). It is also the case that self-healing composites reinforced with natural fibers and nanocellulose will not only display improved mechanical properties but also sustainability (Zheng *et al.*, 2020).

13.4.2 Life Cycle Assessment of Self-Healing Materials

Signed off as one of the more important tools in evaluating environmental impacts throughout the life cycles of self-healing materials from production to disposal, LCA gives information on energy consumption, carbon footprint, and general ecological impact. It has been demonstrated that the self-healing function of materials may significantly minimize this environmental burden by prolonging the life span of products and reducing their overall repair or replacement rate (Santos *et al.*, 2021). This reduced usage of material amount and waste generation is, therefore, in compliance with sustainable development.

13.4.3 Reducing Waste and Increasing Durability

Such a property in materials has enormous implications in terms of waste reduction and improved durability. Through the ability to self-repair after damage, self-healing materials reduce the quantum of wastes emanating from damaged goods and decrease the demand for new raw materials. For example, self-healing coatings on infrastructure prevent corrosion and extend buildings' and bridges' service life, hence minimizing the eco-impact of construction activities Gunático *et al.* (2021). Increased durability in self-healing materials means less usage of resources for maintenance and repair, hence more sustainability.

13.4.4 Challenges in Scalability and Ecological Safety

However, with certain attractive benefits, aspects of scalability and ecological safety become feeble. Some self-healing materials require complicated, hence relatively costly, processes for their production, which becomes problematic when considering their application on a large scale (Zhao *et al.*, 2021a). In addition, an in-depth evaluation of the ecological safety of some of the healing agents and catalysts has to be conducted to show that they will not cause any risk to the environment or health once released or degraded (Zhang *et al.*, 2023b). Continuous research and innovation are needed to find ways of addressing difficulties toward developing cost-effective and environmentally safe self-healing systems.

13.4.5 Contributions to Sustainable Development Goals

Self-healing materials have a vast contribution to a few of the Sustainable Development Goals (SDGs) envisaged by the United Nations. For example, they enhance the usage life of material through durability and reusability, in support of SDG 12: Responsible Consumption and Production (UN, 2021). Self-healing materials are part of SDG 9, with an emphasis on sustainable industrialization and innovation, in that the materials tend to reduce the demand for new materials and eliminate waste (Wong *et al.*, 2024). Importantly, the use of such features with earth-friendly and biodegradable materials in self-healing systems endorses SDG 13: Climate Action through reduced environmental impaction and ensures environmental sustainability (Samir *et al.*, 2022).

13.5 APPLICATIONS OF SELF-HEALING MATERIALS

13.5.1 Aerospace and Automotive Industries

Advancements in self-healing materials have greatly improved the aerospace and automotive industries. In aerospace, self-healing composites are applied to aircraft structures for improved safety and reduced maintenance costs. These materials self-heal autonomously from micro-cracks due to fatigue and environmental stress, thus leading to an extended service life of aero components by 20% (Williams *et al.*, 2007). Similarly, scratch-healing/self-healing polymers in an exterior coating of the car industry can repair scratches and minor damages. If done so, it is expected to reduce by 15% maintenance cost because of reduced vehicle breakdown and increased vehicle longevity (Ngugi *et al.*, 2024).

Figure 13.9 illustrates the integration of self-healing materials in aerospace and automotive components, highlighting their role in enhancing performance, durability, and safety. The airplane and mechanical parts, such as bearings and turbine components, represent typical areas where these advanced materials are applied. In aerospace, components endure significant stress, temperature fluctuations, and mechanical fatigue, making self-healing materials essential to repair micro-cracks or surface damage autonomously and maintain operational integrity. The molecular diagram in the figure demonstrates how self-healing mechanisms function at a microscopic level. When damage occurs, these materials respond by either releasing encapsulated healing agents, reforming dynamic molecular bonds, or activating shape-memory behavior through environmental triggers such as heat or light. This

FIGURE 13.9 Self-healing materials in aerospace and automotive components (Credit: Authors of this chapter).

process restores the material's structural properties without external intervention, ensuring that components can continue to function effectively.

13.5.2 Civil Engineering and Infrastructure

This self-healing functionality has changed much in the phases of maintenance and durability concerning infrastructure in the area of civil engineering. For instance, self-healing concrete is a kind of concrete added to bacteria or chemical agents, which, at cracking, act just like natural skin and start producing calcium carbonate to fill up and heal the cracks (Zhao *et al.*, 2021a). It reduces frequent repairs and hence minimizes costs while increasing the lifespan of structures such as bridges, tunnels, and buildings. Moreover, self-healing asphalt is developed to answer the problems of road degradation, thus able to provide enhanced safety and life expectancy to roads (Zhang *et al.*, 2022).

Figure 13.10 illustrates self-healing concrete, a material designed to repair cracks autonomously, enhancing durability and reducing maintenance. On the left, the healing process is shown, where cracks progressively seal over time. This healing can occur through bacteria-based methods, where embedded bacteria produce calcium carbonate upon exposure to water, or through encapsulated agents that release bonding compounds when cracks form. The sample on the right highlights voids or capsules that facilitate these processes. Self-healing concrete is widely used in infrastructure projects like bridges and roads, improving sustainability and structural reliability by minimizing repairs.

13.5.3 Biomedical and Healthcare Products

Typical biomedical applications of these self-healing materials include wound dressings, implants, and controlled drug delivery systems. In particular, self-healing

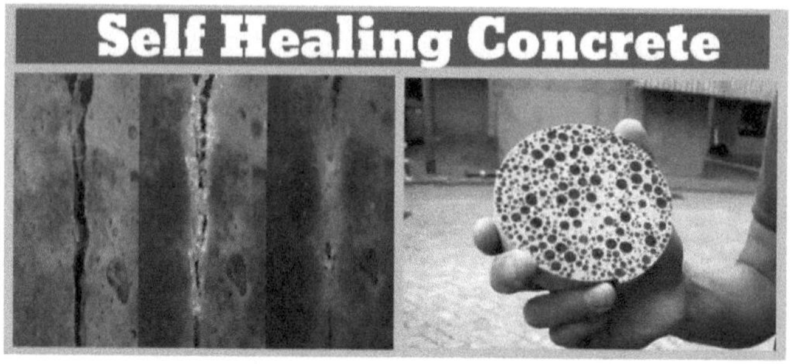

FIGURE 13.10 Self-healing concrete in infrastructure (Credit: Authors of this chapter).

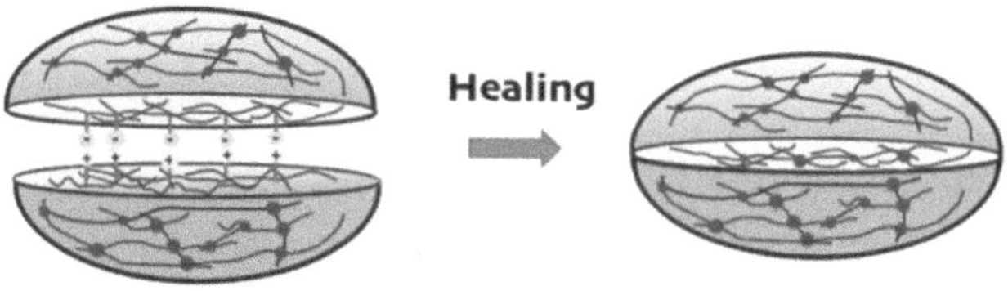

FIGURE 13.11 Self-healing hydrogels (Credit: Authors of this chapter).

hydrogels have proved efficient in wound curing because they can adapt at the wound site to provide a moisturized healing environment and repair themselves mechanically for faster healing (Liu *et al.*, 2024a; Antil *et al.*, 2020). Self-healing polymers can provide long life and functionality to implants by repairing minor damage to their own in implantable devices (Bekas *et al.*, 2016). Liu *et al.* (2024b) state that self-healing materials within drug delivery systems guarantee controlled releases to help attain better treatment outcomes.

In Figure 13.11, the self-healing hydrogel is shown in two distinct stages of recovery. Initially, the hydrogel is damaged, with its structure split into two halves, creating a clear separation. Within the hydrogel, polymer chains, depicted as purple lines, are disconnected due to the break. Healing agents or active sites, represented by small green dots, are visible along the broken polymer chains, preparing to facilitate the repair process. As the healing progresses, these agents form new bonds between the severed polymer chains, indicated by the green connections bridging the gap. Over time, these bonds strengthen, gradually pulling the two halves of the hydrogel back together. In the final stage of the process, the gap is fully closed, and the polymer network is completely restored to its original form. The hydrogel, now reconnected, has successfully undergone self-healing, demonstrating its ability to repair itself without external intervention.

13.5.4 Consumer Electronics and Protective Devices

The new self-healing materials seek to reinforce sustainability in consumer electronics and experiences. For instance, scratch-healing polymers are used in smartphone screens and covers, performing the functions of repairing minor tears to maintain outlook and functionality (Ahmed *et al.*, 2023). Self-healing composite protective devices, like helmets and body armor, absorb impact energy, which causes cracks to heal and therefore protects over several impacts (Alkhatib, 2023).

13.5.5 Energy Production and Storage Devices

As such, self-healing materials are royalty in the development of technologies related to energy production and storage (Wen *et al.*, 2021). In a similar sense, concerning batteries, repairing cracks on electrodes using self-healing polymers and composites will help to increase the battery life cycle and improve their safety and performance. Self-healing coatings can repair minor damage caused by environmental factors on the surface of panels of solar panels, thus maintaining the efficiency and life expectancy of the panels (Fawzy & Abou-Zeid *et al.*, 2017). These innovations mean more sustainable and reliable energy systems.

Figure 13.12 illustrates self-healing materials for energy storage by showing how dynamic covalent and non-covalent bonds, such as hydrogen, coordination, disulfide, and Diels–Alder bonds, enable materials to repair themselves. These bonds are essential for healing key components like self-healing electrodes, electrolytes, and artificial interface layers used in batteries. The figure also emphasizes the role of advanced characterization techniques to study and verify the self-healing capabilities of these materials, enhancing the longevity and performance of energy storage devices.

13.6 CASE STUDIES OF SELF-HEALING MATERIALS

13.6.1 Successful Applications

Self-healing materials have been successfully applied in many engineering applications and therefore have huge potential for improving the durability and sustainability of structures. An example is in the aerospace industry, where self-healing composites are used to repair micro-cracks that arise from cyclic stress (White *et al.*, 2021). In such materials, microcapsules with healing agents release at a time of crack formation and initiate polymerization, which closes the cracks. Another successful application is the one in civil engineering. Qureshi *et al.* (2016) reported that self-healing concrete, by using microencapsulation of healing agents or bacteria precipitating calcium carbonate, significantly enhances the lifecycle of concrete structures. For example, the Delft University of Technology applied self-healing concrete in a range of pilot projects and demonstrated improved durability compared with regular concrete of up to 30%. In the automotive sector, self-healing coatings have been applied in order to preserve the aesthetic and protective qualities of vehicle

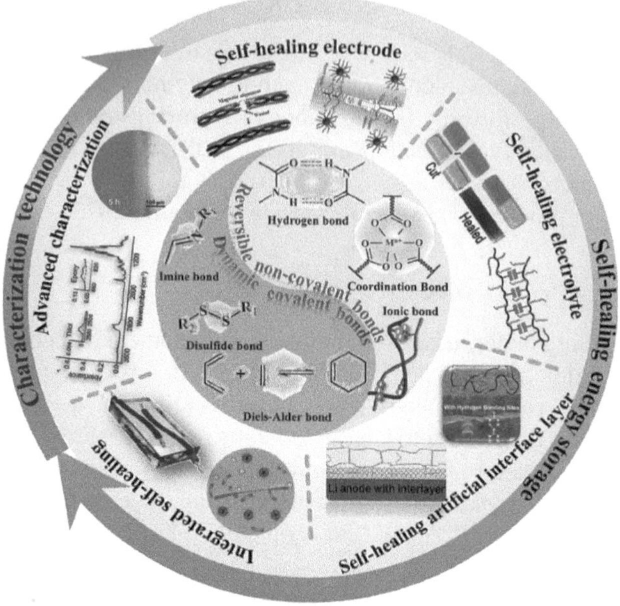

FIGURE 13.12 Self-healing materials in energy storage devices (Wan *et al.*, 2023) (Credit: Creative Commons Attribution 4.0).

surfaces. These coatings can repair minor scratches and dents autonomously; hence, they reduce the cost of maintenance and extend the lifespan of automotive parts (Zhu *et al.*, 2020). Table 13.1 provides an overview of the case studies relating to the successful application of self-healing materials in industries such as aerospace, civil engineering, automotive, biomedical, energy storage, infrastructure, and consumer electronics. Each case study indicates the used material, its specific application, and the outcomes, together with the advantages and challenges of deploying these materials.

13.6.2 Lessons Learned from Failures and Limitations

Although there are bright prospects for application, self-healing materials have faced several challenges. One of the major bottlenecks relates to healing efficiency under certain harsh environments. For example, it is regarded that self-healing polymers applied in marine environments may express reduced efficacy due to continuous actions of salt water and UV radiation (Zhao *et al.*, 2021b). Such findings were reported by Wang *et al.* (2021b). Therefore, this need has been created to come up with more robust mechanisms of self-healing that can demonstrate better resistance to extreme conditions. Another challenge has to do with the long-term stability of the healing agent. At times, micro-encapsulated healing agents would degrade along with time, thus reducing the self-healing capability of the material (Sethi *et al.*,

TABLE 13.1
Case Studies of Self-Healing Applications

Case Study	Material Used	Application	Outcome	Advantages	Challenges	References
Aerospace	Self-Healing Composite	Microcrack Repair in Aircraft Structures	20% Service Life Extension	Improved safety and reduced maintenance costs	Sensitivity to extreme temperature variations	White *et al.* (2021); Chopra *et al.* (2023)
Civil Engineering	Self-Healing Concrete	Structural Durability in Bridges and Tunnels	30% Increase in Durability	Reduced repair frequency and longer lifespan of infrastructure	High initial cost and complex integration	Qureshi *et al.* (2016)
Automotive	Self-Healing Coating	Scratch and Dent Repair on Vehicle Surfaces	Reduced Maintenance Costs	Enhanced aesthetic appeal and protection	Limited healing cycles before coating replacement is needed	Zhu *et al.* (2020); Reddy *et al.* (2020)
Biomedical	Self-Healing Hydrogels	Wound Healing and Drug Delivery Systems	Accelerated Healing and Controlled Release	Adaptation to wound site, reduced infection risks	Limited mechanical strength under high stress	Yang *et al.* (2022); Wang *et al.* (2022a,b)
Energy Storage	Self-Healing Polymers	Battery Electrodes	25% Increase in Cycle Life	Enhanced battery safety and performance	Potential degradation of healing agents over time	Zhao *et al.* (2021a); Jilani *et al.* (2020)

(Continued)

TABLE 13.1 (CONTINUED)
Case Studies of Self-Healing Applications

Case Study	Material Used	Application	Outcome	Advantages	Challenges	References
Infrastructure	Self-Healing Asphalt	Road Surface Durability	40% Reduction in Pothole Formation	Increased safety and reduced maintenance	Sensitivity to temperature fluctuations	Li *et al.* (2018); Zhao *et al.* (2022)
Consumer Electronics	Self-Healing Polymers	Smartphone Screens and Covers	Extended Device Lifespan	Maintained functionality after minor scratches	Reduced transparency and touch sensitivity after repeated healing	Ahmed *et al.* (2023); Zhao *et al.* (2021a)

2024). This problem emphasizes the importance of researching more durable encapsulation techniques and healing agents with longer shelf lives. Moreover, integration into existing manufacturing processes can be rather difficult and expensive for self-healing materials. Initial high costs have limited the wide application of these materials in industries where cost efficiency is paramount (Guadagno *et al.*, 2021). Hence, to some extent, performance has to be weighed against economic feasibility, becoming another direction for further studies. Table 13.2 identifies issues and failure modes of self-healing materials, with common problems such as incomplete healing, slow responses, limited cycles of healing, and their effects on the performance of self-healing materials. Potential solutions to overcome the challenges are also proposed in Table 13.2, supported by relevant literature.

13.6.3 Comparative Analysis of Self-Healing Technologies

Generally, in the area of self-healing technologies, two big groups can be broadly distinguished: intrinsic and extrinsic systems. Examples of intrinsic systems include those using material-inherent properties to self-repair. Extrinsic systems make use of external healing agent integration into the material matrix through carrier systems like microcapsules or vascular networks (Toohey *et al.*, 2007).

Intrinsic self-healing systems, like those that involve reversible Diels–Alder chemistry, have the benefit of recurring without bringing along added constituents. Because of certain distinctive conditions that can be created within these materials, they are capable of repairing themselves autonomously a couple of times, hence making them quite suitable for long-term applications (Yang *et al.*, 2022). Their healing efficiency is, however, often limited by specific conditions required to trigger the mechanism of repair.

Extrinsic systems, such as microcapsule-based healing, are comparatively simpler, wherein the healing agents are released in case of any damage to the material. This has found practical application in many fields including coatings and concrete, with appreciable results (Qureshi *et al.*, 2016). However, once the healing agents get consumed, no further self-repair is possible with the material which needs to be refurnished or replaced.

Recent developments have been toward the design of hybrid self-healing systems that involve intrinsic and extrinsic mechanisms. These hybrid systems are targeted to derive the advantages from both approaches to offer more strong, versatile self-healing capabilities (Liu *et al.*, 2024b). Table 13.3 compares intrinsic versus extrinsic self-healing technologies, including the mechanisms, advantages and disadvantages involved, and main applications. It covers the most recent developments related to hybrid and nanotechnology-enhanced systems that are going to indicate future trends in self-healing materials research.

13.6.4 Insights from Recent Research Breakthroughs

Recent research has introduced several breakthroughs in the field of self-healing materials. One of the recent examples is the designing of self-healing hydrogels with

TABLE 13.2
Challenges and Failure Modes in Self-Healing Material

Challenges and Failure Modes	Description	Examples	Impact on Performance	Potential Solutions	References
Incomplete Healing	The material does not fully recover to its original state after damage	Microcapsule-based healing agents are consumed and can't be replenished	Reduced mechanical strength and durability	Development of hybrid systems combining intrinsic and extrinsic healing	Qureshi *et al.* (2016); Neal *et al.* (2015)
Slow Response Times	The self-healing process takes too long to activate and complete	Thermally activated polymers that require high temperatures to trigger healing	Delayed repair can lead to further damage or failure	Use of faster stimuli-responsive materials (e.g., photo-responsive polymers)	Zhao *et al.* (2022); Hua *et al.* (2023)
Limited Healing Cycles	The material can only self-heal a few times before losing its ability to repair itself	Elastomers with limited reversible bond formation	Cumulative damage over time leads to material failure	Incorporation of more robust dynamic covalent bonds	Yang *et al.* (2021); Zhao *et al.* (2021b)
Reduced Material Strength	Repeated healing reduces the overall strength and durability of the material	Self-healing composites with microcapsules that weaken over time	Lower load-bearing capacity and increased brittleness	Reinforcement with nanomaterials to maintain strength	Li *et al.* (2019); Wen *et al.* (2021)
Failure Under Extreme Conditions	Self-healing mechanisms may fail under extreme temperatures, pressures, or chemical exposure	Marine coatings losing efficacy in salt water environments	Accelerated degradation and corrosion	Development of more resistant and adaptable self-healing agents	Wang *et al.* (2021a,b); Kanu (2019)
Heavy Stress	Excessive mechanical stress can prevent the self-healing process from being effective	Concrete structures exposed to high loads	Cracks that exceed the material's healing capacity	Use of smart materials that can adjust healing properties based on stress level	Thakur *et al.* (2022); Zhu *et al.* (2020)
Economic Feasibility	High production costs and complexity hinder widespread adoption	High initial costs in aerospace applications	Limited adoption in cost-sensitive industries	Scaling up production and reducing material costs	Qamar *et al.* (2020); Chopra *et al* (2023)

TABLE 13.3
Comparative Table of Intrinsic vs. Extrinsic Self-Healing Technologies

Technology	Mechanism	Advantages	Disadvantages	Applications	Recent Developments	References
Intrinsic	Reversible Bonds	Repeated Healing Without External Agents	Often requires specific conditions to activate healing (e.g., temperature, light)	Polymers, Elastomers, Shape Memory Alloys	Development of hybrid intrinsic systems for enhanced performance	Yang *et al.* (2022); Urban (2021)
Extrinsic	Microcapsules or Vascular Networks	Simple Activation, High Healing Efficiency	Limited by the amount of healing agent available; once depleted, no further healing	Coatings, Concrete, Structural Composites	Integration with nanotechnology for improved healing efficiency	Wang *et al.* (2023); Qureshi *et al.* (2016)
Hybrid	Combined Mechanisms (Intrinsic and Extrinsic)	Robust Healing, Enhanced Mechanical Properties	Complexity in design and higher production costs	Advanced Composites, High-Performance Coatings	Ongoing research into bio-inspired hybrid systems for multi-functional applications	Liu *et al.* (2024a); Zhou *et al.* (2022)
Supramolecular Systems	Non-covalent Interactions (e.g., Hydrogen Bonding, Metal-Ligand Coordination)	Self-assembly, Reversible Healing	Limited by environmental stability	Biomedical Applications, Smart Sensors	Recent breakthroughs in bio-inspired designs	Mohamadhoseini and Mohamadnia (2021); Zhou *et al.* (2022)

(Continued)

TABLE 13.3 (CONTINUED)
Comparative Table of Intrinsic vs. Extrinsic Self-Healing Technologies

Technology	Mechanism	Advantages	Disadvantages	Applications	Recent Developments	References
Nanotechnology-Enhanced Systems	Nanoparticles or Nanofibers Integration	Improved Healing and Mechanical Properties	Increased material complexity	Electronics, Aerospace, Renewable Energy Systems	Integration of graphene and carbon nanotubes for enhanced self-healing	Zheng *et al.* (2020); Jilani *et al.* (2020)
Bio-inspired Systems	Mimicking Natural Healing Processes	High Efficiency, Environmentally Friendly	Challenges in scalability and real-world application	Medical Implants, Protective Coatings	Development of hydrogels and biopolymers with rapid healing capabilities	Zhu *et al.* (2020); Hua *et al.* (2023)

rapid and effective healing performance, presenting them with very promising healing potential. The hydrogels that have been designed for biomedical applications are capable of healing within several minutes at room temperature and hold very promising outlooks for applications in wound dressings and tissue engineering (Yang *et al.*, 2022). The other important realization has been the integration of nanotechnology into self-healing materials. Researchers have developed advanced nanocomposite materials with improved mechanical properties and self-healing capability. For instance, graphene-based nanocomposites engineered lately revealed better conductivity and improved healing efficiency that suit flexible electronics and smart sensors (Zhang *et al.*, 2022). Furthermore, over the last few years, a large quantity of bio-inspired self-healing mechanisms has been investigated. Having been inspired by natural systems, like the ability of plants and animals to repair themselves, researchers have come up with materials mimicking these processes. For instance, self-healing polymers with a network topology described by the abalone shell have shown great toughness and efficiency in healing (Song *et al.*, 2021). These developments show that self-healing materials are really promising, and their applications should be able to expand into aerospace, biomedical, civil engineering, and electronic applications, among other fields, in the near future.

13.7 CHALLENGES AND FUTURE DIRECTIONS

As the field of self-healing materials continues to evolve, research is increasingly focused on developing materials with faster healing times, higher healing efficiencies, and the ability to repair under diverse environmental conditions. Multifunctional materials that combine self-healing with other properties, such as conductivity, thermal stability, or self-sensing capabilities, are also an area of intense interest. However, the main challenges when developing self-healing materials are with regard to their mechanical strength versus healing capabilities. On the technical side, certain challenges include quantifying reliable self-healing in various environments and consistent performance for creation to go through cycles of damaging and healing. While economically, these materials promise reduced maintenance costs and long-term life, high upfront investment due to complex manufacturing processes creates a big barrier against general acceptance. Another challenge is how to scale this up from the lab into industrial applications involving process changes that guarantee cost-effectiveness with quality. Future strides in the domain of materials science, in particular, nanotechnology, are envisioned to further increase self-healing properties if integrated with AI. In this light, interdisciplinary concerted actions can expedite goals for moving forward from such known challenges by bringing scientists, engineers, and industry professionals together toward commercialization from next-generation self-healing materials.

13.8 CONCLUSION

Self-healing materials are a major advancement in material science, capable of repairing damage on their own and extending the life of engineered products. They use mechanisms like microencapsulation, vascular networks, and reversible chemical bonds to achieve self-repair. Successfully used in aerospace, automotive, civil engineering, and biomedical sectors, these materials show great promise. However, challenges like environmental stability, long-term durability, and cost need to be addressed. These materials are important for green technology because they reduce waste, lower maintenance costs, and improve sustainability. Future advancements in nanotechnology and artificial intelligence are expected to enhance their capabilities. Collaboration across disciplines is essential to overcome current challenges and develop the next generation of smart, sustainable materials. Continued research is crucial for improving infrastructure resilience and safety in critical applications, benefiting society and the environment.

REFERENCES

Ahmad, S., Hashmi, A.W., Singh, J., Arora, K., Tian, Y., Iqbal, F., Al-Dossari, M., and Khan, M.I. (2024). Innovations in Additive Manufacturing of Shape Memory Alloys: Alloys, Microstructures, Treatments, Applications. *Journal of Materials Research and Technology*, 32, 4136–4197

Ahmed, S., Jeong, J.E., Kim, J.C., Lone, S. and Cheong, I.W. (2023). Self-healing polymers for surface scratch regeneration. *RSC Advances, 13*(50), 35050–35064.

Alkhatib, A.J. (2023). Bio-inspired protective composite structures for automotive applications. *Thin-Walled Composite Protective Structures for Crashworthiness Applications: Recent Advances and Future Developments*, (pp. 87–115).

Antil, P., Kumar Antil, S., Prakash, C., Krolczyk, G., and Pruncu, C. (2020). Multi-objective Optimization of Drilling Parameters for Orthopaedic Implants. *Measurement and Control,* 53(9–10), 1902–1910. https://doi.org/10.1177/0020294020947126

Bekas, D. G., Tsirka, K., Baltzis, D., & Paipetis, A. S. (2016). Self-healing materials: A review of advances in materials, evaluation, characterization, and monitoring techniques. *Composites Part B: Engineering, 87,* 92–119. https://doi.org/10.1016/j.compositesb.2015.09.057.

Chen, H., Zhang, S., Wang, X., Liu, C., & Huang, Y. (2023). Advances in self-healing ceramics: Mechanisms, materials, and industrial applications. *Journal of the American Ceramic Society, 106*(2), 678–694. https://doi.org/10.1111/jace.18790.

Chopra, H., Mohanta, Y., Mahanta, S., Mohanta, T., Singh, I., Avula, S., Mallick, S., Rabaan, A., AlSaihati, H., Alsayyah, A., Alissa, M., Alturaifi, H., AlAlwan, B., Attia, M., Chakraborty, S. & Dhama, K. (2023). Recent updates in nanotechnological advances for wound healing: A narrative review. *Nanotechnology Reviews, 12*(1), 20230129. https://doi.org/10.1515/ntrev-2023-0129.

Fawzy, M., & Abou-Zeid, M. (2017). Technical and economic feasibility of self-healing concrete. *Leadership in Sustainable Infrastructure, Vancouver, Canada, May 31 – June 3, 2017.* The American University in Cairo.

Forintos, N. and Czigany, T. (2019). Multifunctional application of carbon fiber reinforced polymer composites: Electrical properties of the reinforcing carbon fibers – A short review. *Composites Part B: Engineering, 162*, 331–343.

Guadagno, L., Vertuccio, L., Barra, G., Naddeo, C., Sorrentino, A., Lavorgna, M., Raimondo, M., & Calabrese, E. (2021). Eco-friendly polymer nanocomposites designed for self-healing applications. *Polymer, 223,* 123718. https://doi.org/10.1016/j.polymer.2021.123718.

Hager, M. D., Greil, P., Leyens, C., van der Zwaag, S., & Schubert, U. S. (2010). Self-healing materials. *Advanced Materials (Deerfield Beach, Fla.)*, *22*(47), 5424–5430. https://doi.org/10.1002/adma.201003036.

Hot, J., Castelló Lux, K., & Ringot, E. (2023). Investigation of the self-cleaning property of photocatalytic coatings at a laboratory scale. *Photochem*, *3*(4), 461–476. https://doi.org/10.3390/photochem3040028.

Hua, M., Wu, S., Ma, Y., Zhao, Y., Wang, H., Yu, J., & Xu, M. (2023). Recent advances in self-healing materials for flexible electronics. *Journal of Materials Chemistry C, 11*(5), 1234–1250. https://doi.org/10.1039/D2TC05678J.

Irzhak, V. I., Uflyand, I. E., & Dzhardimalieva, G. I. (2022). Self-Healing of Polymers and Polymer Composites. *Polymers, 14*(24), 5404. https://doi.org/10.3390/polym14245404.

Jilani, A., Hussain, S. Z., Khan, A. A. P., Khan, A., Othman, M. H. D., & Ansari, M. O. (2020). Graphene-based material for self-healing: Mechanism, synthesis, characteristics, and applications. In *Self-healing composite materials: From design to applications* (pp. 163–175). Woodhead Publishing. https://doi.org/10.1016/B978-0-12-817354-1.00009-0.

Kanu, N. J., Gupta, E., Vates, U. K., & Singh, G. K. (2019). Self-healing composites: A state-of-the-art review. *Composites Part A: Applied Science and Manufacturing, 121*, 474–486. https://doi.org/10.1016/j.compositesa.2019.04.012.

Kulkarni, S.D., Manjunatha, Chandrasekhar, U., Manjunath, K. V, Prasad, C.D., and Vasudev, H., 2024. Design and optimization of polyvinyl-nitride rubber for tensile strength analysis. *International Journal on Interactive Design and Manufacturing (IJIDeM)*, 18(8), 5893–5908.

Li, W., Dong, B., Yang, Z., Xu, J., Chen, Q., Li, H., Xing, F. and Jiang, Z. (2018). Recent advances in intrinsic self-healing cementitious materials. *Advanced Materials*, *30*(17), 1705679.

Li, G., Xiao, P., Hou, S., & Huang, Y. (2019). Graphene-based self-healing materials. *Carbon, 146*, 371–387. https://doi.org/10.1016/j.carbon.2019.02.011.

Liu, Q., Wang, C., Cheng, M., Hu, L., Zhang, Z., Sun, Q., Wang, S., Fan, Y., Pan, P., & Chen, J. (2024a). Self-healing conductive hydrogels with dynamic dual network structure accelerate infected wound healing via photothermal antimicrobial and regulating inflammatory response. *ACS Applied Materials & Interfaces*, 14(24) 30776–30792. https://doi.org/10.1021/acsami.4c04113.

Liu, Z., Zhong, Y., Li, S., Yu, S., Zhong, J., Yang, Y., & Shen, L. (2024b). Hybrid Self-Repairing Polymer Composites Based on a Mixture of Intrinsic and Extrinsic Self-Healing. *Macromolecular Chemistry and Physics, 255 (18), 2400079.* https://doi.org/10.1002/macp.202400079.

Mohamadhoseini, M., & Mohamadnia, Z. (2021). Supramolecular self-healing materials via host-guest strategy between cyclodextrin and specific types of guest molecules. *Coordination Chemistry Reviews, 432*, 213711. https://doi.org/10.1016/j.ccr.2020.213711.

Neal, J., Mozhdehi, D., & Guan, Z. (2015). Enhancing mechanical performance of a covalent self-healing material by sacrificial noncovalent bonds. *Journal of the American Chemical Society*, 137 14, 4846–50. https://doi.org/10.1021/jacs.5b01601.

Ngugi, M.J. (2024). Self-healing materials: innovations and applications. *Research Invention Journal of Biological and Applied Sciences*, 3(3):57–59.

Norris, C.J., Bond, I.P. and Trask, R.S. (2013). Healing of low-velocity impact damage in vascularised composites. *Composites Part A: Applied Science and Manufacturing, 44*, 78–85.

Panahi, P., Khorasani, S.N., Mensah, R.A., Das, O. and Neisiany, R.E. (2024). A review of the characterization methods for self-healing assessment in polymeric coatings. *Progress in Organic Coatings*, *186*, 108055.

Qamar, I. P.S., Sottos, R.N., & Trask, S.R. (2020). Grand challenges in the design and manufacture of vascular self-healing. *Multifunctional Materials*, 3, 013001. https://doi.org/10.1088/2399-7532/ab69e2.

Qin, J., & Lu, H. (2023). A review of self-cleaning coatings for solar photovoltaic systems: Theory, materials, preparation, and applications. *Environmental Science and Pollution Research, 30*, 91591–91616. https://doi.org/10.1007/s11356-023-28550-5.

Qureshi, T.S., Kanellopoulos, A. and Al-Tabbaa, A. (2016). Encapsulation of expansive powder minerals within a concentric glass capsule system for self-healing concrete. *Construction and Building Materials*, *121*, 629–643.

Reddy, K. R., El-Zein, A., Airey, D. W., Alonso-Marroquin, F., Schubel, P., & Manalo, A. (2020). Self-healing polymers: Synthesis methods and applications. *Nano-Structures & Nano-Objects, 23*, 100500. https://doi.org/10.1016/j.nanoso.2020.100500.

Samir, A., Ashour, F. H., Abdel Hakim, A. A., & Bassyouni, M. (2022). Recent advances in biodegradable polymers for sustainable applications. *npj Materials Degradation, 6*, 68. https://doi.org/10.1038/s41529-022-00277-7.

Santos, A.N., Santos, D.J.D. and Carastan, D.J. (2021). Microencapsulation of reactive isocyanates for application in self-healing materials: a review. *Journal of Microencapsulation*, 38(5), 338–356.

Scheiner, M., Dickens, T., & Okoli, O. (2016). Progress towards self-healing polymers for composite structural applications. *Polymer*, 83, 260-282. https://doi.org/10.1016/J.POLYMER.2015.11.008.

Sethi, S.K., Verma, A., Akubueze, E., Maurya, A.K. (2024). Self-Healing of Polymer Composites: Process and Developments. In: Verma, A., Gupta, H.S., Sethi, S.K. (eds) Hybrid Composite Materials. Springer, Singapore. https://doi.org/10.1007/978-981-97-2104-7_10.

Sharma, A., Shukla, N.K., Belarbi, M.O., Abbas, M., Garg, A., Li, L., Bhutto, J., and Bhatia, A. (2023). Bio-inspired nacre and helicoidal composites: From structure to mechanical applications, Thin-Walled Structures, 192, 111146, 0263-8231, https://doi.org/10.1016/j.tws.2023.111146.

Sharma, S., Nandan, G., Rohatgi, P. K., & Prakash, R. (2019). Recent advances in self-healing materials. *Materials Today: Proceedings, 18*(Part 7), 4729–4737. https://doi.org/10.1016/j.matpr.2019.07.460.

Snyder, R.L., Fortman, D.J., De Hoe, G.X., Hillmyer, M.A. and Dichtel, W.R. (2018). Reprocessable acid-degradable polycarbonate vitrimers. *Macromolecules*, *51*(2), 389–397.

Song, Y., Li, J., Song, G., Zhang, L., Liu, Z., Jing, X., Luo, F., Zhang, Y., Zhang, Y., and Li, X. (2021). Self-healing polyurethane elastomers with high mechanical properties based on synergistically thermo-reversible and quadruple hydrogen bonds. *ACS Appl. Polym. Mater.* 2023, 5, 2, 1302–1311. https://doi.org/10.1021/acsapm.2c01849.

Tay, C. Y., Setyawati, M. I., Xie, J., Parak, W. J., & Chen, X. (2023). Supramolecular self-healing materials: Advances and applications in nanomedicine. *Chemical Society Reviews, 52*(3), 1297–1321. https://doi.org/10.1039/D2CS00845H.

Thakur, A., & Kumar, A. (2022). Self-healing nanocoatings for automotive application. In *Nanotechnology in the automotive industry* (pp. 403–427). Elsevier. https://doi.org/10.1016/B978-0-323-90524-4.00019-0.

Toohey, K.S., Sottos, N.R., Lewis, J.A., Moore, J.S. and White, S.R. (2007). Self-healing materials with microvascular networks. *Nature Materials*, *6*(8), 581–585.

Trask, R.S., Williams, G.J. and Bond, I.P. 2007. Bioinspired self-healing of advanced composite structures using hollow glass fibres. *Journal of the Royal Society Interface, 4*(13), pp.363–371.

Urban, M. W. (2021). Stimuli-responsive self-healing polymers: Chemistry, design, and applications. *Chemical Reviews, 121*(5), 3124–3186. https://doi.org/10.1021/acs.chemrev.0c00783.

Wang, C., Zhang, L., & Zhou, F. (2021a). Recent advances in self-healing coatings for corrosion protection. *Journal of Materials Chemistry A, 9*(18), 11173–11189. https://doi.org/10.1039/D1TA01234G.

Wang, H., Liu, W., & Chen, C. (2021b). Recent advances in self-healing hydrogels for biomedical applications. *Journal of Materials Chemistry B, 9*(18), 4211–4230. https://doi.org/10.1039/D1TB00477E.

Wang, J., Liu, H., & Ma, S. (2022c). Advances in self-healing elastomers: Design, mechanisms, and applications. *Journal of Applied Polymer Science, 139*(35), 52012. https://doi.org/10.1002/app.52012.

Wang, J., Tang, J., Chen, D., Xing, S., Liu, X., & Hao, J. (2023). Intrinsic and extrinsic self-healing fiber-reinforced polymer composites: A review. *Polymer Composites, 44(10), 6304-6323.* https://doi.org/10.1002/pc.27623.

Wang, X., Gao, L., & Wu, X. (2022a). Emerging trends in self-healing multifunctional materials. *Advanced Functional Materials, 32*(27), 2201613. https://doi.org/10.1002/adfm.202201613.

Wang, M., Li, H., & Zhang, Z. (2022b). Self-healing materials for advanced energy storage applications. *Journal of Energy Chemistry, 67*, 81–92. https://doi.org/10.1016/j.jechem.2021.09.011.

Wen, N., Song, T., Ji, Z., Jiang, D., Wu, Z., Wang, Y., and Guo, Z. (2021). Recent advancements in self-healing materials: Mechanicals, performances and features. *Reactive and Functional Polymers*, 168, 105041.

White, T.B., Viana, L.R., Campbell, G., Elverum, C., Bennun, L.A. (2021). Using technology to improve the management of development impacts on biodiversity. *Business Strategy and the Environment*: 1–15. https://doi.org/10.1002/bse.2816

Williams, G., Trask, R, and Bond, I. (2007). A self-healing carbon fibre reinforced polymer for aerospace applications. *Composites Part A: Applied Science and Manufacturing, 38*, 1525–1532.

Wong, P. Y., Mal, J., Sandak, A., Luo, L., Jian, J., & Pradhan, N. (2024). Advances in microbial self-healing concrete: A critical review of mechanisms, developments, and future directions. *Science of The Total Environment, 947*, 174553. https://doi.org/10.1016/j.scitotenv.2024.174553.

Wu, D. Y., Meure, S., & Solomon, D. (2008). Self-healing polymeric materials: A review of recent developments. *Progress in Polymer Science, 33*(5), 479–522. https://doi.org/10.1016/j.progpolymsci.2008.02.001.

Xu, L., Lu, X., & Jiang, F. (2022). Advances in self-healing coatings for corrosion protection. *Corrosion Science, 195*, 110011. https://doi.org/10.1016/j.corsci.2022.110011.

Yang, Y., Xu, L., Wang, J., Meng, Q., Zhong, S., Gao, Y., & Cui, X. (2022). Recent advances in polysaccharide-based self-healing hydrogels for biomedical applications. *Carbohydrate Polymers, 283*, 119161. https://doi.org/10.1016/j.carbpol.2022.119161.

Yedida, V. V. S., & Vasudev, H. (2022). Development of self-healing thermal barrier coatings using thermal spray techniques. *Materials Today: Proceedings, 50*, 1458–1464. https://doi.org/10.1016/j.matpr.2022.05.312.

Yelwa, J.M. and Musa, H. (2024). Innovative Smart Coatings: Advancing Surface Protection and Sustainability Across Industries. *Academia Nano: Science, Materials, Technology*, 1(1), 1–12.

Zhang, H., Wang, L., & Li, X. (2022). Self-healing elastomers: Recent advances in mechanisms and applications. *Advanced Materials Interfaces, 10*(3), 2201789. https://doi.org/10.1002/admi.202201789.

Zhang, J., Liu, W., & Zhao, Y. (2023a). Advances in self-healing coatings for anti-corrosion applications. *Progress in Organic Coatings, 183*, 107078. https://doi.org/10.1016/j.porgcoat.2023.107078.

Zhang, L., Yang, H., & Huang, Y. (2023b). Self-healing polymeric materials for biomedical applications: Current status and future perspectives. *Journal of Biomedical Materials Research Part A, 111*(5), 950-968. https://doi.org/10.1002/jbm.a.37498.

Zhang, S., Li, Y., & Wang, Z. (2023c). Advances in self-healing polymeric materials for aerospace applications. *Advanced Materials Interfaces, 12*(8), 2300019. https://doi.org/10.1002/admi.202300019.

Zhao, H., Li, X., & Chen, Y. (2023d). Self-healing cementitious materials: Mechanisms, recent advances, and future perspectives. *Materials Today Communications, 37*, 106482. https://doi.org/10.1016/j.mtcomm.2023.106482.

Zhao, Q., Liang, Y., Huang, J., & Urban, M. W. (2022a). Sustainable self-healing materials: From molecular design to industrial applications. *Progress in Materials Science, 130*, 101012. https://doi.org/10.1016/j.pmatsci.2022.101012.

Zhao, Q., Qi, H. J., & Xie, T. (2022b). Recent progress in shape memory polymers: New theory and emerging applications. *Materials Today, 49*, 168–186. https://doi.org/10.1016/j.mattod.2022.01.010.

Zhao, C., Zeng, X., & Li, H. (2021a). A review of self-healing materials for electronic devices. *Nano Today, 40*, 101235. https://doi.org/10.1016/j.nantod.2021.101235.

Zhao, X., Li, Y., & Zhang, Q. (2021b). Self-healing polymeric materials: Advances in chemistry and applications. *Journal of Materials Chemistry A, 9*(28), 16157–16176. https://doi.org/10.1039/D1TA03086H.

Zhao, X., Wang, H., Wang, Z., & Xu, X. (2022). Self-healing materials in medical devices: Current status and future perspectives. *Advanced Materials, 34*(18), 2109878. https://doi.org/10.1002/adma.202109878.

Zheng, C., Lu, K., Lu, Y., Zhu, S., Yue, Y., Xu, X., Mei, C., Xiao, H., Wu, Q., & Han, J. (2020). A stretchable, self-healing conductive hydrogel based on nanocellulose-supported graphene towards wearable monitoring of human motion. *Carbohydrate Polymers, 250*, 116905. https://doi.org/10.1016/j.carbpol.2020.116905.

Zhou, Y., Li, J., & Xu, X. (2022). Bio-inspired self-healing materials: From molecular design to advanced applications. *Advanced Functional Materials, 32*(27), 2201612. https://doi.org/10.1002/adfm.202201612.

Zhu, M., Liu, J., Gan, L., & Long, M. (2020). Research progress in bio-based self-healing materials. *European Polymer Journal, 129*, 109651. https://doi.org/10.1016/j.eurpolymj.2020.109651.

14 An In-Depth Analysis and Executive Summary of the Progress and Applications of Self-Healing Materials

Ruchira Srivastava, Ayushi Thakur, Lipsa Das, and Sunil Verma

Description	Abbreviation
Antimicrobial peptides	AMPs
Clonal stem cells	CSCs
Compound annual growth rate	CAGR
Deoxyribonucleic acid	DNA
Polyurethane dispersions	PUD
Self-healing polymers	SHPs

14.1 INTRODUCTION

The structure and basic ideas of self-healing materials are covered in the present chapter using a variety of material classes, inclusive of metals, ceramics, polymers, and concrete [1]. The development of material design has been influenced by improvements in our knowledge of nature, materials, and the requirement of intended applicable areas. Materials were only meant to be passive at first, but as our knowledge of their qualities and their design has grown, so too has our capacity to use these attributes to produce materials with pre-set reaction mechanisms [2]. One of the most thrilling and difficult fields of study is materials science. There are very limited number of studies which are focused on creating and utilizing synthetic materials which are more suitable to human production and survival that includes-self-healing materials [3]. More repairs are not possible after the healing agent in a

DOI: 10.1201/9781003516071-14

particular area displaces with the crack initiation in self-healing that further results in implanted capsules' rupture [4]. Even with the dangers to human health, animals, the ocean, and landfill, only a small percentage of polymeric materials are recycled. One possible, chemistry-driven strategy to reduce the production of continuous, petroleum-derived products is the development of polymers that break down after their intended usage [5]. Particular difficulties still exist in the area of storing and harvesting energy related to the generation and making of suitable self-healing ionic and electrically conductive materials. Given the critical importance of these materials' ability to heal, we focus on delivering a focused look at the most recent advancements in self-healing materials and their applicability in energy harvesting and storage systems [6]. Although there have been descriptions of self-healing polymers in the literature since 1996 and there are known instances from the Roman era, their use in medicine has proven difficult [7]. Different kinetics of self-healing processes may be seen in nature; for example, skin wounds might heal in days, fractured bones can recover in weeks, and bleeding can stop in minutes [8]. Ideally, the systems should have many repair cycles, allowing for the "automatic" mending of newly formed fractures or fissures. The materials' ability to self-heal constantly and frequently would make them ideal for usage in the aerospace and automotive industries as well as allow for the construction of more sustainably designed structures [9]. Adding the capacity for energy storage devices to self-heal has emerged as a viable approach to enhancing their performance and longevity, mostly due to the healing properties of nature. This is very broadly elaborated in Figure 14.1 [10].

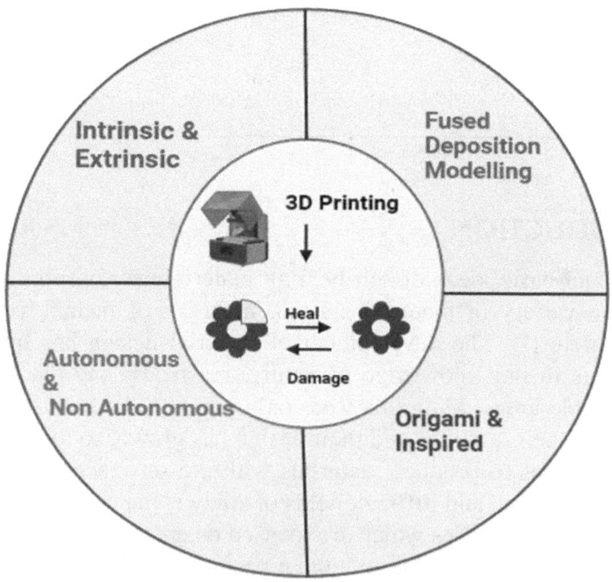

FIGURE 14.1 Mechanisms of self-healing (lab-to-reality). (Credit: Created by authors).

14.2 COMPREHENSIVE OVERVIEW OF THE PROPERTIES OF SELF-HEALING MATERIALS

Materials that can mend damage or resume normal function without external aid are classified as self-healing materials. Some of the fundamental properties of self-healing are shown in Figure 14.2. Because these materials copy the natural healing processes seen in living organisms, they can repair fractures, scratches, and other forms of damage caused by mechanical force, environmental factors, or aging processes [11]. Bionic systems, like skin, have natural mechanisms for repairing damage, which is where the concept of self-healing materials originates. Depending on various domains of interest, the reviewed literature is divided into three basic categories. Some of these areas include the systems and materials employed the experimental methods for assessing the material properties and self-healing capability of the materials or structures, and the characterization methods for assessing the offline and online healing capacity of the suggested systems [12]. The goal of second-generation self-healing materials is to "fully restore" the underlying material structure; first-generation materials aim to "halt" and "fill" damage. This assumption makes it quite possible to differentiate Zeroth generation self-healing materials from conventional composite materials that delay failure [13].

Self-healing materials possess unique properties that enable them to repair damage and go back to normal without external assistance. These characteristics are essential to their efficacy in a broader area of industrial applications. The main characteristics of materials that can mend themselves are outlined below:

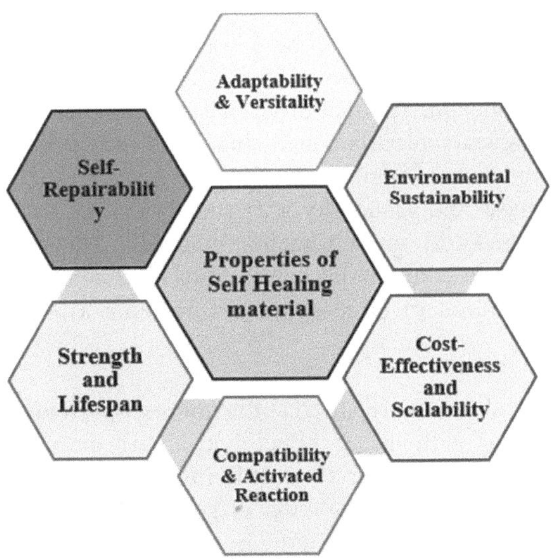

FIGURE 14.2 Basic properties of self-healing materials. (Credit: Created by authors).

- One essential characteristic of self-healing materials is their capacity to self-correct harm. These materials may repair fractures, scratches, and other types of damage brought on by mechanical stress, environmental variables, or aging processes, either by internal mechanisms like polymerization or extrinsic mechanisms like microcapsule-based healing [14].
- It may be possible to engineer self-healing materials that react to particular stimuli or triggers, such as heat, light, moisture, pH shifts, or mechanical stress. By only triggering the repair process when harm occurs, these triggers save energy and resources by activating the healing processes [15].
- Longevity and robustness are characteristics of effective self-healing materials, preserving their mechanical and functional qualities over time. By continually repairing damage, these materials can increase their longevity and reduce the need for routine upkeep or replacement [16].
- Materials with self-healing capabilities have the potential to display an extensive range of characteristics and abilities, making them appropriate for a multitude of uses. Certain performance criteria, such as those for biocompatibility, mechanical strength, flexibility, and thermal stability, can be met by them [17].
- Compatibility with other materials and components in the system or application is a prerequisite for self-healing materials. Whether in vehicle components, electronics, building supplies, or biomedical implants, compatibility guarantees smooth integration and performance in practical settings [18].
- Certain materials with self-repairing capabilities have adaptable qualities that enable them to react quickly to changes in their surroundings or outside stimuli. Because of their versatility, the materials can optimize performance under a variety of operating circumstances by modifying their healing response accordingly [19].
- Through their potential to minimize waste, resource consumption, and the need for regular replacements or repairs, many self-healing materials support environmental sustainability. These materials support the circular economy and sustainable manufacturing practices by increasing product lifespan and reducing material waste [20].
- Cost-effectiveness and scalability are crucial factors to take into account while creating and utilizing self-healing materials. Materials that are more likely to be used in commercial applications across sectors are those that can be manufactured in large quantities utilizing affordable fabrication techniques [21].

Focus on systems that should be repaired at the microscopic and minute level, where damage frequently goes without any notice and allows the material to restore its full mechanical capabilities, rather than utilizing adhesives to achieve a macroscopic repair [22]. Review and calculation of general fundamental material independence principles and processes are done in the context of self-healing materials, taking into account recent significant advancements and upcoming difficulties [23]. Investigating the relationship between material deterioration and material healing helps to understand the kinetics and distortion reversal processes needed to disperse self-healing properties [24]. The final product's characteristics might differ significantly because of the wide range of contents that make up its primary components, cement, water, and different types of aggregates. Therefore, to guarantee that the finished product meets the predetermined specifications, the structural designer must first identify the attributes that are essential for a particular application and then choose the right mixing of the elements of the concrete [25].

14.3 STRATEGIES AND METHODS ADOPTED IN SELF-HEALING MATERIALS

In the last ten years, a lot of attention has been paid to self-healing elastomers having multiple hydrogen bonding due to their intricate chemical compositions, high healing efficacy, quick healing time, and adaptable mechanical characteristics [26]. In the areas of wearable electronics, electronic skins, motion tracking, and health monitoring, self-healing elastomers have the potential to be used by extending their service life and restoring their mechanical qualities quickly. From this angle, we will first discuss the idea of self-healing materials and their categorization. These processes can be broadly classified into two groups: internal self-healing techniques and extrinsic mechanisms [27]. Here are some types of self-healing materials based on their healing mechanisms:

14.3.1 Intrinsic Type Self-Healing

- An increasing number of novel healing systems have developed over time as a result of ongoing scientific and technological advancements as well as the expansion of the field of intrinsic self-healing materials. One such example is the realization of rupture and regeneration based on double or multiple hydrogen bonds.
- Polymerization, for example, reversible cross-linking can repair damage in polymers containing dynamic covalent bonds (such as disulfide bonds and Diels–Alder processes) [28].
- Materials have vascular networks or implanted microchannels that are loaded with therapeutic medicines. When anything is destroyed, the channels break, letting the healing ingredients escape and reacting to fix the harm [29].
- Materials have the ability to return to their original shape after being twisted or damaged [30].

14.3.2 Extrinsic Type Self-Healing

- Agents, including polymers or adhesives, are contained in microcapsules that are scattered over the matrix of materials. The capsules burst when

distortion happens, releasing the restorative chemicals that feed and finish the fractures to repair the material [31].
- Chemical stimuli that stimulate healing systems, such as pH shifts or mechanical stress, may be involved in this [32].
- Chemical markers or integrated sensors that indicate the start of repair processes can be found in autonomous healing materials [33].

14.3.3 HYBRID TYPE SELF-HEALING

- Some materials with self-healing capabilities can be improved in terms of efficacy and versatility by a variety of techniques. For instance, a substance may combine microvascular networks with microcapsules to distribute therapeutic chemicals through various channels [34].
- Hybrid self-healing materials maximize healing efficacy by using the synergistic interactions between several healing processes. These materials have the ability to combine complementary tactics to provide strong and quick damage repair [35].

The development as well as the enhancement of self-healing materials requires the utilization of many methodologies to improve the materials' capacity for healing, their resilience, and their suitability for varied uses. When developing and constructing materials having self-healing characteristics, the following important techniques are used:

14.4 SELECTION OF HEALING MECHANISMS

Researchers determine and include appropriate healing mechanisms relying on the specific requirement of the application and the surrounding circumstances [36]. Polymerization, microencapsulation, shape memory, microvascular networks, and autonomic healing are examples of common processes. Figure 14.3 below clearly depicts some bullet points on criteria of selection of mechanism involved in self-healing.

- When creating materials that can mend themselves, base materials selection is crucial. Researchers use materials such as composites, metals, ceramics, or polymers whose qualities complement the selected healing method [37].
- Self-healing materials include healing agents such as nanoparticles, polymers, adhesives, and resins. These restorative substances are scattered throughout the bulk material or enclosed inside the material matrix. During the healing process, stability, compatibility, and controlled release of healing agents depend on the choice and encapsulation of those substances [38].
- These triggers turn on the healing processes, which save energy and resources by only starting the repair process when harm has already occurred [39].

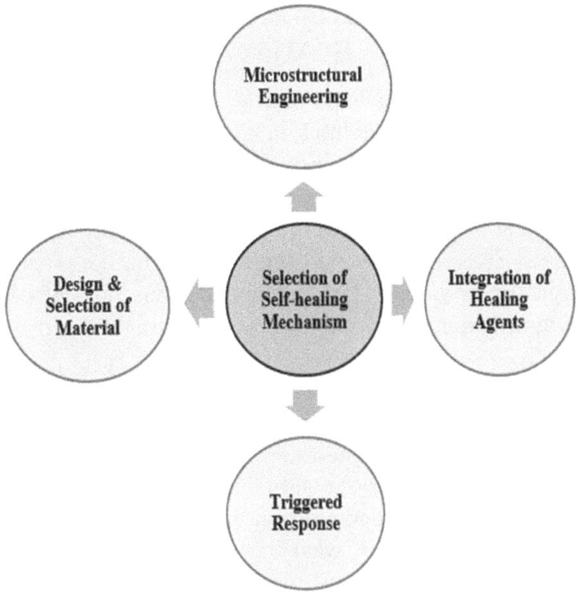

FIGURE 14.3 Selection of mechanism of self-healing (Credit: Created by authors).

- Various research studies have concluded the use of 3D printing and other technologies in the formation of microvascular networks [40].

14.5 CHARACTERIZATION AND EVALUATION

The mechanical characteristics, stability, durability, and healing efficiency of self-healing materials are evaluated by the use of comprehensive characterization methodologies [41].

- Improving manufacturing procedures, increasing output, and taking costs into account are frequently necessary for the enhancement of useful self-healing materials. To get consistent healing outcomes on a large scale, researchers improve material compositions, processing variables, and manufacturing methods [42].
- Materials that have the capability to cure and save themselves must work with current production techniques and be easily incorporated into practical applications. Researchers explore potential applications in the aerospace, automotive, construction, electronics, healthcare, and other industries in collaboration with industry partners, while also addressing specific performance requirements and regulatory standards [43].

14.6 POTENTIAL APPLICABILITY AND FUTURE PROBABILITY OF SELF-HEALING MATERIALS

Researchers have historically innovated new types of materials by imitating nature. Researchers developed a novel method in this technique for creating self-healing polymers (SHPs), which may be applied to address losses caused by damage. Numerous SHPs with diverse attributes have been created for an extensive array of uses. SHPs' applicability and adaptability in fields closely related to our modern and near-future daily lives have opened the door to the flavor of real life. Figure 14.4 makes evident the broader domain of applications for self-healing materials. Some significant applications and potential future paths for self-healing materials are listed below:

- Automobile parts with self-repairing polymers, such as windshields, panels, and bumpers, can last longer and require less maintenance [44].
- The safety and dependability of aviation components can be guaranteed by self-healing composites and coatings, which can stop structural deterioration brought on by collisions, wear and tear, and climatic conditions. In order to minimize downtime and maintenance needs, autonomous repair systems in aircraft structures are able to recognize, find out and fix the damage in real time [45, 46].
- Self-healing concrete extends the life and durability of structures including buildings, bridges, and infrastructure by filling up fractures and halting environmental forces' degradation [47].

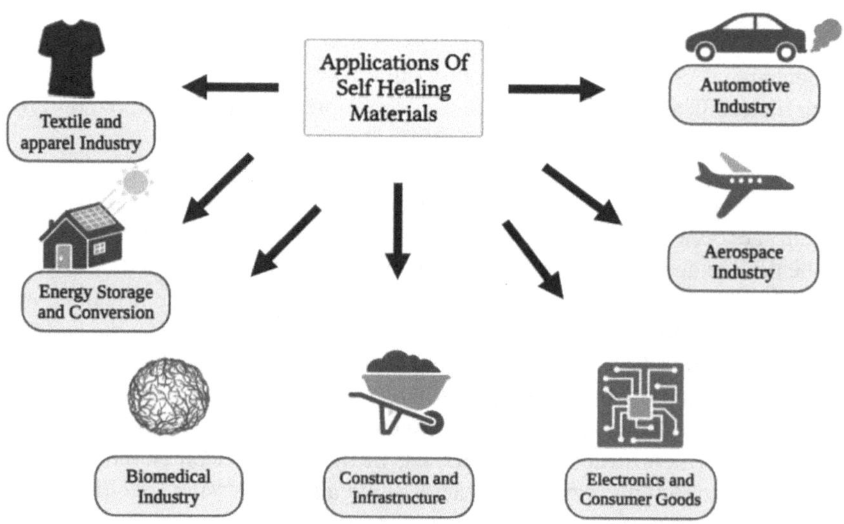

FIGURE 14.4 Potential applicability of self-healing materials. (Credit: Created by authors).

- In order to repair injured tissues and encourage healing, self-healing hydrogels and biomaterials can be used in tissue engineering, drug delivery systems, and regenerative medicine. Medical implants and devices that can mend themselves can increase biocompatibility, lower the chance of problems, and increase the lifespan of the implant [48].
- By stopping leaks and deterioration, self-repairing coatings and membranes in fuel cells and solar panels may preserve performance and efficiency [49].
- Antibacterial fabrics and self-repairing fibers can stop microorganisms from growing and increase the life of products [50].

14.7 FUTURE PROSPECTS

As seen in Figure 14.5, future developments in self-healing materials will probably pay attention to biomimetic materials, especially those with intricate molecular aggregates made of amino acid sequences. When these supramolecular protein-like structures interact in big groups, they will be able to reply back to the outside stimuli in a way that lessens the damage and is capable of reassembling itself into its former form [51]. In the future, self-healing materials have the full capability to completely change the various phases and structure of prevailing industries. The following significant developments and trends are influencing the direction of self-healing materials in the future:

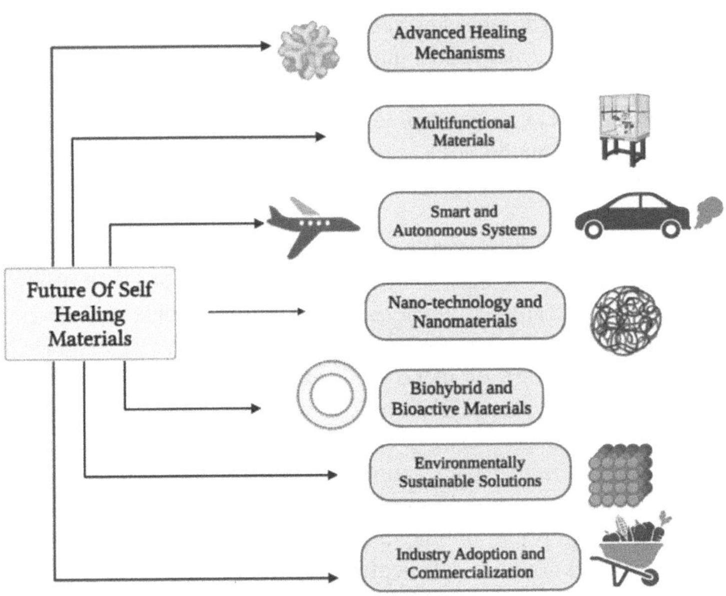

FIGURE 14.5 Future prospects of self-healing materials. (Credit: Created by authors).

- This entails looking at ways to create more dynamic and robust self-repairing materials that are impacted by biological processes, such as how living things heal wounds [52].
- These multifunctional materials open doors to new applications in a wide range of fields as electronics, biomedical and photonics where integrated and advanced functionalities are becoming more desirable [53].
- These self-diagnostic techniques will be capable of identifying deterioration and damage, and will also calculate repairing needs and requirements, and will automatically start the healing process in real time without any requirement for human interference. This will transform maintenance procedures in sectors including infrastructure, automotive, and aerospace [54].
- The procedure and methods involved in self-repairing materials can be improved in ways never before possible because of nanotechnology, which allows for exact control over material characteristics at the nanoscale. Tailored methods for self-healing will be made possible by nanostructured materials, nanoparticles, and nanocomposites, which will also provide greater durability, compatibility with current materials and production processes, and tighter control over healing processes [55].
- Synthetic matrixes can be enhanced with bioactivity and environmental cue responsiveness by adding biological molecules, such as proteins, enzymes, or DNA, which can enable self-repair in biological and biomedical applications [56].
- Two examples of sustainable fabrication methods that will lessen their impact on the environment and promote and publicize the development of circular economy principles in materials production and consumption are green chemistry and additive manufacturing [57].
- Collaborations between government, business, and academia will speed up the transfer of technology, standardize processes, and get regulatory approval [58].

Global issues in sustainability, resilience, and usefulness are what will define the self-healing materials of the future, along with innovation and multidisciplinary cooperation. Self-regenerative and defensive strategies are observed in the realm of living organisms. Self-healing or self-repairing polymeric materials have been suggested as a way to maintain or mend physically damaged or cut samples. These self-healing polymers are a special family of smart materials that can fix internal defects, fractures, or damage caused in any matrix, much like human skin heals from a wound. They can also restore the fractured part's mechanical properties (such as its tensile strength) through an autonomic healing process [59]. Within the domain of living things, tissue and organ maintenance and repair are guaranteed by the basic biological mechanism known as self-healing. Different defensive and self-regenerative mechanisms have developed in organisms to deal with harm and preserve equilibrium [60]. Here are some examples of self-healing mechanisms observed in living organisms:

The fauna, or animals, are investigated. Table 14.1 explains our organizational approach to defining healing in such a huge and diverse set of data. As the first column of Table 14.1 shows, the main groups of creatures, or fauna, will be further separated into subsections. Animals categorized as invertebrates—those lacking a spinal cord or skeleton—and vertebrates—those having a spinal cord—will be referred to as fauna. Moreover, soft tissue wounds and hard tissue wounds (such as those affecting the bone or exoskeleton) will be the two divisions into which the healing process will be divided for both vertebrates and invertebrates [61]. Under the heading "Mechanism(s) in Living Nature," Table 14.1 deconstructs the restorative and protective responses into fundamental mechanisms and processes that are inherent to living things.

TABLE 14.1
Self–Protection Mechanism and Self-Healing Existing in Living Beings (Vertebrates and Invertebrates)

Part	Physical Transformation	Mechanisms In The Natural World	Description Based On Response To Healing And Defense	Ref.
Vertebrates and Invertebrates	Relaxation and contraction of muscles	Actin and myosin, two proteins, interact with each other to slide against one another reversibly.	Reversible tensing (contraction)and relaxing.	[62]
	innate defense mechanism	Barriers, phagocytes, cytokines, (such as mucus or tears).	Physical defenses against dangerous microorganisms and biological reactions.	[63]
	Injury of CNS (central nervous system)	Cellular response: restricted axon regeneration after axonal breakage, glial scarring, inflammation, and blood–brain barrier re-establishment.	Debris removal and quick system stabilization to reduce damage.	[64]
	Injury of PNS (peripheral nervous system)	The biological reaction includes inflammation, the removal of debris by macrophages and Schwann cells, and the joining of the proximal and distal nerve segments via axonal growth.	Removal of waste that leaves less scarring and promotes cell growth.	[65]

(Continued)

TABLE 14.1 (CONTINUED)
Self–Protection Mechanism and Self-Healing Existing in Living Beings (Vertebrates and Invertebrates)

Part	Physical Transformation	Mechanisms In The Natural World	Description Based On Response To Healing And Defense	Ref.
Hard Tissues (Vertebrate)	Breakage of Bone	Cellular response: lamellar bone growth and remodeling, inflammation, and callus development.	Cellular reaction to clear the wound of debris, stabilize the damage, and restructure it (heal).	[66]
Soft Tissue (Vertebrate)	Wound	Response of the cell: coagulation, inflammation, multiplication, and remodeling.	Cellular signaling: transition from transient coagulation to sustained cell proliferation and remodeling.	[67]
	Response to stem cell	To create new tissue, stem cells may divide and develop into any form of cell.	Modified healing through the use of new tissue.	[68]
Hard Tissue (Invertebrate)	Exoskeleton growth	Replacement of the shell as the organism out-grows its existing exoskeleton.	Swapping out the external protective exoskeleton with fresh development on a periodic basis.	[69]
	Wound	Secretion of endoculticles from the cellular epidermis.	Exoskeleton grows continuously from the inside, having a hard outer layer shielding a soft interior layer.	[70]
Soft Tissue (Invertebrate)	Wound	Clotting via plasma protein reactions, blood cell.	Cell division, coagulation between hemocytes and plasma, and clotting across a dense network of hemocytes.	[71]

14.7.1 Cellular Repair Mechanisms

It is required as well as recommended to increase the rate of self-healing material restoration. Although self-healing materials often require several hours to mend, this is very obvious not to change in the near future given the desire for faster self-healing. The study of self-healing material mechanisms should focus on improving the repair speed in order to fulfill the demands of practical application.

Repairing of DNA

In general, cells are delicate and prone to injury. On the other hand, many are capable of healing themselves after being ripped, torn, or even split in two when they sustain damage from an injury or illness, or from regular physiology's everyday wear and tear [72].

Repairing of Protein

Cellular regeneration is demonstrated in a few unicellular model systems. There is a benefit in researching free-living single-celled organisms in that their natural activity may be observed without the effect from any nearby cells as would be the case for cells inside a tissue. This is crucial because, in the event that a cell regenerates inside a tissue context, it is impossible to determine if the cell will produce the required spatial information or if surrounding cells will [73].

14.7.2 Regenerative Processes

Tissue Regeneration

Upon the healing of a wound, the cell must reconstruct any lost or damaged structures. A population of dividing cells may not require an active regeneration mechanism, since during normal cell division they will be creating new structures as they proliferate. Without an active regeneration pathway, cellular structures would continue to grow in number and size until the cell divides, at which point they would be divided between the new cells [74].

Stem Cell Activation

Stem cells are the main topic of discussion when comparing the feasibility of 100% healing versus restoring damaged tissue. On a smaller scale, a process known as physiological regeneration causes blood, skin, and hair cells to all grow back into identical copies of the lost tissue. Just the recent usage of stem cells has shown promise for smaller-scale results in humans' long-term studies of regeneration, which try to replicate the influence of regenerative capacities in salamanders on bigger areas of the body [75].

14.7.3 Immune Response

Inflammatory Response

Materials now in use return to the common self-healing processes seen in the natural environment. But one may counter that there is a potential method for creating materials that can mend themselves for any combination of common processes seen in nature. As an illustration, combining emulsions that make use of micro- and nanospheres with capsules that have functional surfaces might present prospects. In a similar vein, shape memory and muscular control coupled might make use of UV-active smart surfaces that alter through layering form in relation to the Sun on a regular basis to create inflow and effusion of substance required to create layers [76].

Wound Healing

Wound healing is the individual body's very normal physiological reaction to an injury to the skin. However, wound healing is a difficult procedure that requires the cooperation of several cell types and other mediators in order to restore skin integrity. Since hydrogels are made of 3D networks with high water content and flexible rheological characteristics, they are attractive dosage forms for biomedical regenerative therapy. Self-healing hydrogels are especially intriguing for wound care since they may independently reestablish structural damage and regain original functions. Self-healing hydrogels have garnered more attention lately as potential biological materials [77].

14.7.4 Antimicrobial Defense

Antimicrobial Peptides

Novel approaches to tackle the problem of drug-resistant bacteria have been developed in response to their growth. Antimicrobial peptides (AMPs), a class of peptides having antimicrobial activity against microbial infections, are one of the alternative strategies that are garnering more attention [78].

Biofilm Formation

Many living things make biofilms to guard themselves against environmental tragedies. The biofilms help them survive in harsh conditions [79].

14.7.5 Quiescence and Dormancy

Cellular Quiescence

When circumstances improve, quiescent cells can re-enter the cell cycle to multiply and repair tissue because they are more resilient to harm [80].

Dormancy

Dormancy, which is frequently characterized as quiescence on a cellular level and in the context of researching human physiology and illness, is a particularly popular subject in relation to CSCs, or clonal stem cells. In this instance, the criteria of cellular quiescence are established more explicitly by looking at the conditions of the individual cells rather than by inferring them from the ecological state of the organism [81]. Quiescence must be actively maintained in the cell as it is a specifically transient and reversible condition. Thus, in response to different components of signaling, quiescent cells can be awakened to re-enter the cell cycle [82]. Transparency, flexibility, and conductivity—three of the physical attributes of self-healing materials—are not flawless, and the majority of them are still confined to theoretical study domains where they have little practical use.

14.7.6 Global Market of Self-Healing Material and Its Future Trends

The global market for self-healing materials is expected to reach USD 1.61 billion in 2022 and increase at a CAGR (compound annual growth rate) of 25.5% from 2022 to 2032.

- In developing countries, there is growing construction activity in both public and private infrastructure. For instance, these nations' government programs facilitate simple access to raw materials and aid in home manufacturing operations [82].
- Through the application of advanced expertise that supports the expansion of the product portfolio, untapped markets can be captured to increase the customer base, which is anticipated to expand the size of the market for self-healing materials.
- Furthermore, it is anticipated that increased cooperation and early information transfer will accelerate the expansion of the worldwide market [83].
- Self-healing materials depend on polymers to prolong the life of numerous items while also enhancing their functionality. The demand for materials (self-healing) is likely to increase as a result of its ability to replicate biological healing and mend itself when damaged without the need for outside assistance or detection [84].
- Despite their great demand worldwide, these materials' capacity to mend itself is restricted by design. These are significant obstacles to market share [85].

14.8 Self-Healing Opportunities Prevailing in Electronics and Automotive Industry

The automobile and electronics sectors place a high value on lifespan and dependability, hence self-healing materials find extensive use in these domains. Applications include anti-scratch polyurethane coatings for automobile bodies made of self-healing elastomers [86]. Prominent companies in the self-healing material market serve the automobile industry with their state-of-the-art "self-seal technology," which can seal most punctures up to one-quarter of an inch (6 mm) and is perfect for things like car wheel screws and nails. Because it employs a natural rubber sealant compound that preserves all other tire performance without affecting rolling resistance, it remains an ecologically friendly option. The sealant can also be permanently placed into the tire to encapsulate the object and fill the hole [87]. Preventing air leaks and maintaining pressure, will eventually increase sales of self-healing materials. On the other hand, a range of coatings such as encapsulants, silicone gels, and hard, functional, and conformal coatings can protect different types of components against unfavorable environmental factors like shock, vibration, moisture, and dirt [88–92]. These gels combine the self-healing and stress-relieving qualities of a liquid with the dimensional stability of an elastomer [93]. Thanks to manufacturer innovation, self-healing wood finishes now have a softer feel. For instance, the performance of aqueous, environmentally friendly 2k polyurethane coatings has been significantly enhanced with the advent of the new generation of entirely waterborne OH-functional polyurethane dispersions (PUD) that cure with low shrinkage and don't require solvents or cure by-products [94, 95]. These are but a handful of the countless uses for the market share of materials that can mend themselves. The features of this material, such as its weightlessness, improved chemical and physical stability, and processability, are driving the global market for self-healing materials [96–98]. Due to the virus, nearly every nation in the globe is presently under a state of emergency, and all of these industries are closed. The market for electronic devices and materials with self-healing characteristics has therefore reduced [99].

Furthermore, the COVID-19 pandemic has put a halt to new construction projects and the production of automobiles [100]. This has significantly hampered the adoption of self-healing materials, and it will probably take another year for losses to be made up [101]. The ability of materials to repair mechanical friction-induced deformation and the increasing demand for self-healing materials, or polymers, from various end-use sectors reinforce the market's potential, growth, and major trends [102].

Thus, the expanding application of these materials in the building and construction, automotive, electronics, and electrical sectors supports the global trends in the adoption of self-healing materials [103]. Currently, the majority of companies in the semiconductor and electronics industries employ self-healing materials in laptops, desktop computers, mobile phones, and high-end cameras. It is anticipated that the need for materials (self-healing) will rise due to the steady increase in sales of laptops and smartphones as well as design developments [104]. The primary issue preventing the market's expansion and future trends is the high cost of products since self-healing materials are now more expensive than standard materials [105]. It is anticipated that the cost factor will measure future demand for materials with self-healing properties and encourage the use and sales of materials with self-healing properties, provided that it is under control in the next few years.

The utilization and application of self-healing materials are common in biomimicry, electrical devices, construction materials, vehicles, aircraft, military gear, and biological and medical fields. With every year that goes by, the percentage of applications grows as well. For instance, self-healing materials are being employed in the medical sector of dentistry [106]. Its exceptional capacity for self-healing allows for the preservation of the integrity of the body's structure as well as the extension of material life. The importance and position of self-healing materials in medical hydrogels is noteworthy. Certain self-healing hydrogels possess temperature and pH responses, making them highly promising for use in biological science and biotechnology-related fields. Additionally, some self-healing hydrogels can be cultivated to cultivate neural stem cells precursor cells [107].

14.9 CONCLUSION

It is probable that future developments in self-healing research will be influenced by biomimetic materials, namely those containing intricate molecular remains formed from amino acid sequences. When these supramolecular resembling protein structures interact with one another in big groups, they will be able to limit damage and self-assemble back into their original shape in response to external stress.

This research inquires about the self-healing properties of materials from mechanical and thermally induced damage, which is important for the subject of fatigue and fracture. Examined are the development and history of various self-repair systems, such as nano-beam healing components, autonomic self-healing, and passive self-healing, and ballistic self-repair.

The self-healing processes found in these unique materials are heavily inspired by the related fields of crack healing and polymer–polymer interactions. Materials with the property of self-healing have the capability to mend any destruction or deformation, such as fracture development, either fully or partially, and it is expected that this would return the material to its initial functioning. The design and general concepts of self-healing materials are covered in this chapter using a variety of material classifications, like metals, ceramics, concrete and polymers. A summary of significant recent advancements and upcoming obstacles in the domain of self-healing materials is provided, together with a discussion of general, basic, material-independent concepts and mechanisms. For example, it ought to be possible to work with new and latest materials comprised of folded proteins, much like in the area and domain of self-healing nanobeams. This would act as a mechanism of partial absorption of mechanical energy, fold back, and heal itself. It is reasonable to infer that the hunt for materials with original and actual life is similar to the researchers' own. Motivated by this fitness-boosting property, fine-tuned over billions of years of evolution, researchers and technologists have begun adding self-healing properties to artificial materials. New materials with prolonged durability that can repair mechanical integrity and extra functionality after damage have been produced by imitating mechanically triggered chemistry and the storage and transport of liquid reagents.

The public has to be made aware of the usage of self-healing materials, which have not yet been invented or produced. In order for the results to be widely applicable in practice, researchers must continue to concentrate on finding solutions to the issues raised above. In particular, they must concentrate on creating materials with excellent mechanical properties, a quick rate of healing, and a high healing efficiency by choosing inexpensive monomers, straightforward synthesis procedures, and mild reaction conditions.

REFERENCES

1. Hager, M. D., Greil, P., Leyens, C., Van Der Zwaag, S., & Schubert, U. S. (2010). Self-healing materials. *Advanced Materials, 22*(47), 5424–5430.
2. Guimard, N. K., Oehlenschlaeger, K. K., Zhou, J., Hilf, S., Schmidt, F. G., & Barner-Kowollik, C. (2012). Current trends in the field of self-healing materials. *Macromolecular Chemistry and Physics, 213*(2), 131–143.
3. Song, T., Jiang, B., Li, Y., Ji, Z., Zhou, H., Jiang, D.,... Colorado, H. (2021). Self-healing materials: A review of recent developments. *ES Materials & Manufacturing, 14*, 1–19.
4. Toohey, K. S., Sottos, N. R., Lewis, J. A., Moore, J. S., & White, S. R. (2007). Self-healing materials with microvascular networks. *Nature Materials, 6*(8), 581–585.
5. Sumerlin, B. S. (2018). Next-generation self-healing materials. *Science, 362*(6411), 150–151.
6. Chen, D., Wang, D., Yang, Y., Huang, Q., Zhu, S., & Zheng, Z. (2017). Self-healing materials for next-generation energy harvesting and storage devices. *Advanced Energy Materials, 7*(23), 1700890.
7. Menikheim, S. D., & Lavik, E. B. (2020). Self-healing biomaterials: The next generation is nano. *Wiley Interdisciplinary Reviews: Nanomedicine and Nanobiotechnology, 12*(6), e1641.
8. Narayan, R., Laberty-Robert, C., Pelta, J., Tarascon, J. M., & Dominko, R. (2022). Self-healing: An emerging technology for next-generation smart batteries. *Advanced Energy Materials, 12*(17), 2102652.
9. Balazs, A. C. (2007). Modeling self-healing materials. *Materials Today, 10*(9), 18–23.
10. Wan, X., Mu, T., & Yin, G. (2023). Intrinsic self-healing chemistry for next-generation flexible energy storage devices. *Nano-Micro Letters, 15*(1), 99.
11. Wool, R. P. (2008). Self-healing materials: A review. *Soft Matter, 4*(3), 400–418.
12. Bekas, D. G., Tsirka, K., Baltzis, D., & Paipetis, A. S. (2016). Self-healing materials: A review of advances in materials, evaluation, characterization and monitoring techniques. *Composites Part B: Engineering, 87*, 92–119.
13. Brochu, A. B., Craig, S. L., & Reichert, W. M. (2011). Self-healing biomaterials. *Journal of Biomedical Materials Research Part A, 96*(2), 492–506.
14. Khatib, M., Zohar, O., Saliba, W., & Haick, H. (2020). A multifunctional electronic skin empowered with damage mapping and autonomic acceleration of self-healing in designated locations. *Advanced Materials, 32*(17), 2000246.
15. Yang, Y., Wang, H., Zhang, S., Wei, Y., He, X., Wang, J.,... Ji, Y. (2021). Vitrimer-based soft actuators with multiple responsiveness and self-healing ability triggered by multiple stimuli. *Matter, 4*(10), 3354–3365.
16. Fu, X., Tian, L., Fan, Y., Ye, W., Qiao, Z. A., Zhao, J.,... Ming, W. (2021). Stimuli-responsive self-healing anticorrosion coatings: From single triggering behavior to synergetic multiple protections. *Materials Today Chemistry, 22*, 100575.
17. Das, R., Melchior, C., & Karumbaiah, K. M. (2016). Self-healing composites for aerospace applications. In *Advanced composite materials for aerospace engineering* (pp. 333–364). Woodhead Publishing.
18. Yedida, V. V. S., & Vasudev, H. (2022). A review on the development of thermal barrier coatings by using thermal spray techniques. *Materials Today Proceedings*, 50, 1458–1464.
19. Oehlenschlaeger, K. K., Mueller, J. O., Brandt, J., Hilf, S., Lederer, A., Wilhelm, M.,... Barner-Kowollik, C. (2014). Adaptable hetero Diels–Alder networks for fast self-healing under mild conditions. *Advanced Materials, 26*(21), 3561–3566.

20. Xiong, C., Yang, Q., Dang, W., Zhou, Q., Jiang, X., Sun, X.,... Ni, Y. (2023). A multi-functional paper-based supercapacitor with excellent temperature adaptability, plasticity, tensile strength, self-healing, and high thermoelectric effects. *Journal of Materials Chemistry A*, *11*(9), 4769–4779.
21. Kashyap, S. S., Borah, K., Shivam Shailesh Kumar, J., Sarmah, S., Bera, M. K., Basak, P., & Narayan, R. (2024). Fully biobased self-healing and recyclable covalent adaptable networks prepared via a catalyst-free Aza-Michael reaction. *ACS Applied Polymer Materials*, *161*, 100882.
22. Wang, S., Li, L., & Urban, M. W. (2022). Combined reprocessability and self-healing in fluorinated acrylic-based covalent adaptable networks (CANs). *ACS Applied Polymer Materials*, *4*(12), 9360–9367.
23. Peng, K., Zhang, J., Yang, J., Lin, L., Gan, Q., Yang, Z.,... Feng, C. (2022). Green conductive hydrogel electrolyte with self-healing ability and temperature adaptability for flexible supercapacitors. *ACS Applied Materials & Interfaces*, *14*(34), 39404–39419.
24. Fan, C. J., Wen, Z. B., Xu, Z. Y., Xiao, Y., Wu, D., Yang, K. K., & Wang, Y. Z. (2020). Adaptable strategy to fabricate self-healable and reprocessable poly (thiourethane-urethane) elastomers via reversible thiol–isocyanate click chemistry. *Macromolecules*, *53*(11), 4284–4293.
25. Dahlke, J., Zechel, S., Hager, M. D., & Schubert, U. S. (2018). How to design a self-healing polymer: General concepts of dynamic covalent bonds and their application for intrinsic healable materials. *Advanced Materials Interfaces*, *5*(17), 1800051.
26. Priyadarsini, M., Sahoo, D. R., & Biswal, T. (2021). A new generation self-healing composite materials. *Materials Today: Proceedings*, *47*, 1229–1233.
27. Mezzomo, L., Ferrara, C., Brugnetti, G., Callegari, D., Quartarone, E., Mustarelli, P., & Ruffo, R. (2020). Exploiting self-healing in lithium batteries: Strategies for next-generation energy storage devices. *Advanced Energy Materials*, *10*(46), 2002815.
28. Wang, S., & Urban, M. W. (2020). Self-healing polymers. *Nature Reviews Materials*, *5*(8), 562–583.
29. Hansen, C. J. (2022). Microvascular-based self-healing materials. In *Recent advances in smart self-healing polymers and composites* (pp. 141–176). Woodhead Publishing.
30. Saeedi, A., & Shokrieh, M. M. (2019). A novel self-healing composite made of thermally reversible polymer and shape memory alloy reinforcement. *Journal of Intelligent Material Systems and Structures*, *30*(10), 1585–1593.
31. Luo, X., & Mather, P. T. (2013). Shape memory assisted self-healing coating. *ACS Macro Letters*, *2*(2), 152–156.
32. Hornat, C. C., & Urban, M. W. (2020). Shape memory effects in self-healing polymers. *Progress in Polymer Science*, *102*, 101208.
33. Mashkoor, F., Lee, S. J., Yi, H., Noh, S. M., & Jeong, C. (2022). Self-healing materials for electronics applications. *International Journal of Molecular Sciences*, *23*(2), 622.
34. Wang, X., Zhu, J., Zou, F., Zhou, N., Li, Y., & Lei, W. (2022). Ca-Al LDH hybrid self-healing microcapsules for corrosion protection. *Chemical Engineering Journal*, *447*, 137125.
35. Dai, Y., Xiang, Y., & Zhang, G. (2009, December). Self-healing and hybrid diagnosis in cloud computing. In *IEEE international conference on cloud computing* (pp. 45–56). Berlin, Heidelberg: Springer Berlin Heidelberg.
36. Tabaković, A., & Schlangen, E. (2016). Self-healing technology for asphalt pavements. *Self-healing Materials*, 285–306.
37. Dutta, V., Thakur, L., Singh, B., & Vasudev, H. (2022). A study of Erosion–Corrosion behaviour of friction stir-processed chromium-reinforced NiAl bronze composite. *Materials*, *15*, 5401.

38. Reddy, K. R., El-Zein, A., Airey, D. W., Alonso-Marroquin, F., Schubel, P., & Manalo, A. (2020). Self-healing polymers: Synthesis methods and applications. *Nano-Structures & Nano-Objects*, *23*, 100500.
39. Syrett, J. A., Becer, C. R., & Haddleton, D. M. (2010). Self-healing and self-mendable polymers. *Polymer Chemistry*, *1*(7), 978–987.
40. Prashar, G., & Vasudev, H. (2022). A review on the influence of process parameters and heat treatment on the corrosion performance of Ni-based thermal spray coatings. *Surface Reviews and Letters, 29*, 2230001.
41. Keller, M. W., & Sottos, N. R. (2006). Mechanical properties of microcapsules used in a self-healing polymer. *Experimental Mechanics*, *46*, 725–733.
42. Kanu, N. J., Gupta, E., Vates, U. K., & Singh, G. K. (2019). Self-healing composites: A state-of-the-art review. *Composites Part A: Applied Science and Manufacturing*, *121*, 474–486.
43. Thakur, V. K., & Kessler, M. R. (2015). Self-healing polymer nanocomposite materials: A review. *Polymer*, *69*, 369–383.
44. Wu, M., Johannesson, B., & Geiker, M. (2012). A review: Self-healing in cementitious materials and engineered cementitious composite as a self-healing material. *Construction and Building Materials*, *28*(1), 571–583.
45. Majji, B. G. R., Vasudev, H., & Bansal, A. (2022). A review on the oxidation and wear behavior of the thermally sprayed high-entropy alloys. *Materials Today Proceedings, 50*, 1447–1451.
46. Sunitha, K., & Vasudev, H. (2022). A short note on the various thermal spray coating processes and effect of post-treatment on Ni-based coatings. *Materials Today Proceedings, 50*, 1452–1457.
47. Jayabalakrishnan, D., Muruga, D. N., Bhaskar, K., Pavan, P., Balaji, K., Rajakumar, P. S.,... Prabhahar, M. (2021). Self-Healing materials–A review. *Materials Today: Proceedings*, *45*, 7195–7199.
48. Yang, Y., Ding, X., & Urban, M. W. (2015). Chemical and physical aspects of self-healing materials. *Progress in Polymer Science*, *49*, 34–59.
49. Li, W., Dong, B., Yang, Z., Xu, J., Chen, Q., Li, H., & Jiang, Z. (2018). Recent advances in intrinsic self-healing cementitious materials. *Advanced Materials*, *30*(17), 1705679.
50. Seifan, M., Samani, A. K., & Berenjian, A. (2016). Bioconcrete: Next generation of self-healing concrete. *Applied Microbiology and Biotechnology*, *100*, 2591–2602.
51. Zhou, H., Niu, H., Wang, H., & Lin, T. (2022). Self-healing superwetting surfaces, their fabrications, and properties. *Chemical Reviews*, *123*(2), 663–700.
52. Dashofy, E. M., Van der Hoek, A., & Taylor, R. N. (2002, November). Towards architecture-based self-healing systems. In *Proceedings of the first workshop on self-healing systems* (pp. 21–26).
53. Cho, S. H., White, S. R., & Braun, P. V. (2009). Self-healing polymer coatings. *Advanced Materials*, *21*(6), 645–649.
54. Zhang, A., Liu, Y., Qin, D., Sun, M., Wang, T., & Chen, X. (2020). Research status of self-healing hydrogel for wound management: A review. *International Journal of Biological Macromolecules*, *164*, 2108–2123.
55. Roy, N., Bruchmann, B., & Lehn, J. M. (2015). DYNAMERS: Dynamic polymers as self-healing materials. *Chemical Society Reviews*, *44*(11), 3786–3807.
56. Mauldin, T. C., & Kessler, M. R. (2010). Self-healing polymers and composites. *International Materials Reviews*, *55*(6), 317–346.
57. Yuan, Y. C., Yin, T., Rong, M. Z., & Zhang, M. Q. (2008). Self-healing in polymers and polymer composites. Concepts, realization and outlook: A review. *Express Polymer Letters*, *2*(4), 238–250.

58. Jefferson, T., Javierre, E., Freeman, B., Zaoui, A., Koenders, E., & Ferrara, L. (2018). Research progress on numerical models for self-healing cementitious materials. *Advanced Materials Interfaces*, *5*(17), 1701378.
59. Reddy, K. R., El-Zein, A., Airey, D. W., Alonso-Marroquin, F., Schubel, P., & Manalo, A. (2020). Self-healing polymers: Synthesis methods and applications. *Nano-Structures & Nano-Objects*, *23*, 100500.
60. Aïssa, B., Therriault, D., Haddad, E., & Jamroz, W. (2012). Self-healing materials systems: Overview of major approaches and recent developed technologies. *Advances in Materials Science and Engineering*, *2012*. doi:10.1155/2012/854203.
61. Gillott, C. (2005). Insects and humans. *Entomology*, 725–782.
62. Portzehl, H., Caldwell, P. C., & Ru, J. C. (1964). The dependence of contraction and relaxation of muscle fibres from the crab Maia squinado on the internal concentration of free calcium ions. *Biochimica et Biophysica Acta (BBA)-Specialized Section on Biophysical Subjects*, *79*(3), 581–591.
63. Rinkevich, B., & Müller, W. E. (Eds.). (2012). *Invertebrate immunology* (Vol. 15). Springer Science & Business Media.
64. Gillott, C. (2005). *Entomology*. Springer Science & Business Media.
65. Reichert, W. M. (2007). *Indwelling neural implants: Strategies for contending with the in vivo environment*. CRC Press.
66. Singer, A. J., & Clark, R. A. (1999). Cutaneous wound healing. *New England Journal of Medicine*, *341*(10), 738–746.
67. Caplan, A. I. (1991). Mesenchymal stem cells. *Journal of Orthopaedic Research*, *9*(5), 641–650.
68. Parle, E., Dirks, J. H., & Taylor, D. (2016). Bridging the gap: Wound healing in insects restores mechanical strength by targeted cuticle deposition. *Journal of the Royal Society Interface*, *13*(117), 20150984.
69. Ratcliffe, N. A., & Rowley, A. F. (1979). A comparative synopsis of the structure and function of the blood cells of insects and other invertebrates. *Developmental & Comparative Immunology*, *3*, 189–221.
70. Hartmann, H. T., & Kester, D. E. (2002). *Plant Propagation: Principles and Practices*. Prentice Hall.
71. Hartmann, H. T., & Kester, D. E. (1975). *Plant propagation: Principles and practices*. Prentice-Hall.
72. Ghosh, D., Sharman, R., Rao, H. R., & Upadhyaya, S. (2007). Self-healing systems—survey and synthesis. *Decision Support Systems*, *42*(4), 2164–2185.
73. Wu, D. Y., Meure, S., & Solomon, D. (2008). Self-healing polymeric materials: A review of recent developments. *Progress in Polymer Science*, *33*(5), 479–522.
74. Madara, S. R., Sarath Raj, N. S., & Selvan, C. P. (2018, April). Review of research and developments in self healing composite materials. In *IOP conference series: Materials science and engineering* (Vol. 346, p. 012011). IOP Publishing.
75. Asghar, A., Farooq, H., & Imran, A. (2018). Self-healing in emerging cellular networks: Review, challenges, and research directions. *IEEE Communications Surveys & Tutorials*, *20*(3), 1682–1709.
76. De Belie, N., Gruyaert, E., Al-Tabbaa, A., Antonaci, P., Baera, C., Bajare, D.,... Jonkers, H. M. (2018). A review of self-healing concrete for damage management of structures. *Advanced Materials Interfaces*, *5*(17), 1800074.
77. Li, V. C., & Yang, E. H. (2007). Self-healing in concrete materials. In *Self healing materials: An alternative approach to 20 centuries of materials science* (pp. 161–193). Dordrecht: Springer Netherlands.
78. Taylor, D. L., & in het Panhuis, M. (2016). Self-healing hydrogels. *Advanced Materials*, *28*(41), 9060–9093.

79. Kessler, M. R., Sottos, N. R., & White, S. R. (2003). Self-healing structural composite materials. *Composites Part A: Applied Science and Manufacturing, 34*(8), 743–753.
80. Diesendruck, C. E., Sottos, N. R., Moore, J. S., & White, S. R. (2015). Biomimetic self-healing. *Angewandte Chemie International Edition, 54*(36), 10428–10447.
81. Muhammad, N. Z., Shafaghat, A., Keyvanfar, A., Majid, M. Z. A., Ghoshal, S. K., Yasouj, S. E. M.,... McCaffer, R. (2016). Tests and methods of evaluating the self-healing efficiency of concrete: A review. *Construction and Building Materials, 112*, 1123–1132.
82. Luhar, S., & Gourav, S. (2015). A review paper on self healing concrete. *Journal of Civil Engineering Research, 5*(3), 53–58.
83. Srivastava, R., & Srivastava, V. K. (2020). Changing scenario of manufacturing industries due to covid 19. *International Journal for Modern Trends in Science and Technology, 6*(10), 143–146.
84. Bahadur, P. S., Jaiswal, S., Srivastava, R., & Kumar, A. (2021, November). Advanced application of nanotechnology in engineering. In *2021 international conference on technological advancements and innovations (ICTAI)* (pp. 92–95). IEEE.
85. Srivastava, R., Thakur, A., & Rana, A. (2024). The basic understanding of nano materials: A review. In A. Thakur & A. Kumar (Eds.), *Innovations in nanomaterials-based corrosion inhibitors* (pp. 1–20). IGI Global. https://doi.org/10.4018/979-8-3693-3088-3.ch001.
86. Srivastava, R., Arun, J., & Patel, D. K. (2019). Amalgamating the service quality aspect in supply chain management. In *2019 international conference on automation, computational and technology management (ICACTM)* (pp. 63–67). London. https://doi.org/10.1109/ICACTM.2019.8776839.
87. Thakur, A., Vates, U. K., & Mishra, S. (2024). Parametric optimization of 3D-printed PLA part using response surface methodology for mechanical properties. *Smart Science, 12*(1), 91–104.
88. Singh, J., Singh, J. P., Kumar, S., & Gill, H. S. (2023). Short review on hydroxyapatite powder coating for SS 316L. *Journal of Electrochemical Science and Engineering, 13*, 25–39.
89. Singh, J., Kumar, S., & Mohapatra, S. K. (2021). Study on solid particle erosion of pump materials by fly ash slurry using Taguchi's orthogonal array. *Finnish Journal of Tribology, 38*, 31–38.
90. Kumar, P., Singh, J., & Singh, S. (2022). Neural network supported flow characteristics analysis of heavy sour crude oil emulsified by ecofriendly bio-surfactant utilized as a replacement of sweet crude oil. *Chemical Engineering Journal Advances, 11*, 100342.
91. Singh, J. (2021). Slurry erosion performance analysis and characterization of high-velocity oxy-fuel sprayed Ni and Co hardsurfacing alloy coatings. *Journal of Kind Saud University- Engineering Sciences* 1–14. https://doi.org/10.1016/j.jksues.2021.06.009.
92. Prashar, G., & Vasudev, H. (2022). Surface topology analysis of plasma sprayed Inconel625-Al_2O_3 composite coating. *Materials Today Proceedings, 50*, 607–611.
93. Andreoni, G., Barbieri, M., & Colombo, B. (2014). *Developing biomedical devices: Design, innovation and protection*. Springer International Publishing.
94. Idumah, C. I. (2021). Recent advancements in self-healing polymers, polymer blends, and nanocomposites. *Polymers and Polymer Composites, 29*(4), 246–258.
95. Brasileiro, P. P. F., Brandão, Y. B., Sarubbo, L. A., & Benachour, M. (2021). Self-healing concrete: Background, development, and market prospects. *Biointerface Research in Applied Chemistry, 11*, 14709–14725.

96. Sharma, S. N., Prajapati, R., Jaiswal, A., & Dehalwar, K. (2024, June). A comparative study of the applications and prospects of self-healing concrete/biocrete and self-sensing concrete. In *IOP conference series: Earth and environmental science* (Vol. 1326, No. 1, p. 012090). IOP Publishing.
97. Ahmad, S., Hashmi, A. W., Singh, J., Arora, K., Tian, Y., Iqbal, F., Al-Dossari, M., & Khan, M. I. (2024). Innovations in additive manufacturing of shape memory alloys: Alloys, microstructures, treatments, applications. *Journal of Materials Research and Technology*, 32, doi:10.1016/j.jmrt.2024.08.213.
98. Singh, J., Hashmi, A. W., Ahmad, S., & Tian, Y. (2024). Critical review on biodegradable and biocompatibility magnesium alloys: Progress and prospects in bio-implant applications. *Inorganic Chemistry Communications,* 169, 113111.
99. Nik Md Noordin Kahar, N. N. F., Osman, A. F., Alosime, E., Arsat, N., Mohammad Azman, N. A., Syamsir, A.,... Abdul Hamid, Z. A. (2021). The versatility of polymeric materials as self-healing agents for various types of applications: A review. *Polymers, 13*(8), 1194.
100. Idumah, C. I., & Odera, S. R. (2020). Recent advancement in self-healing graphene polymer nanocomposites, shape memory, and coating materials. *Polymer-Plastics Technology and Materials*, *59*(11), 1167–1190.
101. Ammar, M., Haleem, A., Javaid, M., Bahl, S., & Verma, A. S. (2022). Implementing Industry 4.0 technologies in self-healing materials and digitally managing the quality of manufacturing. *Materials Today: Proceedings*, *52*, 2285–2294.
102. Cohades, A., Branfoot, C., Rae, S., Bond, I., & Michaud, V. (2018). Progress in self-healing fiber-reinforced polymer composites. *Advanced Materials Interfaces*, *5*(17), 1800177.
103. Srivastava, R., Thakur, A., & Rana, A. (2024). The basic understanding of nano materials: A review. In: *Innovations in Nanomaterials-Based Corrosion Inhibitors*, edited by Abhinay Thakur and Ashish Kumar, IGI Global, Pennsylvania, USA, pp. 1–20.
104. Bahadur, P. S., Srivastava, R., & Srivastava, V. (2023). Review on the development of smart materials, including shape memory alloys, characterization, and recent applications. In: *Modeling, Characterization, and Processing of Smart Materials*, edited by Ajay Kumar, Parveen Kumar, Ashish Kumar Srivastava, and Vikas Goyat, IGI Global, Pennsylvania, USA, pp. 107–139.
105. Guadagno, L., Raimondo, M., Naddeo, C., Longo, P., & Mariconda, A. (2014). Self-healing materials for structural applications. *Polymer Engineering & Science*, *54*(4), 777–784.
106. Srivastava, R., & Srivastava, V. K. (2020). Changing scenario of manufacturing industries due to Covid 19. *International Journal for Modern Trends in Science and Technology*, *6*(10), 143–146.
107. Aïssa, B., Therriault, D., Haddad, E., & Jamroz, W. (2012). Self-healing materials systems: Overview of major approaches and recent developed technologies. *Advances in Materials Science and Engineering*, *2012*(1), 854203.

15 Role of Deep Learning in Smart Manufacturing

Gurmeet Kaur, Neelofar Sohi, Jashanpreet Singh, and Ashish Kumar

Description	Abbreviation
Active learning	AL
Amazon Web Service	AWS
Artificial intelligence	AI
Auto-encoders	AEs
Bayesian neural networks	BNNs
Bayesian Active Learning by Disagreement	BALD
Cyber-physical systems	CPS
Deep learning	DL
Deep neural networks	DNNs
Differential Privacy	DP
Domain adaptation	DA
Elastic weight consolidation	EWC
Federated learning	FL
Functional encryption	FE
Graph neural networks	GNNs
Generative adversarial networks	GAN
Gated Recurrent Unit	GRU
Homomorphic encryption	HE
Internet of Things	IoT
Learning without forgetting	LwF
Model correlation coefficient	MCC
Neural networks	NNs
Out-of-Distribution Detection	ODD
Privacy-preserving machine learning	PPML
Recurrent neural networks	RNNs
Secure deep learning	safeDL
Secure Sockets Layer	SSL
Transfer learning	TL
Variational auto-encoders	VAEs

DOI: 10.1201/9781003516071-15

15.1 INTRODUCTION

One of the hotspots of modern academic inquiry is artificial intelligence (AI) [1]. Figure 15.1 shows that one area of artificial intelligence called deep learning (DL) is outperforming humans in several tasks [2]. In the game of Go, for example, AlphaGo beat professional players; while in picture recognition, Google Vision outperformed humans [1]. Academics and businesses alike have seen DL's promise, and the technology is finding new uses in fields as diverse as smart manufacturing, healthcare, and personal entertainment. Political digitalisation initiatives, such as Made in China 2025, Advanced Manufacturing Partnership, and Industry 4.0, aim to completely automate and self-adapt contemporary industrial processes. To do this, the manufacturing processes use networked sensors and actuators that bridge the gap between the real and virtual worlds of materials, machines, and products. This is achieved using the principles of cyber-physical systems (CPS) and the Internet of Things (IoT). As a result, there is an urgent need for efficient processing to provide quick reaction times, smart decisions, and global optimisation in the face of an unprecedented amount of data that is constantly being created. Therefore, numerous smart manufacturing applications and other industrial areas like mining and transportation have used data-driven AI methods like deep learning due to their huge potential. According to a survey [3], the AI industry for manufacturing in China is projected to reach a value of $2057 million by 2025. In addition, 83% of the businesses polled believe AI will have an effect or already has. Similarly, 81% of the 86 businesses polled by Microsoft have anticipated AI's increased significance in 2020 [4]. The majority of artificial intelligence applications in smart manufacturing are in the following areas: warehousing, collaborative robots, quality assurance, predictive maintenance, and supply chain

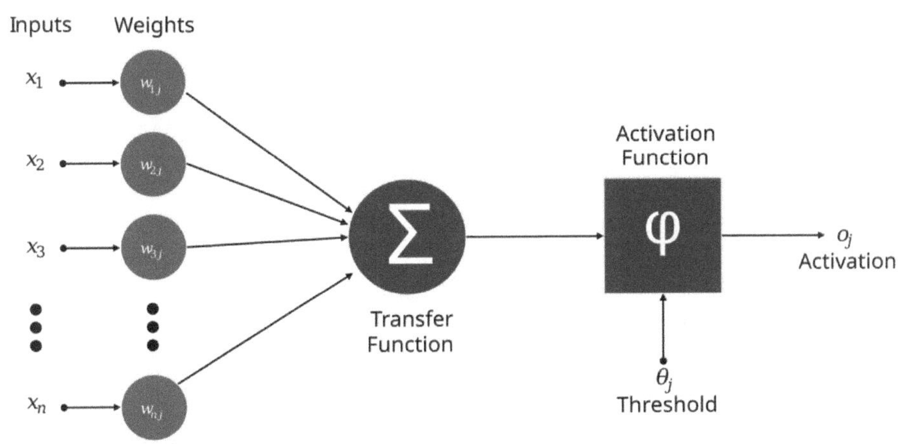

FIGURE 15.1 Illustration of a mathematical model of an artificial neuron [2] (Credit: Creative Commons CC0 1.0 Universal Public Domain Dedication).

management. By implementing early predictive maintenance, both reasonable expenditures and downtime may be reduced. Artificial intelligence paves the way for predictive maintenance to forestall future equipment failure and automated issue identification for industrial machinery [5–7]. Product quality inspections that are automated are the goal of quality assurance. The use of image-based methods, namely visual examination, is prevalent [8–10]. The integrated circuit manufacturing sector has begun to utilise AI technologies to improve yield due to its highly digitalised nature, intricate production procedures, and expensive yield loss costs. In order to increase yield, one must first identify the causes of production losses and then remove them [11–13]. The goal of designing industrial robots essential for increasing production efficiency—is to make them more responsive to their surroundings, rather than relying on predetermined, static positions for every action. The use of vision-guided motion control may accomplish this [14]. On top of that, while keeping people safe, robots are required to function in shared environments with both humans and other robots [15]. Logistics, planning, supplier selection, and warehouse management are some of the many areas that might benefit from supply chain management. Using AI for inventory and demand forecasting allows for more proactive and agile decision-making, which in turn improves planning [16]. By making predictions about their dependability, AI may also assist in supplier selection [17]. The field of logistics focuses on finding the most efficient and cost-effective routes for transporting goods [18]. An overarching goal of AI in warehouse management is to facilitate completely automated warehouse operations, such as inventory forecasting, warehouse robotics, and logistic optimisation [19]. Since deep learning is a data-intensive method, it needs a large amount of data that has already been gathered. Problems with data sampling and storage infrastructures, user expectations for rapid development, unpredictable work environments, and high data pre-processing and labelling costs make this goal difficult to achieve in reality. Hence, real-world data may be sparse, changing in space and time, and lacking labels (for supervised learning). Moreover, industrial data is often secretive; that is, companies are reluctant to share information with outside parties since doing so might hurt their bottom line.

According to this point of view, these challenges may be broadly classified into three areas: AI security, data privacy, and data quality. To help the systems and industrial research communities better understand the issues, this chapter will provide an instructive introduction to neural networks, the basic notion of deep learning, in an effort to make it more accessible to non-experts. In addition, we lay out the cutting-edge methods that will serve as the foundation for overcoming these obstacles. There are a number of surveys on AI for smart manufacturing, but most of them concentrate on specific applications, like predictive maintenance [20], ML-based software development [21], DL techniques and their industrial applications [22–24]. No study that we are aware of outlines the three primary practical obstacles for smart manufacturing that deep learning faces and thoroughly analyses possible solutions to these issues based on recent research findings.

15.2 DEEP LEARNING

Deep neural networks (DNNs) form the basis of a subset of ML methods known as deep learning. Training and inference are two distinct but related parts of machine-learning computation. Training is used to configure ML models utilising training data, whereas inference enables the algorithms to infer features from fresh data [25]. Supervised, unsupervised, and reinforcement learning are the three main categories into which ML algorithms fall in relation to the model training process. In tasks like image-based object identification, supervised learning attempts to link inputs to target outputs (labels). The presence of human-annotated labels, such as item classes, in the training data indicates that it is a "supervised" approach. The two most common applications of supervised learning are regression and data categorisation. Unsupervised learning does not need human innervation and is label-free since it uses the inherent properties of the data. Clustering data and reducing dimensions (from high-dimensional to low-dimensional representations) are common uses. Contrasted with the before, reinforcement learning seeks to teach computers how to behave appropriately. The idea behind it is to enable the machine to get positive reinforcement (feedback) from its surroundings when it acts correctly in response to its current state, and vice versa. Planning and control tasks are common ones for reinforcement learning. Aside from deep learning, there are additional ML algorithms that are sometimes called "traditional" and are used for supervised learning, unsupervised learning, and reinforcement learning [26].

15.2.1 Neural Networks

Derived from more traditional neural networks (NNs), DNNs are the backbone of deep learning. Figure 15.2 shows the basic structure of a neural network, which consists of interconnected mathematical functions called artificial neurons. In typical network architecture, neurons are often arranged in layers and linked so that signals from one layer of neurons are passed on to the next. For example, a layer is formed from the inputs x1, x2, and x3. As NNs are trained, their weights and biases may be adjusted as training parameters. Put simply, a neural network is an input–output structure that uses stacked non-linear transformations [27]. All the layers in a neural network that aren't directly visible to the user are known as hidden layers. Deep neural networks are those that have more than two hidden layers. In every other case, these neural networks are superficial.

15.2.2 Model Training

By carefully selecting the model parameters, the primary goal of training a neural network is to minimise the so-called loss function. In most cases, the objectives of the neural networks are used to determine the loss function. For example, in supervised classification, the loss function may reflect the accuracy with which the

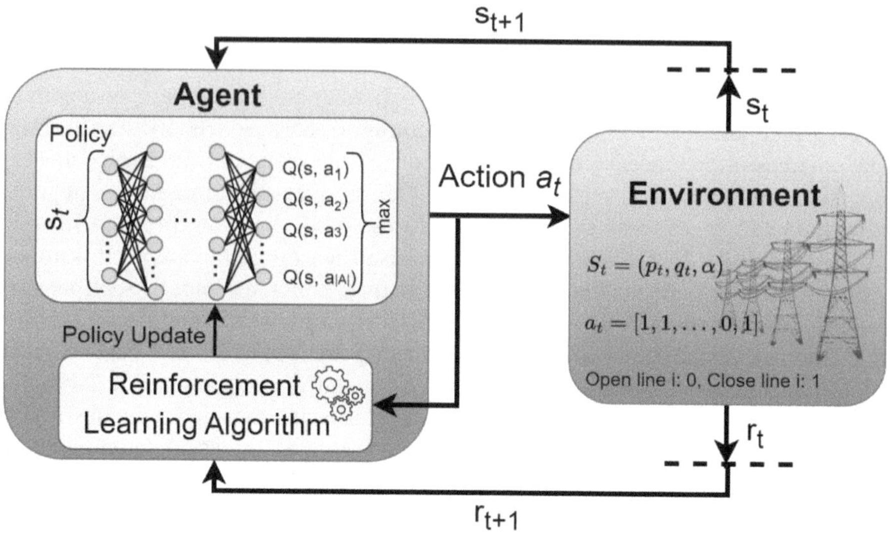

FIGURE 15.2 Schematic structure of deep reinforcement learning [27] (Credit: Creative Commons Attribution 4.0 Licence).

networks accomplish the labels and predictions. Weight sparsity and other numerical regularisations of neural network parameters are also possible in the loss functions. As seen in Figure 15.3, loss functions, on the other hand, are often not convex and could include many local minima. Also, DNNs often store hundreds of thousands, if not millions, of parameters. Thus, analytically seeking a globally minimised loss is almost impossible. And so, DNNs are often taught using gradient descent techniques in an iterative approach. In the direction in which the loss decreases the quickest, the parameters are adjusted slightly on each iteration to account for this. Vector p contains the parameters of the DNN, and p0 represents the parameters that have been changed after each iteration. The learning rate, denoted as α, modifies the step size for every update, Before calculating the loss values and gradients, the data samples are forward-passed to get the outputs, much like inference. The huge number of parameters to optimise and the layer-wise structure of DNNs often lead to the efficient computation of gradients utilising the backpropagation and chain rule procedures. Since the gradients in the final layer are geographically closest to the outputs and loss functions, they may be directly calculated, as shown in Figure 15.4. Based on the gradients calculated in later levels, the gradients in previous layers are determined sequentially, layer by layer.

It is common practice to train a neural network using the whole dataset. However, it's not obvious how to divide up the data samples for each parameter change. It is tedious and causes training procedures to vary to just use one sample every update. However, batch training—which involves updating the whole dataset at once—can be computationally demanding. For this reason, it is common practice to split the training data into smaller chunks, or "mini-batches," that may be used for parameter

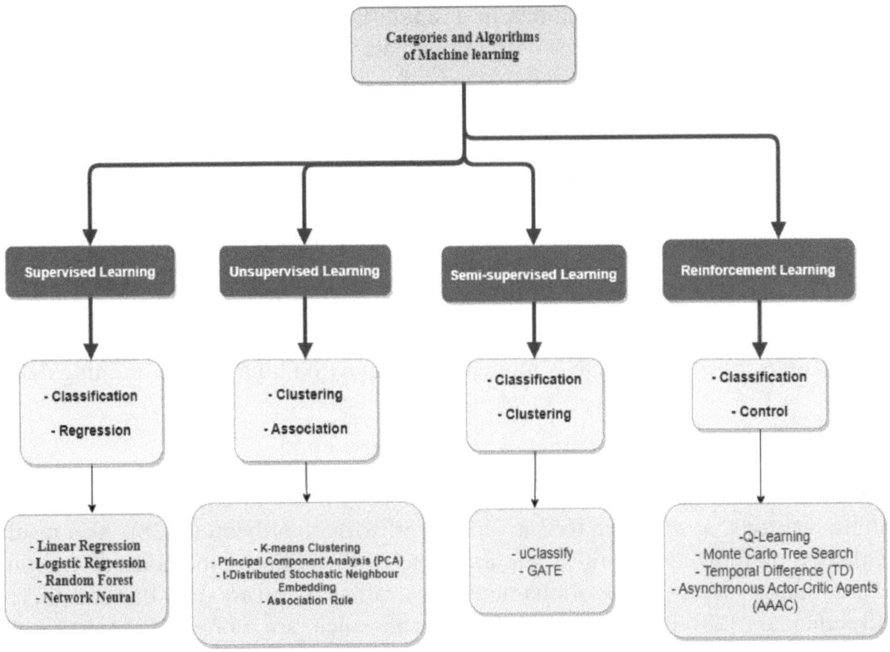

FIGURE 15.3 Model training classification of machine learning [3] (Credit: Creative Commons Attribution 4.0 Licence).

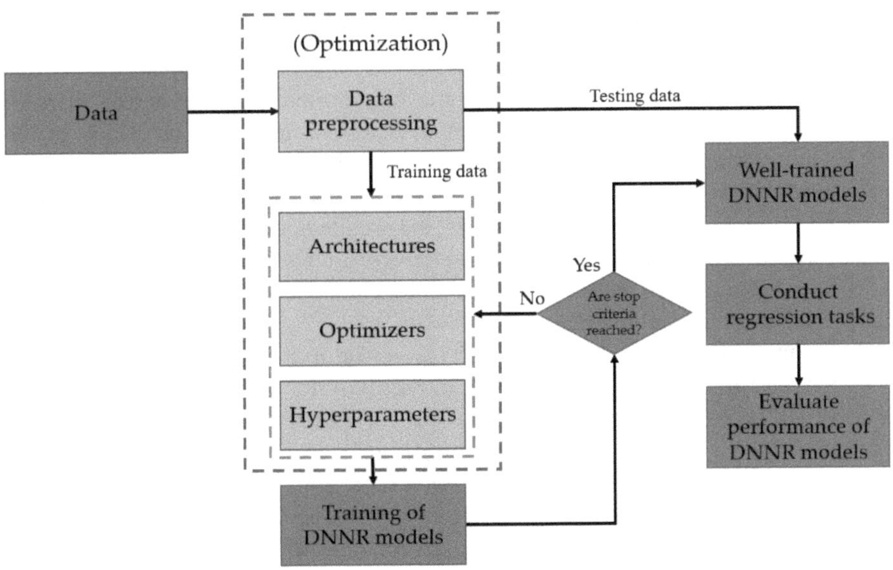

FIGURE 15.4 Example of using deep neural networks [5] (Credit: Creative Commons Attribution 4.0 Licence).

updates and gradient computations in a single training cycle. The average of the gradients for each data sample is used to calculate the gradients using the mini-batch data. A mini-batch technique like this is useful for training convergence as it aims to mimic the gradients calculated using the complete dataset. After many parameter modifications, it is common practice to revisit the complete dataset for an epoch. The goal of training, as previously stated, is to iteratively seek the loss minimum. Neural network training therefore requires tens of thousands to millions of epochs.

The subject of how to control the DNNs' training difficulty remains unanswered. Training time and model accuracy are affected by several aspects such as datasets, hardware platform, optimisation strategy, and network design, while the computational complexity for each epoch may be easily calculated by looking at the number of operations. Specifically, DNN models rely heavily on high-quality training data.

15.2.3 Deep Neural Network Architectures

Any continuous function may be approximated by a shallow neural network with infinite neurons, as stated in the universal approximation theorem [28]. As a result, in theory, shallow NNs can model any dataset. Traditional methods of training rely on human annotations of characteristics rather than raw data that are linked and redundant [28]. Picture object edges, for instance, are more useful features for human recognition than raw pixel values when it comes to completing learning problems; this is because features are bits of qualities or representations of raw data that stand out more.

The complexity of both the learning problems and the data is, however, increasing. This means that the effectiveness of manually annotating features is declining. Not to mention how long they are to train and how bad they are at generalising with complicated raw data, and shallow networks [29]. Fortunately, DNNs are capable of autonomously learning generic, reusable hierarchical features across different data sets. Thus, DL stands out from manual feature engineering as a tool for feature learning or representation learning. Deeper and more complex neural networks have found widespread application as a result of the fast advancements in computing power in the last several decades. In DNNs, the architecture is defined by the amount of neurons, their connection topology, and the learning objectives they are designed to do. In order to make feature learning faster and more accurate, several DNN architectures have been developed in the last few decades.

Convolutional neural networks (CNNs) were the first supervised learning tools that deep learning used to be successful, particularly for image processing applications [30]. The convolutional operators used in classical signal processing are the inspiration for CNNs' convolutional layers (Figure 15.5). Learning spatial-invariant characteristics that may be separated data representations (salient features) is a strong suit of convolution kernels, which take advantage of the interactions between nearby input signals. Used for visual inspection and other image processing tasks, they are computationally more efficient than fully connected layers. [30].

Recurrent neural networks (RNNs) are defined by their reliance on RNN cells and were suggested as a means to learn features characterising the interactions between consecutive input signals [31]. Figure 15.6 shows that RNN cells acquire temporal-invariant characteristics by computing outputs based on current and prior inputs, as opposed to feedforward neural networks that transmit signals sequentially without loops (i.e., by reviewing previous data samples). The network is fed fresh data samples and the results of the previous time step at each time step. Anomaly identification in quality assurance using continuous sensory measures is one common use of RNNs in time series analysis [32].

Layer-level operators, such as fully connected layers, may be used with CNNs and RNNs, which are layer-level architectures. As an example, in order to learn features that are spatially or temporally invariant at several levels, one may stack RNNs or CNNs [7]. For various objectives and data modalities, DNN architectures

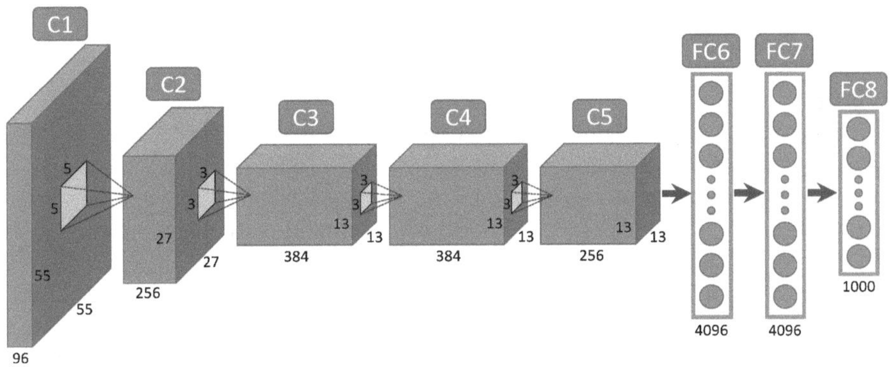

FIGURE 15.5 Conventional recurrent [30] (Credit: Creative Commons Attribution 4.0 Licence).

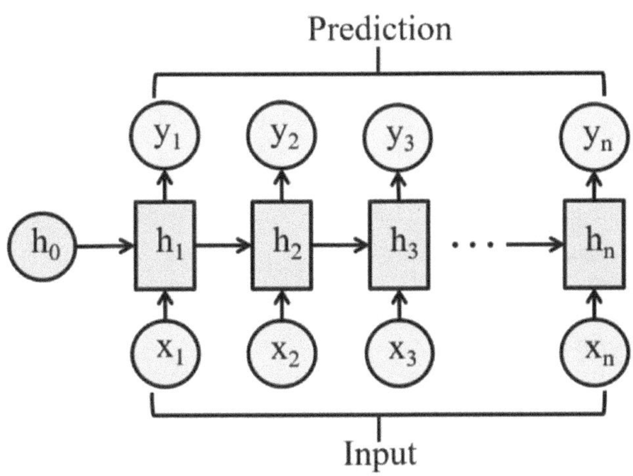

FIGURE 15.6 Architecture of RNNs [31] (Credit: Creative Commons Attribution 4.0 Licence).

are developed using this layer-wise stacking method. Because of their exceptional feature learning capabilities, DNNs are also used for unsupervised learning, which allows them to take advantage of the internal connections found in complicated datasets. Among the many typical methods for dimensionality reduction, auto-encoders (AEs) stand out. Two distinct DNNs, the encoder and the decoder, make up an AE, as seen in Figure 15.7. In order to process the inputs, the encoder converts them to a lower-dimensional space, also known as latent space. You may use the latent values as a generic representation for all data samples since they explain the original data source more effectively than the real raw data. In order to train the decoder to retrieve latent space data as closely as feasible to the original inputs, the loss function is constructed. As a result, encoders may record only the essential details needed for data cleanup.

However, AEs do not need a structured approach to feature extraction since the latent values are sparsely linked and acquired naturally by the network. Since the data samples are assumed to follow underlying probabilistic distributions, namely a Gaussian distribution, Kingma et al. [34] suggested variational auto-encoders (VAEs). The factors of these Gaussian distributions may be used to examine the mathematical relations among the datasets. VAEs are trained to locate these parameters. At the same time, it is possible to transform each latent distribution sample into useful plain data. It is possible to generate new inputs, called generative models, using samples of the latent distributions.

The attention method first suggested for machine translation, is among the most significant contributions to the machine-learning community, alongside the network architectures already discussed [33]. The ability to focus on certain areas of inputs

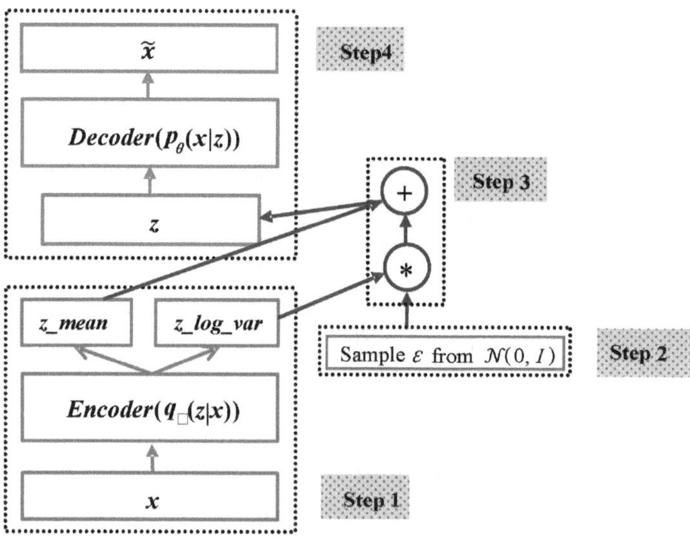

FIGURE 15.7 Architecture of standard VAE [33] (Credit: Creative Commons Attribution 4.0 Licence).

that are more important for task outputs is what makes attention possible for neural models. There has been extensive use of the attention mechanism to boost performance in industrial settings. In order to predict sales, Li et al. [35] use GRU, a form of RNN that incorporates an attention mechanism. Machine prediction was a key consideration when Wang et al. [36] developed CNN. Numerous other network topologies have been put forward, including neural Turing machines and deep belief networks. For a more in-depth look at DNN designs, check out the Neural Network Zoo [37]. Furthermore, selecting the correct model is only one component of an effective industrial deep-learning application. Three major obstacles have emerged from our review of the literature on AI in smart manufacturing, namely, data confidentiality, data quality, and AI safety [38, 39].

15.3 CHALLENGES IN DATA QUALITY

Data quality that "the larger the datasets, the better the performance" applies to deep learning is a commonly held belief. Because state-of-the-art models may include millions of parameters, training a DNN from the start typically takes massive quantities of data samples—perhaps millions. Yet, deep learning engineers seldom have access to well-certified datasets. To begin, establishing new infrastructures is often necessary for dependable data gathering and storage, which is a labour-and cost-intensive process. Second, it takes more work to prepare the data, which includes things like cleaning it up, changing its format, and labelling it. Therefore, methods for building reliable deep-learning models with less-than-ideal datasets are the subject of ongoing investigation.

15.3.1 Data Augmentation

The goal of data augmentation is to prevent overfitting by artificially increasing the data amount and, perhaps, data variety, which in turn improves model generalisation. Methods based on machine learning, methods based on classical signal processing, and methods using handcrafted generated data make up the three main categories of data augmentation techniques. In [40], you will get a much more extensive overview of data augmentation.

Techniques Relying on Signal Processing

When it comes to data augmentation, traditional signal-processing approaches are often based on heuristics. Rotating, flipping, cropping, manipulating colour channels, and injecting noise are typical procedures for photographs, such as those used in visual examination [41]. The time domain, frequency domain, or a combination with local averaging, perturbation, shuffling, etc., may be utilised for data augmentation on time series, for instance, in anomaly detection [42].

Techniques Derived from Machine Learning

Since certain steps in data production cannot be done analytically, it might be challenging to produce samples using the aforementioned approaches that are realistic enough to represent reality. But data-driven modelling of data production processes is where generative models really shine. The training of generative models, which may subsequently create new samples, is often done using real data samples. In their work on fault diagnosis, Gao et al. [43] examine GAN-based data augmentation strategies. They mimicked real-world stream data using many datasets in their studies, and the results demonstrate a 5% improvement in classification accuracy when utilising mixed data.

Data augmentation does not allow for the complete replacement of genuine data with generated datasets. The accuracy gap in training models using actual and augmented samples has to be filled. To start, the enhanced data has to be accurate; in other words, it can't include any samples that don't come from the genuine thing. The second point is to use actual data to test the models that were trained with these synthetic datasets. In this case, it is common practice to use domain adaptation (DA) methods in order to guarantee the model's performance on actual data [44]. When the distributions of the training and testing data do not match, DA may be used as a solution.

15.3.2 Semi-Supervised Learning

When there is insufficient labelled data, one might use semi-supervised learning (SSL), which combines supervised and unsupervised learning techniques [45]. Labelled and unlabelled samples are presumed to have equivalent probability distributions in SSL. Regarding their connections to the labels, the same is true.

Teaching Oneself

For SSL, self-teaching is the most natural approach. Instead of manually labelling unlabelled samples, the model is trained using labelled data and then utilised to predict labels. Proxy labels are a common term for these types of labels. Similar to traditional supervised learning, the models are developed using a combination of labelled and pseudo-labelled data in subsequent training iterations. In order to get the loss function to converge, the pseudo-labels need to be updated with the most recent model after every epoch [46].

Rajani et al. [47] teach a CNN to identify industrial equipment fouling by interpreting ultrasonic data using self-learning algorithms. Their findings demonstrate that, in comparison to training with solely labelled data, semi-supervised learning may decrease the error rate by 15%. Although self-teaching approaches are simple to deploy, if the model's accuracy is poor, to begin with, the mistakes may self-reinforce and spread over epochs. In addition, when the amount of unlabelled data is substantial, it becomes impractical to use.

Generative Models

It is quite likely that numerically near samples will have the same labels if SSL's probability distributions and labelling procedures are consistent for both labelled and unlabelled data. The nearby labelled examples may be used to categorise the unlabelled data. An effort was made to enhance numerical distance comparisons by mapping the samples to feature regions (often latent space) employing unsupervised learning methods, such as generative models. This reasoning is the basis for the M1 and M2 models [48]. Figure 15.8 shows the structure of generative models.

As the number of unlabelled samples increases, these techniques often outperform self-teaching in terms of accuracy. Using sensory time series data and a VAE-based M1 model, Yoon et al. [8] estimate how much longer turbofan engines will last. The findings obtained from testing the VAE-SSL on various prediction models demonstrate a 5% improvement in accuracy. In order to identify bearing anomalies using vibration data, Zhang et al. [49] use the M2 model. The results show that the classification accuracy increases by 3% to 30% when the labelled data fraction varies between datasets.

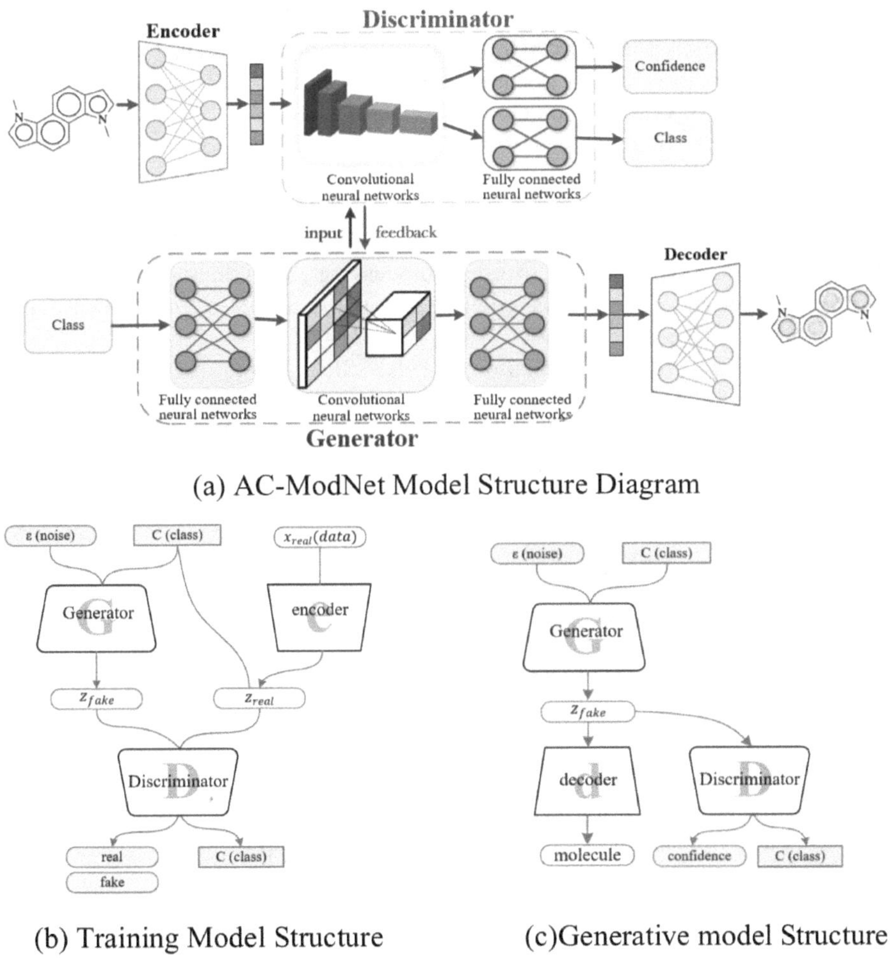

(a) AC-ModNet Model Structure Diagram

(b) Training Model Structure

(c)Generative model Structure

FIGURE 15.8 Structure of generative models [48] (Credit: Creative Commons Attribution 4.0 Licence).

Approaches Based on Graphs

The data samples (the nodes) and the numerical distances (the edges) connecting them, as well as other information (such as communications) indicating their interactions, allow for a graphical representation of datasets. The standard premise upon which graph-based SSL solutions are built is that neighbouring samples in the graphs should have identical labels.

As an alternative to using pre-existing graph embeddings as inputs to DNN classifiers, Yang et al. suggest learning them [50]. For example, node characteristics, which include node attributes and associations with neighbours, are encoded in the embeddings. Graph properties are represented in this way. To train DNNs to learn graph embeddings, one must first force them to forecast node context or characteristics of nearby nodes. This will allow the DNNs to grasp the relationships between nearby nodes. In [51], Chen et al. use graph embedding learning SSL techniques to estimate energy use in steel plants. Model correlation coefficient (MCC) metrics may improve performance by as much as 0.15, where MCC is a measure of the degree to which two variables are comparable to one another.

To classify nodes in a semi-supervised manner, Kipf et al. use graph neural networks (GNNs). Figure 15.9 shows that these networks take in sub-graphs (which consist of nodes and their edges) as inputs and provide predicted labels for the nodes as outputs. GNNs are excellent at capturing the properties of nodes, edges, and subgraphs; they find particular use in graph processing. By reducing the number of incorrect predictions made for the labelled nodes, the model is trained in a supervised manner.

Due to the high expense of obtaining labels, SSL is crucial for industrial deep-learning applications. There is a broader range of possible use cases for graph-based SSL approaches. This is because the datasets may be anything from individual

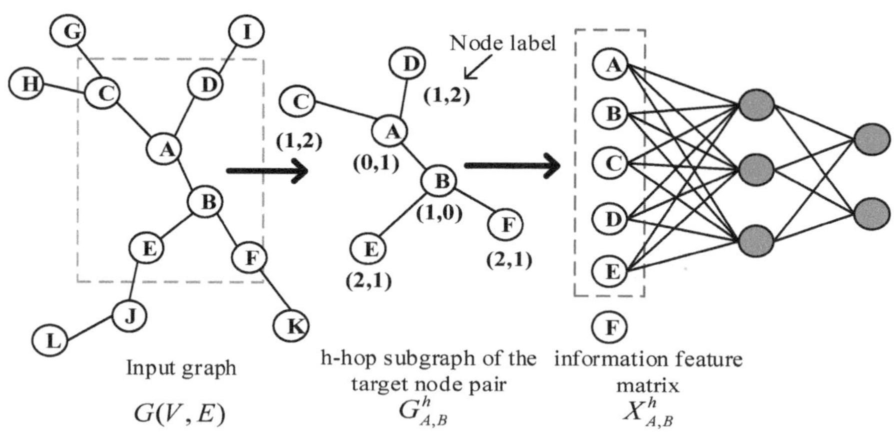

FIGURE 15.9 Neural network graph and subgraph models [7] (Credit: Creative Commons Attribution 4.0 Licence).

instances to genuine explicit graphs, such as communication networks, where the nodes exchange information.

15.3.3 Active Learning

Another class of algorithms used in situations when there is insufficient annotated data is active learning (AL), which is similar to semi-supervised learning. Adapting ML models to use labelled data for training is the first step in active learning; the second is selecting unlabelled samples to ask people (the "oracle") to identify. The retrained model will make use of the freshly annotated data in the subsequent cycle. This cycle continues until either the labelling budget is exhausted or the model's performance reaches a desirable level (convergence).

Minimising the amount of data that has to be labelled while preserving the accuracy of the trained model is the major objective of active learning, and the query method that chooses which samples to label is the greatest obstacle. Discovering examples where the models are blatantly incorrect is the primary objective of the query techniques. Neural networks are often unable to evaluate the accuracy of their predictions. Uncertainty sampling and diversity sampling are the two primary types of query techniques used by deep neural networks.

Uncertainty Sampling

Finding samples about which the models have doubts is the fundamental premise of uncertainty sampling. When it comes to deep learning, most uncertainty sampling query algorithms use Bayesian neural networks (BNNs) [52] since DNNs often can't tell you how uncertain their predictions are. The parameters of BNNs are not fixed numbers but rather statistical distributions, which distinguishes them from traditional neural networks. Predictions on the same inputs might vary after each sampling since inference uses these parameter distributions to calculate predictions.

Conversely, it is reasonable to conclude that the majority of the forecasts are incorrect when the predictions provided by different sampling models do not agree with each other (uncertainty score) [53]. The goal is to annotate these samples that have high uncertainty ratings. In a mathematical sense, Bayesian Active Learning by Disagreement (BALD) chooses samples so that the model parameters and forecasts have the maximum amount of mutual information. However, deep learning suffers from inefficiency and overfitting when retraining with single data samples. Using the samples' mutual information with the model parameters, Kirsch et al. provide BatchBALD, which expands BALD's single-sample selection to a batch of samples [54]. To deal with the emergence of novel sensory measurement patterns induced by deterioration, Martínez-Arellano et al. [55] use an active learning approach based on BNNs for tool condition monitoring. To increase performance, the uncertainty information decides which data to classify and use for retraining.

Diversity Sampling

Through the use of diversity sampling, a portion of the total unlabelled data may be chosen for annotation, with the goal of preserving the model's accuracy when trained with the subset relative to training with the complete dataset. As a result, the amount of labelling data may be significantly decreased. The "long-tail" approach, put out by Geifman et al., uses numerical comparisons to choose samples that are the farthest away from the provided labels [56]. By transforming the query technique into a core-set searching issue, Senser et al. [57] present their core-set method. Their study establishes that finding a dataset's core-set while preserving the DNN training accuracy is the same as a k-centre issue. By minimising the greatest distance from each sample to its nearest centre, a collection of samples is used to choose k-centres in a k-centre issue. It is common practice to combine diversity sampling with uncertainty sampling when drawing samples from a population since the former method is known to provide results with inadequate diversity and informativeness [54]. To diagnose gearbox faults, Chen et al. [58] use a mixed uncertainty/diversity sampling active learning query approach. With a classification accuracy of 90%, their hybrid technique outperforms both diversity sampling and uncertainty sampling. In order to detect software defects, Xu et al. [59] use active learning, integrating diversity and uncertainty metrics. Hybrid approaches also outperform diversity and uncertainty sampling techniques, as they demonstrate.

15.3.4 Transfer Learning

By training a model in a source domain Ds and applying it to a target domain Dt, the idea behind transfer learning (TL) is to alleviate data scarcity by using related datasets. The ML job, related datasets, and model make up a domain. The statistical distributions of the data in Ds and Dt need not be identical. TL speeds up model convergence and helps save the labelled data, much like semi-supervised learning. It is important to acknowledge, however, that TL may result in a negative transfer, which would result in a decrease in the precision of the target model following the transfer. This is particularly true in cases when Ds and Dt strongly disagree. There are four types of transfer learning: instance-, feature-, parameter-, and relational-based [60]. These types are defined by the contents that are transmitted. In particular, deep learning models excel in feature- and parameter-based TL.

Instance-Based Transfer Learning

For instance-based TL to work, models are trained using Ds samples in conjunction with Dt samples. Before transferring, Ds samples should be chosen based on their relevance to the target domain data; this will lessen the likelihood of negative transfer [61]. Statistical similarities between the distributions of Ds and Dt data or the test precision of Dt models on Ds samples are examples of what is often meant by the similarities.

Wang et al. [62] suggest a GAN-style approach to DNN sample selection for Ds, where the "generator" (data selector) chooses Ds samples that fit the distribution of Dt data in order to trick the "discriminator" into thinking they are different from the Dt data, and the "discriminator" makes sure the Ds samples are different from the Dt data as much as possible to improve the data selector's performance. In their instance-based TL application for ball screw deterioration identification, Zhang et al. [63] used reconstruction errors with AEs trained in Dt to choose samples in Ds. Compared to models trained just on short datasets, those trained via transfer learning perform better in experiments conducted on various datasets.

Feature-Based Transfer Learning

Collaborating Ds and Dt is made possible by feature-based transfer learning, which takes use of their shared, or domain-independent, properties. It is challenging to properly analyse and compare DNNs across domains since they learn features automatically from the data. In order for models to learn characteristics that are applicable across domains, additional loss functions are placed on neural networks during training [64].

During training, both Ds and Dt data are used concurrently to construct the target model. The additional loss functions often involve minimising the numerical distances between the layer-wise outputs of the intermediate layers, which are calculated using Ds and Dt data. Accordingly, the model may solve the data scarcity problem in a target domain by learning characteristics that are very comparable to Ds and Dt data. In their work on industrial problem diagnostics of bearings, Zhu et al. [65] utilise this reasoning. When compared to baselines trained with vanilla, transfer learning produces superior performance.

Parameter-Based Transfer Learning

The vast majority of deep-transfer-learning implementations use a parameter-based strategy [66]. Typically, models are trained entirely in Ds, then their parameters are transferred to the target model(s) for initialisation, and finally, using samples from Dt, the models are fine-tuned. Figure 15.10 shows the architecture of parameter-based

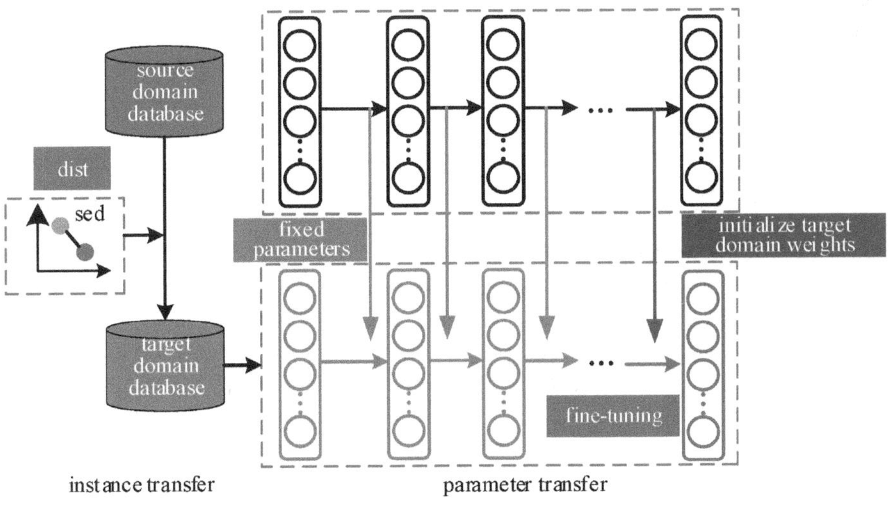

FIGURE 15.10 Parameter-based transfer learning [66] (Credit: Creative Commons Attribution 4.0 Licence).

transfer learning. Layer-sharing parameters should be the same size; there's no need for the source and destination networks to be identical. Using parameters CNN-based computer vision tasks use TL extensively. The ability of the pre-trained CNN models' first layers to capture visual characteristics at a low level which are not heavily reliant on datasets has been reported in the study [67]. Common pretraining sources for CNN layers include ResNet, AlexNet, or VGG which have been trained on massive image recognition datasets like CIFAR or ImageNet [68]. With pre-trained parameters taken from the publicly published trained GoogleNet, Mittel et al. [9] constructed a CNN-based metal fracture identification system and attained a 99.8% accuracy rate.

Since data scarcity is an often-observed problem for deep learning applications, transfer learning has gained increasing attention in the past several years. Most people use TL instead of the other methods. The procedures that choose the items to transmit and avoid a negative transfer, however, vary from instance to situation.

15.3.5 Continual Learning

The emergence of new data patterns is a typical occurrence in deep learning applications. For example, changes in the working environment and/or the age of the equipment might cause different types of mechanical faults. In addition, failure patterns often do not manifest quickly but rather gradually throughout a machine's lifespan. So, gathering massive datasets that include every conceivable class would need years. Users, on the other hand, would want to get the benefits of AI without delay, therefore this is a major turnoff.

With the goal of keeping models up-to-date via retraining with new incoming data while retaining previously acquired information, continuous learning (CL) is sometimes called lifelong learning or incremental learning [69]. Retraining the DNNs from the beginning is obviously going to take a long time and a lot of memory to store all the training data. A few common ways to reduce forgetting include dynamic network designs, memory replay, and regularisation methods. The general idea behind continuous learning is that it's a subset of transfer learning, where the source domain is the original data and the target domain is the one that incorporates additional field samples. Continuous learning is also known as forward transfer.

Regularisation-Based Continual Learning

The parameter changes in regularisation-based continuous learning systems are constrained to remain near the settings prior to retraining. Examples of such algorithms are learning without forgetting (LwF) [70] and elastic weight consolidation (EWC) [71]. By including LwF in object identification tasks, Li et al. [72] train the network to recall previously learnt object classes. New classes cannot be learnt gradually, and regularisation-based approaches are challenging to scale [73].

Memory Reply

Memory respond is based on the simple principle of incorporating both fresh and preserved historical data into the model during retraining to avoid catastrophic forgetting [74]. Memory replay using past data is best understood as instance-based transfer learning according to the transfer learning taxonomy. Since the advent of GANs and VAEs, training deep generative models like these has made effective storing of historical data a reality [75]. In order to facilitate continuous learning for anomaly detection tasks, Wiewel et al. [76] use the VAE decoder for memory replay.

Dynamic Architectures

By combining the two sets of training data, dynamic architecture techniques may increase and reallocate neural network parameters. When retraining for new tasks, PackNet [77] allows DNN parameters with tiny values to change while keeping the big parameters constant. When training to avoid memory loss, progressive neural networks automatically develop new neurons (neural expansion) to handle new tasks, while old neurons are frozen [78]. The appearance of new classes triggers the introduction of new neurons in categorisation tasks. To preserve the acquired information, it is common practice to freeze the initial DNN layers and the output neurons for previous classes [79]. Using neural expansion as its foundation, Zhang et al. developed a system for incremental quality prediction in fluid catalytic cracking [13]. These findings demonstrate that the model is capable of adjusting to its own notion drift.

To keep models up-to-date in dynamic settings, deep constant learning is an essential component. Accordingly, continuous learning methods should be stable (i.e., not drastically degrade performance as the quantity of unknown data rises) and scalable (i.e., not cause the memory footprint to balloon) from an application standpoint. We still have a ways to go before we have a CL system that is both reliable and scalable.

15.4 CHALLENGES IN DATA SECRECY

Although industrial data does not often include personally identifiable information, it is nonetheless subject to strict data confidentiality requirements when it pertains to sensitive company information or even trade secrets, including facts about the manufacturing process or the supplier chain. For the sake of simplicity and cost, both training and inference are frequently conducted directly on the cloud, on servers belonging to third parties or public cloud services. As a result, information may be exposed as it is being sent, shared, and processed, including deep learning models. Direct data disclosure occurs during transmission and processing, whereas indirect data disclosure occurs via model inversion when the training data is inferred from the cloud-stored model. Therefore, to safeguard data after it has left the control of its owners, privacy-preserving machine-learning algorithms have been suggested. While gradient noising is often used to limit indirect data disclose, elimination tactics, encryption techniques, and data cleaning (i.e. differential privacy) are methods for preventing direct data expose. To further secure data during training and inference, new deep-learning architectures have been suggested, which go beyond just removing, encrypting, or sanitising raw data. Federated learning is the most typical method for this. However, it's important to note that decentralised systems may also include privacy-preserving machine learning (PPML) techniques to lessen the chances of data leaks. The primary methods for PPML, such as data encryption, elimination, differential privacy, and DL model architecture methods are discussed. Several of these methods may be used to safeguard raw data or to avoid data disclosure during training and inference. Study [80] details an in-depth analysis of smart manufacturing security.

15.4.1 Elimination Approaches

Taking the easy way out is to filter out any personally identifiable information from raw data. In terms of anonymisation, K-anonymity [81], L-diversity [82], and T-closeness [83] are the three most popular and conventional methods. In order to get K-anonymity, each person in the dataset has to be similar to at least k – 1 other people. Suppression and generalisation are the two approaches to K-anonymity. In datasets, suppression is used to delete values from certain entries, whereas generalisation is used to divide items into categories. In their work on face deidentification, Meden et al. [84] use a generalisation process that averages out faces from the same class to substitute individual faces. As a natural progression from K-anonymity, L-diversity checks that every characteristic in a given class has at least L unique values. Suppression and generalisation are the two primary methods for L-diversity. Two further forms of L-diversity exist, namely entropy l-diversity and recursive L-diversity [85]. Lessening the level of detail in a data representation is the key notion. When the gap between the distributions of the sensitive attributes in the corresponding class and the attribute itself is wider than the threshold, a new measure called T-closeness is introduced as an expansion of L-diversity. Hence, the global table is the limit of what any comparable class's attribute values may provide.

15.4.2 Cryptographic Approaches

Moving encrypted data across computers is one simple way to keep information secret. Both raw data and data used to train and infer models may benefit from encryption's ability to keep information private. Here we describe the two most typical and illustrative methods of encryption, namely functional and homomorphic encryption. Data encryption technologies, however, are laborious and have a significant transmission cost.

Homomorphic Encryption

A common technique for working with encrypted data that does not require decryption is homomorphic encryption (HE). Data is processed in the ciphertext domain, which is encrypted, and the decrypted values are identical to the directly computed outputs from the plaintext domain, which is not encrypted. Consistency is the term used to describe this attribute [86].

Data owners encrypt raw data before sending it via text so that machine-learning models may access and use it. After that, the data owners get the encrypted inference results or trained models, which they may then decrypt locally. Consequently, the raw data is accessible only to the data owners. From a computational standpoint, the major steps in training and inferring using neural networks may be broken down into simple arithmetic operations. Applying encryption solutions to these activities is the core rationale behind building encryption approaches for DNNs. After training in a textual domain, Zhang et al. [87] achieved accuracy equivalent to that of HE-based backpropagation algorithms. It is common practice to estimate the activation functions for efficiency while maintaining the consistency of addition and multiplication. It is inaccurate to use such a function approximation.

One cloud service that Microsoft offers lets users do neural computations in the ciphertext domain [88]. Users are able to immediately execute machine-learning algorithms on encrypted data by uploading it to the cloud using Microsoft Azur [89]. Similar services are offered by Amazon SageMaker via Amazon Web Service (AWS), as mentioned in [90]. In, Zhu et al. [91] provide a framework for training machine-learning algorithms that can support homomorphic encryption.

Functional Encryption

When using functional encryption (FE), the model outputs are immediately available in the plaintext domain, as opposed to the ciphertext domain in homomorphic encryption [92]. In Figure 15.11, the plan with FE has been contrasted. In contrast to HE, however, FE has not yet seen extensive use and is in its early stages.

In functional encryption methods, the data is encrypted such that the key holders may perform a certain function. When doing computations pertaining to deep learning, two main schemes are often used: the inner-product scheme [92] and the quadratic scheme [93]. True to its name, the inner-product scheme permits quadratic

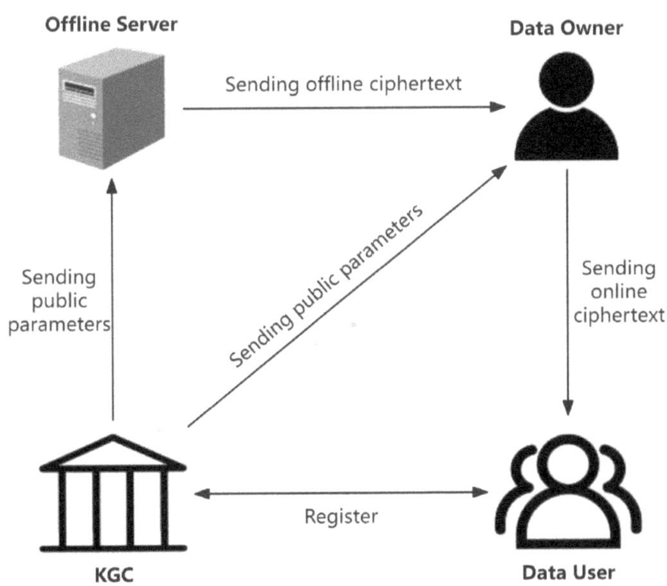

FIGURE 15.11 Functional encryption [86] (Credit: Creative Commons Attribution 4.0 Licence).

and inner-product computations to take place in the ciphertext domain. The inner-product strategy is used in neural network training in CryptoNN, which is proposed by Xu et al. [94]. Polynomial neural networks, an approximation version of normal neural networks [95], are subject to the quadratic scheme used by Ryffel et al. [92].

In order to securely connect data vendors with prospective buyers, Song et al. developed a cloud-based application [96]. At this time, popular cloud providers that offer PPML with functional encryption are not available.

15.4.3 Differential Privacy

The goal of differential privacy (DP) is to reduce the likelihood of sample identification by permuting, or adding noise to, the outputs of particular functions [97].

Data Noising

Google, Apple, Microsoft, and LinkedIn have all used DP-based privacy-preserving techniques to gather user data [98]. On the other hand, DP-based approaches might be harmful to data utilities, particularly when it comes to so-called context-free permutation methods [99]. These methods hypothetically accept all data attributes without discrimination, assuming that attackers have infinite auxiliary knowledge of the data. On the other hand, privacy-preserving solutions that take context into account have been suggested; these methods only use data statistics to switch up privacy-related aspects. Using GAN with raw data as inputs and sanitised data as outputs, Huang et al. [100] suggest learning privacy methods. However, these approaches rely heavily on the target domain and need data utility testing to guarantee that the data are almost equally valuable for ML tasks both before and after cleansing.

The SmartNoise system, built by Microsoft, offers DP modules that customers can utilise to build their own privacy protection solutions. These modules include several private machine-learning model training options [101]. Prior to being input into the deep-learning models, data is DP-sanitised in the energy consumption prediction application described in [102].

Gradient Noising

If an attacker has access to the model designs and parameters, they may avoid data expose via model inference by permuting DNN parameters, such as gradients and model outputs, in addition to noticing the raw data provided above. Assuming that learnt DNN models encode the characteristics of the training data, these permutation approaches rely heavily on differential privacy techniques. To make it less likely that the models would infer the training data, the parameters are randomly changed.

Abadi et al. suggest utilising differential privacy (DP-SGD) to reorder the gradients during training in order to weaken the associations between the training data and model parameters [103]. To hide the impact of individual models on the final output, Papernot et al. suggest introducing noise into the DNN ensemble outputs [104]. Subjecting loss functions to noise is one example of a similar body of work [105]. Tensorflow Privacy and Huawei Mindspore are only two of the many DL frameworks that are DP-SGD compatible.

15.4.4 Federated Learning

During training, dispersed clients were first encouraged to refrain from transmitting raw data in order to facilitate knowledge exchange via federated learning (FL) [106]. In FL, the intermediate parameters—gradients and weights—are uploaded to the central server by each client after they train their local model. For the following round of iterations, the participating clients get the averaged model parameters from the central server.

Parameter exchanges may be fantastic, but they can put an even bigger strain on the system's communication capabilities than exchanging raw data directly. At the same time, model convergence might be hindered by infrequent parameter interchange. The goal is to keep the ratio of communication rate to model convergence constant [107]. Parameter sharing, however, also poses concerns about data confidentiality. The ciphertext domain is used for parameter aggregation and exchange in Privacy Protection Federated Learning, which has been suggested in relation to these [108]. To safeguard businesses' sensitive information, Bagheri et al. [109] suggest using homomorphic encryption in federated learning for equipment prognostics and health monitoring. To further protect against adversary clients inferring other clients' local data, FL may be paired with differential privacy, which entails permuting the local model parameters before sharing [110]. A data processing system for several robots based on federated learning is proposed by Zhou et al. [111]. They use differential privacy techniques to randomly shuffle the data produced by each robot. Their tests show that the system works for various robots and maintains a good balance between privacy and accuracy.

Traditional federated learning, as described above, is also known as vertical federated learning, and it involves each client storing the whole model of their own version locally. Horizontal federated learning, often called split learning, is a method whereby several clients or servers work together to train and infer using a shared set of DNN layers [112]. Consequently, the raw data is not sent between the clients and servers, but rather the outputs of the intermediate layer, which is called the cut layer. The development of horizontal FL has been somewhat slow in comparison to that of vertical FL. Split learning was first suggested by Vepakomma et al. [112] for the purpose of medical image processing using CNNs. Subsequently, in order to safeguard data privacy, Abuadbba et al. used split learning on one-dimensional CNNs and DP noise on the activation functions of the cut layer [113].

15.5 CHALLENGES IN DNN RELIABILITY

In an effort to create more trustworthy and resilient DL systems, the concept of secure deep learning (safeDL) is gaining traction in the academic and professional worlds. At now, healthcare and automotive are the primary areas where the corresponding algorithms are being used. In smart industrial applications, they remain crucial, nevertheless, since erroneous findings may lead to lost output, broken machinery, or even human lives. We will introduce the key algorithmic elements of the DNN in this part, with an emphasis on its dependability. To get a better picture, have a look at [114].

Incorrect model architecture selection and data shift are the two most critical factors affecting deep-learning dependability [114]. The selection of neural architectures is not going to be covered here since it is so reliant on data and domain expertise. On the other hand, NAS is a rapidly expanding field in the field of machine learning [115]. As of right now, smart manufacturing has no uses. When the distributions of the training and testing data are no longer the same, this phenomenon is called a data shift. Biased training data selection and ever-changing surroundings might lead to data shift. Even in commercial settings, these problems manifest often. In their proposal for a framework for the upkeep of machine-learning models, Dreyfus et al. [116] include topics such as the identification and treatment of idea drift in smart manufacturing use cases.

15.5.1 Concept Drift Detection

The statistical features of streaming data may vary with time, a phenomenon known as idea drift. This is because the data is created continually in temporal order. An example, when equipment ages, the vibration intensity may decrease, making it impossible for the DNN model trained on old data to identify abnormalities in predictive maintenance.

It is common practice to look for changes in the distribution of data, mistakes in model predictions, or a combination of the two to identify concept drift [117]. The premise upon which the first group of approaches rests is that idea drift causes a dramatic decline in model accuracy. These techniques operate with a wide range of ML models, including NNs. The second group looks at how the new data compares to the old one statistically. In order to compare the two sets of data, we will first determine their respective distributions [118]. Statistical approaches, however, perform worse with high-dimensional data. In order to forecast software defects, Kabir et al. [119] use model prediction accuracy to identify idea drift. In order to identify idea drift in predictive maintenance, Zenisek et al. [120] model the statistical distributions of the old data using several machine-learning approaches. They think that concept drifts may be used to infer the machine's mistakes. For energy prediction modelling applications, Barmeo et al. [121] employ model accuracy to identify data drift and adjust the model to changing data due to machinery's natural deterioration. Their approach may increase the fit rate of energy estimate by a factor of two when compared to techniques that do not include concept drift detection.

15.5.2 Uncertainty Estimation

For applications where safety is paramount, it is highly recommended that deep-learning models provide uncertainty in their predictions along with the model's confidence. Two sources of uncertainty are the data's inherent and unavoidable unpredictability and non-stationary training and testing data, which means that testing data samples are drawn from different statistical distributions than the training data. One example of this is idea drift. The former describes a state of epistemic doubt while the latter describes a state of aleatory uncertainty [122]. There are three main types of uncertainty estimation approaches used with DNNs: ensembles, uncertainty prediction, and BNNs. In what follows, we provide an overview of all of them.

Instead of using hard and fast values, BNNs use distributions, such as Gaussian, to determine the model parameters, which include biases and weights. Various inference procedures are performed using the same dataset. Parameters of the model are drawn at random from these distributions in every run. Uncertainty may be shown by calculating the variance of the findings after many inferences on the same data sample using various parameter samples [122, 123]. Using sensory time series data, Benker et al. employ BNN to estimate the usable life of the equipment and define the related prediction uncertainty [124]. To fine-tune the forecast and make it more accurate, you may use the estimated uncertainty. Ensembles include training many models with various initialisations on the same data. In a similar vein, the uncertainties are shown by the variances of the discrepancies in their predictions [125] The remaining life for bearings, taking uncertainty into account, is predicted by Nemani et al. using ensembles of LSTMs in [126]. The remaining life for bearings, taking uncertainty into account, is predicted by Nemani et al. using ensembles of LSTMs in [126]. Last but not least, DeVries et al. suggest directly outputting the prediction confidence during training [127]. During model training, the labels are "one-hot encoded," or the prediction probabilities are enforced to be near the label distributions.

15.5.3 Out-of-Distribution Detection

Out-of-distribution detection (ODD) is employed in order to find the unknowns or test samples with different distributions from the training samples. The problem is referred to as open set recognition (OSR) when it is expressly for classification tasks, such as new class samples, and ODD samples are included in the category of data drift.

The introduction of uncertainty-based approaches lays the groundwork for their inclusion in ODD methods. Having said that, these approaches aren't quite pinpoint accurate. Also, epistemic uncertainty and aleatory uncertainty are often intermingled in uncertainties, although ODD samples only generate epistemic uncertainty. The computational efficiency of the original DNNs' intermediate and final layer outputs is used by a different subset of ODD techniques. Hendrycks et al. initially suggest the baseline method, which utilises the softmax distributions; this method depends on their finding that out-of-distribution samples have lower prediction likelihoods than in distribution samples. The prediction probability is determined by the maximum neural outputs prior to the final activation functions [128]. To identify out-of-distribution samples, Lee et al. [129] suggest comparing the training and testing data intermediate layer outputs. Regarding OSR in particular, Bendale et al. provide the distance-based OSR approach, which is predicated on the premise that distribution data belonging to the same class ought to be spatially adjacent to one another [130]. In their work on machinery fault diagnosis, Xu et al. [131] take into account the case when unknown fault types emerge during inference and use a distance-based OSR technique. When out-of-distribution samples are present, the classification accuracy is substantially enhanced with the help of OSR.

15.6 CONCLUSION

In this chapter, we look at the most typical problems with industrial AI for smart manufacturing and how to fix them. Data confidentiality, model dependability, and data quality are the highlighted obstacles. Along with its smart manufacturing applications, a number of potential solutions, ranging from transfer learning to DNN encryption and uncertainty estimates, have been presented and summarised. While some of these algorithms, such as training permutation and out-of-distribution detection, are receiving a lot of attention right now, there are still several that are in the early stages and have not found practical use in business.

The more we dig into the answers for Industry 4.0, the more issues with the AI approaches used in smart manufacturing will become apparent. Some of these are obvious and can be addressed at this meeting. The solutions to the problems highlighted in this study are often more complicated since they cannot manifest in isolation. In cases when data labelling is impractical or impossible to do during runtime, domain adaptation is often used in tandem with semi-supervised learning [132]. Methods like privacy-preserving transfer learning [133], have neglected data secrecy to a large extent. Still unresolved, these concerns may stand in the way of developing strong deep-learning applications for industry.

Plus, in the last two years, there have been proposals for data sharing [134, 135] and adaptable AI [136], both of which are connected to the three difficulties covered in this article. Because organisations may get insight from one another's data via data sharing, it's an easy way to address data shortage. The capacity to access external data is being expanded by more than 70% of data and analytics decision-makers worldwide, according to [134]. Nevertheless, as mentioned in [135], there are obstacles to data privacy, a lack of a standard ontology, and platforms and services that hinder data sharing in the industry. To prevent third parties from gaining access to sensitive company information, it is important to adhere to the data confidentiality standards. To guarantee that all community entities' data can be understood and combined, standard data ontology is being worked on. Collecting, storing, and exchanging data requires hardware facilities, both internal and external, such as fog, cloud, and edge computing [137].

Data may change over time, necessitating retraining to ensure the ML model's performance. With the use of adaptive AI, the model can automatically adapt to new datasets. Our research leads us to believe that adaptable AI must resolve the following technological issues: (1) how to make retraining faster and more accurate; (2) how to plan production such that retraining and normal inference are well-maintained; (3) what kind of hardware architecture works best with adaptive AI? When working with constrained history data and retraining time, deep continuous learning is still unable to attain accuracy levels equivalent to vanilla retraining, as stated in (1). A hybrid retraining technique may be used to ensure performance by proposing methods that balance accuracy, historical data volume, and retraining time. Concerning (2), one should think about whether the production process should hold off till the retraining or whether there's a more efficient approach to advance both at the same time; Concerning (3), it is important to choose computer architectures with care to preserve the optimised solution in (1) and (2), taking into account economic budgets, data secrecy, and particular hardware requirements of the manufacturers.

Finally, machine-learning algorithm developers aren't the only ones who need to work together with domain experts, facility engineers, and third-party organisations to build a reliable AI system for smart manufacturing.

REFERENCES

1. Dash, T.; Chitlangia, S.; Ahuja, A.; et al. A Review of Some Techniques for Inclusion of Domain-Knowledge into Deep Neural Networks. *Scientific Reports* 2022, 12, 1040.
2. Funcs, Artificial Neuron Structure, 31 May 2024. Creative Commons CC0 1.0 Universal Public Domain Dedication. Accessed on 15 August 2024. https://commons.wikimedia.org/wiki/File:Artificial_neuron_structure.svg.
3. Taye, M.M. Understanding of Machine Learning with Deep Learning: Architectures, Workflow, Applications and Future Directions. *Computers* 2023, 12, 91.
4. Ye, F.; Yang, J. A Deep Neural Network Model for Speaker Identification. *Applied Science* 2021, 11, 3603.
5. Chen, C.-H.; Lai, J.-P.; Chang, Y.-M.; Lai, C.-J.; Pai, P.-F. A Study of Optimization in Deep Neural Networks for Regression. *Electronics* 2023, 12, 3071.
6. Singh, J.; Vasudev, H.; Szala, M.; et al. Neural Computing for Erosion Assessment in Al-20TiO2 HVOF Thermal Spray Coating. *International Journal on Interactive Design and Manufacturing* 2024, 18, 2321–2332.
7. Singh, J.; Vasudev, H.; Chohan, J.S. Neural Computing of Slurry Erosion of Al_2O_3-13TiO2 Thermal Spray HVOF Coating for Mining Pump. *International Journal on Interactive Design and Manufacturing* 2024, 18, 2277–2289.
8. Pinto, G.; Wang, Z.; Roy, A.; Hong, T.; Capozzolim, A. Transfer Learning for Smart Buildings: A Critical Review of Algorithms, Applications, and Future Perspectives. *Advances in Applied Energy* 2022, 5(1), 100084.
9. Sharma, D.; Jinwala, D.C. Encrypted Data Ordering with Functional Encryption. In *4th International Conference on Recent Advances in Information Technology (RAIT)*, Dhanbad, India, 2018.
10. Kansal, I.; Popli, R.; Singla, C. Comparative Analysis of Various Machine and Deep Learning Models for Face Mask Detection Using Digital Images. In *9th International Conference on Reliability, Infocom Technologies and Optimization (Trends and Future Directions) (ICRITO)*, Noida, India, 2021, 1–5. https://doi.org/10.1109/ICRITO51393.2021.9596407.
11. Popli, R.; Kansal, I.; Verma, J.; Khullar, V.; Kumar, R.; Sharma, A. ROAD: Robotics-Assisted Onsite Data Collection and Deep Learning Enabled Robotic Vision System for Identification of Cracks on Diverse Surfaces. *Sustainability* 2023, 15, 9314. https://doi.org/10.3390/su15129314.
12. Imoto, K.; Nakai, T.; Ike, T.; Haruki, K.; Sato, Y. A CNN-Based Transfer Learning Method for Defect Classification in Semiconductor Manufacturing. In *Proceedings of the International Symposium on Semiconductor Manufacturing (ISSM'18)*, Tokyo, Japan, 10–11 December 2018, 1–3.

13. Zhang, X.; Zou, Y.; Li, S. Enhancing Incremental Deep Learning for FCCU End-Point Quality Prediction. *Information Science* 2020, 530, 95–107.
14. Alambeigi, F.; Wang, Z.; Hegeman, R.; Liu, Y.H.; Armand, M. A Robust Data-Driven Approach for Online Learning and Manipulation of Unmodeled 3-D Heterogeneous Compliant Objects. *IEEE Transactions on Robotics and Automation* 2018, 3, 4140–4147.
15. Poojari, M.; Hanumanthappa, H.; Prasad, C.D.; Jathanna, H.M.; Ksheerasagar, A.R.; Shetty, P.; Shanmugam, B.K.; Vasudev, H. Computational Modelling for the Manufacturing of Solar-Powered Multifunctional Agricultural Robot. *International Journal on Interactive Design and Manufacturing* 2023, 18(8), 1–12.
16. Singh, J.; Gill, H.S., Vasudev, H.; Computational Fluid Dynamics Analysis on Role of Particulate Shape and Size in Erosion of Pipe Bends. *International Journal on Interactive Design and Manufacturing* 2022, 18(8), 1–16.
17. Kumar, S.; Singh, M.; Singh, J.; et al. Rheological Characteristics of Uni / Bi-Variant Particulate Iron Ore Slurry: Artificial Neural Network Approach. *Journal of Mining Science* 2019, 55, 201–212.
18. Singh, J.; Singh, S. Support Vector Machine Learning on Slurry Erosion Characteristics Analysis of Ni- and Co-alloy Coatings. *Surface Review Letters* 2023. doi:10.1142/S0218625X23400061.
19. Singh, J.; Vasudev, H.; Kumar, R.; et al. Study on Wear Analysis of Ni-20Al2O3 HVOF Micron Layers Using Artificial Neural Network Technique. *Internal Journal on Interactive Design and Manufacturing* 2023. https://link.springer.com/10.1007/s12008-023-01433-2.
20. Ran, Y.; Zhou, X.; Lin, P.; Wen, Y.; Deng, R. A Survey of Predictive Maintenance: Systems. *Purposes and Approaches*. arXiv 2019, arXiv:1912.07383.
21. Singh, J.; Mohapatra, S.K.; Kumar, S. Performance Analysis of Pump Materials Employed in Bottom Ash Slurry Erosion Conditions. *Jurnal Tribologi* 2021, 30, 73–89.
22. Hernavs, J.; Ficko, M.; Berus, L.; Rudolf, R.; Klanˇcnik, S. Deep Learning in Industry 4.0—Brief Overview. *Journal of Production Engineering* 2018, 21, 1–5.
23. Subramaniam, S.; Raju, N.; Ganesan, A.; Rajavel, N.; Chenniappan, M.; Prakash, C.; Pramanik, A.; Basak, A.K.; Dixit, S. Artificial Intelligence Technologies for Forecasting Air Pollution and Human Health: A Narrative Review. *Sustainability* 2022, 14, 9951.
24. Shang, C.; You, F. Data Analytics and Machine Learning for Smart Process Manufacturing: Recent Advances and Perspectives in the Big Data Era. *Engineering* 2019, 5, 1010–1016.
25. Singh, J.; Singh, S.; Vasudev, H.; et al. Neural computing and Taguchi's Method Based Study on Erosion of Advanced Mo2C–WC10Co4Cr Coating for the Centrifugal Pump. *Advanced Materials Processing and Technology* 2024, 10, 2569–2581.
26. Singh, J.; Kumar, S.; Kumar, R. Support Vector Machine-Based Prediction of Wear in Ni-20Cr2O3 Thermal Spray Coating Under Bottom Ash Slurry Conditions. *Jurnal Tribologi* 2023, 39, 84–105.
27. Gholizadeh, N.; Musilek, P. A Generalized Deep Reinforcement Learning Model for Distribution Network Reconfiguration with Power Flow-Based Action-Space Sampling. *Energies* 2024, 17, 5187. https://doi.org/10.3390/en17205187.
28. Deng, L. A Tutorial Survey of Architectures, Algorithms, and Applications for Deep Learning. *Asia Pacific Signal and Information Processing Association Transactions on Signal and Information Processing* 2014, 3, e2.
29. Bengio, Y. *Learning Deep Architectures for AI.* Now Publishers Inc., Delft, The Netherlands, 2009.

30. Song, Z.; Shi, Z.; Yan, X.; Zhang, B.; Song, S.; Tang, C. An Improved Weighted Cross-Entropy-Based Convolutional Neural Network for Auxiliary Diagnosis of Pneumonia. *Electronics* 2024, 13, 2929.
31. Mienye, I.D.; Swart, T.G.; Obaido, G. Recurrent Neural Networks: A Comprehensive Review of Architectures, Variants, and Applications. *Information* 2024, 15, 517.
32. Nanduri, A.; Sherry, L. Anomaly Detection in Aircraft Data using Recurrent Neural Networks (RNN). In *Proceedings of the 12th International Conference on Networking and Services (ICNS'16)*, Lisbon, Portugal, 26–30 June 2016, 5C2-1.
33. Singh, J.; Singh S. A Review on Machine Learning Aspect in Physics and Mechanics of Glasses. *Materials Science and Engineering B* 2022, 284, 115858.
34. Kingma, D.P.; Welling, M. Auto-Encoding Variational Bayes. In *Proceedings of the 1st International Conference on Learning Representations (ICLR'13)*, Scottsdale, AZ, USA, 2–4 May 2013; OpenReview: Scottsdale, AZ, USA, 2013.
35. Li, Y.; Yang, Y.; Zhu, K.; Zhang, J. Clothing Sale Forecasting by a Composite GRU–Prophet Model with an Attention Mechanism. *IEEE Transactions on Industrial Informatics* 2021, 17, 8335–8344.
36. Wang, Y.; Deng, L.; Zheng, L.; Gao, R.X. Temporal Convolutional Network with Soft Thresholding and Attention Mechanism for Machinery Prognostics. *Journal of Manufacturing Systems* 2021, 60, 512–526.
37. van Veen, F. *The Neural Network Zoo*. https://www.asimovinstitute.org/neural-network-zoo/ (accessed on 17 September 2024).
38. Singh, J.; Kumar, S.; Kumar R.; et al. Machine Learning Regression Tools for Erosion Prediction of WC-10Co4Cr Thermal Spray Coating. *Results in Surfaces and Interfaces* 2023, 13, 100156.
39. Singh, J.; Singh, S.; Vasudev, H.; et al. Artificial Neural Network Model for Wear Characteristic Analysis of WC-10Co4Cr and Stellite 6 Thermal Spray Coatings. *Tribology International* 2024, 199, 109924.
40. Lessmeier, C.; Kimotho, J.; Zimmer, D.; Sextro, W. *Condition Monitoring of Bearing Damage in Electromechanical Drive Systems by Using Motor Current Signals of Electric Motors: A Benchmark Data Set for Data-Driven Classification*. https://mb.uni-paderborn.de/en/kat/main-research/datacenter/bearing-datacenter/data-sets-and-download/ (accessed on 17 September 2024).
41. Cui, W.; Zhang, Y.; Zhang, X.; Li, L.; Liou, F. Metal Additive Manufacturing Parts Inspection Using Convolutional Neural Network. *Applied Science* 2020, 10, 545.
42. Gao, J.; Song, X.; Wen, Q.; Wang, P.; Sun, L.; Xu, H. RobustTAD: Robust Time Series Anomaly Detection via Decomposition and Convolutional Neural Networks. arXiv 2020, arXiv:2002.09545.
43. Gao, X.; Deng, F.; Yue, X. Data Augmentation in Fault Diagnosis Based on the Wasserstein Generative Adversarial Network with Gradient Penalty. *Neurocomputing* 2019, 396, 487–494.
44. Bird, J.J.; Faria, D.R.; Ekárt, A.; Ayrosa, P.P. From Simulation to Reality: CNN Transfer Learning for Scene Classification. In *Proceedings of the 10th International Conference on Intelligent Systems (IS'18)*, Varna, Bulgaria, 28–30 August 2020.
45. Chapelle, O.; Scholkopf, B.; Zien, A. Semi-Supervised Learning. *IEEE Transactions on Neural Networks* 2009, 20, 542.
46. Lee, D.H. Pseudo-label: The Simple and Efficient Semi-Supervised Learning Method for Deep Neural Networks. In *Proceedings of the Workshop on Challenges in Representation Learning, ICML 2013*, Vol. 3, Atlanta, GA, USA, 16–21 June 2013.

47. Rajani, C.; Klami, A.; Salmi, A.; Rauhala, T.; Hæggström, E.; Myllymäki, P. Detecting Industrial Fouling by Monotonicity During Ultrasonic Cleaning. In *Proceedings of the International Workshop on Machine Learning for Signal Processing (MLSP'18)*, Aalborg, Denmark, 17–20 September 2018, 1–6.
48. Wei, W.; Fang, J.; Yang, N.; Li, Q.; Hu, L.; Zhao, L.; Han, J. AC-ModNet: Molecular Reverse Design Network Based on Attribute Classification. *International Journal of Molecular Sciences* 2024, 25, 6940.
49. Zhang, S.; Ye, F.; Wang, B.; Habetler, T.G. Semi-Supervised Learning of Bearing Anomaly Detection via Deep Variational Autoencoders. arXiv 2019, arXiv:1912.01096.
50. Yang, Z.; Cohen, W.; Salakhudinov, R. Revisiting Semi-Supervised Learning with Graph Embeddings. In *Proceedings of the International Conference on Machine Learning (ICML'16)*, New York, NY, USA, 19–24 June 2016, 40–48.
51. Chen, C.; Liu, Y.; Kumar, M.; Qin, J.; Ren, Y. Energy Consumption Modelling using Deep Learning Embedded Semi-Supervised Learning. *Computers and Industrial Engineering* 2019, 135, 757–765.
52. Jospin, L.V.; Buntine, W.L.; Boussaid, F.; Laga, H.; Bennamoun, M. Hands-on Bayesian Neural Networks—A Tutorial for Deep Learning Users. arXiv 2018, arXiv:2007.06823.
53. Houlsby, N.; Huszár, F.; Ghahramani, Z.; Lengyel, M. Bayesian Active Learning for Classification and Preference Learning. arXiv 2018, arXiv:1112.5745.
54. Kirsch, A.; Van Amersfoort, J.; Gal, Y. Batchbald: Efficient and Diverse Batch Acquisition for Deep Bayesian Active Learning. In *Proceedings of the 33rd Conference Neural Information Processing Systems (NIPS'19)*, Vol. 32, Vancouver, BC, Canada, 8–14 December 2019, 7026–7037.
55. Martínez-Arellano, G.; Ratchev, S. Towards an Active Learning Approach to Tool Condition Monitoring with Bayesian Deep Learning. In *Proceedings of the ECMS 2019*, Caserta, Italy, 11–14 June 2019, 223–229.
56. Geifman, Y.; El-Yaniv, R. Deep Active Learning over the Long Tail. arXiv 2017, arXiv:1711.00941.
57. Sener, O.; Savarese, S. Active Learning for Convolutional Neural Networks: A Core-Set Approach. In *Proceedings of the International Conference Neural Information Processing Systems (ICLR'18)*, Siem Reap, Cambodia, 13–16 December 2018; ACM: Vancouver, BC, Canada, 2018.
58. Chen, J.; Zhou, D.; Guo, Z.; Lin, J.; Lyu, C.; Lu, C. An Active Learning Method based on Uncertainty and Complexity for Gearbox Fault Diagnosis. *IEEE Access* 2019, 7, 9022–9031.
59. Xu, Z.; Liu, J.; Luo, X.; Zhang, T. Cross-Version Defect Prediction via Hybrid Active Learning with Kernel Principal Component Analysis. In *Proceedings of the 25th International Conference on Software Analysis, Evolution and Reengineering (SANER'18)*, Campobasso, Italy, 20–23 March 2018.
60. Pan, S.J.; Yang, Q. A Survey on Transfer Learning. *IEEE Transactions on Knowledge and Data Engineering* 2009, 22, 1345–1359.
61. Perdana, R.S.; Ishida, Y. Instance-Based Deep Transfer Learning on Cross-domain Image Captioning. In *Proceedings of the International Electronics Symposium (IES'2019)*, Surabaya, Indonesia, 27–28 September 2019, 24–30.
62. Wang, B.; Qiu, M.; Wang, X.; Li, Y.; Gong, Y.; Zeng, X.; Huang, J.; Zheng, B.; Cai, D.; Zhou, J. A Minmax Game for Instance-based Selective Transfer Learning. In *Proceedings of the 25th International Conference on Knowledge Discovery and Data Mining (KDD'19)*, Anchorage, AK, USA, 4–8 August 2019, 34–43.

63. Zhang, L.; Guo, L.; Gao, H.; Dong, D.; Fu, G.; Hong, X. Instance-based Ensemble Deep Transfer Learning Network: A New Intelligent Degradation Recognition Method and Its Application on Ball Screw. *Mechanical Systems and Signal Processing* 2020, 140, 106681.
64. Long, M.; Wang, J. Learning Multiple Tasks with Deep Relationship Networks. In *Proceedings of the 31st Conference on Neural Information Processing Systems (NIPS'17)*, Long Beach, CA, USA, 4–9 December 2017, 1593–1602.
65. Zhu, J.; Chen, N.; Shen, C. A New Deep Transfer Learning Method for Bearing Fault Diagnosis under Different Working Conditions. *IEEE Sensors Journal* 2019, 20, 8394–8402.
66. Zhao, J.; Han, X.; Wang, C.; Yang, J.; Zhang, G. Contrastive-Active Transfer Learning-Based Real-Time Adaptive Assessment Method for Power System Transient Stability. *Sensors* 2024, 24, 5052.
67. Singh, J.; Singh, S.; Singh J.P. Case Study on Implementation of ANN Technique for Analysis of Erosion of Hard Coatings in Different Slurries. In *AIP Conference Proceedings*, 2024.
68. Weiaicunzai. Awesome—Image Classification. https://github.com/weiaicunzai/awesome-image-classification (accessed on 18 September 2024).
69. Parisi, G.I.; Kemker, R.; Part, J.L.; Kanan, C.; Wermter, S. Continual Lifelong Learning with Neural Networks: A Review. *Neural Network* 2019, 113, 54–71.
70. Zenke, F.; Poole, B.; Ganguli, S. Continual Learning through Synaptic Intelligence. In *Proceedings of the 34th International Conference on Machine Learning (ICML'17)*, Sydney, Australia, 6–11 August 2017, 3987–3995.
71. Li, Z.; Hoiem, D. Learning without Forgetting. *IEEE Transactions on Pattern Analysis and Machine Intelligence* 2017, 40, 2935–2947.
72. Li, D.; Tasci, S.; Ghosh, S.; Zhu, J.; Zhang, J.; Heck, L. RILOD: Near Real-Time Incremental Learning for Object Detection at the Edge. In *Proceedings of the 4th Symposium on Edge Computing (SEC'19)*, Arlington, VA, USA, 7–9 November 2019, 113–126.
73. De Lange, M.; Aljundi, R.; Masana, M.; Parisot, S.; Jia, X.; Leonardis, A.; Slabaugh, G.; Tuytelaars, T. Continual Learning: A Comparative Study on How to Defy Forgetting in Classification Tasks. arXiv 2019, arXiv:1909.08383.
74. Lopez-Paz, D.; Ranzato, M. Gradient Episodic Memory for Continual Learning. In *Proceedings of the 31st Conference on Neural Information Processing Systems (NIPS'17)*, Long Beach, CA, USA, 4–9 December 2017, 6467–6476.
75. Shin, H.; Lee, J.K.; Kim, J.; Kim, J. Continual Learning with Deep Generative Replay. In *Proceedings of the 31st Conference on Neural Information Processing Systems (NIPS'17)*, Long Beach, CA, USA, 4–9 December 2017, 2990–2999.
76. Wiewel, F.; Yang, B. Continual Learning for Anomaly Detection with Variational Autoencoder. In *Proceedings of the 44th International Conference on Acoustics, Speech and Signal Processing (ICASSP'19)*, Brighton, UK, 12–17 May 2019, 3837–3841.
77. Mallya, A.; Lazebnik, S. Packnet: Adding Multiple Tasks to A Single Network by Iterative Pruning. In *Proceedings of the 34th Conference on Computer Vision and Pattern Recognition (CVPR'18)*, Salt Lake City, UT, USA, 18–22 June 2018, 7765–7773.
78. Rusu, A.A.; Rabinowitz, N.C.; Desjardins, G.; Soyer, H.; Kirkpatrick, J.; Kavukcuoglu, K.; Pascanu, R.; Hadsell, R. Progressive Neural Networks. arXiv 2016, arXiv:1606.04671.
79. Castro, F.M.; Marín-Jiménez, M.J.; Guil, N.; Schmid, C.; Alahari, K. End-to-End Incremental Learning. In *Proceedings of the 15th European Conference on Computer Vision (ECCV'18)*, Munich, Germany, 8–14 September 2018, 233–248.
80. Tuptuk, N.; Hailes, S. Security of Smart Manufacturing Systems. *Journal of Manufacturing Systems* 2018, 47, 93–106.

81. Samarati, P.; Sweeney, L. Protecting Privacy when Disclosing Information: K-anonymity and its Enforcement through Generalization and Suppression. In *Proceedings of the IEEE Symposium on Research in Security and Privacy (S&P)*, Oakland, CA, USA, 3–6 May 1998.
82. Machanavajjhala, A.; Kifer, D.; Gehrke, J.; Venkitasubramaniam, M. L-diversity: Privacy Beyond K-anonymity. *ACM Transactions on Knowledge Discovery from Data (TKDD)* 2007, 1, 3-es.
83. Li, N.; Li, T.; Venkatasubramanian, S. t-Closeness: Privacy Beyond k-Anonymity and l-Diversity. In *Proceedings of the 2007 IEEE 23rd International Conference on Data Engineering (ICDE'13)*, Istanbul, Turkey, 15–20 April 2007.
84. Meden, B.; Emerši˘c, Ž.; Štruc, V.; Peer, P. K-Same-Net: K-Anonymity with Generative Deep Neural Networks for Face Deidentification. *Entropy* 2018, 20, 60.
85. Gentry, C. Fully Homomorphic Encryption using Ideal Lattices. In *Proceedings of the 41st Annual Symposium on Theory of Computing (STOC'09)*, Washington, DC, USA, 31 May–2 June 2009, 169–178.
86. Zuo, B.; Li, J.; Zhang, Y.; Shen, J. Identity-Based Online/Offline Encryption Scheme from LWE. *Information* 2024, 15, 539.
87. Zhang, Q.; Yang, L.T.; Chen, Z. Privacy Preserving Deep Computation Model on Cloud for Big Data Feature Learning. *IEEE Transactions on Computers* 2015, 65, 1351–1362.
88. Gilad-Bachrach, R.; Dowlin, N.; Laine, K.; Lauter, K.; Naehrig, M.; Wernsing, J. Cryptonets: Applying Neural Networks to Encrypted Data with High Throughput and Accuracy. In *Proceedings of the International Conference on Machine Learning (ICML'16)*, New York, NY, USA, 20–22 June 2016, 201–210.
89. Azure. *Enterprise Security and Governance for Azure Machine Learning*. https://docs.microsoft.com/en-us/azure/security/fundamentals/encryption-overview (accessed on 12 October 2024).
90. Meng, X. *Machine Learning Models that Act on Encrypted Data*. https://www.amazon.science/blog/machinelearning-models-that-act-on-encrypted-data (accessed on 12 October 2024).
91. Zhu, L.; Tang, X.; Shen, M.; Gao, F.; Zhang, J.; Du, X. Privacy-Preserving Machine Learning Training in IoT Aggregation Scenarios. *IEEE Internet Things Journal* 2021, 8, 12106–12118.
92. Ryffel, T.; Dufour-Sans, E.; Gay, R.; Bach, F.; Pointcheval, D. Partially Encrypted Machine Learning using Functional Encryption. In *Proceedings of the 33rd Conference Neural Information Processing Systems (NIPS'19)*, Vancouver, BC, Canada, 8–14 December 2019.
93. Abdalla, M.; Bourse, F.; Caro, A.D.; Pointcheval, D. Simple Functional Encryption Schemes for Inner Products. In *Proceedings of the IACR International Workshop on Public Key Cryptography*, Gaithersburg, MD, USA, 30 March–1 April 2015, 733–751.
94. Boneh, D.; Franklin, M. Identity-based Encryption from the Weil Pairing. In *Proceedings of the Annual International Cryptology Conference (CRYPTO'01)*, Santa Barbara, CA, USA, 19–23 August 2001, 213–229.
95. Singh, J. Slurry Erosion Performance Analysis and Characterization of High-Velocity Oxy-Fuel Sprayed Ni and Co Hardsurfacing Alloy Coatings. *Journal of King Saud University - Engineering Sciences* 2021, https://doi.org/10.1016/j.jksues.2021.06.009.
96. Song, Q.; Cao, J.; Sun, K.; Li, Q.; Xu, K. Try before You Buy: Privacy-preserving Data Evaluation on Cloud-based Machine Learning Data Marketplace. In *Proceedings of the Annual Computer Security Applications Conference (ACSAC'21)*, Austin, TX, USA, 6–10 December 2021, 260–272.
97. Dwork, C.; Roth, A. The Algorithmic Foundations of Differential Privacy. *Foundations and Trends in Theoretical Computer Science* 2014, 9, 3–4.

98. Kenthapadi, K.; Mironov, I.; Thakurta, A.G. Privacy-preserving Data Mining in Industry. In *Proceedings of the WSDM 2019*, Melbourne, Australia, 11–15 February 2019, 840–841.
99. Xu, R.; Joshi, J.B.; Li, C. Cryptonn: Training Neural Networks over Encrypted Data. In *Proceedings of the 39th International Conference on Distributed Computing Systems (ICDCS'19)*, Dallas, TX, USA, 7–9 July 2019, 1199–1209.
100. Huang, C.; Kairouz, P.; Chen, X.; Sankar, L.; Rajagopal, R. Context-aware Generative Adversarial Privacy. *Entropy* 2017, 19, 656.
101. Kopp, A. *Microsoft SmartNoise Differential Privacy Machine Learning Case Studies*. Microsoft Azure White Papers, Microsoft Corporation, Redmond, WA, USA, 2021.
102. Wu, J.; Dang, Y.; Jia, H.; Liu, X.; Lv, Z. Prediction of Energy Consumption in Digital Twins of Intelligent Factory by Artificial Intelligence. In *Proceedings of the 2021 International Conference on Technology and Policy in Energy and Electric Power (ICT-PEP'21)*, Yogyakarta, Indonesia, 29–30 September 2021, 354–359.
103. Abadi, M.; Chu, A.; Goodfellow, I.; McMahan, H.B.; Mironov, I.; Talwar, K.; Zhang, L. Deep Learning with Differential Privacy. In *Proceedings of the Conference on Computer and Communications Security (SIGSAC'16)*, Vienna, Austria, 24–28 October 2016, 308–318.
104. Papernot, N.; Song, S.; Mironov, I.; Raghunathan, A.; Talwar, K.; Erlingsson, U. Scalable Private Learning with PATE. In *Proceedings of the 6th International Conference on Learning Representations (ICLR'18)*, Vancouver, BC, Canada, 30 April–3 May 2018.
105. Gong, M.; Xie, Y.; Pan, K.; Feng, K.; Qin, A.K. A Survey on Differentially Private Machine Learning. *IEEE Computational Intelligence Magazine* 2020, 15, 49–64.
106. Konecny, J.; McMahan, H.B.; Yu, F.X.; Richtárik, P.; Suresh, A.T.; Bacon, D. Federated Learning: Strategies for Improving Communication Efficiency. arXiv 2016, arXiv:1610.05492.
107. Sharma, V.; Singh, J.P.; Raman, R.; Bathla, G.; Saini, A.; Machine Learning Approach for Prediction Analysis of Aluminium Alloy on the Surface Roughness Using CO_2 Laser Machining. *Proceedings of the Institution of Mechanical Engineers, Part E: Journal of Process Mechanical Engineering* 2024, 09544089241231093.
108. Hao, M.; Li, H.; Luo, X.; Xu, G.; Yang, H.; Liu, S. Efficient and Privacy-Enhanced Federated Learning for Industrial Artificial Intelligence. *IEEE Transactions on Industrial Informatics* 2019, 16, 6532–6542.
109. Bagheri, B.; Rezapoor, M.; Lee, J. A Unified Data Security Framework for Federated Prognostics and Health Management in Smart Manufacturing. *Manufacturing Letters* 2020, 24, 136–139.
110. Wei, K.; Li, J.; Ding, M.; Ma, C.; Yang, H.H.; Farokhi, F.; Jin, S.; Quek, T.Q.; Poor, H.V. Federated Learning with Differential Privacy: Algorithms and Performance Analysis. *IEEE Transactions on Information Forensics and Security* 2020, 15, 3454–3469.
111. Zhou, W.; Li, Y.; Chen, S.; Ding, B. Real-Time Data Processing Architecture for Multi-Robots Based on Differential Federated Learning. In *Proceedings of the SmartWorld /SCALCOM/UIC/ATC/CBDCom/IOP/SCI 2018*, Guangzhou, China, 8–12 October 2018, 462–471.
112. Vepakomma, P.; Gupta, O.; Swedish, T.; Raskar, R. Split Learning for Health: Distributed Deep Learning without Sharing Raw Patient Data. arXiv 2018, arXiv:1812.00564.
113. Abuadbba, S.; Kim, K.; Kim, M.; Thapa, C.; Camtepe, S.A.; Gao, Y.; Kim, H.; Nepal, S. Can We Use Split Learning on 1D CNN Models for Privacy Preserving Training? arXiv 2020, arXiv:2003.12365.

114. Sharma, V.; Singh, J.P.; Hussan, K.; Kumar, R. Machine Learning Approach for Predicting the Machining Characteristics of Additive Manufactured ABS Material. *Proceedings of the Institution of Mechanical Engineers, Part C: Journal of Mechanical Engineering Science* 2024, 09544062241283330.
115. Ren, P.; Xiao, Y.; Chang, X.; Huang, P.Y.; Li, Z.; Chen, X.; Wang, X. A Comprehensive Survey of Neural Architecture Search: Challenges and Solutions. *ACM Computing Surveys (CSUR)* 2021, 54, 1–34.
116. Dreyfus, P.A.; Pélissier, A.; Psarommatis, F.; Kiritsis, D. Data-based Model Maintenance in the Era of Industry 4.0: A Methodology. *Journal of Manufacturing Systems* 2022, 63, 304–316.
117. Lu, J.; Liu, A.; Dong, F.; Gu, F.; Gama, J.; Zhang, G. Learning under Concept Drift: A Review. *IEEE Transactions on Knowledge and Data Engineering* 2018, 31, 2346–2363.
118. Singh, J.; Singh, S. A Review on Machine Learning Aspect in Physics and Mechanics of Glasses. *Materials Science and Engineering: B* 2022, 284, 115858.
119. Kabir, M.A.; Keung, J.W.; Bennin, K.E.; Zhang, M. Assessing the Significant Impact of Concept Drift in Software Defect Prediction. In *Proceedings of the 43rd Annual Computer Software and Applications Conference (COMPSAC'19)*, Vol. 1, Milwaukee, WI, USA, 15–19 July 2019, 53–58.
120. Zenisek, J.; Holzinger, F.; Affenzeller, M. Machine Learning-based Concept Drift Detection for Predictive Maintenance. *Computers and Industrial Engineering* 2019, 137, 106031.
121. Bermeo-Ayerbe, M.A.; Ocampo-Martinez, C.; Diaz-Rozo, J. Data-driven Energy Prediction Modeling for Both Energy Efficiency and Maintenance in Smart Manufacturing Systems. *Energy* 2022, 238, 121691.
122. Gal, Y. *Uncertainty in Deep Learning.* Ph.D. Dissertation, Cambridge University Press, Cambridge, 2016.
123. Singh, J.; Kumar, S.; Mohapatra, S.K. Study on Solid Particle Erosion of Pump Materials by Fly Ash Slurry using Taguchi's Orthogonal Array. *Tribologia - Finnish Journal of Tribology* 2021, 38, 31–38.
124. Benker, M.; Furtner, L.; Semm, T.; Zaeh, M.F. Utilizing Uncertainty Information in Remaining Useful Life Estimation via Bayesian Neural Networks and Hamiltonian Monte Carlo. *Journal of Manufacturing Systems* 2021, 61, 799–807.
125. Lakshminarayanan, B.; Pritzel, A.; Blundell, C. Simple and Scalable Predictive Uncertainty Estimation using Deep Ensembles. In *Proceedings of the 31st Conference on Neural Information Processing Systems (NIPS'17)*, Long Beach, CA, USA, 4–9 December 2017, 6402–6413.
126. Nemani, V.P.; Lu, H.; Thelen, A.; Hu, C.; Zimmerman, A.T. Ensembles of Probabilistic LSTM Predictors and Correctors for Bearing Prognostics using Industrial Standards. *Neurocomputing* 2022, 491, 575–596.
127. DeVries, T.; Taylor, G.W. Learning Confidence for Out-of-distribution Detection in Neural Networks. arXiv 2018, arXiv:1802.04865.
128. Hendrycks, D.; Gimpel, K. A Baseline for Detecting Misclassified and Out-of-distribution Examples in Neural Networks. In *Proceedings of the 7th International Conference Neural Information Processing Systems (ICLR'18)*, Montréal, ON, Canada, 3–8 December 2018; OpenReview: San Juan, Puerto Rico, 2018.
129. Lee, K.; Lee, K.; Lee, H.; Shin, J. A Simple Unified Framework for Detecting Out-of-distribution Samples and Adversarial Attacks. In *Proceedings of the 32nd Conference on Neural Information Processing Systems (NIPS'18)*, Montréal, QC, Canada, 3–8 December 2018, 7167–7177.

130. Bendale, A.; Boult, T.E. Towards Open Set Deep Networks. In *Proceedings of the 32nd International Conference on Computer Vision and Pattern Recognition (CVPR'16)*, Las Vegas, NV, USA, 27–30 June 2016, 1563–1572.
131. Xu, J.; Kovatsch, M.; Lucia, S. Open Set Recognition for Machinery Fault Diagnosis. In *Proceedings of the 19th International Conference on Industrial Informatics (INDIN'21)*, Palma de Mallorca, Spain, 21–23 July 2021, 1–7.
132. Berthelot, D.; Roelofs, R.; Sohn, K.; Carlini, N.; Kurakin, A. AdaMatch: A Unified Approach to Semi-Supervised Learning and Domain Adaptation. In *Proceedings of the 9th International Conference on Learning Representations (ICLR'21)*, Online, 3–7 May 2021.
133. Wang, Y.; Gu, Q.; Brown, D. Differentially Private Hypothesis Transfer Learning. In *Proceedings of the European Conference on Machine Learning and Principles and Practice of Knowledge Discovery in Databases (ECML-PKDD'18)*, Dublin, Ireland, 10–14 September 2018, 811–826.
134. Farrall, F.; Mittal, N.; Narra, C.; Tello, J.; Dow, E.M. *Data-Sharing Made Easy*. Technical Report, Deloitte, London, UK, 2021.
135. Otto, B.; Mohr, N.; Roggendorf, M.; Guggenberger, T. *Data Sharing in Industrial Ecosystems*. Technical Report, McKinsey, New York, NY, USA, 2020.
136. Pandey, A.V. *Future of AI in Business: Adaptive Machine Learning Models That Evolve and Improve over Time without Expensive Retraining*. https://www.nvidia.com/en-us/on-demand/session/gtcfall20-a21421/ (accessed on 12 October 2024).
137. Cattaneo, G.; Vanara, F.; Massaro, A. *Advanced Technologies for Industry—ATWATCH technology Focus on Cloud Computing*. Technical Report, EU Commission: Brussels, Belgium, 2020.

16 Zero-Waste Manufacturing

The Role of Sustainable Practices and Circular Economy Solutions

Gursimran Kaur, Shyam Sunder Agrawal, Saloni Raheja, Nitin Goyal, Ankitha Sharma, and Amanjot Singh Syan

Description	**Abbreviation**
Polylactic acid	PLA
Public-private partnerships	PPPs

16.1 INTRODUCTION

The world is facing an approaching environmental crisis caused by unsustainable manufacturing and consumption. The current model of extracting raw materials, producing goods, and then positioning them as waste is not only exhausting the planet's finite resources but also increasing the problem of environmental pollution and climate change.

To address these encounters, a standard shift to sustainable and circular manufacturing practices is critical. The circular reduced model, which highlights the reuse, recycling, and regeneration of resources, offers a promising solution. By participating in production and consumption along with the natural environment, the circular economy proposes to eradicate waste and preserve the value of products as well as materials for as long as possible.

This method is supported by several key principles, including the waste management pyramid of reduce, reuse, and recycle, the concept that one industry's waste is another's resource, and the sustainable commercial model of product–service

DOI: 10.1201/9781003516071-16

systems. The implementation of digital technologies, such as those connected with Industry 4.0, can enable and quicken the transition to a more circular and sustainable industrial ecosystem (Cioffi et al., 2020).

The modification toward zero-waste manufacturing necessitates a comprehensive strategy that talks about the entire product lifecycle, from proposal and production to consumption and end-of-life management. This comprises enhancing resource utilization, improving substantial indirectness, and developing advanced business models that arrange longevity, reuse, and recycling over the outdated linear "take–make–waste" method (Ranta & Saari, 2019; Dam et al., 2020).

Moreover, the circular economy not only suggests environmental benefits, but also has the possibility to drive economic growth and create new job opportunities. By reconsidering production and consumption patterns, manufacturers can reveal new revenue streams, improve operational efficiency, and reduce their dependence on finite resources (Ahmed et al., 2021).

To effectively devise zero-waste manufacturing and circular economy solutions, collaboration across the various stakeholders, including policymakers, industry leaders, researchers, and consumers, is vital. Emerging organized regulatory frameworks, nurturing cross-sectoral partnerships, and encouraging public awareness and engagement are all indispensable elements in driving this conversion (González-Sánchez et al., 2020).

16.1.1 Role of Role of Sustainable Practices Toward a Zero-Waste Manufacturing System

The search for a zero-waste manufacturing system is a vital response to the environmental issues posed by traditional production methods, which frequently result in significant waste generation and resource depletion. Sustainable applications are central to achieving this goal, as they not only reduce waste but also encourage a holistic method of resource management and ecological stewardship.

At the heart of zero-waste manufacturing is the attitude of waste minimization, which can be efficiently addressed through lean manufacturing techniques. Lean manufacturing focuses on the abolition of waste in all forms—whether it be material, energy, or time—by enhancing the production processes (Womack & Jones, 2003). This practice inspires manufacturers to examine every step of their processes to classify inefficiencies and reduce excess. By applying lean principles, the companies can knowingly lower the volume of waste while improving the overall efficiency.

Complementing the lean practices, the implementation of circular economy principles is critical for the development of a zero-waste environment. The circular economy highlights designing products for durability and reparability, allowing the materials to be reused and recycled rather than rejected (Ellen MacArthur Foundation, 2019). This approach inspires the manufacturers to reconsider their product lifecycle, directing on sustainable design that diminishes waste generation from the onset. By embracing circularity, businesses can diminish their dependence on materials and reduce their environmental impact.

Renewable energy incorporation is another dynamic component of sustainable manufacturing practices expected to achieve zero waste. The reliance on fossil energies not only pays to greenhouse gas emissions but also produces waste through pollution and resource extraction (International Energy Agency, 2021). By transitioning to renewable energy sources like solar, wind, and hydropower, manufacturers can meaningfully decrease their carbon footprint and reduce the waste associated with energy production. This shift not only cares the eco-friendly sustainability but expands energy security and decreases exposure to the fossil fuel market fluctuations.

Sustainable sourcing of materials is equally noteworthy in diminishing the waste in manufacturing. Companies are progressively ordering the use of recycled materials and sustainable resources, which diminishes the ecological impression connected with raw material abstraction and dispensation (Kolk & van Tuijl, 2016).

Technological advancements also play a serious role in allowing sustainable practices in the manufacturing sector. Technology like the Internet of Things (IoT), big data analytics, and automation help the real-time monitoring and optimization of production processes. For example, IoT devices can track material usage and waste generation, helping builders make data-driven decisions that expand efficiency and diminish waste (Lee et al., 2018). These technologies can lead to more up-to-date resource management and help with the continuing development toward a zero-waste goal.

16.2 LITERATURE REVIEW

16.2.1 Zero-Waste Manufacturing Industries in India

The concept of zero-waste manufacturing has increased in importance in recent years in India's quickly growing industrial landscape. The manufacturing industry is a main consumer of material and energy resources, creating a considerable quantity of waste that poses a remarkable challenge to the environment. However, the industry also serves as a mainstay to the country's economic prosperity through its GDP and employment.

To address this challenge, the Indian government has been vigorously reassuring the adoption of production principles in order to reduce waste and advance resource utilization (Deshmukh et al., 2017). These ideologies have been predominantly appropriate for the country's small-scale industries, which have tested intense global competition and the need to stay with changing customer demands (Jain & Malik, 2013).

Beyond restructuring the production processes, the application of lean manufacturing has been created to yield added benefits for Indian industries. Through the skilled workers, cumulative job satisfaction, and rapid accomplishment in the manufacturing process, lean practices have the potential to adopt a more engaged and motivated workforce, driving productivity, and efficacy gains.

The acceptance of lean manufacturing principles has not remained without its challenges, however. The change to a leaner, more maintainable production model can be multifaceted, particularly for lesser enterprises that may not have the resources and expertise required (Panizzolo et al., 2012). Nonetheless, the possible rewards of embracing zero-waste manufacturing make a convincing case for Indian industries to endure exploring and refining these strategies, eventually making the country a global leader in sustainable industrial expansion.

16.2.2 Circular Economy: Toward Sustainable Development

The concept of a circular economy has risen as a exceptional method for addressing the global ecological challenges we face today. In a traditional linear economy, resources are removed, used, and then discarded, leading to waste, pollution, and resource depletion. In contrast, the circular economy aims to detach economic growth from the use of raw materials more efficiently through the transition towards more sustainable production and consumption patterns that includes all enterprises and stakeholders (Cheng & Chou, 2018). It is founded on the principles of reducing, reusing, and recycling, with the objective of preserving the value of products and materials for as long as possible (Ranta & Saari, 2019). This includes both biological materials, which can be returned to natural cycles, and technical materials, which can be reused multiple times without producing waste (Ranta & Saari, 2019).

A key aspect of the circular economy is its worldwide and collaborative nature (Konietzko et al., 2020). The Ellen MacArthur Foundation, a leading advocate of the circular economy, has emphasized the necessity of a worldwide systems-level change that extends to the public and private sectors and value chains (Konietzko et al., 2020).

Changing to a circular economy can convey significant environmental and economic benefits. By reducing dependence on natural resources, the circular economy can assist in lessening climate change, loss of biodiversity, and pollution (Sahajwalla & Hossain, 2023). Therefore, the circular economy can create new economic prospects, inspire local job creation, and adopt energy independence (Cheng & Chou, 2018).

However, changing to a circular economy is not without its problems. It involves a fundamental alteration in the way we produce, consume, and dispose of goods, as well as a cooperative effort by all stakeholders. Consumers play a vital role in this transition, as their consumption behavior can either drive or hinder a shift toward a more sustainable and circular financial model (Cheng & Chou, 2018).

The circular economy proposes a talented path toward additional sustainable development, but its achievement will be contingent on the collective efforts of governments, businesses, and consumers to devise principles to make a more sustainable, resilient, and impartial future.

16.2.3 Zero-Waste or Circular Economy

The zero-waste or circular economy has been seen in recent years as a means of meeting environmental challenges, which were caused by the traditional linear economic models centered on the "take–make–waste" method. The flows of materials are intended to be either biological, able to be restored to the biosphere, or technical, accomplished by being recycled various times without producing waste (Andreola et al., 2021). The circular economy is experiential as a solution to the difficulties created by the linear pecuniary model, which has run into significant waste management difficulties (Andreola et al., 2021). It aims to reduce waste by keeping goods and materials in movement for as long as possible, through reducing, recycling, and reprocessing (Vermeşan et al., 2020).

The principle of this economy is preserving efficiency, which is a fundamental idea for the European Union, a major supporter of this economic model (Andreola et al., 2021). Countries that are deficient in natural resources, such as Italy, are particularly interested in adopting circular economy practices to lessen their reliance on imported raw materials (Andreola et al., 2021).

The circular economy model is a waste-conservative approach that emphasizes achieving affluent economic development by saving resources in the form of recycling and other curative practices (Ahmed et al., 2021; Ranta & Saari, 2019).

The application of circular economy principles can take various forms, with the reuse of industrial waste as a resource, the expansion of product–service systems that extend the lifespan of goods, and the implementation of cradle-to-cradle project principles that aim to eradicate waste through the formation of closed-loop systems (Ranta & Saari, 2019).

16.2.4 Public–Private Partnerships (PPPs)

Public–private partnerships play an essential role in the advancing principles of a circular economy, mainly in sectors like support services, education, and recycling. These partnerships influence the strengths of both sectors: governmental authority and regulatory power, on one hand, and private sector novelty and efficacy, on the other. Figure 16.1 illustrates collaborations between public institutions and private companies to promote a circular economy.

First, PPPs carry together the resources and expertise of both sectors to grow and implement policies, regulations, and infrastructure that are in favor of circular practices. Governments deliver the regulatory frameworks and funding, while private entities pay for innovation and effective efficiencies, nurturing a conducive environment for sustainable development.

Second, in education, PPPs play a pivotal role in raising awareness and encouraging the consideration of these economic principles. Cooperative efforts between government agencies, educational institutions, and businesses play a crucial role in programs that educate the public and businesses about sustainable consumption, waste reduction, and the significance of recycling. By promoting these principles in school curricula and public campaigns, PPPs ensure that upcoming generations are fortified with the information and values necessary for a circular economy.

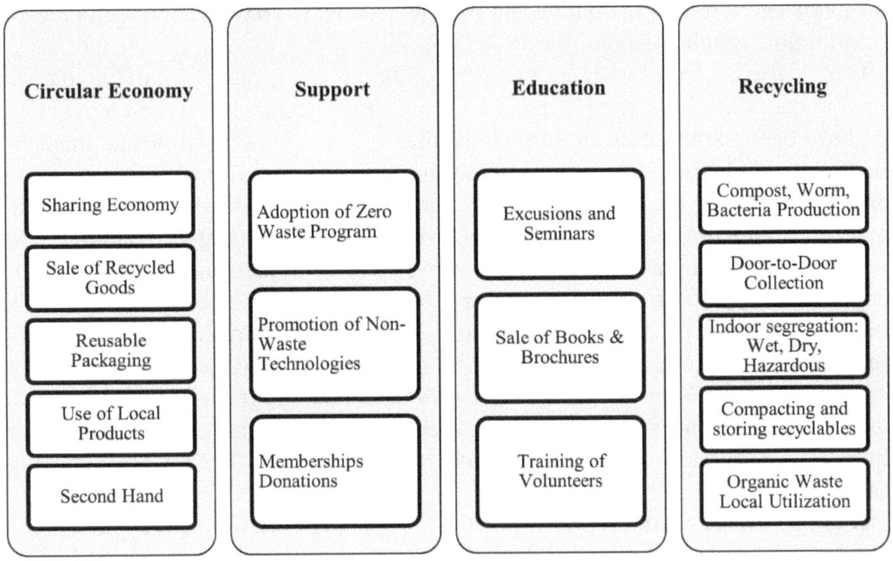

FIGURE 16.1 Public–private partnerships toward circular economy (Credit: created by authors).

Third, in recycling, PPPs drive technical advancements and assets in recycling the infrastructure. This embraces establishing efficient systems for waste collection, sorting, and processing, as well as emerging innovative recycling technologies. By partnering with private firms, governments can influence their proficiency to expand recycling rates and the quality of recycled materials, thereby reducing resource extraction and environmental impact.

16.2.5 Circular Economy: A Sustainable Solution for Zero-Waste

The circular economy has received noteworthy attention in recent years as a practical solution to the global problem. This is a process that sets out to reduce waste and increase the value of resources by keeping them in circulation (Vermeşan et al., 2020; Ranta & Saari, 2019). The traditional linear "take–make–waste" economy is being challenged by this new method, which endorses more sustainable production as well as consumption patterns (Cheng & Chou, 2018).

In a circular economy, the prominence is on reducing waste, reusing, and recycling resources. This method can be practical for biological as well as technical materials (Ranta & Saari, 2019). Biological materials are developed through the renewable sources and natural cycles (Tripathi et al., 2021). Technical materials can be used multiple times without producing waste (Ranta & Saari, 2019). The aims of a circular economy demand a systemic change that comprises the action from various stakeholders, including the public and private sectors, to convert manufacturing and consumption systems (Konietzko et al., 2020).

One of the key enablers of a circular economy is the growth of durable and sustainable materials. For example, polylactic acid (PLA)-based blends and biocomposites have been explored as an auspicious alternative to traditional plastic materials due to their biodegradability and recyclability (Tripathi et al., 2021). These materials can aid in reducing the environmental burden instigated by the linear "take–make–dispose" model and add to the transition toward a more sustainable economy.

Consumers also play a critical role in the positive implementation of a circular economy. The shift from a "throw-away" mentality to conscious and sustainable consumption patterns is essential for substantial economic opportunities, endorsing local job creation, and reducing the environmental hazard (Cheng & Chou, 2018). This transition requires the teamwork of all stakeholders, including policymakers, businesses, and consumers, to generate a comprehensive and rounded approach to attaining a zero-waste circular economy (Konietzko et al., 2020; Cheng & Chou, 2018).

Overall, the circular economy grants a promising solution to the problems of waste management and resource depletion. By adopting a maintainable and restorative approach to production and consumption, we can work toward a future where waste is minimalized, and resources are utilized to their full potential.

16.3 Statistics

India has a major waste management problem. India has the largest population in the world—with 1.43 billion people as of mid-2023—and combined with a growing degree of development, India's daily public waste production has unceasingly grown, surpassing 160,000 metric tons in 2021. With millions of tons of waste ending up

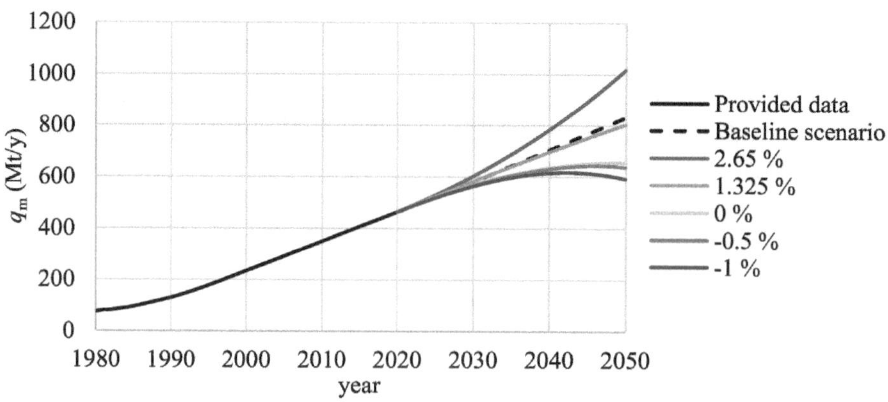

FIGURE 16.2 Projected plastics use change worldwide from 2019 to 2050, by region (Dokl et al. 2024) (Credit: Creative Commons Attribution 4.0 License).

in landfill each year and often dripping back into the environment, enlightening the management system in the country is a top priority for the safe treatment and disposal of waste. Figure 16.2 depicts the anticipated shifts in plastic consumption across different regions, highlighting trends in production and usage.

16.4 SUSTAINABLE WASTE MANAGEMENT: EXPLORING THE CHALLENGES AND OPPORTUNITIES

Waste management is a serious constituent of a sustainable society, as it directly influences the environment, economy, and human health. This chapter provides a complete analysis of the key waste streams, including plastics, tires, paper, metals, and wastewater, and determines the policies and knowledge indispensable to transition toward a more sustainable waste management system (Seadon, 2010; Shams et al., 2014; Chang et al., 2011).

Waste plastics pose a noteworthy challenge due to their non-biodegradable nature and environmental harm caused by inappropriate disposal (Ghosh et al., 2022). The use of plastic has prolonged rapidly across various industries which leads to a continuous increase of plastic waste (Ш е р ш н е в а, 2021). Plastic waste that outflows into the environment can have overwhelming influences on ecosystems, wildlife, and human health (Ilyas et al., 2018). To address this matter, a transition toward a circular economy model, where plastics are intended, shaped, and achieved to allow their reuse and recycling, is important (Hahladakis et al., 2020).

Waste tires also present a challenge, as they are hard to dispose of and can release injurious chemicals into the environment if not following an appropriate and safe process. Tire recycling and the expansion of value-added products from waste tires like rubber-modified asphalt and tire-derived fuel, offer promising solutions to this problem (Lee et al., 2023).

Wastepaper is a significant waste stream that can be efficiently recycled and reused, conducive to preserving efficiency and reducing the environmental impact of paper production. On the other hand, Waste metals can be proficiently improved and re-claimed through healthy recycling programs (Ш е р ш н е в а, 2021).

Wastewater management is also a vital aspect of sustainable waste management, as untreated wastewater can pollute the water bodies and pose a risk to human health as well as the environment. Advances in wastewater treatment technologies and the application of water reuse and reprocessing strategies can minimize the ecological impact of wastewater (Teymourian et al., 2021).

To address these problems, a comprehensive tactic is required, involving policy, technological, and behavioral changes. GIS-based modeling and planning can play a vital role in enhancing waste management systems, improving efficiency, and reducing environmental influences (Chalkias & Lasaridi, 2011). Figure 16.3 represents a GIS framework used for analyzing spatial data and geographical relationships. The development of internet connectivity in current years has considerably improved the possible value of GIS since now it is conceivable to locate and attach the data from many dissimilar GIS databases to produce very rich analytical information on nearly any topic that is related to a physical site (GAO, 2003).

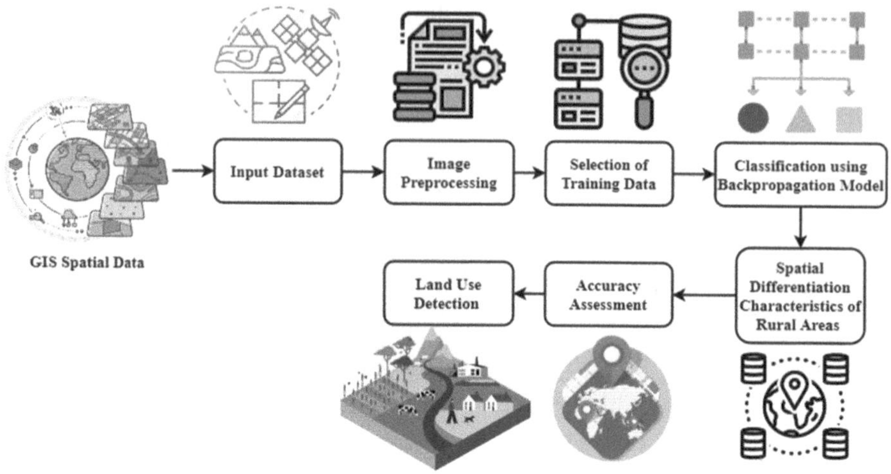

FIGURE 16.3 Geographic information systems model (Chen et al. 2023) (Credit: Creative Commons Attribution 4.0 License).

16.5 BENEFITS OF ZERO-WASTE MANUFACTURING

Zero-waste manufacturing is a progressively popular approach that aims to eradicate waste and exploit the well-organized use of resources throughout the complete production process. This approach offers a range of benefits, both environmental and economic, that make it a good selection for businesses seeking to advance their sustainability and productivity.

One of the primary positives of zero-waste manufacturing is the noteworthy reduction in the environmental impact. By minimizing waste, producers can lower their carbon footprint, preserve natural resources, and decrease their total environmental footprint (Sasane & Adhav, 2020; Bogue, 2014). This not only aids in addressing pressing ecological issues but is also in line with the rising consumer demand for more sustainable products (Bogue, 2014).

In addition to environmental support, zero-waste manufacturing can also deliver substantial monetary advantages for businesses (Samuel et al., 2019; Boyer, 2006). By reducing waste and refining efficacy, manufacturers can lessen their material and disposal costs, which can augment productivity (Samuel et al., 2019). Furthermore, the execution of zero-waste strategies frequently requires the adoption of advanced technologies and processes, which can be central to better-quality efficiency and attractiveness in the market (Bogue, 2014; Boyer, 2006).

One key aspect of zero-waste manufacturing is the acceptance of closed-loop production systems, where waste and by-products are restored into the manufacturing process or repurposed for other aims (Lovins et al., 2006). This approach not only reduces waste but also allows the recovery and reuse of valuable materials, further enhancing the economic and environmental potential of the manufacturing process (Lovins et al., 2006). The benefits of zero-waste manufacturing are wide and compelling (Samuel et al., 2019). By implementing this method, businesses can cut their environmental influence, increase their effectiveness, and place themselves as leaders in sustainable manufacturing practices.

16.6 TOWARD A CIRCULAR ECONOMY: IMPLICATIONS FOR ZERO-WASTE MANUFACTURING

The transition toward a circular economy has noteworthy proposals for the future of industry. A circular economy is a model of sustainable economic development where the generation of waste and pollution is diminished through the upkeep of product and material value for lengthy periods. This viewpoint is distinct from the outdated linear "take–make–waste" model, where the raw materials are used to create goods that are eventually rejected as waste (Ahmed et al., 2021). The core principles of a circular economy—reduce, reuse, and recycle—offer a path to attaining more sustainable manufacturing (Ranta & Saari, 2019). By reducing waste and repurposing materials, producers can decrease their reliance on the determinate natural resources and diminish environmental impact. Furthermore, the adoption of digital technologies, like those enabled by the Industry 4.0, can aid in the application of the circular economy observed across the manufacturing supply chain (Cioffi et al., 2020).

16.7 FUTURE SCOPE OF STUDY IN ZERO-WASTE MANUFACTURING AND CIRCULAR ECONOMY SOLUTIONS

Zero-waste manufacturing and the circular economy portray the transformative tactics intended at diminishing waste generation, exploiting resource efficiency, and nurturing sustainable practices across industries. As global awareness of environmental issues is intensified, and the regulated pressure grows, the study and operation of these principles are require further momentum. This chapter projects the future scope of studying zero-waste manufacturing and its incorporation with circular economy solutions, prominent key drivers, tests, and opportunities.

16.7.1 Drivers of Zero-Waste Manufacturing and Circular Economy

- **Environmental Imperatives**: Climate change, resource depletion, and pollution are persuasive motives for industries to adopt zero-waste practices. By reducing waste generation and recycling materials, companies can lessen environmental impacts and subsidize global sustainability goals (UNEP, 2021).
- **Regulatory Requirements**: Governments worldwide are ratifying stricter environmental regulations, incentivizing businesses to accept sustainable practices. Compliance with principles often dictates the application of zero-waste strategies, positioning them as a critical factor in corporate responsibility and regulatory compliance (European Commission, 2020).
- **Economic Benefits**: Beyond regulatory compliance, zero-waste manufacturing offers economic advantages such as cost savings through reduced material consumption, waste disposal costs, and improved operational

efficiency (Ellen MacArthur Foundation, 2023). Companies implementing these practices can improve effectiveness while reducing their ecological footprint.

- **Technological Advancements**: Quick advancements in technology, such as AI-driven process optimization, advanced recycling skills, and additive manufacturing (3D printing), are allowing more effective resource use and waste reduction in industrial processes (Chang et al., 2022). These inventions are essential in achieving zero-waste goals and attracting the circularity of production systems.

16.7.2 Challenges and Opportunities

- **Complex Supply Chains**: Applying zero-waste strategies across composite supply chains presents challenges related to organization, transparency, and supplier engagement. However, it also offers opportunities for partnership and innovation among stakeholders to improve the material flows and lessen waste jointly (Geng et al., 2021).
- **Behavioral and Cultural Shifts**: Attaining zero waste requires a shift in organizational culture and employee behavior toward sustainability. Education, training, and stakeholder engagement are crucial in nurturing the mindset favorable to waste reduction and preservation efficiency (Heiskanen et al., 2018).
- **Consumer Demand**: Mounting consumer consciousness and preference for eco-friendly products are pressure companies to accept sustainable practices, including zero-waste manufacturing. Meeting consumer expectations for transparency and sustainability can improve a brand's reputation and market attractiveness (Sarkis et al., 2023).

16.7.3 Future Directions in Research and Education

- **Technological Innovation**: Future research in zero-waste manufacturing will emphasize advancing technologies for waste reduction, recycling, and resource recovery. Areas of interest include the expansion of biodegradable materials, closed-loop systems, and sustainable wrapping solutions (Muthu et al., 2022).
- **Policy and Governance**: Research on the policy outlines and governance devices will explore how governments can enable the transition to zero-waste economies through enticements, regulations, and international cooperation (UNEP, 2021).
- **Education and Training**: Educational programs tailored to sustainable manufacturing practices will play a serious role in preparing future generations of professionals to address global ecological trials. Participating in sustainability in engineering, business, and design curricula will adopt novelty and leadership in zero-waste manufacturing (Ellen MacArthur Foundation, 2019).

16.8 CONCLUSION

The change to zero-waste manufacturing and the acceptance of circular economy principles are necessary for focusing on the pressing environmental and economic challenges we face. By applying sustainable practices and advanced solutions, we can generate a more resilient, resource-efficient, and environmentally friendly manufacturing ecosystem that aids both the planet and humankind. Recent research has recognized several key factors that allow the transition toward a circular economy in industry, including product lifecycle management, closed-loop recycling, and ground-breaking business models. By including these strategies, producers can not only lessen waste but also unlock new financial opportunities through the reprocessing and repurposing of material.

Ultimately, the change toward a circular economy in manufacturing signifies a significant challenge, but one that grasps the immense potential to drive maintainable development and alleviate the environmental impacts of industrial production. The future of zero-waste manufacturing and circular economy solutions is promising, determined by environmental imperatives, technological invention, economic benefits, and also evolving consumer preferences. As industries endeavor to realize sustainability goals and regulatory compliance, research and education will play essential roles in advancing knowledge, developing technologies, and shaping policies that encourage a circular economy. By embracing these principles, businesses can not only improve their resilience and competitiveness but also underwrite a more sustainable and equitable future for generations to come.

REFERENCES

Ahmed, A. A., Nazzal, M. A., & Darras, B. M. (2021). Cyber-Physical Systems as an Enabler of Circular Economy to Achieve Sustainable Development Goals: A Comprehensive Review. *Springer Science+Business Media*, 9(3), 955–975. https://doi.org/10.1007/s40684-021-00398-5

Andreola, F., Lancellotti, I., Pozzi, P., & Barbieri, L. (2021). Eco-Compatible Construction Materials Containing Ceramic Sludge and Packaging Glass Cullet. *Multidisciplinary Digital Publishing Institute*, 11(8), 3545–3545. https://doi.org/10.3390/app11083545

Bogue, R. (2014). Sustainable Manufacturing: A Critical Discipline for the Twenty-First Century. *Emerald Publishing Limited*, 34(2), 117–122. https://doi.org/10.1108/aa-01-2014-012

Boyer, J. R. (2006). *What Will It Take to Deliver a Zero-to-Landfill Product?* https://doi.org/10.1115/imece2006-15944

Chalkias, C., & Lasaridi, K. (2011). *Benefits from GIS Based Modelling for Municipal Solid Waste Management.* https://doi.org/10.5772/17087

Chang, N., Pires, A., & Martinho, G. (2011). Empowering Systems Analysis for Solid Waste Management: Challenges, Trends, and Perspectives. *Taylor & Francis*, 41(16), 1449–1530. https://doi.org/10.1080/10643381003608326

Chang, Y., et al. (2022). Advanced Manufacturing Technologies for Sustainability. *Sustainable Production and Consumption*, 32, 123–135.

Chen, L., Hongying, W., & Jing, M. (2023). Image Analysis of Spatial Differentiation Characteristics of Rural Areas Based on GIS Statistical Analysis. *Electronics*, 12(6), 1414. https://doi.org/10.3390/electronics12061414

Cheng, C., & Chou, H. (2018). Applying the Concept of Circular Economy — Using the Cultural Difference of European Consumers as an Example. https://doi.org/10.1109/icasi.2018.8394281

Cioffi, R., Travaglioni, M., Piscitelli, G., Petrillo, A., & Parmentola, A. (2020). Smart Manufacturing Systems and Applied Industrial Technologies for a Sustainable Industry: A Systematic Literature Review. *Multidisciplinary Digital Publishing Institute*, 10(8), 2897–2897. https://doi.org/10.3390/app10082897

Dam, K. V., Simeone, L., Keskin, D., Baldassarre, B., Niero, M., & Morelli, N. (2020). Circular Economy in Industrial Design Research: A Review. *Multidisciplinary Digital Publishing Institute*, 12(24), 10279–10279. https://doi.org/10.3390/su122410279

Deshmukh, G., Patil, C. R., & Deshmukh, M. (2017). *Manufacturing Industry Performance Based on Lean Production Principles*. https://doi.org/10.1109/icnte.2017.7947977

Dokl, M., Anja, C., Damjan, K., Yee, V. F., Vujanović, A., Aviso, K. B., Tan, R. R., Kravanja, Z., & Čuček, L. (2024). Global Projections of Plastic Use, End-of-life Fate and Potential Changes in Consumption, Reduction, Recycling and Replacement with Bioplastics to 2050. *Sustainable Production and Consumption*. https://doi.org/10.1016/j.spc.2024.09.025

Ellen MacArthur Foundation. (2019). *Completing the Picture: How the Circular Economy Tackles Climate Change*. Available at: https://www.hoop-hub.eu/virtual_images/134-6254016ea43c113bc152bb9f06f1ec02.pdf

European Commission. (2020). *Circular Economy Action Plan,* European Commission Report, Brussels, Belgium.

Geng, Y., et al. (2021). Supply Chain Management for Sustainable Development. *Journal of Cleaner Production*, 278, 123–135.

Ghosh, S., Küfer, K., Roy, S. K., & Weber, G. (2022). Carbon Mechanism on Sustainable Multi-Objective Solid Transportation Problem for Waste Management in Pythagorean Hesitant Fuzzy Environment. *Springer Science+Business Media*, 8(5), 4115–4143. https://doi.org/10.1007/s40747-022-00686-w

González-Sánchez, R., Settembre-Blundo, D., Ferrari, A. M., & Muiña, F. E. G. (2020). Main Dimensions in the Building of the Circular Supply Chain: A Literature Review. *Multidisciplinary Digital Publishing Institute*, 12(6), 2459–2459. https://doi.org/10.3390/su12062459

Hahladakis, J. N., Iacovidou, E., & Gerassimidou, S. (2020). Plastic Waste in a Circular Economy. *Elsevier BV*, 481–512. https://doi.org/10.1016/b978-0-12-817880-5.00019-0

Heiskanen, E., et al. (2018). Transition Towards Sustainable Consumption and Production. *Journal of Industrial Ecology*, 32(4), 567–579.

Ilyas, M., Ahmad, W., Khan, H., Yousaf, S., Khan, K., & Nazir, S. (2018). Plastic Waste as a Significant Threat to Environment – A Systematic Literature Review. *De Gruyter*, 33(4), 383–406. https://doi.org/10.1515/reveh-2017-0035

International Energy Agency. (2021). *Renewable Energy Market Update*.

Jain, S. K., & Malik, A. K. (2013). Relevance of Lean Manufacturing in Indian Small Scale Industry. 4(2), 93. https://doi.org/10.1504/ijbcrm.2013.056312

Kolk, A., & van Tuijl, H. (2016). *Sustainable Business: Concepts, Strategies, and Practices*. London: Routledge.

Konietzko, J., Bocken, N., & Hultink, E J. (2020). A Tool to Analyze, Ideate and Develop Circular Innovation Ecosystems. *Multidisciplinary Digital Publishing Institute*, 12(1), 417–417. https://doi.org/10.3390/su12010417

Lee, J., Bagheri, B., & Kao, H.A. (2018). A Cyber-Physical Systems Architecture for Industry 4.0-based Manufacturing Systems. *Manufacturing Letters*, 15, 146–150.

Lee, S., Kim, Y. T., Lin, K. A., & Lee, J. (2023). Plastic-Waste-Derived Char as an Additive for Epoxy Composite. *Multidisciplinary Digital Publishing Institute*, 16(7), 2602–2602. https://doi.org/10.3390/ma16072602

Lovins, A. B., Lovins, L. H., & Hawken, P. (2006). A Road Map for Natural Capitalism. *SAGE Publishing*, 218–234. https://doi.org/10.4135/9781446213704.n17

Muthu, D., G. Suresh Kumar, M. Gowri, M. Prasath, V. Viswabaskaran, V. S. Kattimani, and E. K. Girija. (2022). Rapid Synthesis of Eggshell Derived Hydroxyapatite with Nanoscale Characteristics for Biomedical Applications. *Ceramics International*, 48(1), 1326–1339.

Panizzolo, R., Garengo, P., Sharma, M. K., & Gore, A. (2012). Lean Manufacturing in Developing Countries: Evidence from Indian SMEs. *Taylor & Francis*, 23(10–11), 769–788. https://doi.org/10.1080/09537287.2011.642155

Peiris, M. T. O. V., & Dayarathne, G. L. N. (2022). Application of Life Cycle Framework for Municipal Solid Waste Management: a Circular Economy Perspective from Developing Countries. *Springer Nature*, 3(2), 899–918. https://doi.org/10.1007/s43615-022-00200-x

Ranta, V., & Saari, U. A. (2019). Circular Economy: Enabling the Transition towards Sustainable Consumption and Production. *Springer International Publishing*, 1–12. https://doi.org/10.1007/978-3-319-71062-4_3-1

Riesener, M., Dölle, C., Mattern, C., & Kres, J. (2019). Circular Economy: Challenges and Potentials for the Manufacturing Industry by Digital Transformation. https://doi.org/10.1109/tems-isie46312.2019.9074421

Sahajwalla, V., & Hossain, R. (2023). Rethinking Circular Economy for Electronics, Energy Storage, and Solar Photovoltaics with Long Product Life Cycles. *Springer Nature*, 48(4), 375–385. https://doi.org/10.1557/s43577-023-00519-2

Samuel, A., Oyawale, F. A., & Fayomi, O. (2019). Effects of Waste Management in Beverage Industries: A Perspective. *IOP Publishing*, 1378(2), 022048–022048. https://doi.org/10.1088/1742-6596/1378/2/022048

Sarkis, J., et al. (2023). Green Supply Chain Management. *Journal of Business Research*, 45(2), 189–201.

Sasane, S. S., & Adhav, D. V. (2020). To Eliminate the Muda in HPT in Panel Manufacturing. *International Research Publication House*, V9(02). https://doi.org/10.17577/ijertv9is020054

Seadon, J. K. (2010). Sustainable Waste Management Systems. *Elsevier BV*, 18(16–17), 1639–1651. https://doi.org/10.1016/j.jclepro.2010.07.009

Shams, S., Guo, Z., & Juani, R. (2014). Integrated and Sustainable Solid Waste Management for Brunei Darussalam. https://doi.org/10.1049/cp.2014.1066

Teymourian, T., Teymoorian, T., Kowsari, E., & Ramakrishna, S. (2021). Challenges, Strategies, and Recommendations for the Huge Surge in Plastic and Medical Waste during the Global COVID-19 Pandemic with Circular Economy Approach. *Springer Science+Business Media*, 3(1). https://doi.org/10.1007/s42824-021-00020-8

Tripathi, N., Misra, M., & Mohanty, A. K. (2021). Durable Polylactic Acid (PLA)-Based Sustainable Engineered Blends and Biocomposites: Recent Developments, Challenges, and Opportunities. *American Chemical Society*, 1(1), 7–38. https://doi.org/10.1021/acsengineeringau.1c00011

UNEP. (2021). *Global Environment Outlook: Pathways to a Sustainable Future.*

Vermeşan, H., Mangău, A., & Tiuc, A. E. (2020). Perspectives of Circular Economy in Romanian Space. *Multidisciplinary Digital Publishing Institute*, 12(17), 6819–6819. https://doi.org/10.3390/su12176819

Womack, J. P., & Jones, D. T. (2003). *Lean Thinking: Banish Waste and Create Wealth in Your Corporation*. New York: Simon & Schuster.

Ш е р ш н е в а, Е. Г. (2021). Plastic Waste: Global Impact and Ways to Reduce Environmental Harm. *IOP Publishing*, 1079(6), 062047–062047. https://doi.org/10.1088/1757-899x/1079/6/062047

17 Enhancing Productivity with Lean Tools in the Indian Injection Molding Industry

A Green Technology Approach

Rohit Sharma, Parveen Sharma, and Rakesh Kumar Phanden

Description	Abbreviation
Crucial to quality	CTQ
Current state map	CSM
Defects Per Million Opportunities	DPMO
Define, Measure, Analyze, Improve, and Control	DMAIC
Design of Experiments	DOE
Desirability Function Approach	DFA
Failure Mode and Effect Analysis	FMEA
Future state map	FSM
Necessary non-value adding	NNVA
Non-value adding	NVA
Single Minute Exchange of Dies	SMED
Statistical Process Control	SPC
Small and medium scale enterprise	SME
Total productive maintenance	TPM
Value Addition	VA
Value stream mapping	VSM

DOI: 10.1201/9781003516071-17

17.1 INTRODUCTION

Injection molding is a widely used manufacturing process in which plastic components are produced through the injection of molten material into a mold (Mansur et al. 2016). It is a versatile and efficient process that can produce complex parts with high precision and consistency (Ajitesh et al. 2024). However, like any other manufacturing process, injection molding faces various challenges, such as inefficiencies in production, long lead times, and high defect rates (Beaumont, 2007). These challenges can be addressed through the application of Lean tools, which are proven to improve productivity and efficiency in manufacturing operations (Singh et al. 2022a). Advanced AI and Lean manufacturing tools are a set of principles, methods, and techniques aimed at improving the efficiency and effectiveness of manufacturing processes (Maged et al. 2019; Singh et al. 2022b; Singh and Singh 2023; Singh et al. 2024). Since their original development in the Japanese automobile sector, these tools have found widespread use in a variety of industries.

The primary goal of Lean tools is to eliminate waste, which is any activity that does not add value to the production process. The elimination of waste leads to increased productivity, reduced lead times, and improved quality (Dutta et al. 2022; Majji et al. 2022; Yedida and Vasudev 2022; Kattimani et al. 2023; Poojari et al. 2023; Sharanabasava et al. 2023). There are various Lean tools that can be applied in injection molding to improve productivity (Pereira et al. 2019). One such tool is value stream mapping (VSM), which is a method used to identify and eliminate waste in a production process. To find inefficient regions in the production process, VSM maps the movement of materials and information. By eliminating waste identified in the VSM, the production process becomes more streamlined and efficient, leading to increased productivity and improved quality. The implementation of important Lean concepts and tools in small- and medium-scale enterprise (SMEs) engaged in the operation of plastic injection molding is the subject of this research (Alshammari 2018; Setyabudhi et al. 2019). The major objective is to develop and evaluate various tactics to reduce waste and boost productivity. The application of Lean concepts was done in a methodical manner. Value stream maps were created in their present and future stages to seek ways to reduce waste and its origins and to optimize the relevant manufacturing process. The organization will use the data as a roadmap to map the future state and deploy Lean manufacturing technologies based on the information acquired (Edwin Joseph et al. 2020).

So, the deployment of Lean tools in injection molding can lead to significant improvements in productivity and efficiency. The use of VSM, total productive maintenance (TPM), and DMAIC can help improve productivity, reduce lead times, and improve quality. As a result, companies that implement these Lean tools can gain a competitive advantage by producing high-quality products at lower costs. This research will explore the effectiveness of these Lean tools in improving productivity in injection molding and provide practical recommendations for their implementation.

17.2 LITERATURE REVIEW

A variety of plastic goods, including automobile components, toys, packaging, and medical equipment, are produced using the manufacturing method known as injection molding. Six Sigma is a data-driven approach used to increase quality and decrease variability in industrial operations. In order to boost output, cut waste, and boost profitability, numerous academics have recently concentrated on applying Six Sigma to injection molding operations. The results of research looking at applying Six Sigma to increase productivity in injection molding will be included in this study of the literature.

According to Maged et al.'s (2019) case study part FH447 is an impeller that is positioned on a shaft attached to a compressor. This research used the DMAIC (Define, Measure, Analyze, Improve, and Control) approach of Six Sigma to optimize the injection molding process. The major quantitative properties of a product or a process, known as CTQs (crucial to quality), were employed by the researchers to pinpoint causes of variance. The outcomes demonstrated that the Six Sigma methodology increased injection molding's productivity. Mansur et al. (2016) study shows bush is one of the goods manufactured in the plastic injection division, and the researchers employed the DMAIC approach to pinpoint and address the causes of variability in the process. It includes a few flaws and wastes, including short shoots, stains, sink marks, bubbles, over-cut, under-cut, and scratches. Failure Mode and Effect Analysis (FMEA) results revealed that the Six Sigma technique increased the productivity of the injection molding process by lowering the defect rate and enhancing the process capabilities. A case study by Yadav and Kaushik (2014) for quality improvement in injection molding, this study suggested an integrated Six Sigma methodology. The injection molding procedure was optimized by the researchers using a mix of DMAIC and DPMO (Defects per Million Opportunities). It was discovered that the injection molding company may boost its profit by limiting the rejection rate of latch after applying the Six Sigma DMAIC technique. By reducing the variance in latch-hole diameter, Six Sigma adoption raised the process sigma level from 1.80 to 5.46. According to Alshammari's (2018) paper, the DPMO dropped from 516,500 to 190,000 and the sigma level rose from 1.4 to 2.3. Utilizing Lean Six Sigma, SME, Cause and Effect Matrix, Statistical Process Control (SPC), DOE, and Desirability Function Approach (DFA), more than 50% of process improvement is attributable to this improvement. A Case Study by Pereira and Silva (2019) shows the use of Lean Six Sigma tools, an improvement of around 20% in global OEE was achieved in the instance of this mold industry research involving CNC machines. Lean Six Sigma, VSM, the Pareto diagram, OEE, the 5s methodology, and the Single Minute Exchange of Dies (SMED) approach are all put into practice (Singh 2021; Agung et al. 2018; Singh et al. 2022c; Mahmood et al. 2015). Bhat et al. (2016) reported that the side bend of the plates was the main issue noted throughout the current case study. It will have an immediate impact on the organization's profitability, lengthening the process cycle time and decreasing productivity (Prashar and Vasudev 2022a, 2022b; Sunitha and Vasudev 2022c; Mahmood et al. 2015). As a result, the sigma level for the plate side bend increased from 2.9 to 3.2 by addressing the primary reasons for reworks. The organization's production increased as the sigma level grew.

Previous literature demonstrates that implementing Six Sigma is a successful strategy for enhancing output and quality in the processes involved in injection molding. The research looked at in this study shows that Six Sigma may lead to considerable increases in production efficiency, cycle time reduction, scrap rates reduction, defect reduction, and quality control improvement.

17.3 RESEARCH METHODOLOGY

In this chapter, Lean tools were used for productivity improvement. From finding the problem to implementing the solution these Lean tools used like DMAIC, VSM, and Kaizen and TPM. These Lean tools guide a team logically from issue definition to solution implementation and best practices developed to ensure that solutions are maintained. Proper Lean implementation is a basic requirement in today's manufacturing industry, and this research focuses more on the identification of indicators, practices, tools, and strategies for the application of Lean in manufacturing to improve productivity. Figure 17.1 represents the methodology for the present work.

17.3.1 DMAIC

- *D – Define*
  In the define phase, the problem statement, aims, and objectives had all been determined. The company's high incidence of plastic container rejection, which raised expenses and decreased production, was the stated issue.

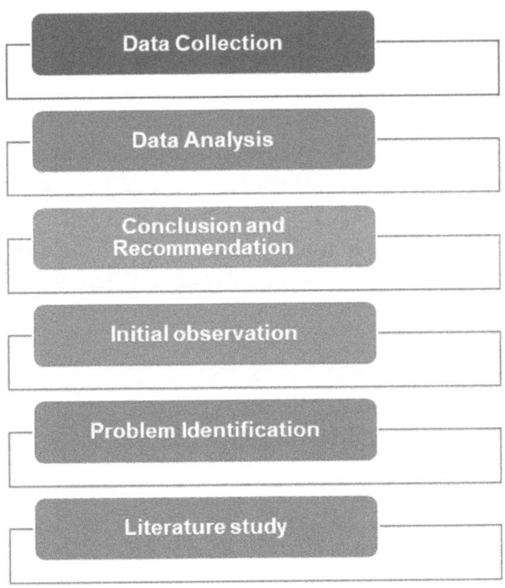

FIGURE 17.1 Methodology for the present work (Credit: Created by authors).

The objective was to decrease the reject rate while boosting productivity by 10% to 20%.

- *M – Measure*

  Measurement is the second stage. A baseline is established during this phase in order to comprehend the process's existing performance. When we wish to compare the improvements achieved, the baseline will be helpful. We had to first comprehend the process the Floor Trap fitting goes through. To depict the steps of the procedure and the interactions between individuals or departments, a deployment chart, also known as a flowchart, will be created. The operator examines the products in his or her lane before deciding whether to accept or reject them. Problematic measurement system difficulties may result from measuring attribute data in ways that rely on human judgment. The operator's judgment is nominal.
- *A – Analyze*

  In this stage, we analyzed various product defect categories from the data gathered from the measure phase, grouping those that could have the most impact on the quality of the final product and looking at the production process stability and capacity. There is the problem that sometimes some kind of debris gets stuck in the barrel during the injection process which generally leads to some kind of bubble formation, improper shapes or sometimes crack on the product which requires regular cleaning on the barrel, due to which a lot of time it being wasted. Some other issues like the hydraulic pump have an abnormal sound, possible causes:
- The filter is choked and causes the pipe to fracture and product sink.
- The inside shaft of the pump is cracked.
- The tank's oil is not at the necessary level for operation.
- *I – Improvement*

In the improvement phase, the project team implemented the solutions and monitored the process using VSM charts to ensure that the improvements were sustained. The next stage is to provide suggested changes for each cause found in the machine after evaluating the underlying causes of the machine's incapacity. The main cause was that some kind of debris got stuck in the barrel which was resolved by proper cleaning and regular maintenance and the bubble formation which was figured out by the proper heating in the barrel.

- *C – Control*

  In order to maintain the changes, the project team set up a system to oversee and manage the process. The method included ongoing process tracking using VSM charts, routine machine maintenance, and instruction for machine workers.

17.3.2 Value Stream Mapping

Value stream mapping is a Lean manufacturing methodology that analyses, depicts, and enhances the information and material movement required to deliver a good or service to a client (Dzanya and Mukada, 2015). It finally offers a roadmap for process optimization and aids in locating waste and inefficiencies in the production process (Bhat and Jose, 2016). The VSM approach entails outlining every stage of the manufacturing process, from receiving the first order to delivering the finished product and identifying the activities that add value and those that don't. Non-value-added activities are those that do not add value and may be reduced or removed, whereas value-added activities enhance the product or service in the eyes of the consumer. A cross-functional team made up of members from many departments, such as production, procurement, shipping, and customer support, should be established in order to construct a VSM. The team must next go through each stage of the manufacturing process, measuring how much time and resources each one requires.

17.4 COMPANY PROFILE

The company Repower Hydraulics is situated in Jammu colony, Ludhiana, registered under the Ministry of Small Enterprises. The shop was registered in 2004 and deals in plastic injection molding 100 gm vertical and hydraulic products.

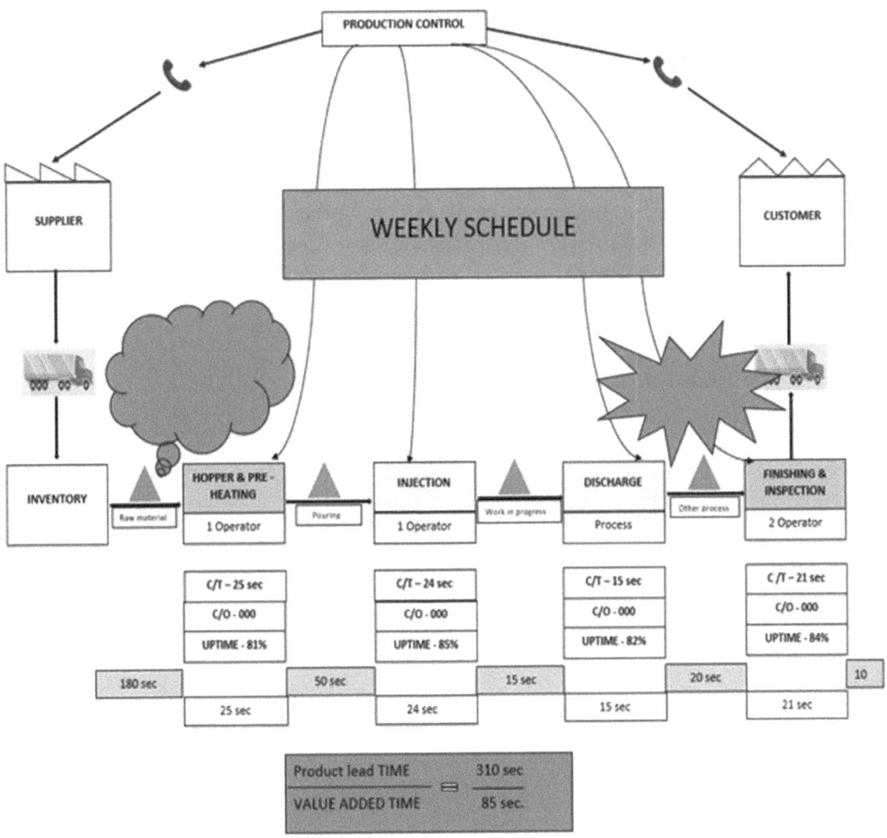

FIGURE 17.2 Current state map (CSM) (Credit: Created by authors).

17.4.1 Current State Map

The present actions involved in the production of the salt and pepper container are depicted in the current state map (CSM) in Figure 17.2. The CSM displays every step of the Shaker production process, including the procurement of raw materials, delivery to the production facility, methods used to produce jars and their cycle periods, as well as delivery and distribution to clients.

By interacting with workshop workers, operators of injection molding machines, and the current supervisor, the information for the present state map was gathered. The data given applied to the workshop's one injection molding machine, which operates for 12 hours each day during a single shift with a single operator.

Customers place phone calls to the workshop's planning department to place orders. The planning division manually communicates its requirements with several providers. The workshop currently holds a raw material inventory in their store for up to one week. Materials are moved via various machinery and procedures from the raw material store to the completed item store. Heated plastic granules are supplied into the machine through the loading hopper as the raw material. The heated and melted plastic is then sent through injection molding, where it is shaped into components. The plastic parts are then discharged from this point and subsequently cut to remove extra material (Rosnah and Othman, 2012). The items undergo further inspection to determine their quality. On the present state map, the value-added time and production lead time were highlighted. Triangles are used to represent inventory storage locations between phases. The two sections make up the timeline at the bottom of the current state map. The lead time for production makes up the first component, while value-added or processing time makes up the second. To calculate the value-added time, the processing times for each process in the value stream are

TABLE 17.1
Activity Classification for Processes

	Process Time (Seconds)			
Process	**Value Addition (VA)**	**Non-Value Adding (NVA)**	**Necessary Non-Value Adding (NNVA)**	**Comments**
Inventory		230		Due to transportation from the warehouse
Hopper and pre-heating	25	45		Heating and WIP are NVA
Injection	24	10		Delays Due to machine stoppage and breakdown
Discharge	15	15		
Finishing and inspection	21	10	1	Delay due to unskilled labor and old equipment

added. This map of the present status shows current positions and offers guidance on any gaps. When certain enhancements or modifications are made, it helps to picture how things might operate. Table 17.1 shows activity classification for processes.

A road plan for improvement is produced as a consequence of the gap area in the current condition. Major issues included the following:

i. The trimming operation requires two workers who utilize outdated hack-saw blades that must be constantly manually sharpened because they are no longer particularly sharp.
ii. Following trimming, inventory builds up since many components stay at this station for a while before going on to the next.
iii. One person handles the inspection procedure.

$$\%\,\text{Non-value Adding (NVA)} = \frac{NVA}{NVA + VA} = \frac{310}{310 + 85} = 78\% \tag{17.1}$$

$$\%\,\text{Productivity} = \frac{Value\ Added\ Time}{Total\ Time} = \frac{85}{85 + 310} = 21\% \tag{17.2}$$

The proportion of time that is not valuable is 78%. This indicates that a lot of superfluous processes are used in the production and processing of salt containers. It demonstrates that time spent on non-value-adding tasks is higher. Then, these become the target of the elimination. This results in components remaining at this station before being examined for a longer period of time, increasing inventory.

17.4.2 Future State Map (FSM)

Value stream mapping is used to identify waste sources and reduce them by putting into practice a future state value stream that may quickly become a reality. The objective is to create a chain of production where each operation is as near to generating just what its customer(s) needs, when they need it, and is linked to its customer(s), either by continuous flow or pull. Figure 17.3 shows the future state map.

$$\%\text{NVA} = \frac{NVA}{NVA + VA} = \frac{107}{107 + 75} = 59\% \tag{17.3}$$

$$\%\ \text{Reduction in NVAT} = \frac{CS\,NVA - FS\,NVA}{CS\,NVA} = \frac{78 - 59}{59} = 32\% \tag{17.4}$$

$$\%\text{Productivity} = \frac{Value\,Added\,Time}{Total\,Time} = \frac{75}{75 + 107} = 41\% \tag{17.5}$$

The future state map was created with the presumption that all problems in the troubled areas would be fully fixed. In actuality, though, the issue could not be entirely fixed. If we compare NVA from the current state map we have reduced the

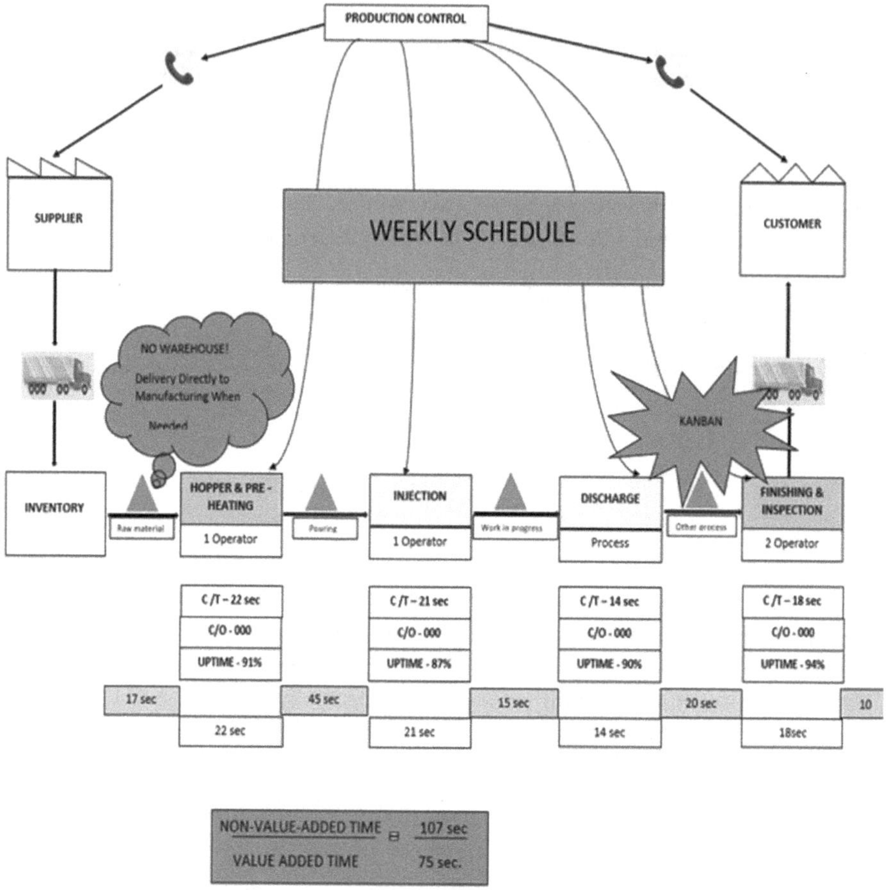

FIGURE 17.3 Future state map (Credit: Created by authors).

non-value-added time to 32%. This shows that there are a lot of improvements in the manufacturing processing of salt and pepper containers. That will help in improving productivity and achieving high quality.

17.5 TECHNIQUES FOR IDENTIFYING AND ELIMINATING WASTE

17.5.1 Pre-Heating

On the machine's top, there is a hopper. A ladder is necessary since it is not easily accessible. Since the molding unit has just one ladder, every time they needed to add materials to the hopper, the operators had to transport it to their equipment. As a result, operators move around within the molding machine in an undesired manner. Unwanted movement arises from this, and operators and equipment remain idle while the process is underway.

- Waste discovered: Waiting while moving
- Cause: Inadequate equipment
- The removal Method: Mobile stairs in place alongside the ladder

17.5.2 Melting

The setup panel isn't working correctly. As a result, configuring the necessary settings takes longer. All parameter values are frequently changed. Assigning values to the parameters must thus be done once again. This significantly lengthens non-value-added operations, which increases setup time.

- Waste discovered: Additional processing time
- Cause: Machine fault and poor maintenance
- The removal method: Total productive maintenance (TPM)

17.5.3 Injection

From the several operations, the injection process takes the longest to complete. During the run, the molten plastic material fails to entirely cover the mold cavity. The machine becomes stuck in the middle of the procedure and needs more time than normal to finish it. Even though the machine's poor state has been updated, the proper course of action has not been followed.

- Waste discovered: Additional processing time
- Cause: Machine fault and poor maintenance
- The removal method: Total productive maintenance (TPM)

17.5.4 Ejection

Normally, it only takes 10 to 12 seconds to expel each product, but owing to a machine malfunction, it now takes 12 to 15 seconds, significantly lengthening the cycle time.

- Waste discovered: Additional processing time
- Cause: Machine fault and poor maintenance
- The removal method: Total productive maintenance (TPM)

17.5.5 Defects in Products

The completed product has flaps of thin excess plastic material, often known as flash, spew, or burrs, along the edges. The end product's appearance may suffer as a result of these flaws. Although they may be removed during the deflashing procedure, this issue significantly increases the processing time. The rejection rate increases when products have strong flashes since they are occasionally even rejected.

- Waste discovered: Additional processing time
- Cause: Machine fault, poor maintenance and old equipment
- The removal method: Total productive maintenance (TPM) and modern equipment

17.6 TOTAL PRODUCTIVE MAINTENANCE (TPM)

The performance of the production process may be enhanced by TPM when utilized in injection molding. A range of tools, including molding machines, molds, and ancillary equipment, are used in the intricate process known as injection molding. Companies may increase the productivity and dependability of their machinery, decrease downtime, and enhance product quality by applying TPM in injection molding. Figure 17.4 shows mold of the study. In injection molding, TPM can be implemented in several ways such as:

i. Daily machine cleaning and inspection: Operators must clean and check their equipment every day to find any problems that can affect the caliber of the molded goods. Cleaning the machine barrel, evaluating the nozzle for wear or damage, and checking the mold for wear or damage are a few examples of what may be done.
ii. In order to save downtime and prevent breakdowns, autonomous maintenance involves empowering machine employees to do common maintenance tasks on their own, such as lubricating and replacing worn components.

FIGURE 17.4 Mold of study (Credit: Captured by authors).

iii. Planned maintenance: To avoid equipment failure and minimize downtime, this entails planning maintenance chores in advance. This may entail carrying out preventive maintenance procedures, checking calibrations, and replacing damaged parts.
iv. Quality maintenance: This entails locating and correcting the fundamental causes of injection molding process quality problems. This may entail locating and fixing problems with the mold or equipment, modifying the process parameters, and enhancing operator training.
v. TPM in administration: To do this, administrative procedures like paperwork and record-keeping must be treated using TPM principles. Implementing visual management tools to monitor equipment performance and maintenance duties can be part of this.

17.7 RESULTS AND DISCUSSION

As indicated in Table 17.2 shows product lead time and value-added time in CSM and VSM comparison, the use of Lean techniques reduced the product lead time by 32% and productivity improved by 95%. All these suggested modifications will result in a considerable increase in workshop productivity, which will lower supply chain costs overall. Now it can supply high-quality salt and pepper containers at a reduced cost than its competitor.

The initial phase in the waste reduction project was to make sure the production area was clear of any extraneous objects, that the workspace and equipment were organized properly, that the production floor and its surrounds were clean, and that any concealed items or issues were identified; 5s are used on the shop floor as a result. There are no longer any required parts, tools, or instructions. Only things that are necessary are kept, prioritized according to needs, and kept in locations that are simple to get to. Everything else is either kept or thrown away. Everything should have a place, and everything should be in its proper position. Each item's location was identified by a label or other marking. Table 17.3 shows a comparison between CSM and VSM uptime.

The fact that keeping things clean has always been part of the job is crucial. It is not an irregular effort started when things get too disorganized. The majority of the workstations maintain uniform and standardized work practices. There is the problem that sometimes some kind of debris gets stuck in the barrel which the injection

TABLE 17.2
Product Lead Time and Value-Added Time in FSM, CSM, and VSM Comparison

Parameter	CSM	FSM
Product Lead Time	310s	107s
Value-Added Time	85s	75s

TABLE 17.3
Comparison Between CSM and VSM Uptime

Variable	Hopper and Pre-Heating	Injection	Discharge	Finishing and Inspection
Uptime	81%	85%	82%	84%
Uptime	91%	87%	90%	94%

process which generally leads to some kind of bubble formation, improper shapes or sometimes crack on the product which requires regular cleaning on the barrel, due to which a lot of time it being wasted. For all these problems total productive maintenance is implemented on the shop floor to overcome the problems.

$$\% \text{ Productivity improvement} = \frac{\textit{Future productivity} - \textit{Current productivity}}{\textit{Current productivity}} = \frac{41-21}{21} = 95\% \quad (17.6)$$

To summarize this case study, it has been shown that the Lean philosophy is a long-term philosophy that requires constant practice in order to reach greater success. Even though the plastic injection molding production process has improved as a result to lower inventory levels and smoother flow following Lean implementation in a short amount of time. By putting in more and more effort to pursue excellence and address problem areas, it is still possible to significantly boost productivity.

17.8 CONCLUSION

The objectives of this project were to increase production, decrease non-value-added time and increase uptime. It is crucial to understand the standard operating procedures and the plastic injection molding process flow. The current state map's creation was essential since it aids in visualizing the real process and provides all the relevant information regarding plastic injection molding operations. The methodical methodology made it possible to find and choose the best Lean tools for process optimization. The large work-in-process inventory that the organization in question was dealing with was a serious issue that was addressed and decreased through the use of Lean concepts. As proof of the improvement, the non-value-adding time was reduced from 32% and productivity increased by 95%.

This study's conclusion is that Lean is a highly extensive and global system that is difficult to execute in tiny doses; it must be done entirely. The entire organization, not just the manufacturing lines, must be ready for change. This transformation will have an impact on the departments of sales, logistics, marketing, and product development, and Lean will not last if one of them does not follow.

17.9 FUTURE SCOPE

This study was restricted to an eight-hour shift on a single injection molding machine. Additionally, it was restricted to being able to produce a single item. To evaluate the outcomes of putting the suggested methods into practice and, if necessary, to make more changes.

REFERENCES

Ajitesh, V.A., Preet Singh, J.I., Singh, S., Dhawan, V., and Sharma, V. (2024). A review on natural fibres: Its applications, chemical and mechanical properties. In *AIP Conference Proceedings*, 2962, 1.

Alshammari, A., Redha, S., Hussain, S., Nazzal, T., Kamal, Z., and Smew, W. (2018). Quality improvement in plastic injection molding industry: Applying Lean Six Sigma to SME in Kuwait, *Proceedings of the International Conference on Industrial Engineering and Operations Management*. Bandung, Indonesia. IEOM Society International, Southfield, MI, pp. 2856–2865.

Antony, J. (2016). *Lean Six Sigma for Small and Medium Sized Enterprises*. CRC Press.

Beaumont, J. P. (2007). *Runner and Gating Design Handbook: Tools for Successful Injection Molding*. Hanser.

Bhat, S. and Jose, M. (2016). Process and productivity improvement through six sigma: A case study at production industry. *Journal of Mechanical Engineering and Automation*, 6(5A), 32–39. https://doi.org/10.5923/c.jmea.201601.07.

Dutta, V., Thakur, L., Singh, B., and Vasudev, H. (2022). A study of erosion–corrosion behaviour of friction stir-processed chromium-reinforced NiAl Bronze composite. *Materials*, 15(15), 5401.

Dwivedi, V., Anas, M., and Siraj, M. (2014). Six sigma; as applied in quality improvement for injection moulding process. *International Review of Applied Engineering Research*, 4(4). http://www.ripublication.com/iraer.htm.

Dzanya, M. and Mukada, C. (2015). Caroline Mukada. Value stream mapping in glide manufacturing: A case study of insti-tools in Zimbabwe. Technology and society. *Special Issue: Design Innovations for Sustainable Development*, 3(2), 6–10. https://doi.org/10.11648/j.ijsts.s.2015030202.12.

Edwin Joseph, R., Kanya, N., Bhaskar, K., Francis Xavier, J., Sendilvelan, S., Prabhahar, M., Kanimozhi, N., and Geetha, S. (2020). Analysis on productivity improvement, using lean manufacturing concept. *Materials Today: Proceedings*, 45, 7176–7182. https://doi.org/10.1016/j.matpr.2021.02.412.

Agung, D., Debora, F., Purba, H.H. (2018). Increased productivity of injection molding with analysis of Overall Equipment Effectiveness (OEE). *International Journal of Research in Engineering, Science and Management,* 1(12), 813–819.

Kattimani, M.A., Venkatesh, P.R., Masum, H., Math, M.M., Bahadurdesai, V.N., Mustafkhadri, S., Prasad, C.D., and Vasudev, H. (2023). Design and numerical analysis of tensile deformation and fracture properties of induction hardened inconel 718 superalloy for gas turbine applications. *International Journal on Interactive Design and Manufacturing (IJIDeM)*, 18(8), 1–11.

Klochkov, Y., Gazizulina, A., and Muralidharan, K. (2019). Lean six sigma for sustainable business practices: A case study and standardisation. *International Journal for Quality Research*, 13(1), 47–74. https://doi.org/10.24874/IJQR13.01-04.

Maged, A., Bhuiyan, N., Kaytbay, S., and Haridy, S. (2019). Continuous improvement of injection moulding using six sigma: Case study. *International Journal of Industrial and Systems Engineering*, 32(2), 243. https://doi.org/10.1504/ijise.2019.10021839.

Mahmood, K., Shevtshenko, E., Mahmood, K., and Shevtshenko, E. (2015). Productivity improvement by implementing lean production approach collaboration enhancing sustainable conceptual model development and implementation for the SME-s in machinery domain view project digital skills for future supply chain management 4.0 (DIGSCM 4.0). *DIGSCM 4.0 View Project Productivity Improvement by Implementing Lean Production Approach*. https://www.researchgate.net/publication/312160611.

Majji, B.G.R., Vasudev, H., and Bansal, A. (2022). A review on the oxidation and wear behavior of the thermally sprayed high-entropy alloys. *Materials Today: Proceedings*, 50, 1447–1451.

Mansur, A., Mu'Alim, and Sunaryo. (2016). Plastic injection quality controlling using the lean six sigma and FMEA method. *IOP Conference Series: Materials Science and Engineering*, 105(1). https://doi.org/10.1088/1757-899X/105/1/012006.

Mishra, A. and Mishra, P. (2015). Six sigma methodology in a plastic injection molding industry: A case study. *International Journal of Industrial Engineering and Technology*, 7(1). http://www.irphouse.com.

Pereira, A.M.H., Silva, M.R., Domingues, M.A.G., and Sá, J.C. (2019). Lean six sigma approach to improve the production process in the mould industry: A case study. *Quality Innovation Prosperity*, 23(3), 103–121. https://doi.org/10.12776/QIP.V23I3.1334.

Poojari, M., Hanumanthappa, H., Prasad, C.D., Jathanna, H.M., Ksheerasagar, A.R., Shetty, P., Shanmugam, B.K., and Vasudev, H. (2023). Computational modelling for the manufacturing of solar-powered multifunctional agricultural robot. *International Journal on Interactive Design and Manufacturing (IJIDeM)*, 8, 1–12.

Prashar, G. and Vasudev, H. (2022a). A review on the influence of process parameters and heat treatment on the corrosion performance of Ni-based thermal spray coatings. *Surface Review and Letters*, 29(1), 2230001.

Prashar, G. and Vasudev, H. (2022b). Surface topology analysis of plasma sprayed Inconel625-Al_2O_3 composite coating. *Materials Today: Proceedings*, 50, 607–611.

Rosnah, M.Y. and Othman, A. (2012). Lean manufacturing implementation in a plastic molding industry. *Applied Science and Engineering Progress*, 5(4), 43–52.

Setyabudhi, A.L. Sanusi and Sipahutar, I. (2019). Application of six sigma methodology to improve the product quality of moldings plastic (Case Study: PT Mega Technology Batam). *IOP Conference Series: Materials Science and Engineering*, 505(1). https://doi.org/10.1088/1757-899X/505/1/012067.

Sharanabasava, H., Raviprakash, M., Prasad, C.D., Ramesh, M.R., Phanibhushana, M.V., Vasudev, H., and Kumar, S. (2023). Microstructure, mechanical and wear properties of SiC and Mo reinforced NiCr microwave cladding. *Advanced Materials Processing Technology*. https://doi.org/10.1080/2374068X.2023.2257937.

Singh, J. (2021). Slurry erosion performance analysis and characterization of high-velocity oxy-fuel sprayed Ni and Co hardsurfacing alloy coatings. *Journal of King Saud University - Engineering Sciences*. https://doi.org/10.1016/j.jksues.2021.06.009.

Singh, J. and Singh, S. (2023). Support vector machine learning on slurry erosion characteristics analysis of Ni- and Co-alloy coatings. *Surface Review and Letters*. https://doi.org/10.1142/S0218625X23400061.

Singh, J., Gill, H.S., and Vasudev, H. (2022a). Computational fluid dynamics analysis on role of particulate shape and size in erosion of pipe bends. *International Journal on Interactive Design and Manufacturing*, 8, 1–16.

Singh, J., Singh, R., Singh, S., Vasudev, H., and Kumar, S. (2022b). Reducing scrap due to missed operations and machining defects in 90PS pistons. *International Journal on Interactive Design and Manufacturing*. https://doi.org/10.1007/s12008-022-01071-0.

Singh, R., Toseef, M., Kumar, J., and Singh, J. (2022c) Benefits and challenges in additive manufacturing and its applications. In *Sustainable Advanced Manufacturing and Materials Processing*, eds. Kaushal, S., Singh, I., Singh, S., and Gupta, A. CRC Press, 137–157. https://doi.org/10.1201/9781003269298-8.

Singh, J., Singh, S., Vasudev, H., and Katiyar, J.K. (2024). Artificial neural network model for wear characteristic analysis of WC-10Co4Cr and Stellite 6 thermal spray coatings. *Tribology International*, 199, 109924.

Sunitha, K. and Vasudev, H., 2022. A short note on the various thermal spray coating processes and effect of post-treatment on Ni-based coatings. *Materials Today: Proceedings*, 50, 1452–1457.

Timans, W., Ahaus, K., and Antony, J. (2014). Six sigma methods applied in an injection moulding company. *International Journal of Lean Six Sigma*, 5(2), 149–167. https://doi.org/10.1108/IJLSS-07-2013-0037.

Yadav, J.S. and Kaushik, P. (2014a). Application of six sigma in injection moulding process. *International Journal of R&D in Engineering*, 1. www.rndpublications.com/journal.

Yadav, J.S. and Kaushik, P. (2014b). Application of six sigma in injection moulding process. *International Journal of R&D in Engineering*, 1. www.rndpublications.com/journal.

Yedida, V.V.S. and Vasudev, H. (2022). A review on the development of thermal barrier coatings by using thermal spray techniques. *Materials Today: Proceedings*, 50, 1458–1464.

18 Interpretive Structural Modeling Approach for Sustainable Supply Chain Planning

Vishal Jagota, Rajnish Kler, Samrat Ray, and Rajan Vohra

Description	Abbreviation
Dependence Power	DeP
Driving power	DrP
Interpretive structural modeling	ISM
Matriced Impacts Croises-Multiplication Applique and Classment	MICMAC
Structural Self-Interaction Matrix	SSIM
Sustainable supply chain	SSC

18.1 INTRODUCTION

The supply chain is a phenomenon that connects various organizations in carrying out various processes that are required to supply products to customers [1]. The systemic and strategic integration of everyday business operations and strategies into a supply chain network aims to increase the long-term success of businesses. The rising acceptance of the ISO 14001 environmental standards has enhanced the role of supply chain management in corporate environmental practices [2]. Sustainable supply chain planning can be understood as the conceptualization of entire supply chain activities in such a way that it ensures the minimum possible environmental load along with reasonable efficiency of the supply chain network. For businesses to successfully compete in the global market, a rising number of company activities and jobs are being outsourced, while components, sub-assemblies, services, and finished goods are being imported [3]. Currently, one of the main concerns of organizations is how sustainability can be successfully incorporated into their supply chain network [4]. Sustainability practices have the ability to take care of environmental

 DOI: 10.1201/9781003516071-18

concerns and therefore it becomes crucial to integrate them into supply chain planning [5]. The implementation of sustainability in the supply chain network will not only improve the brand image of the organization but also provide a competitive edge in the business [6]. Therefore, this study focuses on identifying the key strategies that can be utilized by organizations for better adoption of sustainability in their supply chain planning.

18.2 LITERATURE REVIEW

An extensive literature search was carried out to find various strategies that can improve the implementation of sustainability in the supply chain planning of organizations. Online sources such as Scopus and Web of Science were used for literature searches. A list of the strategies along with their relevant references is presented in Table 18.1.

Strong support from top management is expected to achieve the goal of SSC planning [7, 8]. Sustainability can be adopted to gain an advantage over competitors and therefore it should be seen as not only a business opportunity but also as an essential practice by the companies [9, 10]. It is the responsibility of suppliers that they maintain a strong relationship with their clients to ensure a strong level of faith and commitment that is ultimately beneficial for the business [11]. Customers should know why it is important to look for environmentally benign products and therefore training should be conducted to make them aware of sustainability [12]. It is always better

TABLE 18.1
Strategies for Sustainable Supply Chain Planning

Code	Strategies	References
S1	Continuous support from top management	[7,8]
S2	Considering sustainability as a business opportunity	[9]
S3	Considering sustainability as an essential practice on a strategic level	[10]
S4	Develop strong client/supplier relationship	[11]
S5	Training of customers about the environmental issues	[12]
S6	Considering the sustainability approaches at the beginning of supply chain planning	[13]
S7	Use company-specific environmental principles, rules and standards	[14]
S8	Perform sustainability in cross-functional teams	[15]
S9	Education and training for each individual in the supply chain	[16]
S10	Apply sustainability-based methods or tools for support	[17, 18]
S11	Include an environmental expert in the supply chain management team	[19]
S12	The proactive role of the government in supporting sustainability	[20]
S13	The presence of an environmental champion in companies	[21]

to consider eco-friendly ideas at the beginning of a process and therefore logistic practitioners should be encouraged to consider sustainability during the planning stage itself [13]. The requirements of various types of companies are different and depend upon the type of their business. Therefore, company-specific environmental rules and regulations should be established [14]. Sustainability cannot be achieved only by a single department of the organization and hence it should be practiced in cross-functional teams across the different departments of the companies [15]. It always remains a challenge for individuals to understand how sustainability can be adopted at the practical level and therefore the individuals involved in supply chain activities should be provided with education and training for implementing various eco-friendly practices [16]. The presence of an environmental expert in the team of supply chain practitioners can not only encourage the team but also ensure the efficient execution of various environmentally benign plans [17–19]. The attitude of the government is a key factor when it comes to the adoption of sustainability. The active role of the government can help in the successful adoption of sustainability in organizations [20, 21].

18.3 RESEARCH METHODOLOGY

A flowchart of the research methodology is given in Figure 18.1. The research is conducted in three main steps which are as follows:

18.3.1 Identification of the Strategies for Sustainable Supply Chain Planning

The strategies for sustainable supply chain planning was obtained using a rigorous literature survey and through the opinion of industry experts. Initially, 15 strategies were identified through a literature search but later two strategies were eliminated due to low significance after discussion with industry experts. Thus, a total of 13 strategies were finalized for this study.

18.3.2 Analysis of the Strategies Using the ISM Approach

An ISM approach was utilized in this research for analyzing the strategies for sustainable supply chain planning. It is an interpretive method in which the judgment of the experts determines how different factors are associated with each other. ISM was initially thought of as an approach for group learning but later recognized as a method to be used individually [22]. It is a qualitative technique that provides simplified solutions through structural mapping of complex inter-relationships among various elements [23, 24]. ISM is used in a variety of applications such as green supply chains [25, 26], sustainability in the oil and gas sector [27], and cellular manufacturing systems [28]. Various advantages that lead toward the selection of the ISM method for this study were [29]: i) the process is systematic and all the pairwise relations between different factors are either formed only by the responses

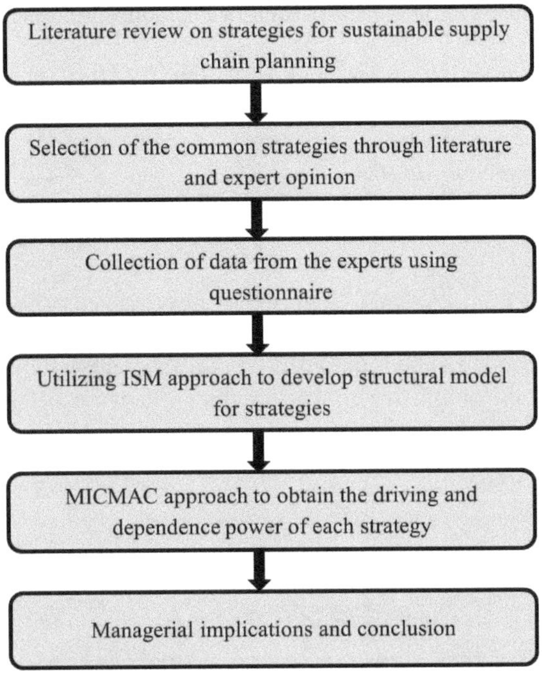

FIGURE 18.1 Flow diagram of the research methodology (Credit: Created by authors).

or by including transitive inference, ii) the relational queries are reduced by 50 to 80% through the use of transitive inference, iii) the respondents are not required to have any knowledge about underlying process and they just need to respond to the relational queries, iv) it develops a structural model that can be interpreted easily, and v) it takes only a single query at any given time, which improves the quality of response. The common steps of ISM are as follows [25, 30]:

Step 1: List down various strategies (criteria) to be examined.
Step 2: Establish a contextual relationship for each pair of the strategies.
Step 3: Develop a Structural Self-Interaction Matrix (SSIM) which represents a pairwise relationship among the strategies.
Step 4: Develop a reachability matrix by using SSIM and check the transitivity of contextual relations. It is a conjecture in ISM which states that if X has a relation with Y and Y has a relation with Z then there must be a relation between X and Z.
Step 5: Partition of the final reachability matrix into different levels.
Step 6: Construct a diagraph using the relationships given in the final reachability matrix and then eliminate the transitive links from the diagraph.
Step 7: Add the statements for replacing the nodes of the diagraph to transform it into an ISM model.

Step 8: Check the developed ISM model for any conceptual inconsistency.

Each step mentioned above is presented in Figure 18.2. The execution of each step corresponding to the current study is explained in the next sub-sections.

- *Contextual relationship (CR)*
  A CR of the type "influence" was chosen after the discussion with the experts. It represents that one variable has a positive influence on the other.
- *SSIM*
  On the basis of the selected CR, the experts were asked to give their opinion about the type of relationship between each pair of strategies (a and b) and also the direction of the relationship. Four different notations are used to show the relationship between the strategies which are as follows:

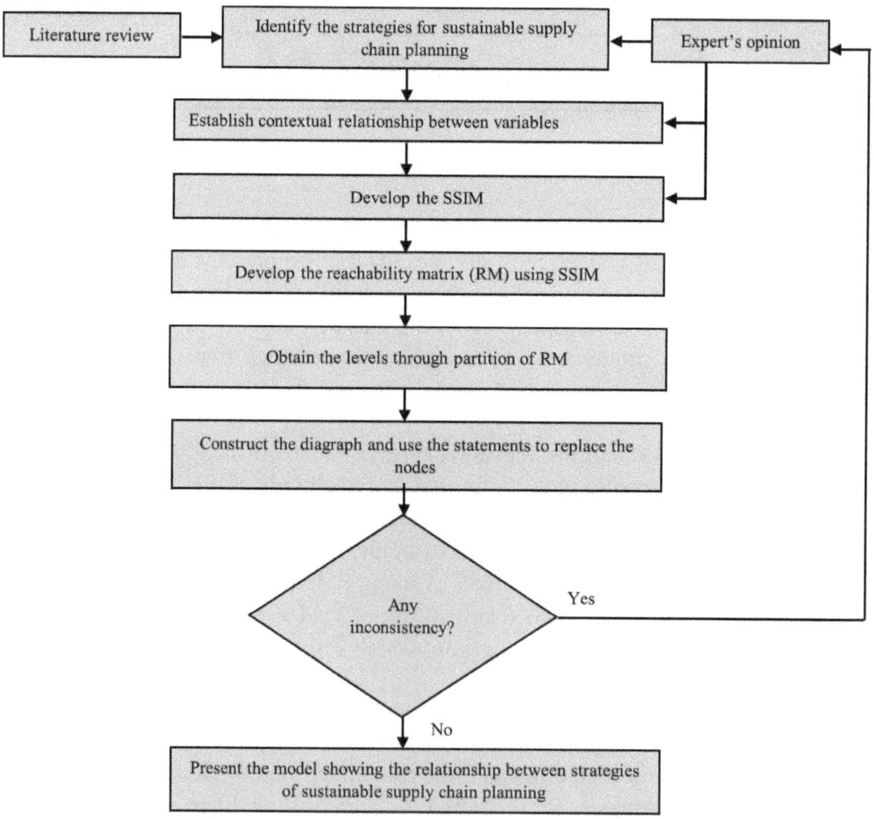

FIGURE 18.2 Basic steps of the ISM approach.

I: Factor a influences factor b; J: Factor b influences factor a; K: Factors a and b influence each other; and L: Factors a and b are not related to each other. The developed SSIM for the strategies for SSC planning is given in Table 18.2.

- *Initial RM*
  In this case, SSIM is used to construct the initial RM. The data available in each block of SSIM is converted into binary form (i.e., 1 and 0) to obtain the initial RM. This conversion into binary form is carried out by following certain rules [30] which are given below:
- If the entry in a block (a, b) of SSIM is I, then insert "1" in the cell (a, b) and "0" in the cell (b, a) of the initial RM.
- If the entry in a cell (a, b) of SSIM is J, then insert "0" in the cell (a, b) and "1" in the cell (b, a) of initial RM.
- If the entry in a cell (a, b) of SSIM is K, then insert "1" in both the cells (a, b) and (b, a) of initial RM.
- If the entry in a cell (a, b) of SSIM is L, then insert "0" in both the cells (a, b) and (b, a) of the initial RM.

The initial RM constructed using the above four criteria is given in Table 18.3.

Final RM was constructed by including transitivity as mentioned in Step 4 of the ISM method. The modified entries after considering transitivity are represented with a star (*) sign as given in Table 18.4.

- *Level partitions*

TABLE 18.2
SSIM for the Strategies for SSC Planning

	S13	S12	S11	S10	S9	S8	S7	S6	S5	S4	S3	S2
S1	K	J	I	L	I	I	I	L	I	I	I	I
S2	K	J	I	L	I	I	I	L	I	I	I	
S3	J	J	I	I	I	I	I	I	I	L		
S4	J	L	L	L	L	I	L	L	J			
S5	J	J	L	L	L	L	L	L				
S6	J	L	J	L	J	L	K					
S7	J	J	J	I	J	L						
S8	L	L	J	L	L							
S9	J	J	L	I								
S10	L	L	L									
S11	L	L										
S12	L											

TABLE 18.3
Initial RM

	S1	S2	S3	S4	S5	S6	S7	S8	S9	S10	S11	S12	S13
S1	1	1	1	1	1	0	1	1	1	0	1	0	1
S2	0	1	1	1	1	0	1	1	1	0	1	0	1
S3	0	0	1	0	1	1	1	1	1	1	1	0	0
S4	0	0	0	1	0	0	0	1	0	0	0	0	0
S5	0	0	0	1	1	0	0	0	0	0	0	0	0
S6	0	0	0	0	0	1	1	0	0	0	0	0	0
S7	0	0	0	0	0	1	1	0	0	1	0	0	0
S8	0	0	0	0	0	0	0	1	0	0	0	0	0
S9	0	0	0	0	0	1	1	0	1	1	0	0	0
S10	0	0	0	0	0	0	0	0	0	1	0	0	0
S11	0	0	0	0	0	1	1	1	0	0	1	0	0
S12	1	1	1	0	1	0	1	0	1	0	0	1	0
S13	1	1	1	1	1	1	1	0	1	0	0	0	1

In this step, all the factors under consideration are presented in the form of reachability set, antecedent set, intersection set, and different levels. Level I is always given the top place in the ISM model and the successive levels are provided below Level I accordingly. In the present study, the levels of all strategies are obtained in six iterations, as given in Table 18.5. These levels are used to construct the diagraph and final ISM model.

- *Development of the ISM model*
  A diagraph was constructed (shown in Figure 18.3) using the relationships given in the final RM. Further, the transitive links were removed from the diagraph. Nodes of the diagraph were replaced with relationship statements to develop the ISM model for strategies. The developed ISM model is shown in Figure 18.4.

18.3.3 MICMAC Analysis

The abbreviation MICMAC stands for "Matriced Impacts Croises-Multiplication Applique and Classment". This is used to find out the significant factors based on the driving power (DrP) and dependence power (DeP) of each strategy. (DrP) of a strategy is given by the row sum of that factor in the final RM whereas the column sum of a factor in the final RM indicates its (DeP). (DrP) and (DeP) of each strategy is given in Table 18.4. In this analysis, the factors under investigation are divided into four clusters which are as follows:

TABLE 18.4
Final RM

	1	2	3	4	5	6	7	8	9	10	11	12	13	Driving
1	1	1	1	1	1	1*	1	1	1	1*	1	0	1	**12**
2	1*	1	1	1	1	1*	1	1	1	1*	1	0	1	**12**
3	0	0	1	1*	1	1	1	1	1	1	1	0	0	**9**
4	0	0	0	1	0	0	0	1	0	0	0	0	0	**2**
5	0	0	0	1	1	0	0	1*	0	0	0	0	0	**3**
6	0	0	0	0	0	1	1	0	0	1*	0	0	0	**3**
7	0	0	0	0	0	1	1	0	0	1	0	0	0	**3**
8	0	0	0	0	0	0	0	1	0	0	0	0	0	**1**
9	0	0	0	0	0	1	1	0	1	1	0	0	0	**4**
10	0	0	0	0	0	0	0	0	0	1	0	0	0	**1**
11	0	0	0	0	0	1	1	1	0	1*	1	0	0	**5**
12	1	1	1	1*	1	1*	1	1*	1	1*	1*	1	1	**13**
13	1	1	1	1	1	1	1	1*	1	1*	1*	0	1	**12**
Dependence	**4**	**4**	**5**	**7**	**6**	**9**	**9**	**9**	**6**	**10**	**6**	**1**	**4**	**80**

* indicates transitive relation

TABLE 18.5
Level Partitions and Iterations

	Set S. No.	Reachability set	Antecedent set	Intersection set	Level
Iteration I	**1**	1, 2, 3, 4, 5, 6, 7, 8, 9, 10, 11, 13	1, 2, 12, 13	1, 2, 13	
	2	1, 2, 3, 4, 5, 6, 7, 8, 9, 10, 11, 13	1, 2, 12, 13	1, 2, 13	
	3	3, 4, 5, 6, 7, 8, 9, 10, 11	1, 2, 3, 12, 13	3	
	4	4, 8	1, 2, 3, 4, 5, 12, 13	4	
	5	4, 5, 8	1, 2, 3, 5, 12, 13	5	
	6	6, 7, 10	1, 2, 3, 6, 7, 9, 11, 12, 13	6, 7	
	7	6, 7, 10	1, 2, 3, 6, 7, 9, 11, 12, 13	6, 7	
	8	8	1, 2, 3, 4, 5, 8, 11, 12, 13	8	I
	9	6, 7, 9, 10	1, 2, 3, 9, 12, 13	9	
	10	10	1, 2, 3, 6, 7, 9, 10, 11, 12, 13	10	I
	11	6, 7, 8, 10, 11	1, 2, 3, 11, 12, 13	11	
	12	1, 2, 3, 4, 5, 6, 7, 8, 9, 10, 11, 12, 13	12	12	
	13	1, 2, 3, 4, 5, 6, 7, 8, 9, 10, 11, 13	1, 2, 12, 13	1, 2, 13	
Iteration II	**1**	1, 2, 3, 4, 5, 6, 7, 9, 11, 13	1, 2, 12, 13	1, 2, 13	
	2	1, 2, 3, 4, 5, 6, 7, 9, 11, 13	1, 2, 12, 13	1, 2, 13	
	3	3, 4, 5, 6, 7, 9, 11	1, 2, 3, 12, 13	3	
	4	4	1, 2, 3, 4, 5, 12, 13	4	II
	5	4, 5	1, 2, 3, 5, 12, 13	5	
	6	6, 7	1, 2, 3, 6, 7, 9, 11, 12, 13	6, 7	II
	7	6, 7	1, 2, 3, 6, 7, 9, 11, 12, 13	6, 7	II
	9	6, 7, 9	1, 2, 3, 9, 12, 13	9	
	11	6, 7, 11	1, 2, 3, 11, 12, 13	11	
	12	1, 2, 3, 4, 5, 6, 7, 9, 11, 12, 13	12	12	
	13	1, 2, 3, 4, 5, 6, 7, 9, 11, 13	1, 2, 12, 13	1, 2, 13	

(Continued)

TABLE 18.5 (CONTINUED)
Level Partitions and Iterations

	Set S. No.	Reachability set	Antecedent set	Intersection set	Level
Iteration III	**1**	1, 2, 3, 5, 9, 11, 13	1, 2, 12, 13	1, 2, 13	
	2	1, 2, 3, 5, 9, 11, 13	1, 2, 12, 13	1, 2, 13	
	3	3, 5, 9, 11	1, 2, 3, 12, 13	3	
	5	5	1, 2, 3, 5, 12, 13	5	III
	9	9	1, 2, 3, 9, 12, 13	9	III
	11	11	1, 2, 3, 11, 12, 13	11	III
	12	1, 2, 3, 5, 9, 11, 12, 13	12	12	
	13	1, 2, 3, 5, 9, 11, 13	1, 2, 12, 13	1, 2, 13	
Iteration IV	**1**	1, 2, 3, 13	1, 2, 12, 13	1, 2, 13	
	2	1, 2, 3, 13	1, 2, 12, 13	1, 2, 13	
	3	3	1, 2, 3, 12, 13	3	IV
	12	1, 2, 3, 12, 13	12	12	
	13	1, 2, 3, 13	1, 2, 12, 13	1, 2, 13	
Iteration V	**1**	1, 2, 13	1, 2, 12, 13	1, 2, 13	V
	2	1, 2, 13	1, 2, 12, 13	1, 2, 13	V
	12	1, 2, 12, 13	12	12	
	13	1, 2, 13	1, 2, 12, 13	1, 2, 13	V
Iteration VI	**12**	12	12	12	VI

Cluster I: Autonomous factors:

These factors are weak (DrP) as well as weak (DeP). They cause less effect on the model as they are comparatively less connected to the model. As evident from their names, they are independent in nature.

Cluster II: Dependent factors

These factors largely depend on the other factors of the system and also have less (DrP). Thus, these factors are considered as the least significant factors.

Cluster III: Linkage factors

These factors form a bridge between driving and dependent factors. These factors possess strong (DrP) and high dependency. These are considered unstable factors and any action on them does not influence other factors.

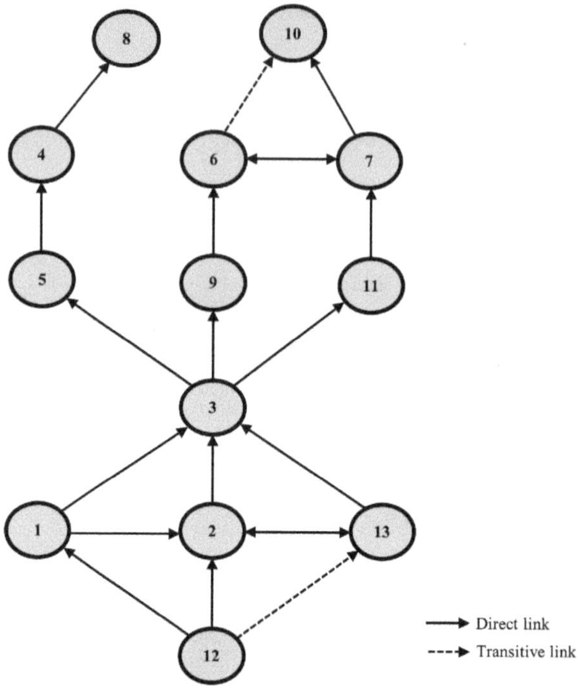

FIGURE 18.3 Diagraph for the ISM model (Credit: Created by authors).

Cluster IV: Driving factors

These factors are also known as independent factors because they have low dependency on the other factors. Also, they possess high (DrP) and are therefore considered to have a high influence on the system under investigation.

A diagram (known as a power matrix) showing the (DrP) and (DeP) of all the strategies is presented in Figure 18.5.

18.4 DISCUSSION ON FINDINGS

ISM approach divides the strategies into six different levels as given by Table 18.5. These levels are allocated in the ISM model as given in Figure 18.4. Two strategies, "perform sustainability in cross-functional teams" (S8) and "apply sustainability-based methods or tools for support" (S10), are assigned to Level I and given the top position in the ISM model. "Proactive role of the government to support sustainability" (S12) is assigned to Level VI and assigned a place at the bottom of the ISM model. The rest of the ten factors are divided within the four intermediate levels, i.e., Level II to Level V. The factors assigned to the bottom place have a significant effect on the factors which are provided at the top of the model. Thus, it can be clearly observed from the model that the "proactive role of the government to

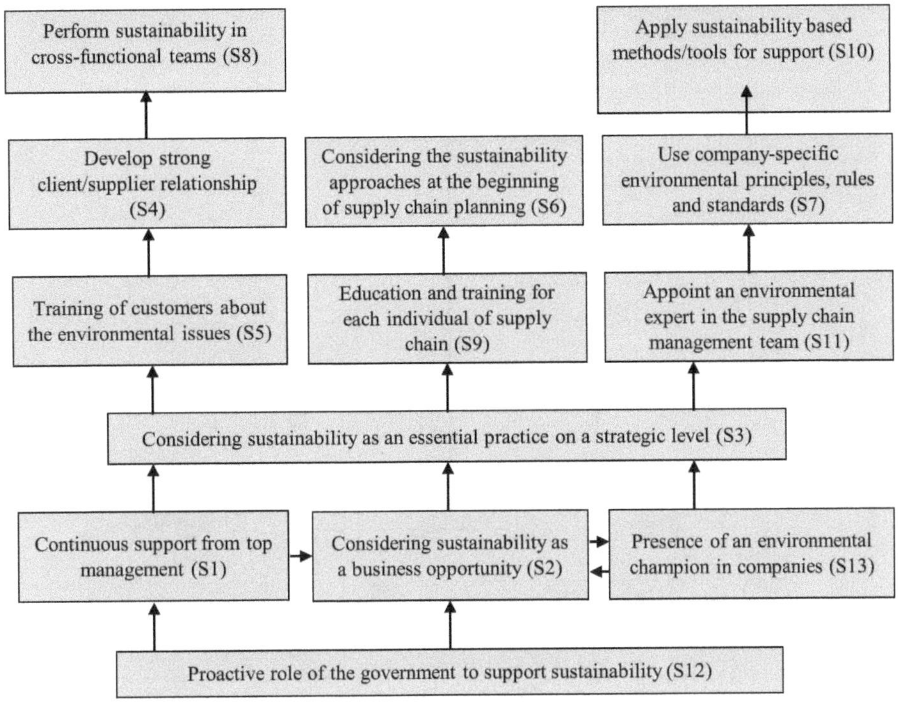

FIGURE 18.4 ISM model (Credit: Created by authors).

support sustainability" (S12) has the highest influence on the strategies for sustainable supply chain planning.

MICMAC analysis represents the strategies in four different clusters as shown in the power matrix (Figure 18.5). A discussion on each factor under different clusters is provided as follows:

- Cluster I: "Training of customers about the environmental issues" (S5), "education and training for each individual of the supply chain" (S9) and "appointing an environmental expert in the SCM team" (S11) are the strategies under cluster I and have autonomous nature. These factors are self-directed and cannot be controlled by outside forces. In this cluster, all three factors have the same dependence power of six but "appointing an environmental expert in the supply chain management team" (S11) has the highest driving power of five whereas "training customers about the environmental issues" (S5) has the least driving power of three. Thus, "appointing an environmental expert in the supply chain management team" (S11) is the most significant factor in cluster I.
- Cluster II: "Develop strong client/supplier relationship" (S4), "consider the sustainability approaches at the beginning of supply chain planning" (S6), "use company-specific environmental principles, rules and standards" (S7),

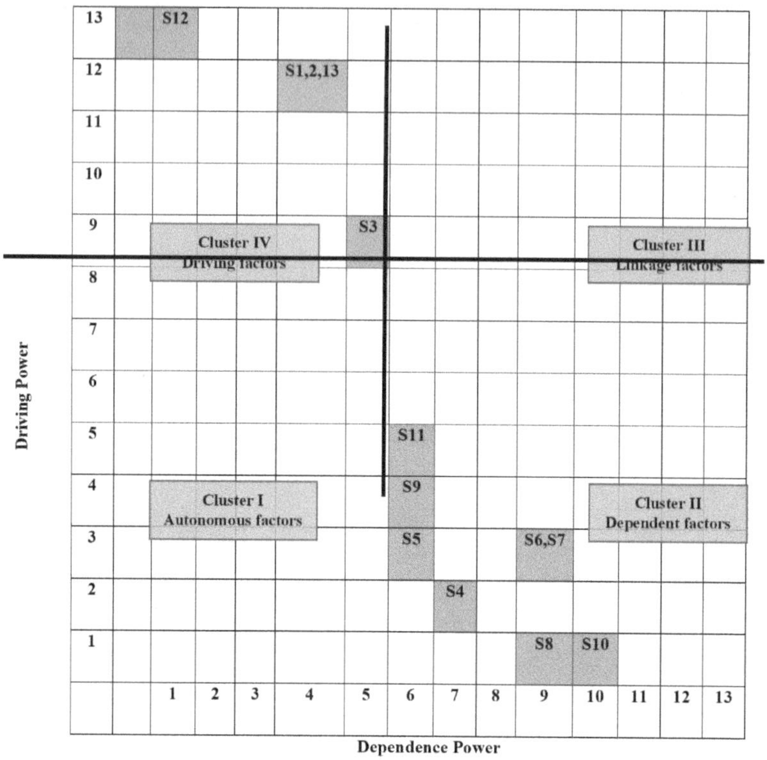

FIGURE 18.5 Power matrix showing driving and dependence power of strategies (Credit: Created by authors).

"perform sustainability in cross-functional teams" (S8) and "apply sustainability-based methods or tools for support" (S10) are the strategies under cluster II. The factors in this cluster show high dependency on the other factors of the system and also have less driving power. "Apply sustainable based methods or tools for support" (S10) has the highest (DeP) of ten and the least (DrP) of one along with (S8). S6 and S7 have the same (DeP) and (DrP) of 9 and 3, respectively. S6 and S7 are the significant factors in cluster II having the highest (DrP) of 3.

- Cluster III: In this case, the factors generally have strong (DrP) and high dependency. They are assumed to be unstable factors and do not have much effect on the other factors of the system. In this study, it is interesting to observe that there is no strategy that falls in cluster III. Therefore, it can be understood that all the selected strategies in this study show a stable nature.
- Cluster IV: "Continuous support from top management" (S1), "considering sustainability as a business opportunity" (S2), "considering sustainability as an essential practice on a strategic level" (S3), "proactive role of the government to support sustainability" (S12) and "presence of an environmental champion in companies" (S13) are the strategies in cluster IV. Generally,

the factors of this cluster possess high (DrP) and low dependency. S1, S2, and S13 have similar (DrP) and (DeP) of 12 and 4, respectively. "Proactive role of the government to support sustainability" (S12) has the highest (DrP) of 13 and the least (DeP) of one whereas "considering sustainability as an essential practice on a strategic level" (S3) possesses the least (DrP) of nine and maximum (DeP) of five in cluster IV.

With an overall observation of the power matrix (Figure 18.5), it is clear that the "proactive role of the government to support sustainability" (S12) is the most influential strategy among all having the highest (DrP) of 13 and the least (DeP) of 1.

18.5 MANAGERIAL INSIGHTS OF THE RESEARCH

This research reveals that the implementation of sustainability in supply chain planning can be improved by considering a set of strategies. The role of the top management is a key factor in the growth of an industry whether it is a small or a large industry. Therefore, the consideration of environmentally friendly practices in industries is significantly affected by the approach and attitude of the top management of these industries toward various environmental needs. Sustainability should not be considered as a burden but as a business opportunity. If sustainability is accepted as a part of the business, then a corporate culture can be imagined where eco-friendly practices will have a space along with profit-making psychology. Only the acceptance of the philosophy of sustainability will not be enough until it is executed at a strategic level in industries. The government also need to have a proactive role not only in the implementation of sustainability but also in the overall growth of corporate culture. The government can make strict environmental policies to be adopted by these industries but at the same time, the government also needs to provide financial support because lack of financial resources is a major constraint, especially for SMEs.

All the strategies mentioned above in this section have the ability to drive the other strategies. For example, the financial support from the government can encourage the top management to recruit some key personnel such as the appointment of an environmental expert in the SCM team and the selection of an environmental champion in the company. Environmental experts can provide continuous support to bring the environmental criteria in the very beginning of the supply chain planning of products and services whereas an environmental champion can encourage everyone from the top management to the lower staff to include the environmental criteria as a routine practice. The environmental champion can also keep a check on different ecodesign activities taking place in cross-functional teams and can suggest necessary improvements wherever possible.

A proactive role of the government and a positive attitude of the top management toward environmental needs may lead to the organization of various conferences and workshops for improving the knowledge and training of employees for a better understanding of various environmental requirements. Proper knowledge will help the supply chain team to use the company-specific environmental principles, rules, and standards, and at the same time they can conceptualize the sustainability practices in the early stages of supply chain planning. It will also help the supply chain practitioners to learn and use the different sustainability methods and tools. Companies can also invite their clients such as suppliers and customers to these workshops to give them exposure to different eco-friendly products and services and their importance to society.

18.6 CONCLUSIONS

As per the current scenario worldwide, organizations are realizing the importance of sustainability to be incorporated into their business activities. In the current study, an ISM approach is utilized to analyze the strategies for implementing sustainability in the supply chain planning of companies. This research can be useful for decision makers to identify the mutual relationships between different strategies and to identify the critical ones that can lead to significant improvement in sustainable supply chain planning. The results obtained in this study conclude that "continuous support from top management", "considering sustainability as a business opportunity", "considering sustainability as an essential practice on a strategic level", "proactive role of the government to support sustainability", and "presence of an environmental champion in companies" are the most significant strategies toward sustainable supply chain planning in organizations. These strategies have the ability to drive and influence the other strategies to reach the goal. Although the proactive role of the government comes out to be the most crucial strategy, the organizations do not have any control over this strategy. Therefore, it is better that the organizations focus on the other key strategies such as "considering sustainability as a business opportunity". It will help the organizations to improve their brand image as well as to gain an advantage over their competitors. There exist some limitations in the current study. The results of this research are dependent on the experience, perception and beliefs of the experts associated with this study. Therefore, some elements of biases may be associated with the outcomes of the research. However, the inclusion of some more experts in the study will easily eliminate this limitation. In this study, a traditional ISM approach has been used. However, some hybrid methods such as TISM–DEMATEL or PLS–SEM can be used and their results can be verified with the outcomes of this study. Also, some fuzzy-based methods such as fuzzy AHP or fuzzy TOPSIS can be used that can handle the vagueness and uncertainty in the opinion of the experts.

REFERENCES

1. A. Kaur, G. Singh, V. Kukreja, S. Sharma, S. Singh, and B. Yoon, "Adaptation of IoT with Blockchain in Food Supply Chain Management: An Analysis-Based Review in Development, Benefits and Potential Applications," *Sensors*, vol. 22, no. 21. MDPI AG, p. 8174, Oct. 25, 2022. doi: 10.3390/s22218174.

2. J. Sarkis, "A Strategic Decision Framework for Green Supply Chain Management," *Journal of Cleaner Production*, vol. 11, no. 4, pp. 397–409, Jun. 2003, doi: 10.1016/S0959-6526(02)00062-8.
3. P. Centobelli, R. Cerchione, E. Esposito, R. Passaro, and Shashi, "Determinants of the Transition Towards Circular Economy in SMEs: A Sustainable Supply Chain Management Perspective," *International Journal of Production Economics*, vol. 242. Elsevier BV, p. 108297, Dec. 2021. doi: 10.1016/j.ijpe.2021.108297.
4. M. Yang, M. Movahedipour, J. Zeng, Z. Xiaoguang, and L. Wang, "Analysis of Success Factors to Implement Sustainable Supply Chain Management Using Interpretive Structural Modeling Technique: A Real Case Perspective," *Mathematical Problems in Engineering*, vol. 2017, pp. 1–14, 2017, doi: 10.1155/2017/7274565.
5. R. D. Raut, B. Narkhede, and B. B. Gardas, "To Identify the Critical Success Factors of Sustainable Supply Chain Management Practices in the Context of Oil and Gas Industries: ISM Approach," *Renewable and Sustainable Energy Reviews*, vol. 68, pp. 33–47, Feb. 2017, doi: 10.1016/j.rser.2016.09.067.
6. D. Andalib Ardakani and A. Soltanmohammadi, "Investigating and Analysing the Factors Affecting the Development of Sustainable Supply Chain Model in the Industrial Sectors," *Corporate Social Responsibility and Environmental Management*, vol. 26, no. 1, pp. 199–212, Jan. 2019, doi: 10.1002/csr.1671.
7. E. Ünal, A. Urbinati, and D. Chiaroni, "Managerial Practices for Designing Circular Economy Business Models: The Case of an Italian SME in the Office Supply Industry," *JMTM*, vol. 30, no. 3, pp. 561–589, Apr. 2019, doi: 10.1108/JMTM-02-2018-0061.
8. C.-H. Wang, "How Organizational Green Culture Influences Green Performance and Competitive Advantage: The Mediating Role of Green Innovation," *JMTM*, vol. 30, no. 4, pp. 666–683, Jun. 2019, doi: 10.1108/JMTM-09-2018-0314.
9. O. Chkanikova and O. Mont, "Corporate Supply Chain Responsibility: Drivers and Barriers for Sustainable Food Retailing: Sustainable Food Retailing," *Corporate Social Responsibility and Environmental Management*, vol. 22, no. 2, pp. 65–82, Mar. 2015, doi: 10.1002/csr.1316.
10. D. Bagdonienė, A. Daunorienė, and A. Simanavičienė, "Integration of Sustainable Development Principles into the Balanced Scorecard," *Intellectual Economics*, vol. 5, pp. 460–476, 2011.
11. L. Domingo, M. Buckingham, E. Dekoninck, and H. Cornwell, "The Importance of Understanding the Business Context When Planning Eco-design Activities," *Journal of Industrial and Production Engineering*, vol. 32, no. 1, pp. 3–11, Jan. 2015, doi: 10.1080/21681015.2014.1000398.
12. C. M. Dües, K. H. Tan, and M. Lim, "Green as the New Lean: How to Use Lean Practices as a Catalyst to Greening Your Supply Chain," *Journal of Cleaner Production*, vol. 40, pp. 93–100, Feb. 2013, doi: 10.1016/j.jclepro.2011.12.023.
13. R. K. Sharma, P. K. Singh, P. Sarkar, and H. Singh, "A Hybrid Multi-Criteria Decision Approach to Analyze Key Factors Affecting Sustainability in Supply Chain Networks of Manufacturing Organizations," *Clean Technologies and Environmental Policy*, vol. 22, no. 9, pp. 1871–1889, Nov. 2020, doi: 10.1007/s10098-020-01926-8.
14. A. Diabat, D. Kannan, and K. Mathiyazhagan, "Analysis of Enablers for Implementation of Sustainable Supply Chain Management – A Textile Case," *Journal of Cleaner Production*, vol. 83, pp. 391–403, Nov. 2014, doi: 10.1016/j.jclepro.2014.06.081.
15. V. Yadav, R. Jain, M. L. Mittal, A. Panwar, and M. K. Sharma, "An Appraisal on Barriers to Implement Lean in SMEs," *JMTM*, vol. 30, no. 1, pp. 195–212, Jan. 2019, doi: 10.1108/JMTM-12-2017-0262.

16. M. Delmas and M. W. Toffel, "Stakeholders and Environmental Management Practices: An Institutional Framework," *Business Strategy and the Environment*, vol. 13, no. 4, pp. 209–222, Jul. 2004, doi: 10.1002/bse.409.
17. M. Marconi and C. Favi, "Eco-design Teaching Initiative Within a Manufacturing Company Based on LCA Analysis of Company Product Portfolio," *Journal of Cleaner Production*, vol. 242, p. 118424, Jan. 2020, doi: 10.1016/j.jclepro.2019.118424.
18. J. Monteiro, F. J. G. Silva, S. F. Ramos, R. D. S. G. Campilho, and A. M. Fonseca, "Eco-Design and Sustainability in Packaging: A Survey," *Procedia Manufacturing*, vol. 38, pp. 1741–1749, 2019, doi: 10.1016/j.promfg.2020.01.097.
19. D. Millet, L. Bistagnino, C. Lanzavecchia, R. Camous, and T. Poldma, "Does the Potential of the Use of LCA Match the Design Team Needs?," *Journal of Cleaner Production*, vol. 15, no. 4, pp. 335–346, Jan. 2007, doi: 10.1016/j.jclepro.2005.07.016.
20. Q. Zhu and Y. Geng, "Drivers and Barriers of Extended Supply Chain Practices for Energy Saving and Emission Reduction Among Chinese Manufacturers," *Journal of Cleaner Production*, vol. 40, pp. 6–12, Feb. 2013, doi: 10.1016/j.jclepro.2010.09.017.
21. P. K. Singh and P. Sarkar, "A Framework Based on Fuzzy AHP-TOPSIS for Prioritizing Solutions to Overcome the Barriers in the Implementation of Ecodesign Practices in SMEs," *International Journal of Sustainable Development & World Ecology*, vol. 26, no. 6, pp. 506–521, Aug. 2019, doi: 10.1080/13504509.2019.1605547.
22. A. Diabat, D. Kannan, and K. Mathiyazhagan, "Analysis of Enablers for Implementation of Sustainable Supply Chain Management – A Textile Case," *Journal of Cleaner Production*, vol. 83, pp. 391–403, Nov. 2014, doi: 10.1016/j.jclepro.2014.06.081.
23. H. Pfohl, P. Gallus, and D. Thomas, "Interpretive Structural Modeling of Supply Chain Risks," *International Journal of Physical Distribution & Logistics Management*, vol. 41, no. 9, pp. 839–859, Oct. 2011, doi: 10.1108/09600031111175816.
24. R. H. Watson, "Interpretive Structural Modeling—A Useful Tool for Technology Assessment?," *Technological Forecasting and Social Change*, vol. 11, no. 2, pp. 165–185, Jan. 1978, doi: 10.1016/0040-1625(78)90028-8.
25. A. Diabat and K. Govindan, "An Analysis of the Drivers Affecting the Implementation of Green Supply Chain Management," *Resources, Conservation and Recycling*, vol. 55, pp. 659–667, 2011.
26. K. Mathiyazhagan, K. Govindan, A. NoorulHaq, and Y. Geng, "An ISM Approach for the Barrier Analysis in Implementing Green Supply Chain Management," *Journal of Cleaner Production*, vol. 47, pp. 283–297, May 2013, doi: 10.1016/j.jclepro.2012.10.042.
27. R. Raut, B. E. Narkhede, B. B. Gardas, and H. T. Luong, "An ISM Approach for the Barrier Analysis in Implementing Sustainable Practices: The Indian Oil and Gas Sector," *Benchmarking: An International Journal*, vol. 25, no. 4, pp. 1245–1271, May 2018, doi: 10.1108/BIJ-05-2016-0073.
28. A. Sharma and R. K. Sharma, "Modelling and Analysis of Enablers for Successful Implementation of Cellular Manufacturing System," *International Journal of Process Management and Benchmarking*, vol. 8, no. 1, pp. 103–123, 2018.
29. R. Attri, N. Dev, and V. Sharma, "Interpretive Structural Modelling (ISM) Approach: An Overview," *Research Journal of Management Sciences*, vol. 2, no. 2, pp. 3–8, 2013.
30. G. Kannan, S. Pokharel, and P. Sasi Kumar, "A Hybrid Approach Using ISM and Fuzzy TOPSIS for the Selection of Reverse Logistics Provider," *Resources, Conservation and Recycling*, vol. 54, no. 1, pp. 28–36, Nov. 2009, doi: 10.1016/j.resconrec.2009.06.004.

19 Comparative Study of Hybrid Vehicles with Traditional Vehicles

Kamalkishor Maniyar, Gitanjali Kale, Ankur Salunke, Avinash Salunke, Roshan Marode, Pallavi Jadhav, Sunilkumar Harsur, Prashant Patil, Pushparaj Warke, Nitin Sherje, Lalitkumar Wadhwa, and Praveen Kumar Verma

Description	**Abbreviation**
Auxiliary Power Module	APM
Electric vehicle	EV
Fuel-cell electric vehicle	FCEV
High voltage	HV
Hybrid and electric vehicle	HEV
Internal combustion engine	ICE
Low voltage	LV
Plug-in hybrid electric vehicle	PHEV

19.1 INTRODUCTION

Propulsion is provided by the combination of engine systems, while traction is provided by the motor. A car with an electric motor that runs entirely on battery power is called an all-electric vehicle [1]. The charging stations for this car must be as far apart as the distance it can travel between charges. If the electric motor is powered by both the heat engine and the battery, the car is said to be hybrid [2]. This vehicle is actually a hybrid fuel-cell automobile. Hybrid or all-electric electric cars may reduce pollution locally, but their impact is not the same on a global scale [3–8]. Reducing hazardous emissions and fuel consumption in comparison to a standard vehicle is the ultimate goal of hybrid electric vehicles [9]. Energy efficiency and environmental preservation have accelerated the development of EVs. But the marketing of electric

DOI: 10.1201/9781003516071-19

vehicles was a disaster [10]. The main problem was that their high cost and restricted availability prevented them from satisfying the demands of the clients. As a result, HEV and FCEV are currently developing rapidly [11]. In the coming decades, new technologies will assist promote the development of batteries and enhanced propulsion [12]. BEVs will mostly be developed for small cars in a specialized market, such as community transportation. HEV will grow faster and be able to meet consumer demand. The primary difficulty with HEVs is figuring out how to best use the many energy sources to deliver the best performance at the lowest cost [13]. Figure 19.1 illustrates the schematic diagram of a HEV.

In the long term, FCEV might replace conventional vehicles. The start torque of the starter generators in hybrid electric vehicles has a major effect on the HEV's emissions [14]. Due to the vehicle's space constraints, they are frequently made in a flat shape. To optimize the SG's output torque and efficiency, the stator windings are built as concentrated windings; yet, this will result in a noticeably higher torque ripple [15]. This study investigates the toque ripple of a focused winding SG. When calculating the torque ripple, an average approach and a WO dimensional Finite Element Analysis (FEA) approach are used. In order to minimize torque ripple, three separate SGs with unique topologies are made [16]. Hybrid electric vehicles, or HEVs, integrate multiple energy sources. In this work, electric motors present in the propulsion systems of hybrid and electric vehicles are used to provide antilock braking [17]. The study illustrates how HEV antilock braking may be easily achieved through the use of iterative learning control for various driving conditions [18]. In order to regulate the antilock performance of HEVs during deceleration, three models were created: a vehicle model, a slip ratio model, and a speed observer. The motor torque is adjusted through an iterative process of learning to ensure that the tire slip ratio remains consistent with the maximal traction coefficient during braking [19]. Figure 19.2 shows the electric vehicle power equipment.

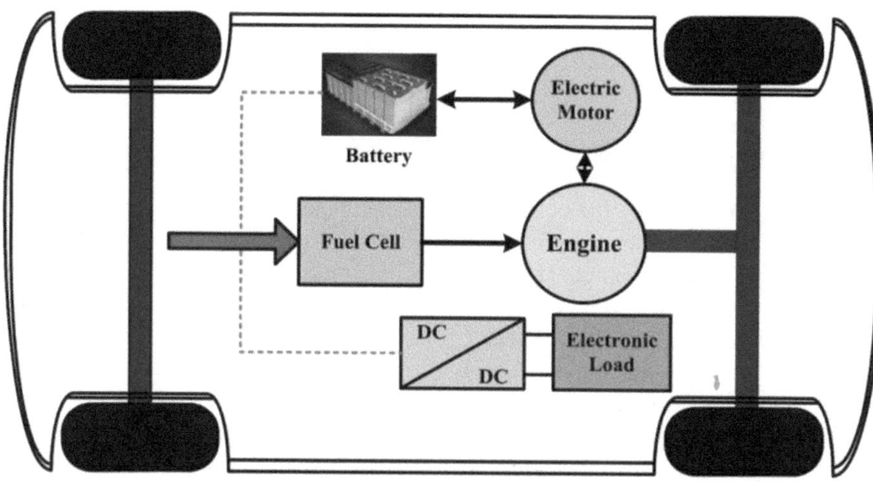

FIGURE 19.1 Schematic diagram of a HEV [14]. (Credit: Creative Commons Attribution 4.0 License)

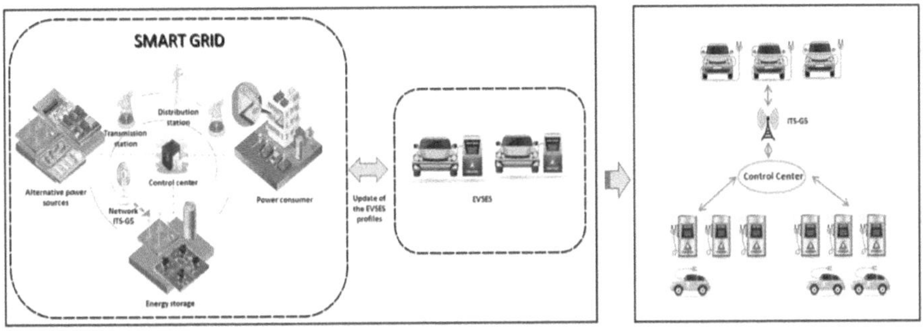

FIGURE 19.2 System overview of electric vehicle power equipment [20]. (Credit: Creative Commons Attribution 4.0 License).

19.2 HYBRID ELECTRIC VEHICLES

This case study offers a general literature survey of HEVs. The basic costs and advancements in the field of HEVs are analyzed to illustrate the role this technology plays in the current automotive market [21]. There is also a full discussion of the advantages and disadvantages of this automotive technology. Additionally, the study is focused on market estimates for HEVs, both current and prospective, with a particular focus on legislative actions that are helping to increase the market share of environmentally friendly cars [22]. The views of researchers and automakers will be taken into account in order to predict which technical advancements will have the greatest future influence [23]. Plug-in hybrid electric vehicles, or PHEVs, are hybrid electric vehicles that may power their propulsion systems by utilizing and storing energy from the electric grid [24]. A plug-in hybrid might be able to go from using energy derived from petroleum to using electricity generated by a number of other sources using this simple and practical accessory for the standard hybrid electric vehicle [25]. There will be large and mostly positive impacts on the amount of petroleum utilized in transportation energy, greenhouse gas emissions, CO_2 emissions, and electrical grid construction and operation as a result of this [26–30]. Using plug-in hybrid electric cars is one of the best strategies to improve the transportation and stationary energy sectors' near-term sustainability [31]. Figure 19.3 shows the plug-in hybrid electric vehicle (PHEV).

This overview covers the basic design elements of PHEVs, such as the grid, energy management systems, drivetrain components' operation, and the architecture of the vehicle. Most cars on the roads today are powered by internal combustion engines or ICEs. ICEs are the main cause of pollution in cities, notwithstanding the enormous development they have experienced during the previous century. HEVs have the potential to significantly increase efficiency and reduce hazardous emissions, despite the fact that pure electric vehicles are presently far behind internal combustion engine vehicles in terms of performance [33]. Figure 19.4 shows the internal combustion engine.

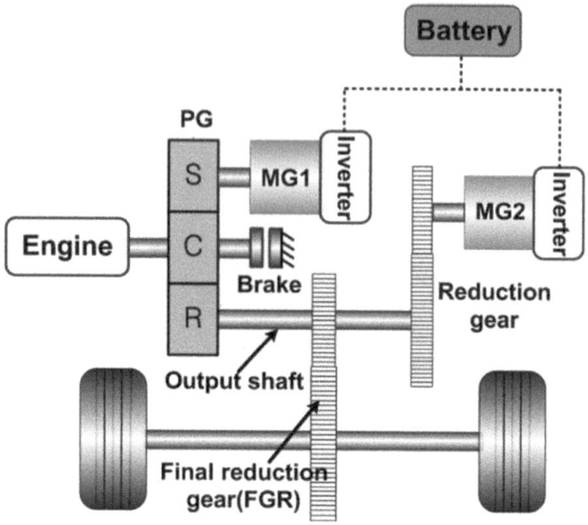

FIGURE 19.3 PHEV system's intended vehicle configuration [32]. (Credit: Creative Commons Attribution 4.0 License).

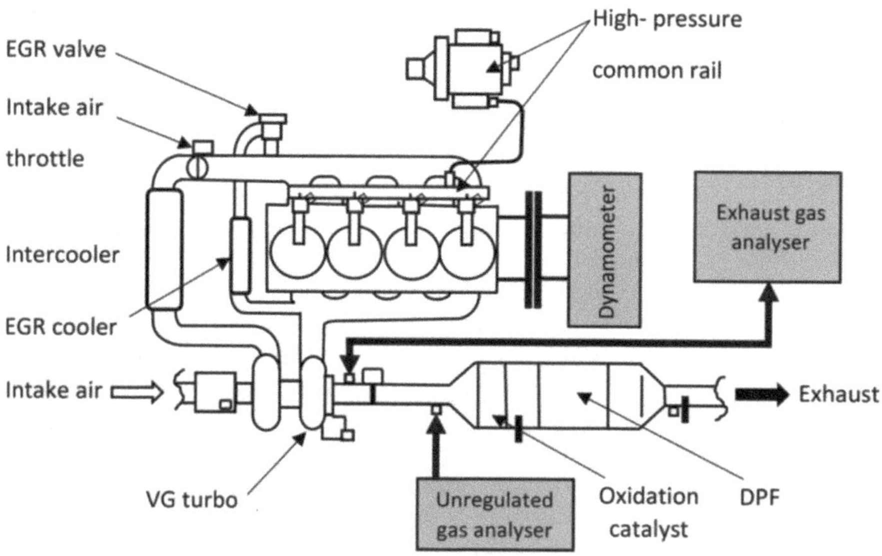

FIGURE 19.4 The experimental system of internal combustion engine [34]. (Credit: Creative Commons Attribution 4.0 License).

19.3 DESIGN AND DEVELOPMENT OF A SMALL HYBRID ELECTRIC VEHICLE

In this chapter, the design and development of small HEVs for personal urban transportation is discussed. Ideal power management strategies and architectural decisions are highlighted, with a focus on the electric powertrain system. The corporate industry is seeing a significant shift toward hybrid electric vehicles owing to their proven benefits of lower pollution emissions and fuel efficiency. Fuel efficiency is still highly valued in both the commercial and military sectors, despite the fact that military vehicles have completely different drivers than hybrid electric cars. Because hybrid electric vehicles can generate a significant quantity of electricity inside the vehicle to meet the demands of the warfighter and increase mobility, survivability, and lethality for his mission, the military needs them [35]. The army recognizes the advantages of hybrid electric propulsion for combat and tactical vehicles. Nevertheless, the deployment of hybrid electric military vehicles has lagged behind that of commercial vehicles due to technical difficulties. Dynamics theory states that for all 30 driving cycles, the maximum vehicle traction power requirements are less than the vehicle's actual traction power. Based on statistical analysis, the optimal driving ratio for a sedan HEV (hybrid electric vehicle) is 55%:28%:17% in the city, suburbs, and on the highway. On series, parallel, and PRIUS HEV models, the optimal driving cycle is selected from a list of 30, and it is simulated using the optimum ratio of the road condition [36].

Motor duty cycles are dynamic efficiency test cycles for motor propulsion systems of series, defined by the ideal torque load ratio and speed load ratio model findings. In order to replace the alternator used in most automobiles, a compact DC/DC converter is required for the architectures of electric and hybrid electric vehicles (EV/HEV). Power transfer between the vehicle's low voltage (LV) and high voltage (HV) DC buses is facilitated by the small DC/DC converter, also referred to as the Auxiliary Power Module (APM) of the vehicle. The APM HV interface is coupled to the HV DC bus, which consists of several power conversion devices (traction inverters) and energy storage systems (HV batteries) [37, 38]. Figure 19.5 illustrates the DC/DC converter.

The vehicle's LV DC bus supplies power to additional electronic control modules and 12V accessory loads, and is coupled to the low-voltage battery and APM LV

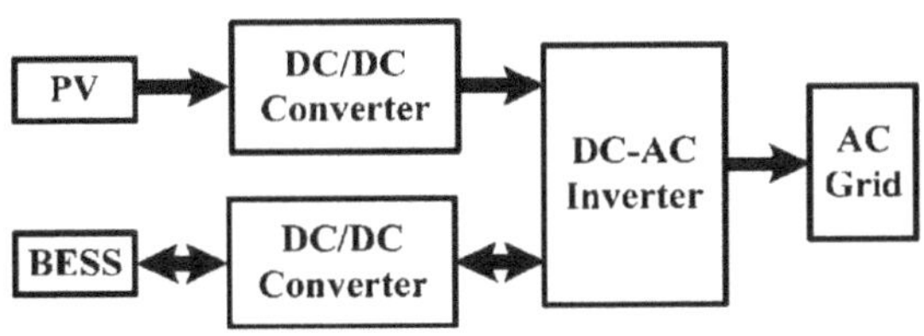

FIGURE 19.5 Typical power flow diagrams for a PV generator and BESS in their conventional setup, with two power converters [39]. (Credit: Creative Commons Attribution 4.0 License).

interface. The performance, overall efficiency, and electric range of the vehicle are therefore significantly influenced by the design and functionality of the APM. An extensive series hybrid electric vehicle model is used in the development and testing of a SCS. The EMM control system can now carry out local powertrain efficiency maximization in real-time without requiring extra processing while driving thanks to a control map that was produced through off-line optimization. The examination of the individual efficiency diagrams for the PS and SS is a critical component of this optimization [40]. It was shown that the EMM control, as opposed to thermostat control, increases fuel economy over a range of typical drive cycles. Designers have been able to develop SCSs thanks to insights acquired from analyzing the transient response of the engine using the dynamic nature of the vehicle model [41]. The suggested study provides an innovative approach to identifying current trends and best practices in the concept of electric mobility cars and significant technological developments. The DLR Vehicle Concept Database contains information on over 200 electric concept and series cars that have been displayed at auto shows or are currently available for purchase in a number of foreign locations [40]. The technical details of each car are collected for five key technologies, leading to complex claims and detailed analyses of advancements in state-of-the-art research for hybrid motors—an average electric motor that has more than doubled in power [41]. Assuming a 1,0 kW/kg increase in machine power density between 2006 and 2012 and an average battery capacity of 19 Wh per kg vehicle mass for BEVs in 2012, the NEDC test cycle can accommodate an electric range of around 150 km [42]. The employment of optimization algorithms for the electrical energy management of cars is a novel concept in the transportation industry. To accurately duplicate the physical behavior of the vehicle under consideration in this context, it is required to identify the performances starting with an accurate electrical device model [43]. Figure 19.6 shows the challenges and opportunities with electric vehicles.

Starting from a model that considers a variety of factors, such as the battery and all electrical loads, is especially important. In the transportation sector, using optimization algorithms for automobile electrical energy management is a new idea. Starting with a realistic electrical device model, it is necessary to identify the performances in order to accurately replicate the physical behavior of the vehicle under discussion in this context [44]. It is especially crucial to start with a model that accounts for a wide range of parameters, including the battery and all electrical loads [45]. As carbon emissions increase and fossil fuel energy sources diminish, technology is shifting toward systems whose main design principle is renewable energy [46]. Due to the benefits of reduced fuel consumption, lower operational costs, less noise emissions, lower emissions, smaller engine capacities, and longer engine longevity, hybrid electric vehicles are among the most popular technologies of the present day [47]. A dynamic analysis of the behavior of a hybrid electric vehicle is necessary to accurately assess its performance, utilization, and emission level. The two distinct categories of simulation tools that are available for electric and hybrid cars are stable state and dynamic tools. A dynamic model that encompasses the electric motor, batteries, and motor controller is necessary for a comprehensive analysis of the vehicle's performance. The core controller of this sedan consists of a 1/0 board, an adapter,

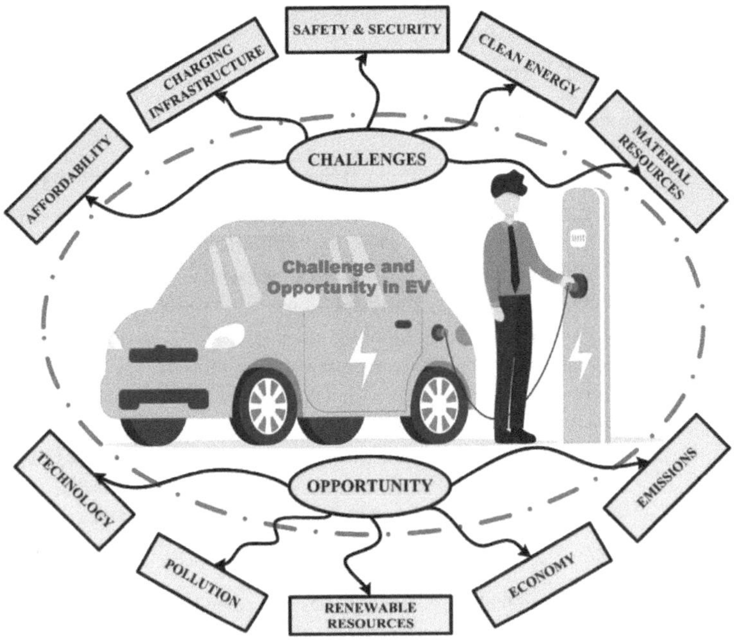

FIGURE 19.6 EVs for providing a variety of possibilities and challenges [14]. (Credit: Creative Commons Attribution 4.0 License).

and a motor-driving microcontroller. First, a power analysis tool is used to scale the vehicle's components [48].

Many technologies are in the works for the next generation of automobiles. Certain technological obstacles still need to be overcome, particularly in the case of fuel-cell automobiles. There are still important problems to be overcome in the areas of weight, volume, and cost in order to achieve the required performance and efficiency. The value to the customer as a function of pricing is the most important factor, but other considerations include the product's dependability, safety, durability, and ability to be manufactured. It is not so much the technology as it is the economics that prevent "More Electric Vehicle" development. The value of a hybrid or plug-in hybrid vehicle must exceed its cost [49–52]. **Table 19.1** summarizes a comparison of ICE, HV, and EV.

19.4 CONCLUSION

The development and expansion of a nation's economy are significantly influenced by both public and private transportation. Market expansion and resilience are enabled by an efficient transportation infrastructure. It promotes and sustains the robust economic expansion that boosts global competitiveness. A double fall in social bonds and a reduction in production are among the primary drivers of inefficient systems.

TABLE 19.1
Comparison of ICE, HV, and EV

Parameter	Internal Combustion Engine (ICE)	Hybrid Vehicle	Electric Vehicles (EVs)
Fuel Efficiency	Lower fuel efficiency due to energy losses in the combustion process.	Provides improved fuel efficiency by electrical power and combustion engine.	It provides higher energy efficiency by using an electrical motor.
Powertrain	Relies solely on an internal combustion engine powered by fossil fuels.	Combines an internal combustion engine with an electric motor and a battery.	Utilizes an electric motor powered by rechargeable batteries.
Emissions	Emits greenhouse gases and pollutants during combustion.	Emissions depend on the driving mode (electric or combustion engine).	Produces zero tailpipe emissions, but emissions depend on the electricity generation source.
Range	Dependent on fuel tank capacity, but generally offers longer range without refueling.	Range varies depending on the size of the fuel tank and battery capacity.	Range is determined by battery capacity and driving conditions.
Cost	Generally has a lower purchase price, but fuel and maintenance costs can be higher.	Cost can vary depending on the type (mild, full, plug-in), but typically higher than ICE vehicles and lower than EVs.	Higher initial purchase price but can have lower operating costs due to lower fuel and maintenance expenses.
Maintenance	Requires regular maintenance, such as oil changes, filter replacements, and tune-ups.	Maintenance requirements are similar to ICE vehicles, but the presence of an electric motor and battery may add some additional maintenance considerations.	Generally has fewer moving parts and less maintenance required compared to ICE vehicles. Battery maintenance and replacement can be a factor.
Examples	Ford Mustang Gt Mahindra Xuv 700 Tata Nexon	Toyota Prius Maruti Suzuki Grand Vitara Volvo Xc90	Mahindra Xuv 400 Tesla Model S BMW I3

This study presents a comparison of many hybrid electric vehicles. Extensive literature survey investigations have been carried out in order to meet the research goal. According to a survey of the literature, research into hybrid vehicles is currently progressing more quickly than that of electric and internal combustion engine vehicles. The response characteristics and criteria are used to assess a hybrid car's performance. Furthermore, there is a greater impact on the response factors. This study contributes to the definition of hybrid car performance characteristics and helps determine the ideal configuration. Hybrid electric vehicle studies are expected to become more common in the coming years because of advancements in battery technology, control strategies, and government support for automakers and owners. This study gave a general overview of HEVs with an emphasis on classification, history, benefits and drawbacks, vehicle types, and energy management techniques.

REFERENCES

1. Chan CC, The state of the art of electric and hybrid vehicles, *Proceedings of the IEEE*, vol. 90, no. 2, 247–275, Feb. 2002, doi: 10.1109/5.989873.
2. Cheng S, Pei Y, Chai F, and Cui S, The torque pulsation analysis of a starter generator with concentrated windings based hybrid electric vehicles, in *IEEE International Electric Machines and Drives Conference, 2003. IEMDC'03*, vol. 1, Madison, WI, USA, 218–221, 2003, doi: 10.1109/IEMDC.2003.1211266.
3. Kumar S, Singh J, and Mohapatra SK, Role of particle size in assessment of physico-chemical properties and trace elements of Indian fly ash, *Waste Management & Research Journal for a Sustainable Circular Economy*, vol. 36, 1016–1022, 2018.
4. Singh J, Gill HS, and Vasudev H, Computational fluid dynamics analysis on role of particulate shape and size in erosion of pipe bends, *International Journal of Interactive Design and Manufacturing*, vol. 17, no. 5, 1–16, 2022.
5. Kumar P, Singh J, and Singh S, Neural network supported flow characteristics analysis of heavy sour crude oil emulsified by ecofriendly bio-surfactant utilized as a replacement of sweet crude oil, *Chemical Engineering Journal Advances [Internet]*, vol. 11, 100342, 2022, https://doi.org/10.1016/j.ceja.2022.100342.
6. Singh R, Toseef M, Kumar J, et al., Benefits and challenges in additive manufacturing and its applications, in Kaushal S, Singh I, Singh S, et al., editors, *Sustainable Advanced Manufacturing and Materials Processing* [Internet], 1st ed., Boca Raton: CRC Press, 137–157, 2022.
7. Kattimani MA, Venkatesh PR, Masum H, et al., Design and numerical analysis of tensile deformation and fracture properties of induction hardened inconel 718 superalloy for gas turbine applications, *International Journal of Interactive Design and Manufacturing*, vol. 18, no. 8, 1–11, 2023.
8. Thakur L, Vasudev H, Singh J, et al., *Artificial Intelligence and Machine Learning in the Thermal Spray Industry: Practices, Implementation, and Challenges*, CRC Press, 2023.
9. Mi C, Lin H, and Zhang Y, Iterative learning control of antilock braking of electric and hybrid vehicles, *IEEE Transactions on Vehicular Technology*, vol. 54, no. 2, 486–494, March 2005, doi: 10.1109/TVT.2004.841552.
10. Sânchez-Repila D and William Poxon JE, Hybrid electric vehicles: current concepts and future market trends, *Buran*, vol. 23, 5–30, 2006.
11. Bradley TH and Frank AA, Design, demonstrations and sustainability impact assessments for plug-in hybrid electric vehicles, *Renewable and Sustainable Energy Reviews*, vol. 13, no. 1, 115–128, 2009.
12. Concari C, Franceschini G, and Toscani A, Design and optimization of a hybrid-electric vehicle for advanced urban mobility, in *IEEE Industry Applications Society Annual Meeting*, IEEE, 2008.

13. Khalil G, Challenges of hybrid electric vehicles for military applications, in *IEEE Vehicle Power and Propulsion Conference,* IEEE, 2009.
14. Kumar M, Panda KP, Naayagi RT, Thakur R, Panda G, Comprehensive review of electric vehicle technology and its impacts: Detailed investigation of charging infrastructure, power management, and control techniques, *Applied Sciences*, vol. 13, no. 15, 8919, 2023.
15. Li W, et al., Compilation of dynamic efficiency test cycle for motor propulsion system on hybrid electric vehicle, in *IEEE International Conference on Intelligent Computing and Intelligent Systems*, Vol. 1, IEEE, 2010.
16. Hasan, SMN, et al., Features and challenges for Auxiliary Power Module (APM) design for hybrid/electric vehicle applications, in *IEEE Vehicle Power and Propulsion Conference*, IEEE, 2011.
17. Shabbir W, Arana C, and Evangelou SA, Series hybrid electric vehicle supervisory control based on off-line efficiency optimization, in *IEEE International Electric Vehicle Conference*, IEEE, 2012.
18. Frieske B, Kloetzke M, and Mauser F, Trends in vehicle concept and key technology development for hybrid and battery electric vehicles, in *World Electric Vehicle Symposium And Exhibition (EVS27)*, IEEE, 2013.
19. Saravade N, Sedani C, Maniyar K, Nandkishore D, Deshmukh S, Rupanar S, Gidhad B, and Surase S, Study on heavy machinery utilizing the most recent technologies for excavation of road construction and mining, *Journal of Mines, Metals and Fuels*, vol. 72, no. 7, 667–673, 2024, https://doi.org/10.18311/jmmf/2024/45327.
20. Farhani R, El Hillali Y, Rivenq A, Boughaleb Y, Hajjaji A, Assignment approach for electric vehicle charging using traffic data collected by Sumo, *World Electric Vehicle Journal*, vol. 14, no. 2, 40, 2023.
21. Patil GH, Mishra E, Maniyar K, Rawoot M, and Satarkar MFAR, Study on Prediction of lightning impulses on transmission line tower due to harsh field and variable working condition, environmental condition and tower footing resistances, *Journal of Mines, Metals and Fuels*, vol. 72, no. 7, 763–769, 2024, https://doi.org/10.18311/jmmf/2024/45322.
22. Maniyar KK, Harsur SK, Shinde SN, Shaaikh JA, Chandan MN, and Rupanar A, An experimental investigation of machining parameters on aluminum composites, *Journal of Polymer and Composites*, vol. 12, no. 6, 8–13, 2024.
23. Maniyar KK and Harsur SK, Experimental findings of the EDM process parameters for metal matrix composites, *Journal of Polymer and Composites*, vol. 12, no. 6, 14–18, 2024.
24. Maniyar KG and Ingole DS, Multi response optimization of EDM process parameters for aluminium hybrid composites by GRA, *Materials Today: Proceedings*, vol. 5, 1983643, 2018, https://doi.org/10.1016/j.matpr.2018.06.347.
25. Maniyar KG and Ingole DS, Investigation of EDM process parameters for hybrid metal matrix composites, *Institute of Physics Conference Series: Materials Science and Engineering*, vol. 377, 012204, 2018, https://doi.org/10.1088/1757-899X/377/1/012204.
26. Maniyar KG and Ingole DS, Mathematical modeling and optimization of EDM process parameters for aluminium hybrid composites, *Materials Today: Proceedings*, vol. 5, 27700–27709, 2018, https://doi.org/10.1016/j.matpr.2018.10.004.
27. Satyavathi Yedida VV and Vasudev H, A review on the development of thermal barrier coatings by using thermal spray techniques, *Materials Today: Proceedings* [Internet], vol. 50, 1458–1464, 2022, https://linkinghub.elsevier.com/retrieve/pii/S2214785321058144.
28. Singh J, Singh JP, Kumar S, et al., Short review on hydroxyapatite powder coating for SS 316L, *Journal of Electrochemical Science and Engineering*, vol. 13, 25–39, 2023.

29. Dutta V, Thakur L, Singh B, et al., A study of Erosion–Corrosion behaviour of friction stir-processed chromium-reinforced NiAl Bronze composite, *Materials*, vol. 15, 5401, 2022, https://www.mdpi.com/1996-1944/15/15/5401.
30. Majji BGR, Vasudev H, and Bansal A, A review on the oxidation and wear behavior of the thermally sprayed high-entropy alloys, *Materials Today: Proceedings*, vol. 50, 1447–1451, 2022.
31. Maniyar KG and Ingole DS, Influence of EDM control factors for aluminium hybrid composites, *Lecture Notes on Multidisciplinary Industrial Engineering*, 275–284, 2018, https://doi.org/10.1007/978-3-319-76276-0_27.
32. Son H and Kim H, Development of near optimal rule-based control for plug-in hybrid electric vehicles taking into account drivetrain component losses, *Energies*, vol. 9, no. 6, 420, 2016.
33. Maniyar KG, An experimental analysis of aluminium hybrid metal matrix composites, *Materials Today: Proceedings*, 2024, https://doi.org/10.1016/j.matpr.2024.01.021.
34. Puškár M, Živčák J, Král Š, Kopas M, and Lavčák M, Analysis of biodiesel influence on unregulated gaseous emissions of diesel motor vehicles, *Applied Sciences*, vol. 11, no. 10, 4646, 2021.
35. Maniyar KG, Marode RV, and Chikalthankar SB, Optimization of EDM process parameters on MRR and TWR of Tungsten Carbide by Taguchi Method, *International Journal of Engineering and Advanced Technology*, vol. 5, 112–116, 2016, https://doi.org/10.1016/j.matpr.2024.01.021.
36. Maniyar KG, Analysis of EDM process parameters on material removal rate of WC, *2nd National Conference on Recent Innovations in Science, Engineering*, vol. 5. no. 9, 37–39, 2017.
37. Maniyar KG, Agrawal SK, and Ingole DS, Optimization of multiple performance characteristics in EDM: A critical literature review, *International Journal of Innovative Technology and Exploring Engineering*, vol. 5, no. 10, 27–33, 2016.
38. Poojari M, Hanumanthappa H, Prasad CD, et al., Computational modelling for the manufacturing of solar-powered multifunctional agricultural robot, *International Journal on Interactive Design and Manufacturing*, vol. 18, no. 8, 1–12, 2023.
39. Ye D, Martinez S, A Three-Port DC-DC converter with partial power regulation for a photovoltaic generator integrated with energy storage, *Electronics*, vol. 13, no. 12, 2304, 2024.
40. Maniyar KG, Deore HV, Design and testing of carbon fiber composite material with comparative analysis, *International Research Journal of Engineering and Technology*, vol. 7, no. 8, 1120–1122, 2020.
41. Maniyar KG, Ingole DS, Current research development in machining of hybrid metal matrix composites: A review, *International Journal on Recent and Innovation Trends in Computing and Communication*, vol. 4, no. 7, 150–154, 2016.
42. Maniyar K, Jadhav P, Biradar R, Gawande J, An experimental study of process parameters for nano composites, *NanoWorld Journal*, vol. 10, no. S1, S209–S211, 2024.
43. Maniyar K, Jadhav P, Biradar R, Gawande J, Effect of nano-lubricants on journal bearing for performance enhancement, *NanoWorld Journal*, vol. 10, no. S1, S206–S208, 2024.
44. Maniyar K, Suhane S, Parametric analysis of electric discharge machining for nano composites, *NanoWorld Journal*, vol. 10, no. S1, S63–S65, 2024.
45. Maniyar K, Puri A, Pawar P, Patil V, Jadhav P, Patil G, et al., The comprehensive study for machining of metal matrix composites, *American Institute of Physics Conference Proceedings*, vol. 3196, 040005–5, 2024. https://doi.org/10.1063/5.0227955.

46. Maniyar K, Warke P, Jadhav P, Waware S, Harsur S, Biradar R, An extensive study of mechanical properties and tribological characteristics for hybrid composites, *American Institute of Physics Conference Proceedings*, vol. 3203, 030038–8, 2024. https://doi.org/10.1063/5.0224994.
47. Maniyar K, Pawar P, Kolhe P, Ingale M, Patil P, Sarode G, The comprehensive study of synthesis and characterization of hybrid nano fluids, *American Institute of Physics Conference Proceedings*, vol. 3149, 030048, 2024. https://doi.org/10.1063/5.0224988.
48. Ghutepatil PR, Khapare S, Joshi RB, Maniyar KG, A critical review of nanotechnology in various fields of engineering, *American Institute of Physics Conference Proceedings*, vol. 3149, 030002, 2024. https://doi.org/10.1063/5.0224984.
49. Thamke V, Saravade N, Banarase S, Maniyar K, Seismic performance and cost analysis of AAC blocks vs. fly ash bricks: A comparative study utilizing STAAD pro, *American Institute of Physics Conference Proceedings*, vol. 3196, 060006, 2024. https://doi.org/10.1063/5.0227895.
50. Banarase S, Banarase M, Saravade N, Thamke V, Maniyar K, Evaluation of concrete building strengthening through non-destructive testing methods - A case study, *American Institute of Physics Conference Proceedings*, vol. 3196, 040001, 2024. https://doi.org/10.1063/5.0227927.
51. Jadhav PA, Maniyar K, Evaluation and characterization of aluminum silicon carbide metal matrix composites, *American Institute of Physics Conference Proceedings*, 2024. https://doi.org/10.1063/5.0227945.
52. Farhani R, El Hillali Y, Rivenq A, Boughaleb Y, Hajjaji A, Assignment approach for electric vehicle charging using traffic data collected by Sumo, *World Electric Vehicle Journal*, vol. 14, no. 2, 40, 2023.

20 Building a Sustainable Future

The Power of Energy-Efficient Materials

Hitesh Kumar, Gaurav Prashar, and Hitesh Vasudev

Description	Abbreviation
Active thermal storage systems	ATSS
Green house gases	GHG
Heating, cooling, and air conditioning	HVAC
Passive thermal storage system	PTSS
Phase change materials	PCM
Thermal energy storage	TES
Volatile organic compound	VOC

20.1 INTRODUCTION

The global building and construction sector utilizes more than 3 billion tons of raw materials annually, accounting for 40% of total global consumption. Annually, construction activities account for approximately 25% of the global wood harvest, and 40% of the mined stone, and gravel. This sector is responsible for generating 50% of the worldwide emissions of greenhouse gases and chemicals that contribute to acid rain. Hence climate change represents a significant theme and challenge in both historical and contemporary contexts. A multitude of areas globally have encountered severe weather phenomena, including floods and heat waves, with an increasing frequency observed in recent years. The Paris Climate Agreement aims to restrict global warming to a maximum of 1.5°C by implementing strategies to decrease energy consumption and carbon emissions [1]. Governments across the globe have set ambitious climate targets to tackle this objective. For instance, the United States aims for a 50–52% reduction in greenhouse gas emissions from 2005 levels by 2030 [2]. Similarly, China has unveiled an action plan to achieve a CO_2 peak before 2030 [3], while the European Union is advocating for a reduction of greenhouse

DOI: 10.1201/9781003516071-20

gas emissions to at least 55% below 1990 levels by 2030 [4]. Simultaneously, the worldwide urban demographic has undergone significant expansion since 1950. In 2018, approximately 55% of the global population resided in urban environments. The projection indicates that this figure is anticipated to rise to 68% by the year 2050 [5], as per a forecast from the United Nations. The ongoing rapid urbanization is expected to lead to significant new construction activities. The global building inventory is projected to double by 2060, which translates to constructing a complete New York City every month over the next four decades [6]. As reported by the US Energy Information Administration (EIA), the building sector represented 39% of primary energy consumption in the United States in 2017. Over the decades, it has become evident that significant quantities of CO_2 are emitted into the atmosphere as a result of the burning of fossil fuels and conventional fuel sources. The atmospheric concentrations exhibit an exponential increase, as illustrated in Figure 20.1. This figure presents the variations from 2016 to 2022, with the red line representing the seasonal forecasted cycle. The data indicates that the projected CO_2 levels for the forthcoming years are expected to continue following this upward trend. The increase in greenhouse gas emissions contributes to the acceleration of climate change and global warming, resulting in significant adverse effects on both the ecosystem and human populations [7].

Renewable plants represent a viable solution; however, their effectiveness is significantly influenced by weather conditions, solar radiation levels, and wind speed. A further solution involves the implementation of home energy management through the integration of local resources, as well as bio-sourced, green, and geo-materials within the building's structure. This solution is designed to be more eco-friendly,

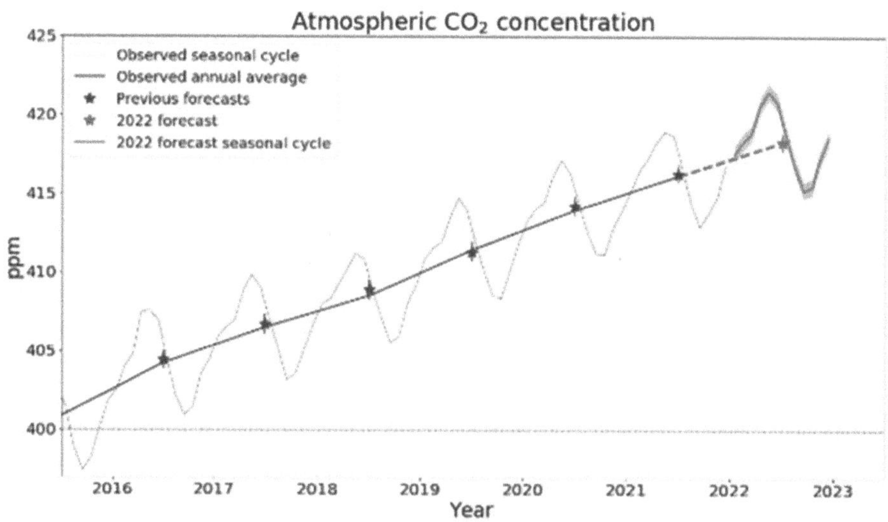

FIGURE 20.1 The continuous rise in CO_2 levels can be attributed to multiple sources of emissions [7]. (Credit: Creative Commons Attribution 4.0 License).

minimizing pollutants and reducing negative environmental impact while promoting lower greenhouse gas emissions. Hence to enhance energy efficiency and achieve carbon reduction goals, it is essential to conduct a thorough analysis of buildings. The development process has led to the emergence of cutting-edge and novel building materials, significantly transforming the construction sector. These developments, however, now exhibit significant negative consequences on the environment, affecting ecosystems and human health. The extraction procedure followed by material processing, their transportation, and lastly assemblage of construction materials represent the main applications of energy. Consequently, it is essential to reduce GHG emissions in order to mitigate negative ecological impacts [8]. Energy-efficient materials can contribute positively to building projects by providing both ecological benefits and financial savings due to their advantageous properties. The production of these materials necessitates reduced energy consumption, resulting in lower harmful emissions and pollutants linked to building materials. Additionally, the thermal properties, including heat retention and storage, play a significant role in enhancing the comfort of an interior environment. It is essential to incorporate energy-efficient characteristics alongside other factors when selecting suitable materials from the outset of the design process [9].

20.2 ENERGY-EFFICIENT MATERIALS: SELECTION

The planning and execution of new constructions, along with the upgrading of existing structures to enhance energy efficiency, are significantly influenced by the selection and application of materials. When selecting materials, it is essential to consider factors such as compatibility with climate and environment, selected material strength, its durability, material cost-effectiveness, resistance to fire and local availability. To achieve sustainability from the beginning of the design process, it is essential to develop environmentally intelligent designs that require innovation across all levels—products, structures, and communities. The selection of materials is often influenced by several key factors, one of which is the cost associated with the materials. Innovative approaches are required to tackle the significant challenge of affordable building materials. Examples of targeted interventions include phase change materials, reflective interior coatings, durable outdoor coatings, and thermal insulation materials [10].

20.2.1 Methodology of Selection of Energy-Efficient Materials

The main aspects of the proposed methodology are:

1. The parameters under review must be taken into account at each of the three lifecycle stages of a construction system: engineering (during the development of detailed documentation and work method statements for individual process cycles); construction (execution of works on the construction site in accordance with the designs, including implementing solutions to save water, electric power, heat, compressed air, and other resources when

drafting the construction site layout plans; using prefabricated transformable container units for temporary construction camps, and conducting quality control); and operation.

2. Identifying the key factors that influence the energy efficiency of a construction system throughout its lifecycle stages.
3. Identifying materials and technologies that optimize energy efficiency while adhering to environmental standards [11].

20.3 FACTORS AFFECTING ENERGY-EFFICIENT MATERIALS

The variables encompass a range of factors, including the environmental impact, suitability of design, cost considerations, sources for components, manufacturing processes, and transportation methods. This indicates that to aid architects in recognizing and prioritizing relevant criteria, as well as to facilitate the assessment of trade-offs among technical, and performance factors a structured approach to material selection is necessary [12]. Additional factors that influence energy-efficient materials include cost-effectiveness, thermal and mechanical properties, environmental impact, as well as aesthetic and design considerations (in Figure 20.2).

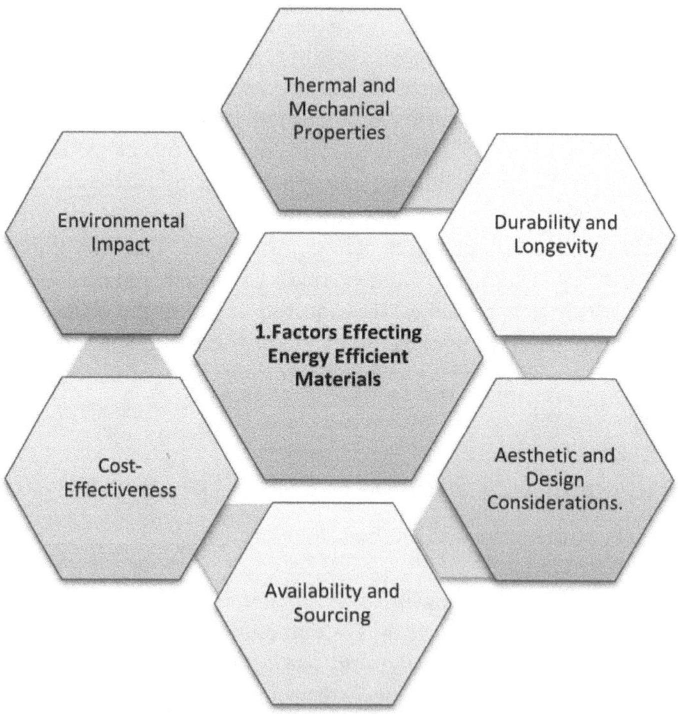

FIGURE 20.2 Factors affecting energy-efficient materials (Credit: Created by authors).

20.3.1 Thermal Properties Affecting Energy-Efficient Materials

To reduce environmental impact, enhance sustainability, and lower energy consumption, the use of energy-efficient construction materials is crucial. The thermal properties of these materials significantly influence their overall performance and effectiveness. The fundamental thermal characteristics of energy-efficient materials are as follows:

Density

A minor fraction of the pores present in various construction materials may become occupied by liquid water, humid air, dry air, or alternative gases. The material's density can be significantly affected by the proportion of voids to the solid phase and by the presence of liquid water, humid air, or dry air within those voids. The density significantly influences the overall performance, moisture resistance, structural integrity, and thermal properties of energy-efficient building materials. Enhancements in energy efficiency can be achieved through the application of both low-density and high-density materials, tailored to their specific applications and the distinct needs of the architectural design.

Thermal Conductivity

Watts per meter-kelvin (W/m•K) is a unit that quantifies a material's ability to conduct heat, commonly referred to as thermal conductivity. Materials that exhibit low-temperature conductivity and serve effectively as insulators include aerogels, foam, and fiber-glass. They improve energy efficiency and reduce heat transfer, making them suitable for minimizing the demand for heating and cooling in buildings.

Thermal Resistance (R-Value)

Thermal resistance, the reciprocal of thermal conductivity, is quantified in square meters-kelvin per watt (m^2•K/W) and reflects a material's capacity to withstand heat transfer. Higher R-values suggest improved insulating properties. In the context of roofs, walls, and floors, materials with elevated R-values are frequently employed to minimize thermal loss during winter months and mitigate heat gain in the summer season. Rigid foam and cellulose insulation represent a selection of these materials [13].

Specific Heat Capacity (c)

The amount of thermal energy required to raise the temperature of a unit mass of a substance by 1°C is referred to as its specific heat capacity (J/kg°C). Materials such as stone, brick, and concrete exhibit high specific heat capacity due to their enhanced capability to absorb and retain heat, contributing to the stabilization of interior temperatures. This feature minimizes the requirement for cooling and heating by leveraging the thermal mass effect [13].

Thermal Diffusivity (α)

Thermal diffusivity quantifies the rate of heat diffusion within a material. It is determined by dividing thermal conductivity by the product of density and specific heat capacity, yielding units of m^2/s. Materials with low thermal diffusivity exhibit a reduced rate of temperature change, thereby enhancing thermal stability. This characteristic is essential for mitigating abrupt temperature fluctuations and enhancing occupant comfort.

Emissivity (ε)

The emissivity of a material, varying between 0 and 1, reflects its capacity to emit thermal radiation. Higher values signify increased emissivity. The application of low-emissivity (low-e) window coatings effectively reduces heat transmission by reflecting infrared radiation while permitting the passage of visible light. This function reduces energy expenses related to cooling and heating, while also contributing to the maintenance of comfortable indoor temperatures [13].

Thermal Expansion Coefficient

The extent to which a material undergoes expansion or contraction due to variations in temperature is quantified by the thermal expansion coefficient. Materials exhibiting low thermal expansion coefficients demonstrate the ability to endure a diverse range of temperatures while maintaining their structural integrity and functionality. This characteristic is essential for the performance of building envelope materials, as considerable expansion or contraction can lead to air leaks and gaps, ultimately diminishing energy efficiency [13].

20.3.2 Mechanical Properties Affecting Energy-Efficient Materials

The energy efficiency of construction materials is significantly affected by their mechanical properties. Durable and resilient materials such as concrete, steel, and wood offer essential structural stability, enabling the integration of energy-efficient elements like insulation while maintaining overall structural integrity. Materials exhibiting elasticity and flexibility, capable of withstanding thermal expansion and contraction, encompass rubber and various polymers. This mitigates the development of fractures and voids, which may reduce energy efficiency. Durable materials, such as ceramics and stones, maintain their insulating characteristics over time due to their inherent resistance to degradation. The application of robust and malleable substances, including metals such as aluminum and copper, which can be shaped into energy-efficient parts and endure vibrations and impacts, has the potential to enhance the longevity and strength of structures. Structural elements subjected to significant loads and stretching pressures necessitate materials that exhibit elevated tensile and compressive strengths. The shear strength and fatigue resistance are critical for maintaining the integrity of joints and connections when subjected to dynamic loads, whereas creep resistance is essential for preserving the shape and functionality of components under extended stress conditions. Impact-resistant materials, including tempered glass and certain composites, safeguard the structural integrity of buildings during unforeseen events while also enhancing energy efficiency by preserving the integrity of insulating layers. By evaluating these mechanical properties, builders and architects can choose materials that enhance energy efficiency while maintaining the durability, structural stability, and safety of structures [14].

20.3.3 Environmental Impact on Energy-Efficient Materials

The environmental implications of energy-efficient materials are intricate, influencing both their production processes and application, alongside their overall sustainability profile. The lifecycle of these materials, encompassing extraction, production, installation, and disposal, reveals significant differences in their energy consumption and emission profiles over time. Materials that exhibit low embodied energy, such as bamboo and recycled steel, contribute to a reduced negative impact on the environment during the manufacturing process. Furthermore, the acquisition of materials is essential; utilizing locally sourced and responsibly obtained resources minimizes transportation emissions and promotes ecological equilibrium. Energy-efficient materials such as aluminum and glass can be recycled and maintained, which reduces their environmental impact by extending their lifespan and minimizing waste. Considering toxicology is essential; selecting non-toxic, environmentally friendly options, such as low-VOC paints and natural fiber insulation, contributes to reducing environmental pollution and creating a healthier living space. Recognizing and addressing these environmental factors can significantly enhance the sustainability and ecological integrity of construction projects through the careful selection and implementation of energy-efficient materials.

20.3.4 Impact of Cost-Effectiveness on Energy-Efficient Materials

The economic viability of energy-efficient materials significantly influences their adoption and long-term sustainability in construction initiatives. While these materials might have a higher upfront cost compared to traditional options, they generally demonstrate a return on investment over time. Substantial energy savings over time can be realized by effectively minimizing heating and cooling expenses through the implementation of advanced insulation materials and high-efficiency window systems. The durability and minimal upkeep of these materials significantly boost their economic viability by lowering expenses related to replacement and maintenance. By optimizing initial expenditures alongside tax credits, rebates, and various financial incentives, the application of energy-efficient materials can lead to significant cost reductions and enhance accessibility. Employing cost-effective, energy-efficient materials can enhance a property's value and appeal to environmentally conscious buyers or tenants. The cost-effectiveness of energy-efficient materials is crucial for advancing sustainable construction practices, as they integrate potential financial incentives with long-term savings and initial expenditures [15].

20.3.5 Impact of Durability and Longevity on Energy-Efficient Materials

The durability and lifespan of energy-efficient materials significantly influence the effectiveness and sustainability of construction endeavors. The enduring preservation of structural integrity and thermal performance is characteristic of specific materials, such as reinforced concrete and engineered wood. This quality minimizes the necessity for frequent replacements and repairs, thereby decreasing resource consumption and waste generation. This robust characteristic guarantees a steady degree of energy efficiency, leading to ongoing cost reductions for both cooling and heating. Moreover, the extended durability of these materials reduces the ecological impact associated with their production, logistics, and end-of-life management. Consequently, employing robust, energy-efficient materials enhances a home's value and marketability, appealing to environmentally aware tenants and buyers. Therefore, the significance of longevity and durability is paramount in optimizing both the environmental and financial benefits of employing energy-efficient materials in sustainable construction.

20.3.6 Impact of Availability and Sourcing on Energy-Efficient Materials

The availability and purchasing methods of energy-efficient materials play a crucial role in determining their prevalence and sustainability in construction projects. Choosing materials that are sourced sustainably and supplied locally reduces the environmental footprint associated with long-distance transportation, thereby decreasing emissions and supporting ecological balance. The availability of these materials enhances the efficiency of the supply chain, potentially leading to lower costs and expedited project timelines. Employing sustainable sourcing methods, such as obtaining certifications for ethical harvesting and production practices, can enhance the environmental credentials of energy-efficient products. Prioritizing the utilization of locally available resources not only strengthens regional economies but also fortifies community resilience. Emphasizing the utilization of locally sourced materials enhances the durability of community structures and fosters the development of regional enterprises. By prioritizing the importance of availability and sustainable sourcing practices, architects and builders can effectively reduce the environmental impact of their projects while also significantly enhancing the long-term sustainability of the construction sector.

20.3.7 Impact of Aesthetic and Design Considerations on Energy-Efficient Material

The aesthetics and configuration of energy-efficient materials play a crucial role in determining the overall effectiveness of sustainable construction initiatives. Maximizing energy efficiency remains paramount; however, incorporating visually appealing materials and design elements not only improves the aesthetic value of a building but also contributes to its functional performance. To develop spaces that are visually appealing and environmentally sustainable, architects and designers need to meticulously align energy efficiency requirements with the intended aesthetic outcomes. Materials that enhance energy efficiency while reducing environmental impact—like recycled metals, sustainable woods, and eco-friendly finishes—provide a wide range of design options and optimize creative expression. Moreover, strategic design decisions, such as optimizing natural light and airflow, contribute to occupant comfort while also improving energy efficiency. By seamlessly integrating energy-efficient materials with aesthetic and architectural considerations, architects and designers can create sustainable buildings that minimize energy consumption while providing residents with inspiring and enjoyable environments.

20.4 UTILIZATION OF ENERGY-EFFICIENT CONSTRUCTION MATERIALS IN THE DESIGNING OF COASTAL URBAN ENVIRONMENTS

Considering the principles of sustainable development, the incorporation of energy-efficient building materials in coastal urban areas can significantly address the demands for environmentally friendly and health-conscious structures, while also aligning with the community's focus on ecological preservation and reduced energy consumption. Therefore proactive advancement of energy-efficient building practices and the ongoing enhancement of renewable energy application are leading to an increased integration of renewable energy in the architectural design of marine cities [16]. The city's marine characteristics necessitate research into materials and application techniques for environmental protection that are appropriate for marine city architecture. This approach aligns with the Scientific Outlook on Development and adheres to the worldwide pattern of resource conservation (Figure 20.3) [17].

The role of energy-efficient windows and doorways is to regulate the energy exchange between the interior and exterior, contributing to the energy efficiency of marine city structures. Regarding the energy efficiency of doors and windows, it is essential for energy-efficient marine city structures to enhance the thermal efficiency of window units and glazing systems. The introduction of energy-efficient doors and windows has significantly elevated the overall quality of construction projects aimed at energy conservation in marine urban environments [18]. In energy-efficient marine city structures, the materials used for doors and windows play a crucial role in heat transfer and conduction, significantly impacting the overall energy performance of these buildings. The doors and windows constitute a significant portion of the total energy consumption in marine city buildings (Figure 20.4). Consequently, the extensive and judicious implementation of energy-efficient doors and windows can significantly enhance the energy conservation standards of marine city structures, foster the ongoing advancement and evolution of energy-saving practices in these buildings, and offer inhabitants a more comfortable, sustainable, and eco-friendly living environment.

The combination technology of solar energy with marine city buildings can be categorized into two main types: photo-thermal systems and photovoltaic systems. Photo-thermal integration represents a method for transforming solar energy into

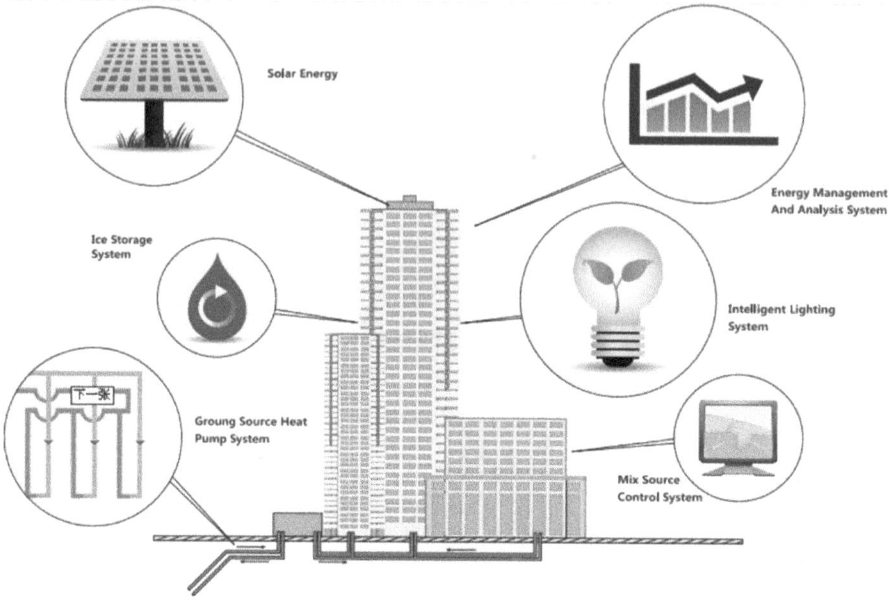

FIGURE 20.3 Diagram of sustainable construction technique [16]. (Credit: Creative Commons Attribution 4.0 License).

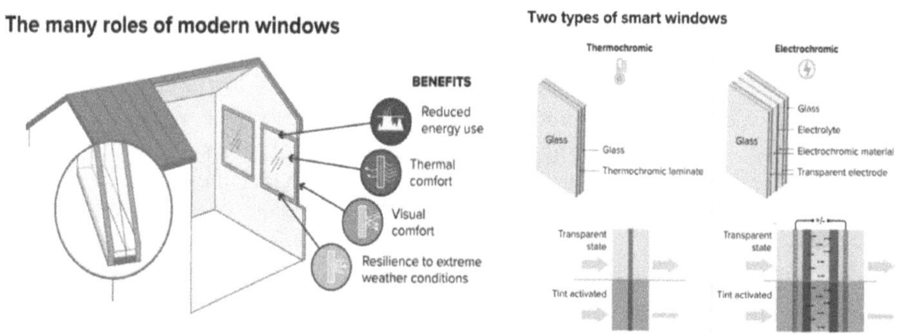

FIGURE 20.4 Utilization of energy-efficient window systems [16]. (Credit: Creative Commons Attribution 4.0 License).

thermal energy for practical applications. In the context of marine city buildings, various utilization strategies include solar air collector systems for heating, solar water heaters for hot water supply, passive solar rooms that leverage integrated heat collection and storage principles, and solar heating systems designed to improve ventilation within these structures. Currently, the utilization of solar water heaters for water supply is prevalent. Photoelectric integration involves converting solar energy into electrical energy, which is stored in batteries for use during nighttime, catering to indoor lighting and other requirements. The module consists of multiple monocrystalline or polycrystalline silicon cells arranged in series and parallel configurations. By employing solar cell modules for electricity generation, we can create rooftop components for marine city buildings, effectively merging solar power generation with urban architecture [16].

20.5 EMERGING ENERGY-EFFICIENT MATERIALS

Over the years, the significant rise in global energy consumption rates, coupled with environmental issues and the depletion of fossil fuel reserves, has underscored the critical need for effective energy utilization. Consequently, it is essential to elucidate the advancement of a novel approach to energy conservation that can be realized through thermal energy storage (TES) systems. These systems are designed to store thermal energy in mediums such as phase change materials (PCM) including rock or water, and subsequently release it as required. TES systems are increasingly being adopted across various industries due to their effectiveness in enhancing energy efficiency and promoting the utilization of renewable energy sources. The TES systems are categorized into three primary types: sensible storage, latent storage, and thermo-chemical storage [19]. Figure 20.5 illustrates the passive thermal storage system (PTSS), which includes multiple applications, such as the incorporation of PCM into building walls, floors, and roofs. Integration into insulation, façade, shutter, and shading systems is also possible [20–21]. Conversely, PCM in active thermal storage systems (ATSS), as illustrated in Figure 20.6, facilitates the conversion of mechanical or electrical energy, allowing the system to store and release heat as required. This process enhances heat transfer efficiency and minimizes energy consumption of structures. Typically, in this mode, the PCM is employed within the heating, cooling, and air conditioning (HVAC) sector, as well as in aquifers, boreholes, snow storage, pits, or tanks located in proximity to buildings. The PCM must exhibit stability, as these materials are intended for long-term application. This indicates that the materials must maintain their integrity over time and should remain unaffected by ambient air or moisture in order to preserve their properties as they age.

The building sector is widely recognized as one of the most detrimental to the environment. It is responsible for the extraction of over 40% of natural resources in industrialized nations, accounts for a minimum of 70% of electricity consumption,

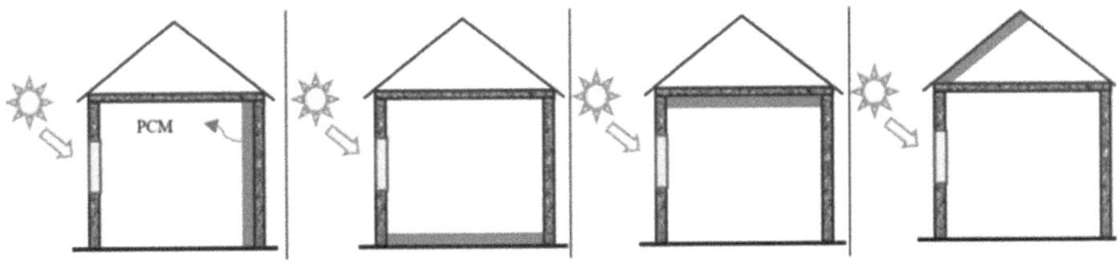

FIGURE 20.5 Passive heating utilizing phase change materials (PCM) integrated within the walls, floor, ceiling, and roof [7]. (Credit: Creative Commons Attribution 4.0 License).

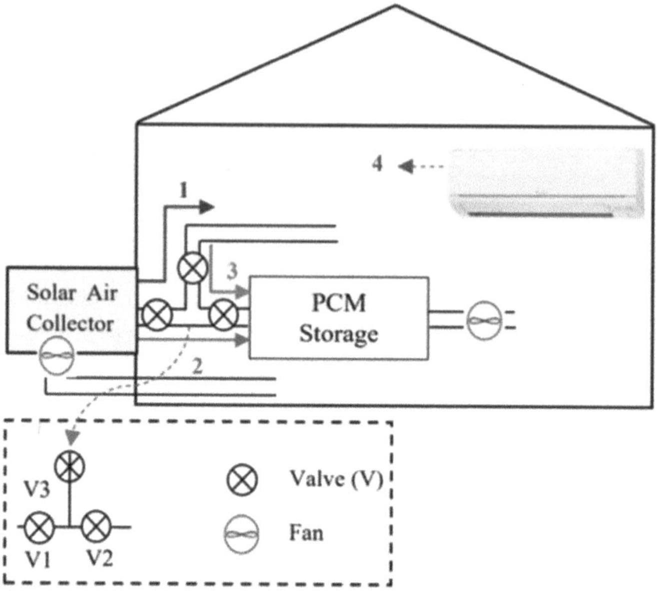

FIGURE 20.6 Representation of the various pathways for supplying PCM with heating and cooling energy [7]. (Credit: Creative Commons Attribution 4.0 License).

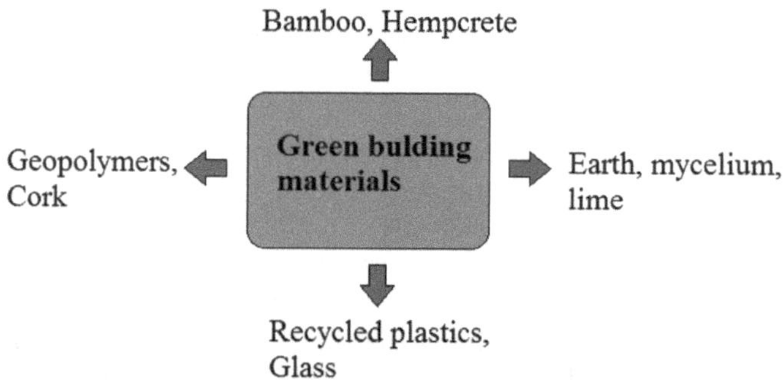

FIGURE 20.7 Emerging green building materials. (Credit: Created by authors).

utilizes 12% of potable water, and generates a significant amount of waste, with disposal rates ranging from 45% to 65% directed to landfills [22–26]. The building sector is responsible for 30% of greenhouse gas emissions and numerous harmful emissions resulting from its operations, material extraction, and transportation [27–31]. The global population is projected to increase from 6.5 billion in 2005 to approximately 9.0 billion by 2035, which is anticipated to further elevate these figures. Therefore to determine the environmental sustainability of the building, various certification codes are utilized to validate its green status. Leadership in Energy & Environmental Design (LEED) is the most prevalent standard in the field. Green building materials are categorized as environmentally friendly, exhibiting minimal or no impact on the environment and posing no harm to human health throughout their production cycle. Their composition includes renewable resources instead of non-renewable resources. Consequently, it is essential that these materials originate from natural sources and possess the ability to endure heat, humidity, and various environmental factors. Numerous green building materials are available, as shown in Figure 20.7.

20.6 POSSIBLE INSIGHTS FOR THE FUTURE

The domain of energy-efficient materials presents a promising and growing landscape for research and innovation. As advancements in technology unfold and the significance of sustainability becomes increasingly apparent, numerous avenues are emerging that warrant additional research and development efforts. Material science advancements offer a significant chance to create innovative materials that improve energy efficiency and promote environmental sustainability [32–36]. Enhancing energy efficiency in buildings can be achieved by exploring innovative composite materials, leveraging advancements in nanotechnology, and drawing inspiration from biomimetic designs. Moreover, it is increasingly essential to conduct comprehensive lifecycle assessments to thoroughly evaluate the environmental impacts of energy-efficient materials throughout their entire lifespan [37–39]. Future research efforts should focus on a comprehensive assessment of factors like embodied energy, carbon footprint, and recyclability to enhance the accuracy of sustainability evaluations for these materials. Furthermore, addressing the challenges related to the cost-effectiveness and accessibility of energy-efficient materials is a vital focus for upcoming research endeavors. The emphasis of targeted initiatives could be on creating cost-effective alternatives and exploring techniques to enhance production, thereby ensuring these materials are accessible to a broader array of stakeholders.

The potential for energy savings through the application of AI in building systems remains inadequately explored. The current body of research predominantly emphasizes particular elements of building performance, including design, construction, and operational control. The findings suggest a wide spectrum of possible savings, ranging from 2% to 60% [40–44]. Furthermore, there is still an absence of a systematic and standardized approach for quantifying savings from AI. The process by which AI can effectively minimize energy usage and lower carbon emissions in buildings remains ambiguous. Buildings are intricate assemblies, comprising numerous elements, including walls, windows, and HVAC and lighting systems [45]. Construction projects typically entail meticulous planning, thorough analysis, systematic development, and precise execution, each demanding significant expertise, resources, and workforce commitment [46–48]. Construction projects frequently present risks that can impact the health and safety of workers on site. The implementation of AI can significantly lower expenses throughout different phases of the construction process, minimize risks, and improve health and welfare outcomes [49–53]. Furthermore, the relationships between occupants and building elements exhibit nonlinearity, making them challenging to model with conventional rule-based control systems. Utilizing sophisticated algorithms like deep learning and reinforcement learning, the AI model is capable of autonomously learning from operational data [54]. It can adapt and enhance itself through ongoing live data, thereby optimizing objective functions and elevating performance.

We assert that artificial intelligence has the potential to enhance energy efficiency and lower carbon emissions via two primary methodologies: (a) Artificial intelligence enhances the scalability of optimal technologies and methodologies. The potential to enhance technology scaling and accelerate adoption is substantial, particularly through the reduction of construction and labor expenses. Consequently, this can result in broader adoption of efficient technologies; (b) AI has the potential to enhance and refine design, construction, and operation throughout the entire life-cycle of buildings, yielding further savings [55].

20.7 CONCLUSION

This chapter explores the properties and potential for energy efficiency of energy-efficient materials. The main points are summarized below:

- Materials that enhance energy efficiency offer several benefits, such as reduced energy usage, lower operational expenses, and greater environmental stewardship, making them crucial for the advancement of sustainable building practices.

This category encompasses a diverse range of materials, each offering unique benefits toward sustainability and energy efficiency, including advanced insulation solutions and environmentally conscious finishes.

- A variety of materials, including environmentally conscious finishes and advanced insulation solutions, are included in this category, with each serving a specific function in promoting sustainability and energy efficiency.

Incorporating these materials into building projects not only enhances resilience and adaptation but also aligns with sustainability initiatives addressing resource exhaustion and climate change.

- To foster a more sustainable future and accelerate the adoption of energy-efficient materials, collaboration among stakeholders in the construction sector is essential.

Ecological energy-saving buildings represent the future of architectural design, serving as a crucial strategy for cities to attain sustainable social and economic growth. These structures not only address the scarcity of domestic energy resources but also enhance air quality, minimize pollution, and safeguard arable land. Furthermore, they significantly elevate the living standards of the population. Consequently, designers must expand their focus beyond mere shape and aesthetics in architectural design for marine cities. It is essential to incorporate principles of ecological energy efficiency, integrating concepts of sustainability, environmental protection, and energy conservation into the design process. Emphasizing the utilization of renewable energy sources, such as solar and wind energy, is crucial while enhancing the comfort of residents' living environments. This approach aims to minimize energy resource consumption, safeguard the ecological environment, and achieve a balanced coexistence between humanity and mother nature.

REFERENCES

1. UNFCCC. 2015. *The Paris Agreement.* United Nations Framework Convention on Climate Change.
2. White House. 2021. *Fact Sheet: President Biden Sets 2030 Greenhouse Gas Pollution Reduction Target Aimed at Creating Good-Paying Union Jobs and Securing U.S. Leadership on Clean Energy Technologies.* The White House.
3. State Council. 2021. *Action Plan for Carbon Dioxide Peaking Before 2030.* The State Council of the People's Republic of China.
4. European Commission. 2020. *Communication from the Commission to the European Parliament, the Council, the European Economic and Social Committee and the Committee of the Regions: Stepping up Europe's 2030 Climate Ambition Investing in a Climate-Neutral Future for the Benefit of Our People.* European Commission.
5. UN DESA. 2018. *World Urbanization Prospects: The 2018 Revision.* United Nations Department of Economic and Social Affairs, Population Division.
6. Gates, B. and Gates, M. 2019. *Our 2019 Annual Letter: We Didn't See This Coming.* https://www.gatesnotes.com/2019-Annual-Letter?WT.mc_id=02_12_2019_05_AL2019_GF-GFO_&WT.tsrc=GFGFO.
7. Nasr, Y., El Zakhem, H., Hamami, A.E.A., El Bachawati, M., and Belarbi, R. 2023. Comprehensive Review of Innovative Materials for Sustainable Buildings' Energy Performance. *Energies* 16(21): 7440.
8. Hussain, M.A. and Kamal, M.A. 2015. Energy Efficient Sustainable Building Materials: An Overview. *Key Engineering Materials* 650: 38–50.
9. Yüksek, Í. 2015. The Evaluation of Building Materials in Terms of Energy Efficiency. *Periodica Polytechnica* 59(1): 45–58.
10. Garg, A., Kumar, A., Piprralia, S., and Kumar, P. Sustainable Building Materials for Energy Efficiency. In *International Conference on Urban Sustainability: Emerging Trends, Themes, Concepts and Practices.*
11. Singh, J., Mohapatra, S.K., and Kumar, S. 2021. Performance Analysis of Pump Materials Employed in Bottom Ash Slurry Erosion Conditions. *Jurnal Tribologi* 30: 73–89.
12. Ogunkah, I. and Yang, J. *Investigating Factors Affecting Material Selection: The Impacts on Green Vernacular Building Materials in the Design-Decision Making Process.* ISSN 2075-5309.
13. Johra, H. *Thermal Properties of Building Materials - Review and Database.* https://www.researchgate.net/publication/357355874. ISSN 1901-726X.
14. Zhang, X. and Xiang, X. 2023. *Mechanical Properties of Advanced Materials and Structures for Energy Absorption.* doi: 10.3389/fmats.2023.1343791.
15. Almusaed, A., Yitmen, I, Myhren, J.A., and Almssad, A. 2024. Assessing the Impact of Recycled Building Materials on Environmental Sustainability and Energy Efficiency: A Comprehensive Framework for Reducing Greenhouse Gas Emissions. *Buildings* 14(6): 1566.

16. Zhang, L., Yuan, J., and Kim, C.S. 2023. Application of Energy-Saving Building's Designing Methods in Marine Cities. *Energy Reports* 9: 98–110.
17. Chiu, Y.R., Liaw, C.H., Hu, C.Y., Tsai, Y.L., and Chang, H.H. 2009. Applying GIS-Based Rainwater Harvesting Design System in the Water-Energy Conservation Scheme for Large Cities. In *2009 13th International Conference on Computer Supported Cooperative Work in Design*, 722–727. IEEE.
18. Poojari, M., et al. 2023. Computational Modelling for the Manufacturing of Solar-Powered Multifunctional Agricultural Robot. *International Journal on Interactive Design and Manufacturing (IJIDeM)* 18(8): 1–12.
19. Raoux, S. 2009. Phase Change Materials. *Annual Review of Materials Research* 39(1): 25–48.
20. Saffari, M., de Gracia, A., Ushak, S., and Cabeza, L.F. 2017. Passive Cooling of Buildings with Phase Change Materials Using Whole-Building Energy Simulation Tools: A Review. *Renewable and Sustainable Energy Reviews* 80: 1239–1255.
21. Akeiber, H., Nejat, P., Majid, M.Z.A., Wahid, M.A., Jomehzadeh, F., Famileh, I.Z., Calautit, J.K., Hughes, B.R., and Zaki, S.A. 2016. A Review on Phase Change Material (PCM) for Sustainable Passive Cooling in Building Envelopes. *Renewable and Sustainable Energy Reviews* 60: 1470–1497.
22. Singh, J., Satish, K., and Mohapatra, S.K. 2021. Study on Solid Particle Erosion of Pump Materials by Fly Ash Slurry Using Taguchi's Orthogonal Array. *Tribologia - Finnish Journal of Tribology* 38(3–4): 31–38.
23. Franzoni, E. 2011. Materials Selection for Green Buildings: Which Tools for Engineers and Architects? *Procedia Engineering* 21: 883–890.
24. Singh, J. 2021. Slurry Erosion Performance Analysis and Characterization of High-Velocity Oxy-Fuel Sprayed Ni and Co Hardsurfacing Alloy Coatings. *Journal of King Saud University - Engineering Sciences*. https://doi.org/10.1016/j.jksues.2021.06.009.
25. Singh, J., Satish, K., and Mohapatra, S.K. 2018. Role of Particle Size in Assessment of Physico-Chemical Properties and Trace Elements of Indian Fly Ash. *Waste Management & Research* 36(11): 1016–1022. http://journals.sagepub.com/doi/10.1177/0734242X18804033.
26. Singh, J., Gill, H.S., and Vasudev, H. 2022. Computational Fluid Dynamics Analysis on Role of Particulate Shape and Size in Erosion of Pipe Bends. *International Journal on Interactive Design and Manufacturing (IJIDeM)* 18(8): 1–16.
27. Castro-Lacouture, D., Sefair, J.A., Flórez, L., and Medaglia, A.L. 2009. Optimization Model for the Selection of Materials Using a LEED-Based Green Building Rating System in Colombia. *Building and Environment* 44(6): 1162–1170.
28. Kumar, P., Singh, J., and Singh, S. 2022. Neural Network Supported Flow Characteristics Analysis of Heavy Sour Crude Oil Emulsified by Ecofriendly Bio-Surfactant Utilized as a Replacement of Sweet Crude Oil. *Chemical Engineering Journal Advances* 11(June): 100342. https://doi.org/10.1016/j.ceja.2022.100342.
29. Pulselli, R.M., Simoncini, E., Pulselli, F.M., and Bastianoni, S. 2007. Energy Analysis of Building Manufacturing, Maintenance and Use: Em-Building Indices to Evaluate Housing Sustainability. *Energy and Buildings* 39(5): 620–628.
30. Singh, R., Toseef, M., Kumar, J., and Singh, J. 2022. Benefits and Challenges in Additive Manufacturing and Its Applications. In *Sustainable Advanced Manufacturing and Materials Processing*, ed. Sarbjeet Kaushal, Ishbir Singh, Satnam Singh, and Ankit Gupta. Boca Raton: CRC Press, 137–57.
31. Singh, J. 2023. Support Vector Machine Learning on Slurry Erosion Characteristics Analysis of Ni- and Co-Alloy Coatings. *Surface Review and Letters*. https://doi.org/10.1142/S0218625X23400061.

32. Goia, F. 2016. Search for the Optimal Window-to-Wall Ratio in Office Buildings in Different European Climates and the Implications on Total Energy Saving Potential. *Solar Energy* 132: 467–492.
33. Niachou, A., Papakonstantinou, K., Santamouris, M., Tsangrassoulis, A., and Mihalakakou, G. 2001. Analysis of the Green Roof Thermal Properties and Investigation of Its Energy Performance. *Energy and Buildings* 33: 719–729.
34. Sacks, R., Girolami, M., and Brilakis, I. 2020. Building Information Modelling, Artificial Intelligence and Construction Tech. *Developments in the Built Environment* 4: 100011.
35. Singh, J. 2022. Reducing Scrap Due to Missed Operations and Machining Defects in 90PS Pistons. *International Journal on Interactive Design and Manufacturing (IJIDeM)*. https://doi.org/10.1007/s12008-022-01071-0.
36. Pan, Y. and Zhang, L. 2023. Integrating BIM and AI for Smart Construction Management: Current Status and Future Directions. *Archives of Computational Methods in Engineering* 30: 1081–1110.
37. Dutta, V., Thakur, L., Singh, B., and Vasudev, H. 2022. A Study of Erosion–Corrosion Behaviour of Friction Stir-Processed Chromium-Reinforced NiAl Bronze Composite. *Materials* 15(15): 5401. https://www.mdpi.com/1996-1944/15/15/5401.
38. Blanco, J.L., Fuchs, S., Parsons, M., and Ribeirinho, M.J. 2018. *Artificial Intelligence: Construction Technology's Next Frontier*. McKinsey & Company.
39. Abioye, S.O., et al. 2021. Artificial Intelligence in the Construction Industry: A Review of Present Status, Opportunities and Future Challenges. *Journal of Building Engineering* 44: 103299.
40. Kattimani, M.A., et al. 2023. Design and Numerical Analysis of Tensile Deformation and Fracture Properties of Induction Hardened Inconel 718 Superalloy for Gas Turbine Applications. *International Journal on Interactive Design and Manufacturing (IJIDeM)* 18(8): 1–11.
41. Majji, B.G.R., Vasudev, H., and Bansal, A. 2022. A Review on the Oxidation and Wear Behavior of the Thermally Sprayed High-Entropy Alloys. *Materials Today: Proceedings* 50: 1447–1451.
42. Prashar, G. and Vasudev, H. 2022. A Review on the Influence of Process Parameters and Heat Treatment on the Corrosion Performance of Ni-Based Thermal Spray Coatings. *Surface Review and Letters* 29(1): 2230001.
43. Yang, T., A. Bandyopadhyay, Z. O'Neill, J. Wen, and B. Dong. 2022. From Occupants to Occupants: A Review of the Occupant Information Understanding for Building HVAC Occupant-Centric Control. *Building Simulation* 15: 913–932.
44. Kiliccote, S., Olsen, D., Sohn, M.D., and Piette, M.A. 2016. Characterization of Demand Response in the Commercial, Industrial, and Residential Sectors in the United States. *WIREs Energy and Environment* 5: 288–304.
45. Sharanabasava, H., et al. 2023. Microstructure, Mechanical and Wear Properties of SiC and Mo Reinforced NiCr Microwave Cladding. *Advances in Materials and Processing Technologies* 10(4). doi:10.1080/2374068X.2023.2257937.
46. Sunitha, K. and Vasudev, H. 2022. A Short Note on the Various Thermal Spray Coating Processes and Effect of Post-Treatment on Ni-Based Coatings. *Materials Today: Proceedings* 50: 1452–1457. https://linkinghub.elsevier.com/retrieve/pii/S2214785321058132.
47. Fernandez, N.E.P., et al. 2017. *Impacts of Commercial Building Controls on Energy Savings and Peak Load Reduction*. Pacific Northwest National Laboratory (PNNL).
48. Merabet, G.H., et al. 2021. Intelligent Building Control Systems for Thermal Comfort and Energy-Efficiency: A Systematic Review of Artificial Intelligence-Assisted Techniques. *Renewable and Sustainable Energy Reviews* 144: 110969.

49. Srivastava, S. 2023. *AI in Construction—How Artificial Intelligence is Paving the Way for Smart Construction.* https://appinventiv.com/blog/ai-in-construction/.
50. Yedida, V., Satyavathi, V., and Vasudev, H. 2022. A Review on the Development of Thermal Barrier Coatings by Using Thermal Spray Techniques. *Materials Today: Proceedings* 50: 1458–64.
51. Saka, A.B., et al. 2023. Conversational Artificial Intelligence in the AEC Industry: A Review of Present Status, Challenges and Opportunities. *Advanced Engineering Informatics* 55: 101869.
52. Kor, M., Yitmen, I., and Alizadehsalehi, S. 2023. An Investigation for Integration of Deep Learning and Digital Twins Towards Construction 4.0. *Smart and Sustainable Built Environment* 12: 461–487.
53. Singh, J. and Singh, S. 2022. A Review on Machine Learning Aspect in Physics and Mechanics of Glasses. *Materials Science and Engineering: B* 284: 115858.
54. Regona, M., Yigitcanlar, T., Xia, B., and Li, R.Y.M. 2022. Opportunities and Adoption Challenges of AI in the Construction Industry: A PRISMA Review. *Journal of Open Innovation: Technology, Market, and Complexity* 8: 45.
55. Ding, C., Ke, J., Levine, M., and Zhou, N. 2024. Potential of Artificial Intelligence in Reducing Energy and Carbon Emissions of Commercial Buildings at Scale. *Nature Communications* 15(1): 5916.

21 Application of OFAT Approach to Examine the Effect of Green Cutting Process Parameters for Aluminum Material

Vikas Sharma, Sanjay Kumar, Jaiinder Preet Singh, Vinay Saini, Ashiwani Kumar, Mukesh Kumar, Rajeev Kumar, Ashish Goyal, and Shashi Bahl

Description	Abbreviation
Heat-affected zone	HAZ
Laser beam machining	LBM
One Factor at a Time	OFAT

21.1 INTRODUCTION

Laser beam machining (LBM) is considered an advanced machining process to remove material from a workpiece without any physical contact with workpiece material [1]. Due to its ability to produce precise and exact machining results on a variety of materials, LBM has attracted substantial attention in a number of sectors. In comparison to conventional cutting methods, the laser machining process has less heat-affected zones, less tool wear, and greater flexibility. LBM has several benefits, such as producing small details and complex forms [2]. In industrial contexts, it is frequently used for surface treatment, welding, micro-machining, drilling, and cutting. A laser beam is focused onto the workpiece's surface and vaporizes the material to form the required shape or cut [3, 4]. The depth and width of the cut may be controlled by varying parameters including scanning speed, laser power, intensity, and duration of pulse.

DOI: 10.1201/9781003516071-21

The traditional machining methods, like turning, drilling, and milling, play a crucial role in a wide range of precision, scientific, and industrial applications [5–8]. The turning process is particularly useful for shaping cylindrical parts, mainly by reducing their diameter. On the other hand, milling is all about removing unwanted material from various shapes, allowing for the creation of complex features such as channels, pockets, and edges. These processes are essential for crafting the precise components in many industries.

There are many eco-friendly techniques used in traditional machining [9–13]. One innovative method is green machining with cryogenic cooling [14]. This approach involves using liquid nitrogen or other highly cooled gases under high pressure to lower the temperature of the machining area [15–18]. Due to this, the generation of a cushion of gas takes place which helps to reduce the frictional heat generated between the tool and the workpiece. This technique offers several benefits, including better chip removal, lower temperatures, improved surface quality of the finished part, and extended tool life. Figure 21.1 depicts the application of green techniques in traditional and non-traditional machining processes.

Advanced or non-conventional machining processes are recognized as alternatives to traditional machining techniques due to their unique features [19–22]. Methods such as spark erosion machining, abrasive water jet machining, laser beam machining, and ultrasonic machining are among the most advanced techniques in the green machining revolution [23–25]. These processes are known for their precision machining of a wide range of engineering materials and can create intricate shapes and tiny components. Moreover, it is essential to reveal that some of these advanced methods can create environmental challenges. The use of certain fluids and consumables may lead to toxic waste, contamination, thermal damage, and the

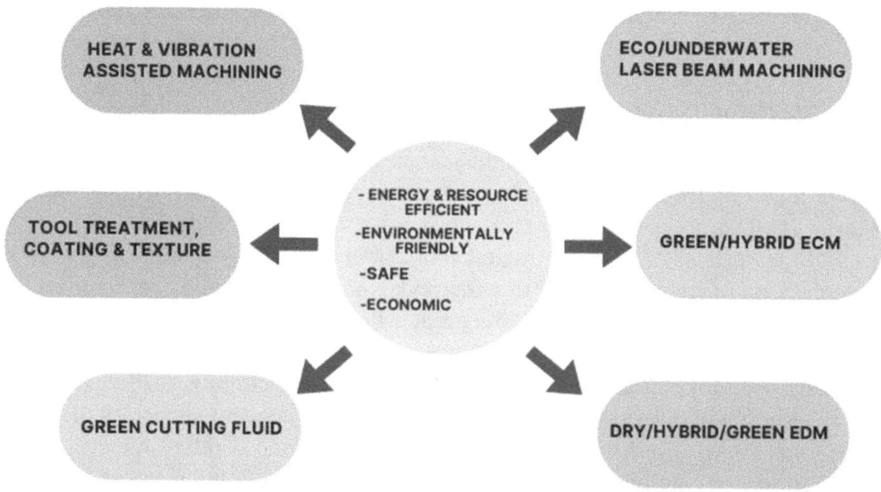

FIGURE 21.1 Green techniques in traditional and non-traditional machining (Credit: Created by authors).

release of harmful fumes and gases. To address these issues, various green or sustainable machining strategies have been developed. In this section, we will take a closer look at these innovative approaches aimed at making machining more environmentally friendly.

Laser machining, which stands for "light amplification by stimulated emission of radiation," utilizes a powerful laser beam to cut through various engineering materials and shape them into specific forms. However, there are some drawbacks to this process, such as thermal damage, the formation of micro-cracks, high energy consumption, and the release of harmful particles and gases. To enhance the sustainability of laser machining, various research has been revealed on several innovative solutions which include underwater laser machining, which creates a wet environment during the melting and vaporization stages. This approach significantly reduces the re-deposition of molten material, minimizes thermal damage, and decreases the emission of gases and particulates, making the process safer and more environmentally friendly. Moreover, advancements in laser methods have been introduced like fiber and disk lasers which are energy-efficient alternatives to traditional CO_2 lasers.

21.2 LITERATURE REVIEW

21.2.1 Effect of Reinforced Particles on Quality Characteristics

Cristobal et al. [25] implemented CO_2 laser cutting of PMMA sheets and revealed the impact of gas pressure, focal point position, cutting speed and laser power on quality characteristics. The optimal operational parameters were identified which resulted in improved kerf profile deviation. Moreover, an artificial neural network model predicted accurate kerf value deviation values. Rao et al. [26] conducted a study comparing kerf width values of aluminum alloy (7050) reinforced with a 20% B4C composite and a 20% Al_2O_3 composite alloy. The results showed that the B4C composite exhibited a smaller kerf width compared to the Al_2O_3 composite. Additionally, the study found that complicated cut profiles had a larger kerf taper as compared to normal cut profiles. Kumar et al. [27] utilized the Box–Behnken design and wire electric discharge machine to investigate the behavior of pure titanium. Their study focused on optimizing process variables using response surface methodology and desirability. Sharma et al. [28] examined the normal and complex kerf quality using Nd:YAG laser-machined nickel alloy for kerf characteristics. The results showed that the complex contours yielded different outcomes for kerf taper and kerf deviation compared to the straight-cut profiles. However, the kerf width remained consistent for both profiles. Mathew et al. [29] carried out investigations on the laser cutting of a 2 mm thick composite sheet using pulsed Nd:YAG technology. A second-order response model was formulated with the assistance of CCD experimental design, allowing the analysis of the effects of different laser input parameters, including ablation gas and laser pulse types with different spans, on both heat-affected zone (HAZ) and kerf properties.

21.2.2 Machinability Analysis of Various Materials

Yilbas et al. [30] studied the stresses at different corner points in laser cutting using an alumina tile with a triangular cutting profile. The research revealed that high-stress concentration lay in between the workpiece profile. Cicala et al. [31] conducted research on various alloy materials including stainless steel. The research revealed that higher laser pulse frequency and reduced nozzle speed have a significant impact on quality. Dubey et al. [32] applied a response surface methodology to identify significant levels of various parameters on laser cutting of aluminum alloy work material. Venkata et al. [33] conducted deep research on various composite materials based on various SiC and Al_2O_3-reinforced particles. They found that AMC exhibited superior machinability in terms of cutting force, radial force, and surface roughness compared to the other composites. Riveiro et al. [34] studied the various laser cutting variables with a focus on variable wave (continuous-pulse) modes on Al-Cu alloy and compared the results for various wave modules. Caydas et al. [35] used laser machining with the help of various statistics tools such as GRA and factorial approach to optimize the quality characteristics. Yang et al. [36] investigated the laser welding process on various materials having different physical properties. It was revealed that brass material had greater welded strength over stainless steel. Pandey and Dubey [37] conducted laser machining of titanium alloy sheets to obtain optimized values of Kt and Rz. A multi-objective optimization was applied for the regression models for quality characteristics, whereas the objective functions were generated using a genetic algorithm. Santosh et al. [38] utilized the Taguchi array to determine the parameters for laser machining, analyzing results through response tables and graphs. Surface topography was examined using scanning electron microscopy, whereas the thermodynamic properties were investigated using differential scanning calorimetry (DSC). X-ray diffraction (XRD) identified various phases present. Yan et al. [39] established a model for predicting morphological changes in laser-assisted machining of composite materials. An experimental result confirmed the model's accuracy, showing that laser-assisted machining reduces subsurface damage. This research provided extensive results for minimizing damage during the machining of metal matrix composites. Chong Ng et al. [40] examined the laser processing parameters to optimize the outcome responses. The 3D surface profiles were created to demonstrate the feasibility of a machine-learning regression model. Sivasankara et al. [41] fabricated hemp, pineapple, and palm fiber eco-friendly composites for ballistic and architectural uses. The various composites were ranked using AHM and TOPSIS with confirmation of thermal stability (palm: 276°C, pineapple: 241°C, hemp: 166°C). It was examined that Pineapple-palm composites excelled in strength, whereas the hemp–pineapple composites showed superior water absorption. Abbas et al. [42] applied response surface methodology (RSM) and evaluated the material removal rate (MRR) and tool wear rate (TWR) for the composite material. It was also found that optimization with RSM-COPRAS and machine learning identified the optimum machining settings such as Ton 60 μs, Toff 60 μs, SV 7 V, I 12 A, and brass tool, yielding an MRR of 1.11 g/s and TWR of 0.0114 g/s. Jyotiprasad et al. [43] examined the stress-strain behavior of polycrystalline carbon (graphite fiber) laminates under thermal and mechanical loads. It was further analyzed that variation of stress-strain resulted in different orientations whereas some configurations outperformed longitudinal and transverse stacking.

21.3 EXPERIMENTAL PROCEDURE AND MATERIALS

Using a One-Factor-at-a-Time approach, a series of experimental runs were designed to machine the aluminum material. The different work samples were cut using high high-precision cutting machine. The various aluminum material samples of the same dimension were finished further so that the absorption of the laser could be suitable for the surface. Using OFTA, a total of 27 experimental runs were conducted (as shown in Table 21.1) to compare the quality characteristics values. The machining of work specimens was conducted using a CO_2 laser cutting machine (Trulaser, 3000 W). The schematic of the laser cutting setup is shown in Figure 21.2. The various laser-cutting variables used in the process were laser nitrogen gas pressure, nozzle diameter, cutting speed, and laser power.

TABLE 21.1
Laser Cutting Variables Defined Range

Run No.	Laser speed	Cutting power	Gas Pressure	Orifice Diameter
Units	mm/min	watt	bar	mm
1	1100–1500	500–1300	3–11	1.0–1.8

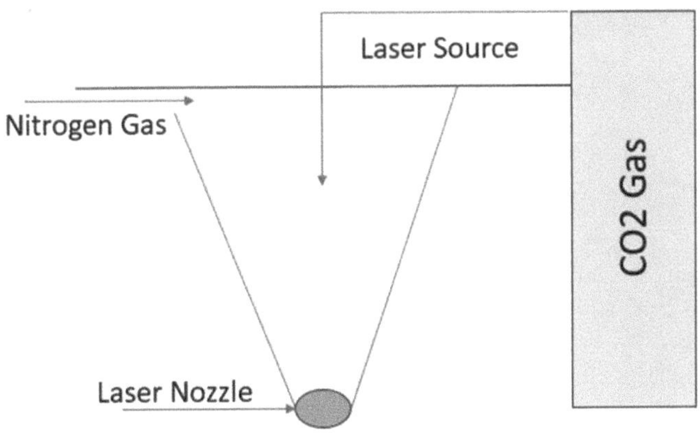

FIGURE 21.2 Schematic of laser setup (Credit: Created by authors).

21.3.1 Relationship Between Kerf Taper and Process Variables

The variation in kerf taper values was investigated by analyzing the impact of top and bottom kerf widths. The reduced values of kerf taper may be obtained with the help of various optimized laser cutting variables. The various researchers executed the work to analyze the effect of normal and complex cut profiles on kerf taper. In reference to Equation 21.1, this research assesses the implications of laser cutting parameters on the kerf taper. Equation 21.1 employs the workpiece material thickness, along with the values of top and bottom kerf width, to analyze their impact on the kerf taper. The kerf taper values can be calculated using the equation:

$$K_t = \{(K_{wt} - K_{wb}) \times 180\} / 2\pi t \quad (21.1)$$

Where t denotes the thickness of the workpiece material, K_{wt} is the top surface profile kerf width, and K_{wb} is the bottom kerf profile width.

21.4 RESULTS AND DISCUSSIONS

21.4.1 Influence of Laser Cutting Speed and Laser Power Settings

Figure 21.3(a) illustrates the effect of laser nozzle speed, with the nozzle speed range displayed on the x-axis ranging from 1100 to 1500 mm/min. It was identified that incremental values of cutting speed have shown an effect on the values of kerf taper values which rose from 0.27 mm to 0.39 mm. A dramatic variation has been examined in the case of cutting speed within the range of 1300 to 1900 mm/min which resulted in higher kerf taper values from 0.3 to 0.39 mm. The possible reason for the phenomenon is that rapid cutting speed lowers the amount of laser energy accessible to the profile section due to less beam interaction with the cutting surface. As a result, there is a slower rate of temperature rise over the surface and a greater increase in kerf taper due to the spread of diffusion energy transfer in the vicinity of the material.

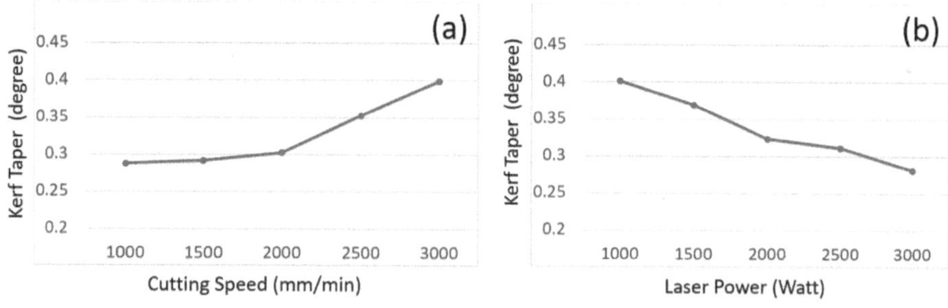

FIGURE 21.3 Trends of parameters in terms of speed and power on kerf taper (Credit: Drawn by authors).

In Figure 21.3(b), the detailed analysis of laser cutting power on the kerf taper has been shown using a graphical representation. In this research, it was revealed that incremental laser power influences the kerf taper angle. The higher laser power lowered the kerf taper values in a significant pattern. Moreover, with the higher values of laser cutting power ranging from 500 watts to 1300 watts, a significant decrease in kerf taper has been observed from 0.43 to 0.27 mm. When the laser power was raised to higher values of 1300 watts, the heat supplied in the vicinity also increased which resulted in a wider kerf width. Due to the high laser power, the laser beam penetrates easily into the material. The significance of wider kerf increases as it reduces the kerf taper values resulting in lower taper formation on the cut surface.

21.4.2 Analysis of Gas Pressure and Nozzle Diameter Effects

The gas pressure is an important parameter to consider as it has a large effect on the surface cut quality. Moreover, it becomes more significant to analyze the gas pressure influence with respect to various types of gases such as argon, nitrogen, compressed air etc. on quality characteristics. In the present research, nitrogen gas has been used with a variation of gas pressure to analyze the influence on the kerf taper. Moreover, the research revealed the inversely proportional relation between kerf taper and gas pressure, higher gas pressure values resulted in lower kerf taper values as depicted in Figure 21.4(a).

It is analysed that Kerf taper values was predicted lower i.e. 0.32 degree which may be occurred due to increase in gas pressure (up to 11 bar). The lower kerf taper values were reported due to the removal of molten material with the high gas pressure from the vicinity of the kerf profile. This resulted in a decrease in the kerf values with respect to the top and bottom kerf profiles. On the other hand, lower gas pressure constrains the molten material due to the presence of an oxide layer. The effect of nozzle diameter has been depicted in Figure 21.4(b). It has been examined that with the increase in nozzle diameter, there is no significant effect that has been seen on the kerf taper. It may be due to the turbulence effect which is possibly compensated by the oxide layer formation in the vicinity of the kerf profile.

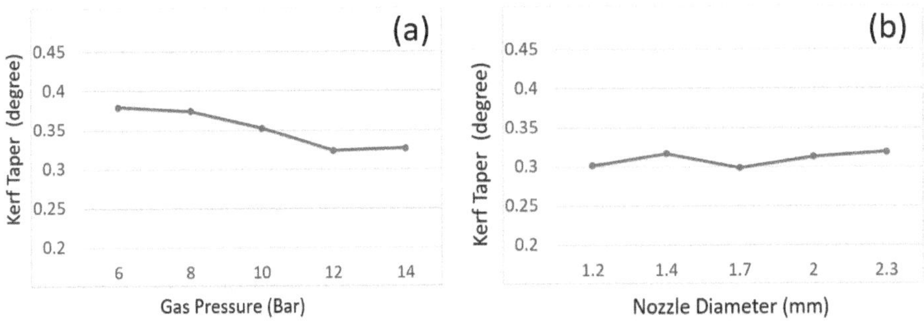

FIGURE 21.4 Trends of gas pressure (a) and orifice diameter (b) on kerf taper (Credit: Drawn by authors).

21.4.3 Microstructure Analysis

The machined specimens were examined for microstructural analysis for the top surface profile and bottom surface profile kerf width. The thermal behavior of aluminum material toward laser absorption makes it important to analyze the output characteristics with the variation of laser cutting parameters. Due to the melting temperatures of the aluminum and different reinforced particles, the bottom and top kerf widths were analyzed with large variations in values. Based on the experimental results, the critical factors influencing kerf taper include nitrogen gas pressure, laser power and nozzle speed. The microstructure of the machined specimens may change due to the thermal cutting action of laser beam machining. There are many factors which contribute to the stress concentration, which initiates the process of crack formation.

The variation in top and bottom kerf width resulted in variable values of kerf taper. The results derived from optical micrographs predicted low values of bottom kerf width as compared to top kerf width. Moreover, the clustering of reinforced particles may also result in variable heat-absorbing mechanisms in the profile of the kerf zone. In Figure 21.3(a), the bottom kerf width values were observed in the range of 0.26 to 0.37 mm whereas in Figure 21.3(b) the top kerf width was observed with higher values in the range 0.38 to 0.47 mm. It has been investigated whether various reinforced particles show different behavior with respect to laser heat absorbing behavior. Due to the low clustering of particles, a high variation was found, which resulted in higher viscosity levels and surface tension of liquid material. The stress concentration also resulted in high variation in kerf taper values, which may be caused due to the presence of reinforced particles.

21.5 CONCLUSION

In this current study, an analysis of different laser cutting parameters was conducted to assess kerf taper values, taking into account both bottom and top kerf measurements. With the help of one factor at a time, it has been examined laser cutting speed, laser power and gas pressure shown a significant effect on kerf taper. The effect of nozzle diameter does not significantly affect the kerf taper values. The variable laser heat absorbing mechanism has been confirmed using variable top and bottom kerf width values for respective kerf taper values. The agglomeration effect of reinforced particles has been considered for the output characteristics. From the optical micrographs, the kerf values were investigated to examine the effect on the kerf taper values.

REFERENCES

1. Dubey, A. K., & Yadava, V. (2008). Laser beam machining—A review. *International Journal of Machine Tools and Manufacture*, *48*(6), 609–628.
2. Wang, K., & Sun, W. (2018). A multiscale multi-permeability poroplasticity model linked by recursive homogenizations and deep learning. *Comput. Methods Appl. Mech. Engrg.*, 334, 337–380. https://doi.org/10.1016/j.cma.2018.01.036.

3. Sharanabasava, H., Raviprakash, M., Prasad, C.D., Ramesh, M.R., Phanibhushana, M.V., Vasudev, H., & Kumar, S. (2023). Microstructure, mechanical and wear properties of SiC and Mo reinforced NiCr microwave cladding. *Adv. Mater. Process. Technol.*, 10(4). https://doi.org/10.1080/2374068X.2023.2257937.
4. Sunitha, K., & Vasudev, H. (2022). A short note on the various thermal spray coating processes and effect of post-treatment on Ni-based coatings. *Mater. Today Proc.*, 50, 1452–1457. https://doi.org/10.1016/j.matpr.2021.09.017.
5. Faludi, J., Bayley, C., Bhogal, S., & Iribarne, M. (2015). Comparing environmental impacts of additive manufacturing vs traditional machining via life-cycle assessment. *Rapid Prototyp. J.*, 21, 14–33. https://doi.org/10.1108/RPJ-07-2013-0067.
6. Liang, X., Liu, Z., & Wang, B. (2019). State-of-the-art of surface integrity induced by tool wear effects in machining process of titanium and nickel alloys: A review. *Meas. J. Int. Meas. Confed.*, 132, 150–181. https://doi.org/10.1016/j.measurement.2018.09.045.
7. Prashar, G., & Vasudev, H. (2022). A review on the influence of process parameters and heat treatment on the corrosion performance of Ni-based thermal spray coatings. *Surf. Rev. Lett.*, 29, 2230001.
8. Singh, J. (2021). Slurry erosion performance analysis and characterization of high-velocity oxy-fuel sprayed Ni and Co hardsurfacing alloy coatings. *J. King Saud Univ. - Eng. Sci.*, 1–14. https://doi.org/10.1016/j.jksues.2021.06.009.
9. Singh, J., Singh, R., Singh, S., Vasudev, H., & Kumar, S. (2022). Reducing scrap due to missed operations and machining defects in 90PS pistons. *Int. J. Interact. Des. Manuf.* https://doi.org/10.1007/s12008-022-01071-0.
10. Kumari, S., Sonia, P., Singh, B., Abhishek, K., & Saxena, K.K. (2020). Optimization of surface roughness in EDM of pure magnesium (Mg) using TLBO. *Mater. Today Proc.*, 26, 2458–2461.
11. Singh, J., & Singh, S. (2023). Support vector machine learning on slurry erosion characteristics analysis of Ni- and Co-alloy coatings. *Surf. Rev. Lett.* https://doi.org/10.1142/S0218625X23400061.
12. Pramanik, A., Basak, A.K., Littlefair, G., Debnath, S., Prakash, C., Singh, M.A., Marla, D., & Singh, R.K. (2020). Methods and variables in electrical discharge machining of titanium alloy – A review. *Heliyon*, 6, e05554. https://doi.org/10.1016/j.heliyon.2020.e05554.
13. Kattimani, M.A., Venkatesh, P.R., Masum, H., Math, M.M., Bahadurdesai, V.N., Mustafkhadri, S., Prasad, C.D., & Vasudev, H. (2023). Design and numerical analysis of tensile deformation and fracture properties of induction hardened Inconel 718 superalloy for gas turbine applications. *Int. J. Interact. Des. Manuf.*, 18(8), 1–11.
14. Özbek, N.A., & Özbek, O. (2022). Effect of cryogenic treatment holding time on mechanical and microstructural properties of Sverker 21 steel. *Mater. Test.*, 64, 1809–1817. https://doi.org/10.1515/mt-2022-0207.
15. Jakhar, C., Saroha, A., Antil, P., Ahlawat, V., Rani, A., Buddhi, D., & Kumar, V. (2023). Deep cryogenic treated high carbon steel blades: Tribological, morphological, and economic analysis. *Surf. Rev. Lett.*, 30. https://doi.org/10.1142/S0218625X22410025.
16. Altan Özbek, N., & Özbek, O. (2022). Investigation of the effects of shallow cryogenic treatment on the mechanical and microstructural properties of 1.2436 tool steel. *J. Mater. Mechat. A*, 3, 151–162. https://doi.org/10.55546/jmm.1113194.
17. Wu, Y., Deng, B., Li, X., Li, Q., Ye, T., Xiang, S., Zhao, M.-C., & Atrens, A. (2024). In-situ EBSD study on twinning activity caused by deep cryogenic treatment (DCT) for an as-cast AZ31 Mg alloy. *J. Mater. Res. Technol.*, 30, 3840–3850.

18. M'Saoubi, R., Axinte, D., Soo, S.L., Nobel, C., Attia, H., Kappmeyer, G., Engin, S., & Sim, W.M. (2015). High performance cutting of advanced aerospace alloys and composite materials. *CIRP Ann. - Manuf. Technol.*, 64, 557–580. https://doi.org/10.1016/j.cirp.2015.05.002.
19. Li, J., Shen, Z., Xu, W.Y.T., Lam, W.Y.H., Hsung, R.T.C., Pow, E.H.N., Kosuge, K., & Wang, Z. (2019). A compact dental robotic system using soft bracing technique. *IEEE Robot. Autom. Lett.*, 4, 1271–1278. https://doi.org/10.1109/LRA.2019.2894864.
20. Kurtz, S., Ong, K., Lau, E., Mowat, F., & Halpern, M. (2007). Projections of primary and revision hip and knee arthroplasty in the United States from 2005 to 2030. *J. Bone Jt. Surg.*, 89, 780–785. https://doi.org/10.2106/JBJS.F.00222.
21. Ahuja, N., Batra, U., & Kumar, K. (2020). Experimental investigation and optimization of wire electrical discharge machining for surface characteristics and corrosion rate of biodegradable Mg alloy. *J. Mater. Eng. Perform.*, 29, 4117–4129. https://doi.org/10.1007/s11665-020-04905-8.
22. Singh, J., Gill, H.S., & Vasudev, H. (2022). Computational fluid dynamics analysis on role of particulate shape and size in erosion of pipe bends. *Int. J. Interact. Des. Manuf.*, 18(8), 1–16.
23. Dutta, V., Thakur, L., Singh, B., & Vasudev, H. (2022). A study of erosion–corrosion behaviour of friction stir-processed chromium-reinforced NiAl bronze composite. *Materials (Basel)*, 15, 5401. https://doi.org/10.3390/ma15155401.
24. Yedida, V.V.S., & Vasudev, H. (2022). A review on the development of thermal barrier coatings by using thermal spray techniques. *Mater. Today Proc.*, 50, 1458–1464.
25. Cristobal, L., & Ivan, L. (2021). Kerf profile analysis and neural network-based modeling of increasing thickness PMMA sheets cut by CO2 laser. *Opt. Laser Technol.*, 144, 107386.
26. Rao, R., & Yadava, V. (2009). Multi-objective optimization of Nd:YAG laser cutting of thin super alloy sheet using grey relation a analysis with entropy measurement. *Opt. Laser Technol.*, 41, 922–930.
27. Kumar, A., Kumar V., & Kumar J. (2013). Multi-response optimization of process parameters based on response surface methodology for pure titanium using WEDM process. *Int. J. Adv. Manuf. Tech.*, 68, 2645–2668.
28. Sharma, A., Yadava, V., & Rao, R. (2010). Optimization of kerf quality characteristics during Nd:YAG laser cutting of nickel based super alloy sheet for straight and curved cut profiles. *Opt. Lasers Eng.*, 48, 915–925.
29. Mathew, J., & Goswami, L. (1999). Parametric studies on pulsed Nd: YAG laser cutting of carbon fibre reinforced plastic composites. *J. Mater. Proc. Technol.*, 89, 198–203.
30. Yilbas, S., Akhtar, S., & Karatas, C. (2014). Laser cutting of triangular geometry into alumina tiles: Morphological changes and thermal stress analysis. *Mach. Sci. Technol.*, 18, 424–447.
31. Cicala, E., Soveja, A., & Sallamandb, P. (2008). The application of the random balance method in laser machining of metals. *J. Mater. Proc. Technol.*, 196, 393–401.
32. Dubey, A., & Yadava, V. (2007). Simultaneous optimization of multiple quality characteristics in laser beam cutting using Taguchi method. *Int. J. Precis. Eng. and Mfg.*, 8, 10–15.
33. Venkata, B., Ganguly, R., Srinivasarao, G., & Sahoo, K., (2013) Machinability of aluminum metal matrix composite reinforced with in-situ ceramic composite developed from mines waste colliery shale. *Mater. Manuf. Proc.*, 10, 1082–1089.
34. Riveiro, A., Quintero, F., & Lusquinos, F. (2010). Parametric investigation of CO2 laser cutting of 2024-T3 alloy. *J. Mater. Proc. Technol.*, 210, 1138–1152.

35. Caydas, U., & Hascalik, A. (2008). Use of grey relational analysis to determine optimum laser cutting parameters with multi-performance characteristics. *Opt. Laser Tech.*, 40, 987–994.
36. Yang, L., & Shengsun, H. (2014). Dissimilar welding of H62 Brass-316L stainless steel using continuous-wave Nd: YAG laser. *Mater. Manuf. Proc.*, 29, 916–921.
37. Pandey, A., & Dubey, A. (2012). Simultaneous optimization of multiple quality characteristics in laser cutting of titanium alloy sheet. *Opt. Laser Technol.*, 44, 1858–1865.
38. Santosh, S., Srivatsan S., & Vignesh Pandian, R. (2023). Unravelling the effect of CO2 laser machining parameters on the surface and shape memory characteristics of CuAlFeMn quaternary shape memory alloy. *Opt. Laser Technol.*, 163, 109306.
39. Yan, A., Zhang, H., Deng, B., Peng, F., Yan, R., & Cui, F. (2023). Analytical modeling of subsurface damage in laser-assisted machining of metal matrix composites based on the reinforcement fracture probability. *J. Manuf. Proc.*, 109, 300–312.
40. Ng, C.-K., Chen, C., Yang, Y., Zhang, F., Ju, B.-F., & Chen, Y.-L. (2024). Femtosecond laser micro-machining of three-dimensional surface profiles on flat single crystal sapphire. *Opt. Laser Technol.*, 170, 110205.
41. Rallabandi, S.R., Pilla, D.P., Dowluru, S., Palli, S., Sharma, N., Sharma, S.K., & Sharma, R.C. (2024). Critical evaluation of epoxy-hemp-pineapple-palm fiber composites using hybrid AHM-TOPSIS technique for sustainable structural applications. *JCIE*, 47, 325–336. https://doi.org/10.1080/02533839.2024.2308250.
42. Abbas, T., Sharma, N., Alsuhaibani, A., Sharma, S., Soliman, S., & Sharma, R.C. (2023). Processing of Al/SiC/Gr hybrid composite on EDM by different electrode materials using RSM-COPRAS approach. *Metals*, 13, 1125. https://doi.org/10.3390/met13061125.
43. Jyotiprasad, G., Dowluru, S., Rallabandi, R., Sharma, N., et al. (2023). Thermal-structural and macromechanical behavior analysis of graphite-based laminated composite. SAE Technical Paper, 01:5041, doi:10.4271/2023-01-5041.

Index

FSC
www.fsc.org
MIX
Papier | Fördert
gute Waldnutzung
FSC® C083411

Kolibri 360 GmbH
Ferdinand-Jühlke-Straße 7
99095 Erfurt, Deutschland
produktsicherheit@kolibri360.de